Revit Architecture 2025 For Novices

(Learn by Doing)

CADSoft Technologies

CADSoft Technologies

Revit Architecture 2025 for Novices: Learn By Doing

CADSoft Technologies

ISBN 978-1-64057-160-0

NOTICE TO THE READER

Publisher does not warrant or guarantee any of the products described in the text or perform any independent analysis in connection with any of the product information contained in the text. Publisher does not assume, and expressly disclaims, any obligation to obtain and include information other than that provided to it by the manufacturer.

The reader is expressly warned to consider and adopt all safety precautions that might be indicated by the activities herein and to avoid all potential hazards. By following the instructions contained herein, the reader willingly assumes all risks in connection with such instructions.

The publisher makes no representation or warranties of any kind, including but not limited to, the warranties of fitness for particular purpose or merchantability, nor are any such representations implied with respect to the material set forth herein, and the publisher takes no responsibility with respect to such material. The publisher shall not be liable for any special, consequential, or exemplary damages resulting, in whole or part, from the reader's use of, or reliance upon this material.

www.cadsofttech.com

DEDICATION

To students, who are dedicated to learning new technologies

THANKS

To employees of CADSoft Technologies for their cooperation

Online Training Program Offered by CADSoft Technologies

CADSoft Technologies provides effective and affordable virtual online training on various software packages including Computer Aided Design, Manufacturing and Engineering (CAD/CAM/CAE), computer programming languages, animation, architecture, and GIS. The training is delivered 'live' via Internet at any time, any place, and at any pace to individuals as well as the students of colleges, universities, and CAD/CAM/CAE training centers. The main features of this program are:

Training for Students and Companies in a Classroom Setting

Highly experienced instructors and qualified Engineers at CADSoft Technologies conduct the classes under the guidance of Prof. Sham Tickoo of Purdue University Northwest, USA. This team has authored several textbooks that are rated "one of the best" in their categories and are used in various colleges, universities, and training centers in North America, Europe, and in other parts of the world.

Training for Individuals

CADSoft Technologies with its cost effective and time saving initiative strives to deliver the training in the comfort of your home or work place, thereby relieving you from the hassles of traveling to training centers.

Training Offered on Software Packages

CADSoft Technologies provides basic and advanced training on the following software packages:

CAD/CAM/CAE: *CATIA, Creo Parameteric, SOLIDWORKS, Autodesk Inventor, Solid Edge, NX, AutoCAD, AutoCAD LT, Customizing AutoCAD, AutoCAD Electrical, and ANSYS*

Architecture and GIS: *Autodesk Revit (Architecture/Structure/MEP), AutoCAD Civil 3D, AutoCAD Map 3D, Autodesk Navisworks, Bentley Staad.Pro, Oracle Primavera P6, MS Project, ArcGIS, and Raster Design.*

Animation and Styling: *Autodesk 3ds Max, 3ds Max Design, Autodesk Maya, Autodesk Alias, Pixologic ZBrush, and CINEMA 4D*

Computer Programming: *C++, VB.NET, Oracle, AJAX, and Java*

For more information, please visit the following link:
https://www.cadsofttech.com

Table of Contents

Chapter 2: Starting an Architectural Project

Chapter 3: Creating Architectural Walls

Chapter 4: Using Basic Building Components-I

Chapter 5: Using the Editing Tools

Chapter 6: Working with Datum Plane and Creating Standard Views

Chapter 7: Using Basic Building Components-II

Chapter 8: Using Basic Building Components-III

Chapter 9: Adding Site Features

Chapter 10: Using Massing and Family Tools

Chapter 11: Adding Annotations and Dimensions

Chapter 12: Creating Project Details and Schedules

Chapter 13: Creating and Plotting Sheets

Chapter 14: Creating 3D Views

Chapter 15: Rendering Views and Creating Walkthroughs

Chapter 16: Using Advanced Features

This Page is Intentionaly Left Blank

Preface

Revit Architecture 2025 for Novices

Autodesk Revit 2025is a Building Information Modeling software developed by Autodesk. This software helps AEC professionals to develop high quality and accurate building design.

The **Autodesk Revit 2025 for Architecture** textbook introduces the users to the spectacular realm of one of the most powerful software in the architectural quiver. This textbook is a gateway to power, skill, and competence in the field of architectural and interior presentations, drawings, and documentations.

This textbook is specially meant for professionals and students of architecture and interior design, facilities planners, and CAD professionals who are associated with the building construction and allied fields in the construction industry.

Special emphasis has been laid to explain new concepts, procedures, and methods in Revit by using sufficient text and graphical examples. The accompanying tutorials and exercises, which relate to the real-world projects, help you understand the usage and abilities of the tools available in Autodesk Revit.

The main features of this textbook are as follows:

- **Project-based Approach**
  The author has adopted the project-based approach and the learn-by-doing approach throughout the textbook. This approach helps the users learn the concepts and procedures easily.

- **Real-World Designs as Projects**
  The author has used real-world building designs and architectural examples as projects in this textbook so that the users can correlate them to the real-time designs.

- **Tips and Notes**
  Additional information related to various topics is provided to the users in the form of tips and notes.

- **Learning Objectives**
  The first page of every chapter summarizes the topics that will be covered in that chapter. This will help the users to easily refer to a topic.

- **Self-Evaluation Test, Review Questions, and Exercises**
  Every chapters ends with Self-Evaluation test so that the users can assess their knowledge of the chapter. The answers to Self-Evaluation Test are given at the end of the chapters. Also, the Review Questions and Exercises are given at the end of the chapters and they can be used by the instructors as test questions and exercises.

Symbols Used in the Textbook

Note

The author has provided additional information to the users about the topic being discussed in the form of notes.

Tip

Special information and techniques are provided in the form of tips that helps in increasing the efficiency of the users.

Formatting Conventions Used in the Textbook

Please refer to the following list for the formatting conventions used in this textbook.

- Names of tools, buttons, options, browser, palette, panels, and tabs are written in boldface.

  Example: The **Wall: Architecture** tool, the **Modify** button, the **Build** panel, the **Architecture** tab, the **Properties** palette, the **Project Browser**, and so on.

- Names of dialog boxes, drop-downs, drop-down lists, list boxes, areas, edit boxes, check boxes, and radio buttons are written in boldface.

  Example: The **Options** dialog box, the **Wall** drop-down of the **Build** panel in the **Architecture** tab, the **Name** edit box of the **Name** dialog box, the **Chain** check box in the **Options Bar**, and so on.

- Values entered in edit boxes are written in boldface.

  Example: Enter **Brick Wall** in the **Name** edit box.

- Names of the files are italicized.

  Example: *c14_Club_tut2.rvt*

- The methods of invoking a tool/option from the ribbon, **File** menu, or the shortcut keys are given in a shaded box.

  Ribbon: Architecture Tab > Build > Wall drop-down > Wall
  Shortcut Keys: WA

- When you select an element or a component, a contextual tab is displayed depending upon the entity selected. In this textbook, this contextual tab is referred to as **Modify | (Elements / Components)**.

Naming Conventions Used in the Textbook

Please refer to the following list for the naming conventions used in this textbook.

Tool

If you click on an item in a panel of the ribbon and a command is invoked to create/edit an object or perform some action, then that item is termed as **tool**.

Ribbon: Modify | (Elements / Components) > Modify > Move
Shortcut Key: MV

For example:
Wall: Architectural tool, **Window** tool, **Railing** tool
Filled Region tool, **Trim/Extend to Corner** tool, **Rotate** tool
Link Revit tool, **Detail Line** tool

If you click on an item in a panel of the ribbon and a dialog box is invoked wherein you can set the properties to create/edit an object, then that item is also termed as **tool**, refer to Figure 1.

For example:
Load Family tool, **Materials** tool, **Project Units** tool
Design Options tool, **Visibility/Graphics** tool

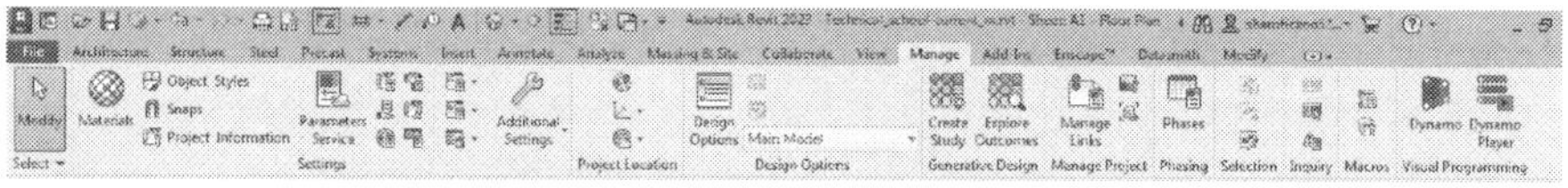

Figure 1 *Tools in the ribbon*

Button

The item in a dialog box that has a 3d shape is termed as **button**. For example, **OK** button, **Cancel** button, **Apply** button, and so on. If an item in a ribbon is used to exit a tool or a mode then that item is also termed as a button. For example, **Modify** button, **Finish Edit Mode** button, **Cancel Edit Mode** button, and so on; refer to Figure 2.

Figure 2 *Choosing the **Finish Edit Mode** button*

Dialog Box

In this textbook, different terms are used for referring to the components of a dialog box. Refer to Figure 3 for the terminology used.

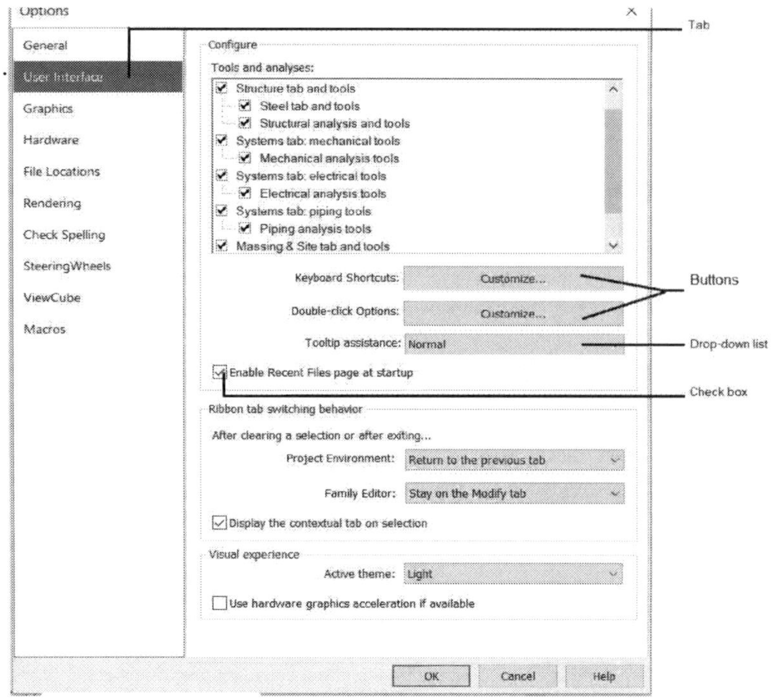

***Figure 3** Different components of a dialog box*

Drop-down

A drop-down is the one in which a set of common tools are grouped together for performing an action. You can identify a drop-down with a down arrow on it. The drop-downs are given a name based on the tools grouped in them. For example, **Wall** drop-down, **Component** drop-down, **Region** drop-down, and so on, refer to Figure 4.

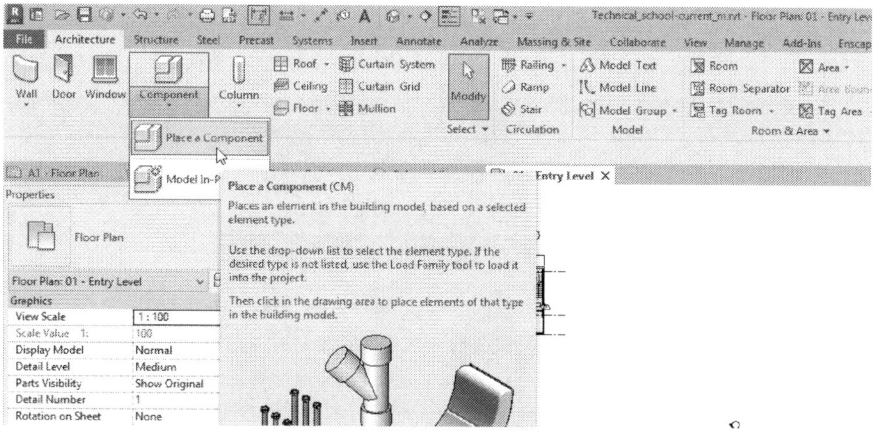

Figure 4 *Choosing a tool from a drop-down*

Drop-down List

A drop-down list is the one in which a set of options are grouped together. You can set a parameter using an option from this drop-down list. You can identify a drop-down list with a down arrow on it. For example, **Type Selector** drop-down list, **Units** drop-down list, and so on; refer to Figure 5.

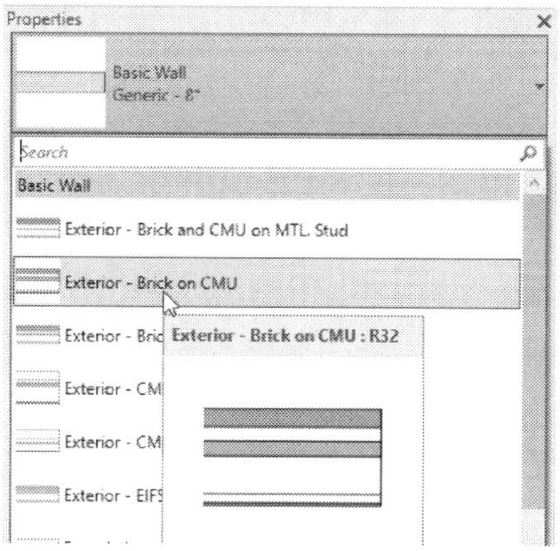

Figure 5 *Selecting an option from the* ***Type Selector*** *drop-down list*

Options

Options are the items that are available in shortcut menus, drop-down lists, dialog boxes, flyouts, and so on. For example, choose the **Zoom In Region** option from the shortcut menu displayed on right-clicking in the drawing area; refer to Figure 6.

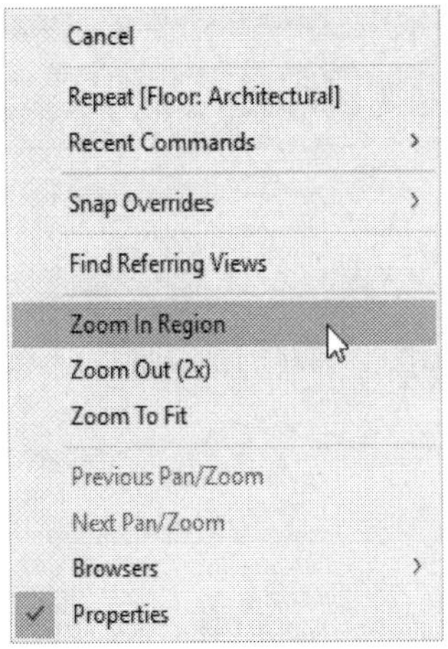

Figure 6 *Choosing an option from the shortcut menu*

Some New and Enhanced Features in Revit 2025

In Revit 2025, some of the enhanced and new features are discussed next:

Measure in 3D

In Revit 2025, you can measure elements while working in a 3D view.

View on Sheet Indicator

The project browser has an option to indicate which views have been placed on sheets. If a view is placed on a sheet then a blue colored square icon appears next to the view name in the **Project Browser**.

Swap Views on Sheets

Views that are placed on sheets can now be swapped for other views. TTo swap a view, select it in the sheet and then select the sheet to swap the view with, from the ribbon.

Filter by Sheet for Schedules

The filter by sheet will look at the objects that are visible on the sheet that the schedule is placed on. As a schedule can be placed on multiple sheets, the filter will adapt to each sheet and list only the objects visible in the views placed on each sheet.

Cut CAD Models

In Revit 2025, there is an option to have imported or linked CAD models cut with the view settings. When a model is linked, a check box in its instance properties can be used to enable cutting in views.

Performance Improvement

In Revit 2025, there have also been improvements in performance for working with large models.

Tag Groups and Linked Models

In Revit 2025, groups and linked models have standard parameters applied and they have the ability to add custom parameters. These parameters can also be tagged and included in schedules.

Duplicate Material Assets

In Revit 2025, you can now duplicate materials and assets simultaneously.

Fill Pattern in 3D Families

In Revit 2025, while working in a 3D family, you can create fill patterns on selected views for added detailing value.

Pick Working Plane

In Revit 2025, you can select your working planes using the **Pick Plane** tool.

This Page is Intentionaly Left Blank

Chapter 1

Introduction to Autodesk Revit 2025 for Architecture

Learning Objectives

After completing this chapter, you will be able to:
- *Understand the basic concepts and principles of Revit 2025 for Architecture*
- *Understand different terms used in Revit*
- *Know the parametric behavior of Revit*
- *Start the Revit 2025 program*
- *Use different components of the User interface screen of Revit*
- *Access the Revit 2025 Help*
- *Know worksharing using Revit Server*
- *Know about Autodesk Construction Cloud*

INTRODUCTION

Welcome to the realm of Revit, a powerful building modeler that has changed the outlook of the building industry about computer aided designs. Autodesk Revit is a design and documentation platform that enables you to use a single, integrated building information model to conceptualize, design, and finally document a project. Its integrated parametric modeling technology is used to create the information model of a project, and to collect and coordinate information across all its representations. In Autodesk Revit, drawing sheets, 2D views, 3D views, and schedules are a direct representation of the same building information model. Using its parametric change engine, you can modify a design at any stage of a project. The change in the project is automatically made and represented in all its views, resulting in the development of better designs, along with an improved coordination. The use of Revit provides a competitive advantage and a higher profitability to architects and building industry professionals.

Autodesk Revit AS A BUILDING INFORMATION MODELER

The history of computer aided design and documentation dates back to the early 1980s when architects began using this technology for documenting their projects. Realizing its advantages, information sharing capabilities were developed, especially to share data with other consultants. This led to the development of object-based CAD systems in the early 1990s. Before the development of these systems, objects such as walls, doors, windows were stored as a non-graphical data with the assigned graphics. These systems arranged the information logically, but were unable to optimize its usage in a building project. Realizing the advantages of the solid modeling tools, the mechanical and manufacturing industry professionals began using the information modeling CAD technology. This technology enabled them to extract data based on the relationship between model elements.

In 1997, a group of mechanical CAD technologists began working on a new software for the building industry. The Building Information Modeling (BIM) provided an alternative approach to building design, construction, and management. This approach, however, required a suitable technology to implement and reap its benefits. In such a situation, the use of parametric technology with the Building Information Modeling approach was envisaged as an ideal combination. They developed a software that was suitable for creating building projects. This software was earlier known as Autodesk Revit Architecture, and has now been changed to Autodesk Revit.

Autodesk Revit is a building design and documentation platform in which a digital building model is created using the parametric elements such as walls, doors, windows, and so on. All the building elements have inherent relationship with one another, which can be tracked, managed, and maintained by the computer.

BASIC CONCEPTS AND PRINCIPLES

Autodesk Revit enables you to envisage and develop a building model with actual 3D parametric building elements. It provides a new approach to the architectural thought and the implementation process. In a way, it replicates the way architects conceive a building. For example, 2D CAD platforms mostly use lines to represent all elements, as shown in Figure 1-1. However, in Autodesk Revit, you can create a building model using 3D elements such as walls, floors, doors, and windows, as shown in Figure 1-2.

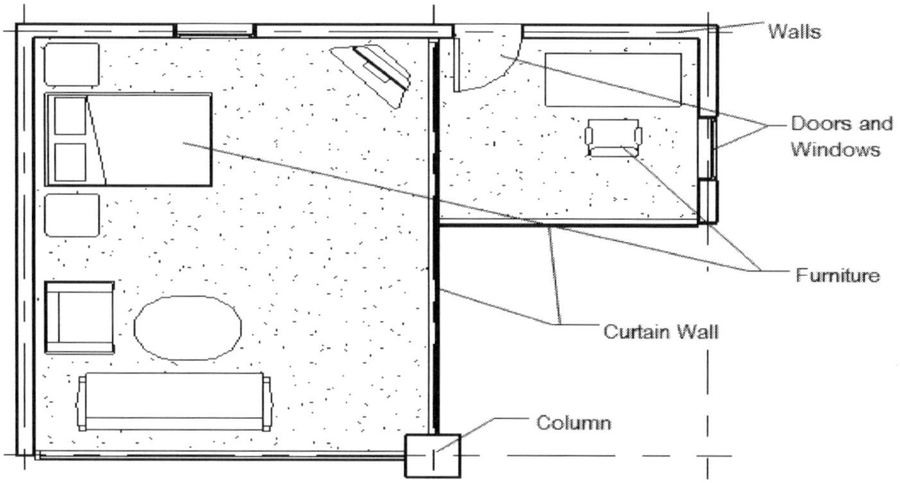

Figure 1-1 *CAD project created using 2D lines and curves*

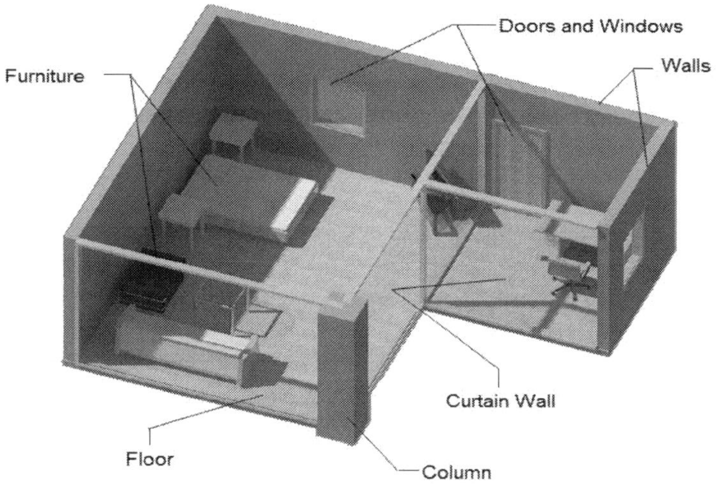

Figure 1-2 *Autodesk Revit project created using parametric building model*

Using these 3D elements, you can visualize the architectural or interior project with respect to its scale, volume, and proportions. This enables you to study design alternatives and develop superior quality design solutions. Autodesk Revit automates routine drafting and coordination tasks and assists in reducing errors in documentation. This, in turn, saves time, improves the speed of documentation, and lowers the cost for users.

Understanding the Parametric Building Modeling Technology

A project in Autodesk Revit is created using the in-built parametric building elements. The term parametric refers to the relationship parameters between various building elements. Some relations are applied automatically to building elements while creating them, and some are user defined. For example, doors, which have an inherent parametric relationship with walls cannot be created without first creating a host wall. A door always moves with the host wall. Similarly, floors too are parametrically linked to walls. When you move walls, the floor extents are also modified automatically. Each building element has in-built bidirectional associativity with many other elements in the project.

A building information model is created using different interdependent parametric building elements such as walls, floors, roof, ceiling, stairs, ramps, curtain walls, and so on. As they are bidirectionally associated elements, any change made in one element is automatically adopted by others. The integrated building information model thus created contains all the data for a project. You can then create project presentation views such as plans, sections, elevations, and so on for documentation. As you modify the model while working in certain views, Autodesk Revit's parametric change engine automatically updates other views. This capability is, therefore, the underlying concept in Autodesk Revit.

Autodesk Revit's parametric change engine enables you to modify design elements at any stage of the project development. As changes are made immediately and automatically, it saves the time and effort in coordinating them in all other associated views which for most projects is an inevitable part of the design process. Revit's capability to coordinate between various aspects of the building design provides immense flexibility in the design and development process along with an error-free documentation.

Revit also provides a variety of in-built parametric element libraries that can be selected and used to create a building model. It also provides you with the flexibility of modifying properties of these elements or create your own parametric elements based on the project requirement.

Terms Used in Autodesk Revit

Before using Revit, it is important to understand the basic terms used for creating a building model.

Creating a Building Model Using Parametric Building Elements

Another classification of categories of elements followed in Revit is based on their usage. Revit uses five classes of elements: host, component, annotation, view, and datum. Hosts are the element categories that form the basic structure of a building model and include model elements such as walls, floor, roof, and ceiling. Components are the elements that are added to host elements or act as stand-alone elements such as doors, windows, and furniture. Annotations are the 2D, view-specific elements that add content to the project documentation such as dimensions, tags, text notes, and so on. Views represent various orientations of a building model such as plans, elevations, sections, 3D views, and so on. Datum refers to the reference elements that assist you in creating a building model, which include grids, levels, reference planes, and so on.

There is no specific methodology available for creating a building model in Revit. It provides you with the flexibility of generating building geometry based on the project requirement, design complexity, and other factors. However, the following steps describe a general procedure that may be followed for creating an architectural building model using the in-built parametric elements provided in Revit.

The first step is to create the exterior walls of a building at the predefined lowest level (level 1). Next, create interior walls at that level and add components to the building model. Then, define the upper levels based on the story height of the building. You can also link the control height of the walls to the levels and extend the exterior walls to their full height. Next, create floors and roof using the defined levels. Add the site topography to the building model and then add site components to complete the building project. You can then create drawing sheets with the desired views for its presentation. Revit also provides tools to create rendered 3D views and walkthroughs. Figure 1-3 shows an example of a building section with various building elements and annotations.

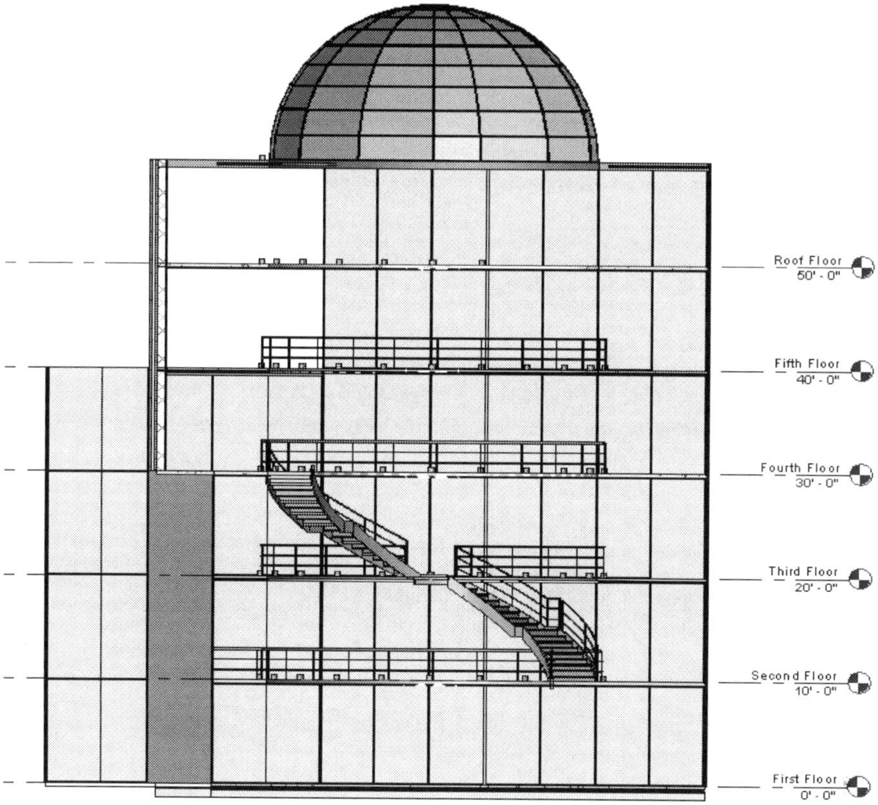

Figure 1-3 *Building section showing building elements and levels*

Visibility/Graphics Overrides, Scale, and Detail Level

Revit enables you to control the display and graphic representation of a single element or the element category of various elements in project views by using the visibility and graphics overrides tools. You can select a model category and modify its linetype and detail level. This can also be done for various annotation category elements and imported files. These settings can be done for each project view based on its desired representation. You can also hide an element or an element category in a view using the **Hide in View** and **Isolate** tools. You can override the graphic representation of an element or an element category in any view using the **Visibility/ Graphics** tool.

The scale is another important concept in a Revit project. You can set the scale for each project view by selecting it from the available list of standard scales such as 1/16"=1'0", 1/4"=1'0",

1"=1'0", 1/2"=1'0", and so on (for imperial). As you set a scale, Revit automatically sets the detail level appropriate for it. There are three detail levels provided in an Revit project: **Coarse**, **Medium**, and **Fine**. You can also set the detail level manually for each project view. Each detail level has an associated linetype and the detail lines associated with it. The details of annotations such as dimensions, tags, and so on are also defined by the selected scale.

Extracting the Project Information

A single integrated building information is used to create and represent a building project. You can extract project information from a building model and create area schemes, schedule, and cost estimates, and then add them to the project presentation.

Revit also enables you to export the extracted database to the industry standard Open Database Connectivity (ODBC) compliant relational database tables. The use of the building information model to extract database information eliminates the error-prone method of measuring building spaces individually.

Creating an Architectural Drawing Set

After creating the building model, you can easily arrange the project views by plotting them on drawing sheets. Drawing sheets can also be organized in a project file based on the established CAD standards followed by the firm. In this manner, the project documentation can easily be transformed from the conceptual design stage to the design development stage and finally to the construction document stage. The project view on a drawing sheet is only a graphical representation of the building information model and therefore, any modification in it is immediately made in all the associated project views, keeping the drawings set always updated.

Creating an Unusual Building Geometry

Revit also helps you conceptualize a building project in terms of its volume, shape, and proportions before working with the actual building elements. This is possible by using the **Massing** tool, which enables you to create quick 3D models of buildings and conduct volumetric and proportion study on overall masses. It also enables you to visualize and create an unusual building geometry. The same massing model can then be converted into a building model with individual parametric building elements. It provides continuity in the generation of building model right from sketch design to its development.

Flexibility of Creating Special Elements

Revit provides a large number of in-built family types of various model elements and annotations. Each parametric element has the associated properties that can be modified based on the project requirement.

Revit also enables you to create the elements that are designed specifically for a particular location. The in-built family editor enables you to create new elements using family templates. This provides you with the flexibility of using in-built elements for creating your own elements. For example, using the furniture template, you can create a reception desk that is suitable for a particular location in the design.

Creating Structural Layouts

Revit's structural tools enable you to add structural elements to a building model. An extensive in-built library of structural elements has been provided in Revit. You can add structural columns, beams, walls, braces, and so on to the project. Thus, structural consultants can also incorporate their elements in the basic architectural building model and check for inconsistency, if any.

Working on Large Projects

In Revit, you can work on large projects by linking different building projects together. For a large project that consists of a number of buildings, you can create individual buildings as separate projects and then link all of them into a single base file. The database recognizes the linked projects and includes them in the project representation of the base file.

For example, while working on a large campus of an educational institution, you can create separate project files for academic building, administration area, gymnasium, cafeteria, computer centre, and so on, and then link them into the base site plan file. In this manner, large projects can be subdivided and worked upon simultaneously.

Working in Large Teams and Coordinating with Consultants

Worksets in Revit enable the division of the building model into small editable set of elements. The worksets can be assigned to different teams working on the same project and then their work can easily be coordinated in the central file location. The effort required to coordinate, collaborate, and communicate the changes between various worksets is taken care of by computer. Various consultants working on a project can be assigned a workset with a set of editable elements. They can then incorporate their services and modify the associated elements.

For example, a high rise commercial building project can be divided into different worksets with independent teams working on exterior skin, interior walls, building core, toilet details, finishes, and so on. The structural consultants can be assigned the exterior skin and the core workset in which they can incorporate structural elements. Similarly, the rest of the teams can work independently on different worksets.

STARTING Autodesk Revit

You can start Autodesk Revit 2025 by double-clicking on its shortcut icon located on the desktop. Alternatively, you can start Autodesk Revit 2025 from the taskbar. To do so, choose the **Start** button; a menu is displayed. Choose **Autodesk > Revit 2025,** as shown in Figure 1-4; the interface screen will be displayed, as shown in Figure 1-5.

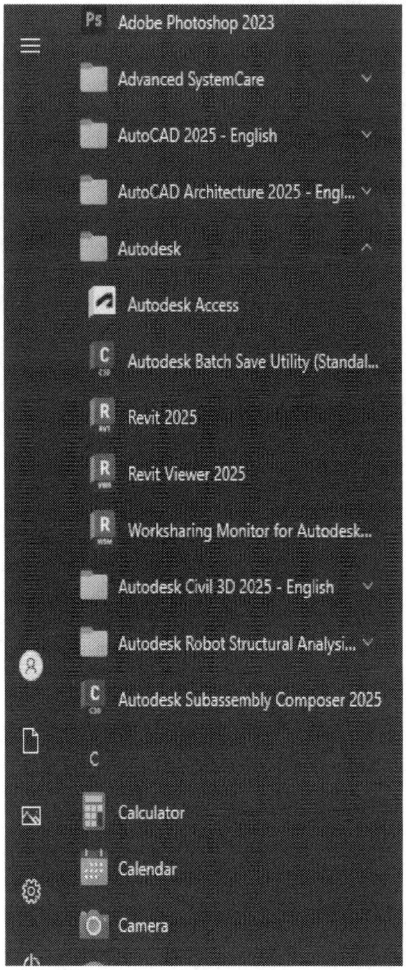

Figure 1-4 *Starting Revit 2025 from the taskbar*

Note
The path for starting Revit depends on the operating system being used.

The Revit interface screen has two sections: **Models**, and **Families**. The options in the **Models** section are used to open an existing model, a new model, and an existing template. The options in the **Families** section are used to open a new or an existing family. You can also invoke the Conceptual Mass environment from this section to create a conceptual mass model.

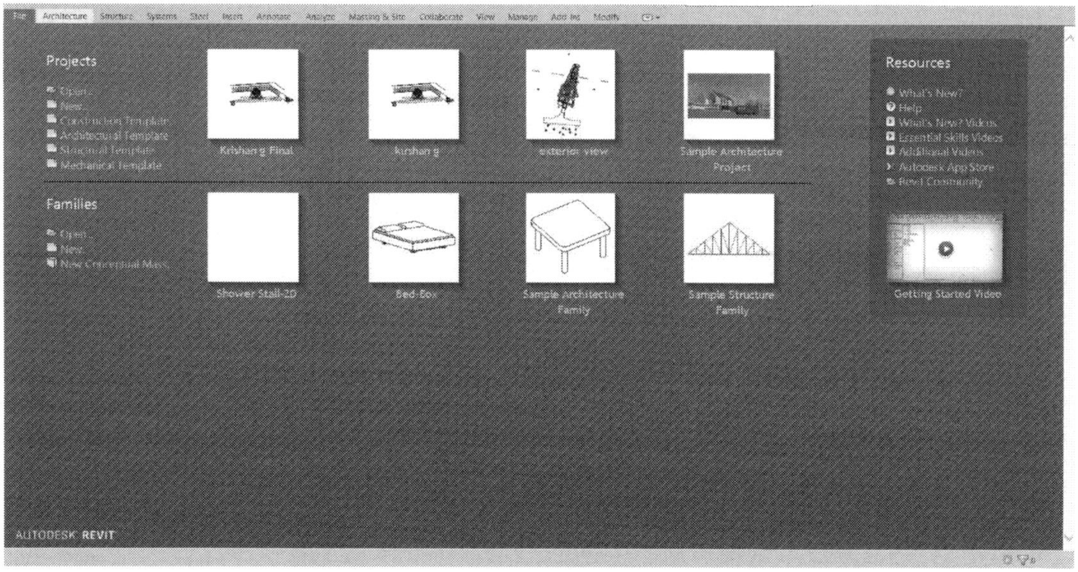

Figure 1-5 *The interface of Autodesk Revit 2025*

In the interface screen you can choose the **What's New?** option to get information about the new tools and features in Revit 2025.

When you click on the **What's New?** option, you are directed to the **Autodesk Revit 2025** page that has a list of videos of newly added features in Revit. You can click on a link to view the corresponding video.

You can choose the **Autodesk App Store** option from the **InfoCenter** to access various add-ons that can be used to enhance the productivity of Revit. On choosing this option, the **Autodesk App Store** page will be displayed. In this page, various links are available as add-ons which can be used in Revit applications. In the **Revit Community** option of the **InfoCenter** section, you can access information related to various communities and their contribution in the form of articles, tutorials, and videos.

In addition, you can choose the **Help** option from the **InfoCenter** to get help on various tools. When you choose this option, you will be directed to *https://help.autodesk.com/view/RVT/2025/ENU/*. Also, the Autodesk Revit 2025 page with the **Welcome to Revit 2025 Learning** area will be displayed. To access information related to additions and enhancements in **Revit 2025** release, you can expand the **What's New** node from the left pane and then click on the **What's**

New link. On doing so, the **What's New** page will be displayed with various links. You can visit the links to learn about enhancements in Revit 2025.

The new interface screen of Autodesk Revit 2025 has been enhanced to provide better experience for navigating BIM 360 projects. Now, you can access your projects stored in BIM 360 directly from the interface screen of Revit 2025 by clicking on the **Autodesk Docs** button.

Some other features are also available in the interface screen of Autodesk Revit 2025. These include **Community Forum** and **Customer Support** options. When you click on the **Community Forum** option, Revit 2025 directs you to *https://forums.autodesk.com/t5/revit-products/ct-p/2003* page. On this page, you can access Autodesk's Revit Forums to share your knowledge, ask questions, collaborate on ideas, and explore popular Revit topics. When you click on the **Customer Support** option, you are directed to *https://knowledge.autodesk.com/contact-support* page. This page provides you the information on purchasing, downloading, installing and troubleshooting of various softwares. This page also assists you in managing your Autodesk account and your software license.

In the **Models** section, choose the **Open** option; the **Open** dialog box will be displayed. Browse to the desired location in the dialog box and select the file. Now, choose the **Open** button to open the file.

To open a new model file, choose the **New** option from the **Models** section. Alternatively, choose **New > Project** from the **File** menu; the **New Project** dialog box will be displayed. In this dialog box, you can select the desired template from the **Template file** drop-down list or you can browse the other template files by using the **Browse** button from the **Template File** area. When you choose the **Browse** button, the **Choose Template** dialog box will be displayed. In this dialog box, make sure the **Project** radio button is selected, and then choose the **OK** button; a new project file will open and the interface screen will be activated.

USER INTERFACE

Autodesk Revit has ribbon interface. The ribbon which contains task-based tabs and panels, streamlines the architectural workflow and optimizes the project delivery time. In Revit, when you select an element in the drawing area, the ribbon displays a contextual tab that comprises of tools corresponding to the selected element. The interface of Autodesk Revit is similar to the interfaces of many other Microsoft Windows based programs. The main parts in the Revit interface are **Ribbon**, **Options Bar**, **Project Browser**, **Drawing Area**, **Status Bar**, and **View Control Bar**, as shown in Figure 1-6.

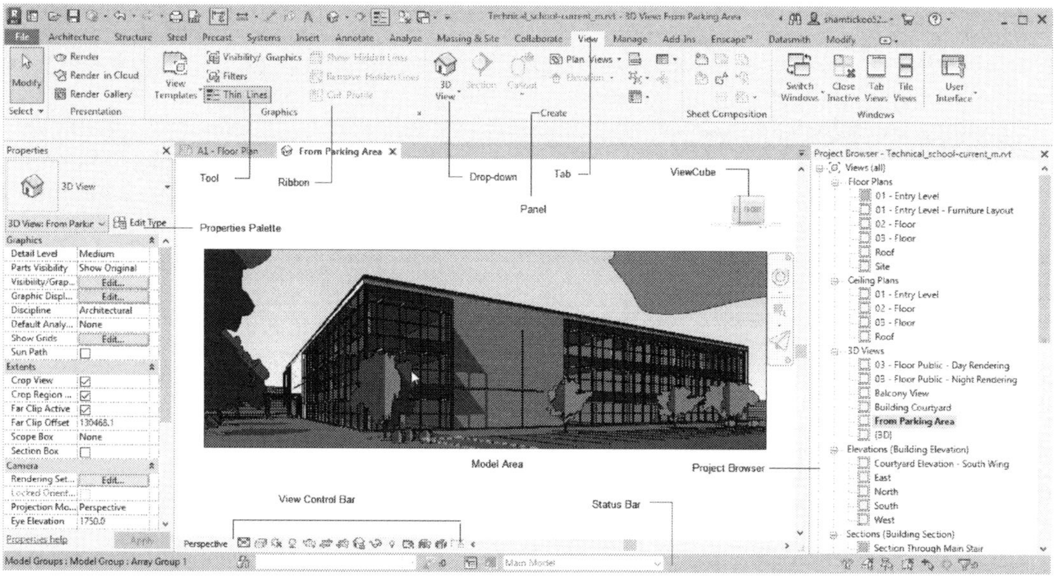

Figure 1-6 The Autodesk Revit 2025 user interface screen

Invoking Tools

To perform an operation, you can invoke the required tools by using any one of the following two options:

Ribbon: You can invoke all necessary tools from the ribbon.

Shortcut Keys: Some tools can also be invoked by using the keys on the keyboard.

Title Bar

The Title bar, docked on the top portion of the user interface, displays the program's logo, program's name of the current project, and the view opened in the viewing area. **Project 1- Floor Plan: Level 1** is the default project and view.

Ribbon

The ribbon, as shown in Figure 1-7, is an interface that is used to invoke tools. When you open a file, the ribbon is displayed at the top in the screen. It comprises of task-based tabs and panels, refer to Figure 1-7, which provide all the tools necessary for creating a project. The tabs and panels in the ribbon can be customized according to the need of the user. This can be done by moving the panels and changing the view states of the ribbon (changing the ribbon view state is discussed later in this chapter). The ribbon contains buttons, drop-downs, panels, tabs, and tools.

Figure 1-7 Different components of a ribbon

Status Bar

The **Status Bar** is located at the bottom of the interface screen. When the cursor is placed over an element or component, the **Status Bar** displays the name of the family and type of the corresponding element or components. It also displays prompts and messages to help you use the selected tools.

View Control Bar

The **View Control Bar** is located at the lower left corner of the drawing window, as shown in Figure 1-8. It can be used to access various view-related tools. The **Scale** button shows the scale of the current view. You can choose this button to display a flyout that contains standard drawing scales. From this flyout, you can then select the scale for the current view. The **Detail Level** button is used to set the detail level of a view. You can select the required detail level as **Coarse**, **Medium**, and **Fine**. Similarly, the **Visual Style** button enables you to set the display style. The options for setting the display style are: **Wireframe**, **Hidden Line**, **Shaded**, **Consistent Colors**, **Realistic**, and **Raytrace**.

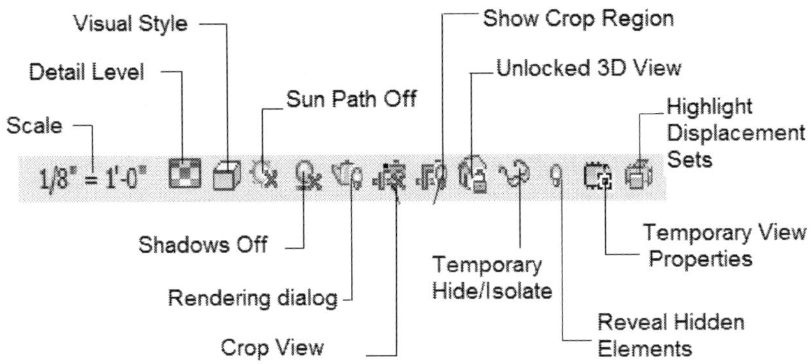

*Figure 1-8 The **View Control Bar***

Options Bar

The **Options Bar** provides information about the common parameters of component type and the options for creating or editing of building elements. The **Options Bar** changes its appearance based on the type of component selected or being created. You can also modify the properties of the component by entering a new value in the edit box for the corresponding parameter in the **Options Bar**. For example, the **Options Bar** for the **Wall** tool displays various options to create a wall, as shown in Figure 1-9.

*Figure 1-9 The **Options Bar** with different options to create a wall*

Type Selector

The **Type Selector** drop-down list is located in the **Properties** palette of the currently invoked tool. For example, if you invoke the **Wall** tool, all the properties of the wall will be displayed in the **Properties** palette. In the **Properties** palette, you can use the **Type Selector** drop-down list to select the required type of the wall. The options in the **Type Selector** drop-down list keep on changing based on the current function of the tool or the elements selected. When you place an element or a component in a drawing, you can use the **Type Selector** drop-down list to specify the type of element or component. You can also use this drop-down list to change the existing type of a selected element to a different type. In Revit, you can add the **Type Selector** drop-down list to the **Quick Access Toolbar**. To do so, right-click on the down arrow on the right in the **Type Selector** drop-down list in the **Properties** palette; a flyout will be displayed. Choose the **Add to Quick Access Toolbar** option from the flyout.

Drawing Area

The Drawing Area is the actual modeling area where you can create and view the building model. It covers the major portion of the interface screen. You can draw various building components in this area using the pointing device. The position of the pointing device is represented by the cursor. The Drawing Area also has the standard Microsoft Windows functions and buttons such as close, minimize, maximize, scroll bar, and so on. These buttons have the same function as that of the other Microsoft Windows-based programs.

Project Browser

The **Project Browser** is located below the ribbon. It displays project views, schedules, sheets, families, and groups in a logical, tree-like structure, as shown in Figure 1-10, and helps you open and manage them. To open a view, double-click on the name of the view; the corresponding view will be displayed in the drawing area. You can close the **Project Browser** or dock it anywhere in the drawing area.

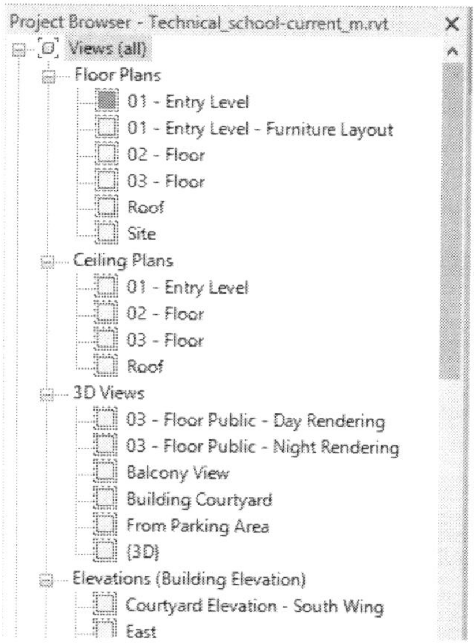

*Figure 1-10 The **Project Browser***

Note

*If the **Project Browser** is not displayed on the screen, choose the **View** tab from the ribbon and then click on the **User Interface** drop-down from the **Windows** panel. Next, select the **Project Browser** check box from the flyout displayed.*

The **Project Browser** can be organized to group the views and sheets based on the project requirement. For example, while working on a large project with a number of sheets, you can organize the **Project Browser** to view and access specific sheets.

Note

*In the **Project Browser**, you can expand or collapse the view listing by selecting the '+' or '-' sign, respectively. The current view in the drawing window is highlighted in bold letters. The default project file has a set of preloaded views.*

Keyboard Accelerators

In Revit, accelerator keys have been assigned to some of the frequently used tools. These keys are shortcuts that you can type through the keyboard to invoke the corresponding tool. Accelerator keys corresponding to a tool appear as a tooltip when you move the cursor over the tool. In Revit 2025, you can export all commands (even if they do not have shortcut keys assigned) to a XML file. You can further edit the XML file to assign shortcut keys to commands, and then import them back to be used in Revit.

Tip

*As you get accustomed to using Revit, you will find these **Keyboard Accelerators** quite useful because they save the effort of browsing through the menus.*

Properties Palette

The **Properties** palette, as shown in Figure 1-11, is a modeless interface, which displays the type and element properties of various elements and views in a drawing. The **Properties** palette is dockable and resizable, and it supports multiple monitor configurations. The **Properties** palette is displayed in the Revit interface by default and it shows the instance properties of an active view. When you select an element from a drawing, the **Properties** palette displays its instance properties. You can also access the **Type Properties** of the selected element from the **Properties** palette. To do so, choose the **Edit Type** button from the palette; the **Type Properties** dialog box will be displayed. In this dialog box, you can change the **Type Properties** of the selected element. In the **Properties** palette, you can assign a type to a selected element in a drawing from the **Type Selector** drop-down list. In Revit, you can toggle the display of the **Properties** palette in its interface. Choose the **Properties** button in the **Properties** panel of the **Modify** tab to hide it. Similarly, you can choose the **Properties** button to display the palette if it is not visible in the interface.

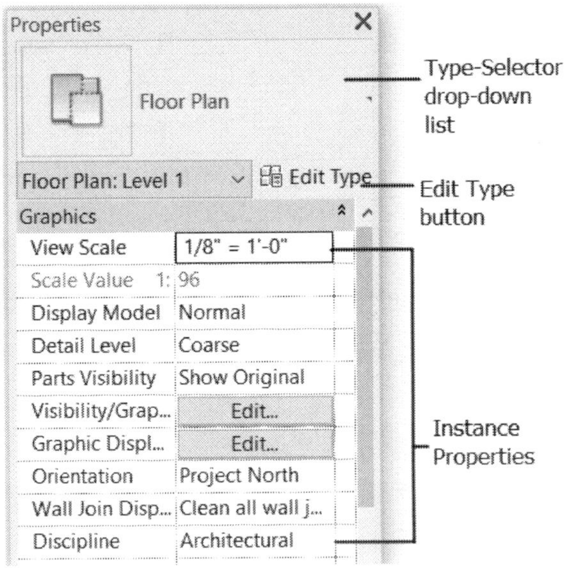

*Figure 1-11 The **Properties** palette*

DIALOG BOXES

Some Revit tools when invoked display a dialog box. A dialog box is a convenient method of accessing and modifying the parameters related to that tool. For example, when you choose **Save As > Project** from the **File** menu, the **Save As** dialog box will be displayed, as shown in Figure 1-12. A dialog box consists of various parts such as dialog label, radio buttons, text or edit boxes, check boxes, slider bars, image box, and tool buttons, which are similar to other windows-based programs. Some dialog boxes contain the [**...**] button. On choosing such buttons, another related dialog box will be displayed. There are certain buttons such as **OK**, **Cancel**, and **Help**, which appear at the bottom of most of the dialog boxes. The names of the buttons imply their respective functions. The button with a dark border is the default button.

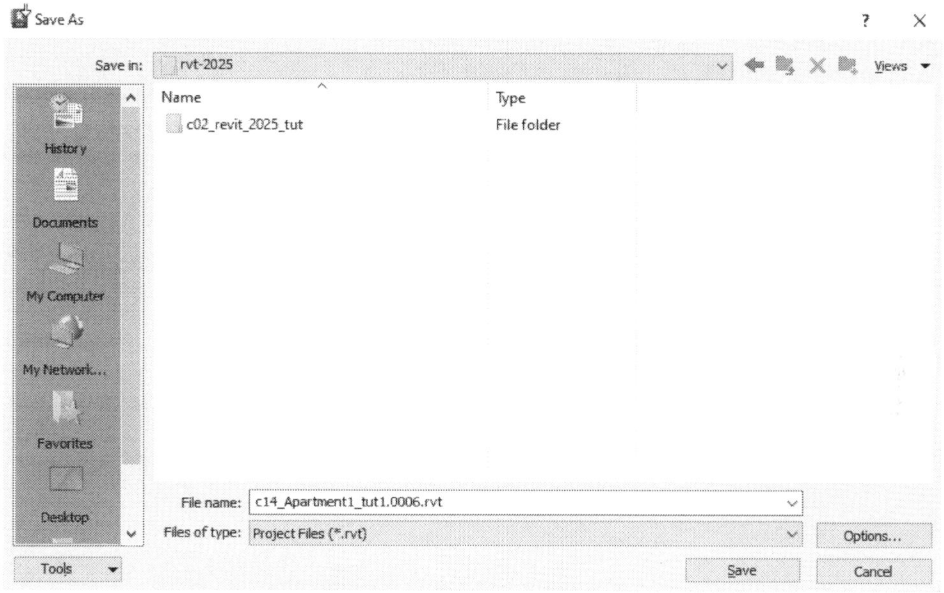

*Figure 1-12 The **Save As** dialog box*

MULTIPLE DOCUMENT ENVIRONMENT

The multiple document environment feature allows you to open more than one project at a time in a single Revit session. This is very useful when you want to work on different projects simultaneously and make changes with reference to each other.

Sometimes you may need to incorporate certain features from one project into the other. With the help of multiple document environments, you can open multiple projects and then use the **Cut**, **Copy**, and **Paste** tools from the **Clipboard** panel of the **Modify (type of element)** tab to transfer the required components from one project to another. These editing tools can also be invoked by using the CTRL+C and CTRL+V keyboard shortcuts.

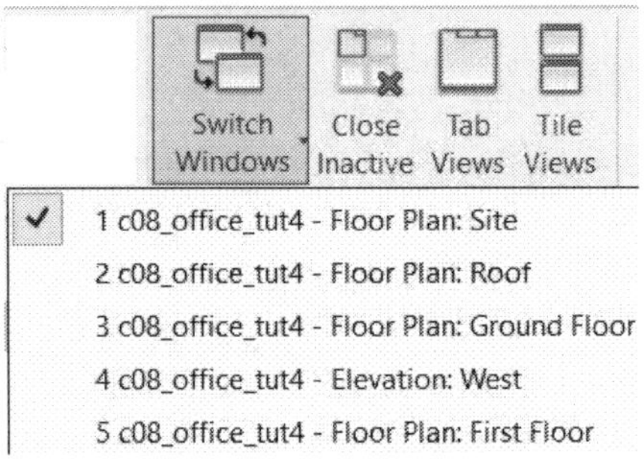

***Figure 1-13** Selecting an option from the **Switch Windows** drop-down*

To access the opened projects, choose the **Switch Windows** drop-down from the **Windows** panel of the **View** tab; a menu will be displayed showing the name of different project files opened, as shown in Figure 1-13. Like other Microsoft Windows-based programs, you can select and view the opened projects using the **Tile** and **Tab** tools from the **Windows** panel of the **View** tab. Figure 1-14 shows the tab view of projects.

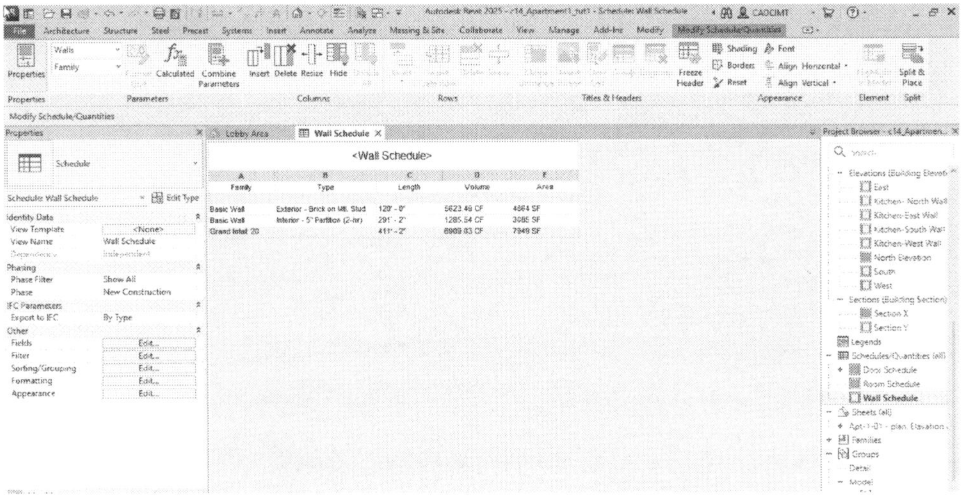

***Figure 1-14** The Tab view of the projects*

INTEROPERABILITY OF Autodesk Revit

The models or geometries created in Revit can easily be exported to AutoCAD based programs, such as 3ds Max and Max Design in the DWG file format. This enables you to visualize and create photorealistic exterior and interior renderings for your project designs. You can also transfer drawings from Revit to Trimble SketchUp to visualize your projects in a better way.

Revit follows a wide range of industry standards and supports various CAD file formats such as DWG, DXF, DWF, DGN, FBX, and SAT. For image files, it supports JPG, TIFF, BMP, PNG, AVI, PAN, IVR, and TGA file formats. Besides these, the formats that are supported by Revit include ODBC, HTML, TXT, gbXML, XLS, and MDB. Revit is compatible with any CAD system that supports the DWG, DXF, or DGN file format. Revit can import the models and geometries as

ACIS solids. This enables designers to import models from AutoCAD Architecture and AutoCAD MEP (Mechanical, Electrical, and Plumbing) software and to link and import 3D information to Revit. This feature makes Revit an efficient, user-friendly, and compatible software.

In Revit, you can directly link the files into 3ds Max and load selected views in it. You can also override material in 3ds Max and retain its settings when you reload Revit link file. Also, in 3ds Max, you can add high level of details to the curved objects to make them smooth. Microstation is interoperable with Revit. Therefore, the Microstation files can be imported to the Revit project. In addition to this, mapping functionality for levels, lines, line weights, patterns, and texts and fonts is added to export DGN workflow.

Revit provides fully certified IFC (Industry Foundation Class) import and export based on buildingSMART® IFC data exchange standards. For importing IFC files Revit supports IFC files based on the following buildingSMART International (bSI) data exchange standards: IFC2x3, IFC2x2, and IFC2x. For import (link only), Revit also supports IFC files based on the bSI IFC4 standard.

For export, Revit supports the following standards: IFC4, IFC2x3, and IFC2x2. When you export a Revit building information model to IFC format, the information can be used directly by other stakeholders, such as structural and building services engineers. For example, if a building information models is created in Revit and saved as *.rvt* file format, you can export the building model using the IFC format to an IFC-certified application such as Tekla, Open Building and others file format. The exported model can be opened and worked on in the non-native application. Similarly, in Revit you can import an IFC file, create a RVT file, and work on the building model in Revit.

IFC file uses architecturally meaningful data to describe real-world building objects. These data include parameters that have meaningful values. In Revit 2025, the standard Revit elements have corresponding IFC scheme. These do not require any specific user action to export them. (For example, Revit walls export as ifcWalls.) Other Revit families (such as escalators) require that you map them to IFC scheme before exporting

BUILDING INFORMATION MODELING AND Autodesk Revit

Building Information Modeling (BIM) is defined as a design technology that involves creation and use of coordinated, internally consistent, and computable information about a building project in design and construction. BIM covers spatial relationships, geographic information, quantities, and properties of building components. Using this technology, you can demonstrate the entire life cycle of a building project starting from the process of construction, facility operation, and information about quantities and shared properties of elements. BIM enables the circulation of virtual information model from the design team to contractors and then to the owner, thereby adding changes and their knowledge to update the model at each stage of transfer. The ability to keep information up-to-date and make it available in an integrated digital environment enables the architects, owners, builders, and engineers to have clear vision of the project before the commencement of actual construction. It enables them to make better and faster decisions as well as to improve the quality and profitability of projects. Autodesk Revit is a specially designed platform based on BIM. Revit is the best example of the BIM technology. Revit's parametric model represents a building as an integrated database of coordinated information. In Revit, change anywhere is change everywhere. Any change made in your project at any stage is reflected in the entire project, and also, due to the parametric behavior of elements, the project is updated automatically according to the changes made anywhere in the project. Also, the integration of Revit with the available in-built commercial tools such as solar studies, material takeoffs, greatly simplifies the project design and reduces the time consumed by these analyses, thereby enabling faster decision making.

Autodesk Revit 2025 HELP

Autodesk Revit provides help to easily understand various tools and methods used in it. In Autodesk Revit 2025, you can access online help documentation. To access the help feature, click on the down arrow on the right of the **InfoCenter**; a flyout will be displayed. Next, choose the **Help** option, as shown in Figure 1-15. Various options to access the help are discussed next.

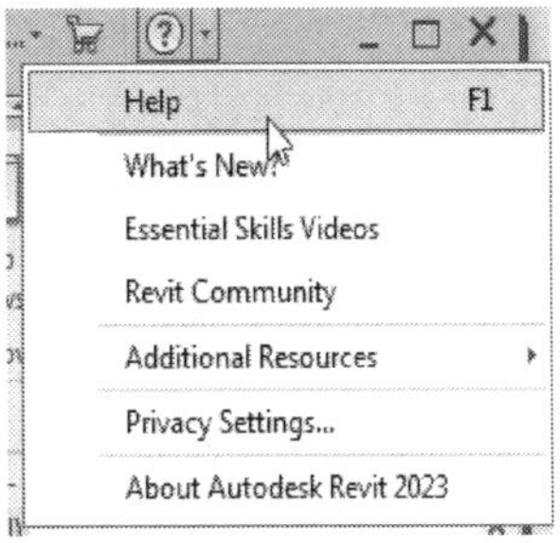

Figure 1-15 *A drop-down menu displaying help options*

Using the Revit 2025 Help

You can access Autodesk Revit 2025 help when you are online. To do so, choose the **Help** tool from the **InfoCenter**; the **Autodesk Revit 2025** page will be displayed, as shown in Figure 1-16. In this page, there are several tabs that contain information of help topics. These tabs are useful to understand the basic concepts of Revit.

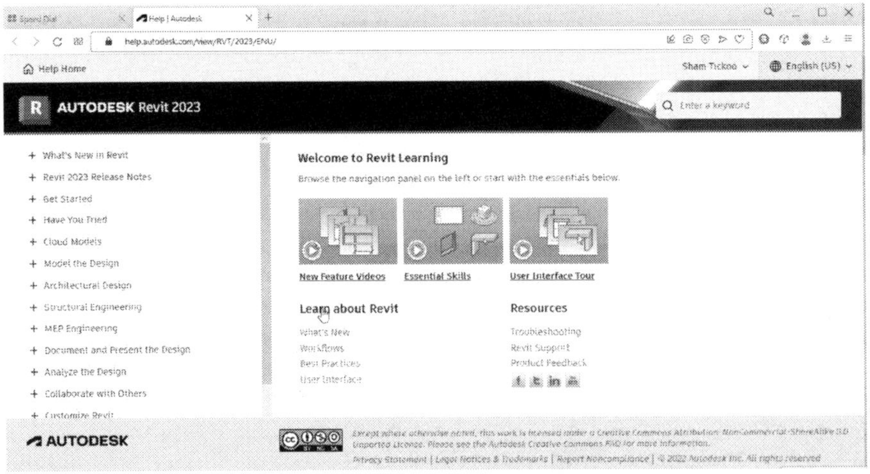

Figure 1-16 *The **Revit Help** page*

WORKSHARING USING REVIT SERVER

Worksharing is a method of distributing work among team involved in a project, and accomplishing it within the stipulated period of time. In worksharing, each person involved in the project is assigned a task that has to be accomplished by proper planning and by coordinating with the other members of the team.

In a large scale building project, worksharing helps in finishing a project in time and meeting the quality requirements that are set during the process. Generally, in a large scale building project, worksharing is based on the specialization of work. The professionals such as structural engineers, architects, interior architects, and MEP engineers are involved in their respective fields to accomplish the project. So, the distribution of work at the primary stage is made on the basis of the area of specialization. Each professional has his own set of work to perform for the accomplishment of the project.

You can apply server-based worksharing with the help of Revit Server as it is a server based application. Revit Server uses a central server and multiple local servers for collaborating across a Wide Area Network (WAN). The central server hosts the central model of a workshared project and remain accessible to all the team members over the Wide Area Network. Similarly, the local server is accessible to all team members in a Local Area Network (LAN). The local server hosts a local updated copy of the central model. In the Worksharing environment, the team members are not aware of the local server, as it is transparent in their daily operations. Refer to Figure 1-17 for the network model of Revit Server.

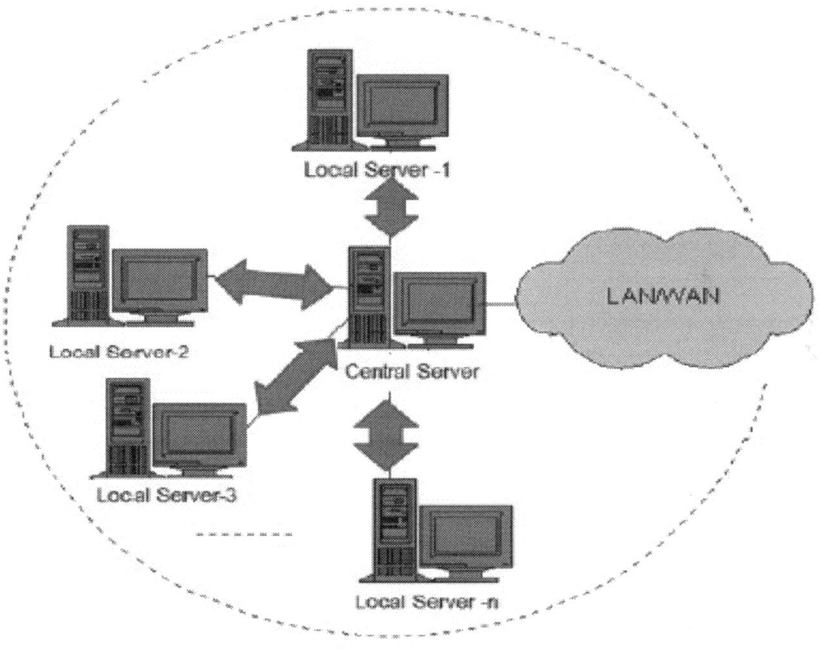

Figure 1-17 *The network model of Revit Server*

In Worksharing environment, a team member starts working on the local model of the central model. The local model will be saved in the computer of the team member. As the team member works, the local server requests updated information from the central model on the central server using available network capacity to transfer the data over the WAN. The updated version of the model is stored on the local server, so the updates are readily available when a team member requests them.

AUTODESK CONSTRUCTION CLOUD

Autodesk Construction Cloud™ is used as a cloud-based construction management and collaboration solution developed by Autodesk. Autodesk Construction Cloud solution connects building data, project work flows, and project teams throughout the entire building life cycle, from design to operations. Autodesk Construction Cloud is built on a integrated platform and common data environment that will empower general contractors, specialty trades, designers and owners to drive better business outcomes. Using the Autodesk Construction Cloud solution, project teams will now have a comprehensive construction management platform in which all data will be located centrally. As such, it will simplify collaboration, predict project changes, and will provide guidance that will be driven by data for improvement of the organization. The Autodesk Construction Cloud offers three main products namely Autodesk Build, Autodesk Quantity, and Autodesk BIM Collaborate. As part of the unified Autodesk Construction Cloud, every product is reinforced by Autodesk Docs, Insights, and Administration.

Autodesk Docs cloud platform supports the common data environment for every product of Autodesk Construction Cloud. It provides the users with uninterrupted navigation of data across the project teams and throughout the project workflow. Autodesk Docs is a single, cloud-based platform for streamlining document management by storing and managing all project documents and data from design through construction in a single platform.

Insights platform is used to delivers analytical report from the data collected and analyzed from the project. It is also used to export that data to the user with the application of artificial intelligence to identify and mitigate risk.

Chapter 2

Starting an Architectural Project

Learning Objectives

After completing this chapter, you will be able to:

- *Start a new architectural project*
- *Set units of various measurement parameters of a project*
- *Understand the concept of snaps, dimensions, and object snaps*
- *Save a project*
- *Modify parameters and settings of a project*
- *Close a project and exit Revit 2025*
- *Open an existing project*
- *Explore the building model using viewing tools*
- *Use navigation tools*

INTRODUCTION

In Revit, you can work on structural, architectural, and MEP (Mechanical, Electrical and Plumbing) projects on a single platform. The chapters in this textbook are specially written for professionals in architectural and interior design field. In this chapter, you will learn about the tools and the processes involved in starting up a new architectural project.

STARTING A NEW ARCHITECTURAL PROJECT

Shortcut Key:	CTRL+N
File menu:	New > Project

In Revit, a project is considered as a single database that contains all information related to building design. Starting from geometry to construction data, each project file contains complete information of the building design. In a building design, the three dimensional models drawn using this software are called BIM (Building Information Model). BIM is a process involving the generation and management of digital representation of physical and functional features of different building elements.

To start an architectural project, choose **New > Project** from the **File** menu, as shown in Figure 2-1. On doing so, the **New Project** dialog box will be displayed, as shown in Figure 2-2. Using this dialog box, select an existing *.rte* template file format that can be used in the new project.

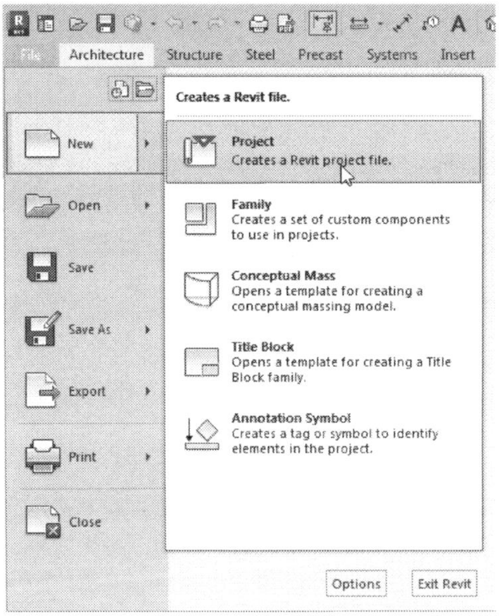

Figure 2-1 *Choosing the **Project** option*

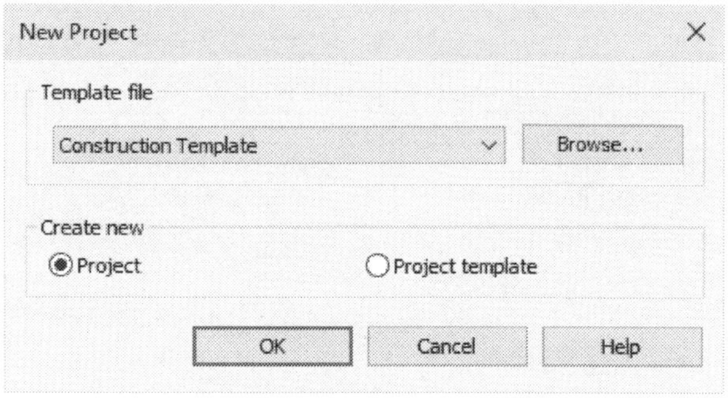

***Figure 2-2** The **New Project** dialog box*

When you start a new Revit project, you need to apply a template to it. Revit comes with several templates which are pre-installed ranging from architectural templates to structural templates. These template files are required to start a project and are designed to save your time while setting up a new project. You can have as many different templates as per your company standards. If you are working on different types or projects, whether it is of different disciplines such as plumbing or architectural or even if it is categorized by purpose such as commercial or residential, you are can set up your templates according to your requirement.

A template file can be defined as a template which has various project parameters such as units and views, already saved in it. On using the template file, the new project file will adopt the same parameters as the template file. The difference between a template file and a project file is that the template file has a *.rte* extension, whereas the project file has a *.rvt* extension. You can either select any of the template files provided in Revit or create your own file. Any project file can be saved as a template file.

PROJECT UNITS

Ribbon:	Manage > Settings > Project Units
Shortcut Key:	UN (By Default)

Units are important parameters in a project. The default selection of units helps you open project with the specified/selected unit system. However, you can change the default unit set system. To set units, choose the **Project Units** tool from the **Settings** panel of the **Manage** tab; the **Project Units** dialog box will be displayed, as shown in Figure 2-3. Project units are grouped into seven disciplines: **Common**, **Structural**, **HVAC**, **Electrical**, **Piping**, **Energy**, and **Infrastructure**. Each discipline has a set of measurement parameters. You can select any of these disciplines from the **Discipline** drop-down list of this dialog box. Select the **Common** option from the **Discipline** drop-down list; various measurement parameters such as **Angle**, **Area**, **Cost per Area**, **Distance**, **Length**, **Mass Density**, **Rotation Angle**, **Slope**, **Speed**, **Time**, **Volume**, and **Currency** will be displayed in the **Project Units** dialog box. The **Format** column in the dialog box displays the current unit format for the corresponding parameter. You can preview and select the possible digit grouping and decimal separators from the **Decimal symbol/digit grouping** drop-down list located at the lower left corner of the dialog box, refer to Figure 2-3. Some of the options for settings measurement units are discussed next.

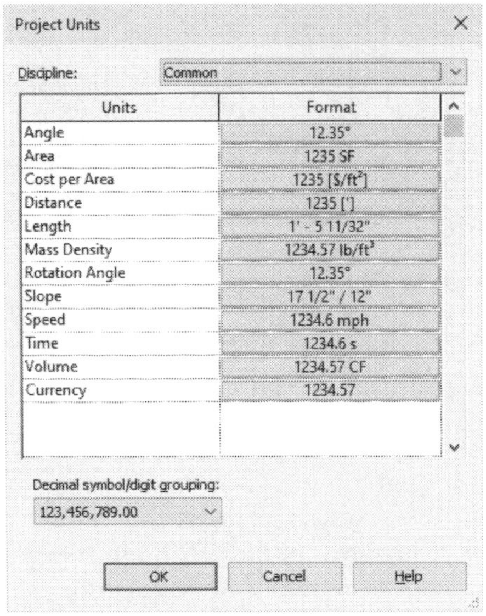

*Figure 2-3 The **Project Units** dialog box*

Angle Unit

The unit for angle can be set by using the **Units** drop-down list in the **Format** dialog box of the **Angle** parameter.

Area Unit

To assign a unit for measuring the areas of building elements, click on the **Format** column for the **Area** parameter; the **Format** dialog box will be displayed. In this dialog box, you can set the unit for measuring an area by selecting an option from the **Units** drop-down list. This drop-down list contains various options such as **Square feet**, **Square meters**, **Acres**, and so on. By default, the **Square feet** option is selected in this drop-down list. The settings for rounding, rounding increment, and units can be done by selecting the desired option from the respective drop-down list and edit boxes.

Length Unit

To assign a unit for measuring the lengths of building elements in your project, click on the **Format** column corresponding to the **Length** parameter; the **Format** dialog box will be displayed, as shown in Figure 2-4. This dialog box displays different units of length and their settings. You can select the desired unit from the **Units** drop-down list in the dialog box. After selecting the desired unit, you can specify the rounding value for the selected unit. To do so, select the desired option from the **Rounding** drop-down list in the **Format** dialog box. Note that by default, the **Rounding increment** edit box in the **Format** dialog box is inactive. To make it active, select an option from the **Units** drop-down list and then select the **Custom** option from the **Rounding** drop-down list. Note the **Custom** option remains inactive for the **Feet and fractional inches**, **Fractional inches**, and **Meters and Centimeters** options of the **Units** drop-down list. The default value in the **Rounding increment** edit box is 1. You can change this value by entering a value in this edit box. Similarly, the **Unit symbol** drop-down list will be inactive for the **Feet and fractional inches**, **Fractional inches**, and **Meters and Centimeters** options of the **Units** drop-down list. From the **Unit symbol** drop-down list, you can select the measurement symbol that will be added to the unit of length. For example, if you select '**m**' as the measurement symbol, then '**m**' will be added to all metric length measurements. In case, you select **Feet and fractional inches** from the **Units** drop-down list in the **Format** dialog box, then you need to select the **Suppress spaces** check box to remove spaces in between the values when a length string is expressed in feet and fractional inches to denote a particular measurement.

Figure 2-4 *The* ***Format*** *dialog box*

Mass Density Unit

The mass density of building elements is required for structural analysis. In Revit, you can assign a unit for measuring mass density. To assign the unit of mass density, invoke the **Project Units** dialog box. In the **Format** column of this dialog box, choose the button corresponding to the **Mass Density** parameter; the **Format** dialog box will be displayed. In this dialog box, you can select different units from the **Units** drop-down list. Also, you can assign a unit symbol for the selected unit. To do so, click on the **Unit symbol** drop-down list and then select any of the options displayed.

Slope Unit

To specify the unit for the slope measurement, click in the **Format** column for the **Slope** parameter; the **Format** dialog box will be displayed. In this dialog box, you can specify the desired unit settings by selecting the required option from the **Units** drop-down list. The default selected option for the Imperial unit setting in the drop-down list is **Rise / 12"**.

Speed Unit

To specify the unit for speed, click in the **Format** column corresponding to the **Speed** parameter: the **Format** dialog box will be displayed. In this dialog box, you can select the desired unit of speed from the **Units** drop-down list. This drop-down list contains various options such as **Feet per minute, Meters per second**, **Feet per second**, **Kilometers per hour**, and **Miles per hour**.

Time Unit

To specify the unit for time, click in the **Format** column corresponding to the **Time** parameter; the **Format** dialog box will be displayed. In this dialog box, you can select the desired unit of time from the **Units** drop-down list. This drop-down list contains various options such as **Milliseconds**, **Seconds**, **Minutes**, and **Hours**.

Volume Unit

The units for volume can be set similar to that of the length and area. You can set the unit for the volume measurement by selecting any of the options from the **Units** drop-down list in the **Format** dialog box of the **Volume** parameter.

Currency Unit

The currency unit is used to set the unit of currency for its usage in the cost and estimation schedules. To set the unit of currency, invoke the **Project Units** dialog box and then choose the button displayed in the **Format** column corresponding to the **Currency** parameter; the **Format** dialog box will be displayed. From this dialog box, you can select the required type of currency symbol from the **Unit symbol** drop-down list.

Tip

*As soon as you change the units and choose the **OK** button to close the **Format** dialog box, the numbers and units shown for each measurement parameter in the **Project Units** dialog box are modified to the new settings. You can modify these settings and format them as per your requirement.*

Note

*You can format the display of units represented on the screen using the **Project Units** dialog box. The actual values for these units in the project may be different. For example, if you set the wall length rounding to the nearest 1', the wall may show this rounded value, but the actual length of the wall might be in fractional feet.*

SNAPS TOOL

Ribbon: Manage > Settings > Snaps

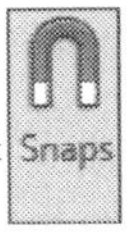

The **Snaps** tool is one of the most productive tools available while creating and editing elements in a building model. This tool represents the ability of the cursor to snap or jump to the preset increments or specific object properties of various elements such as endpoint, midpoint, and so on. Invoke the **Snaps** tool from the **Settings** panel of the **Manage** tab; the **Snaps** dialog box will be displayed, as shown in Figure 2-5. This dialog box has three areas: **Dimension Snaps**, **Object Snaps**, and **Temporary Overrides**.

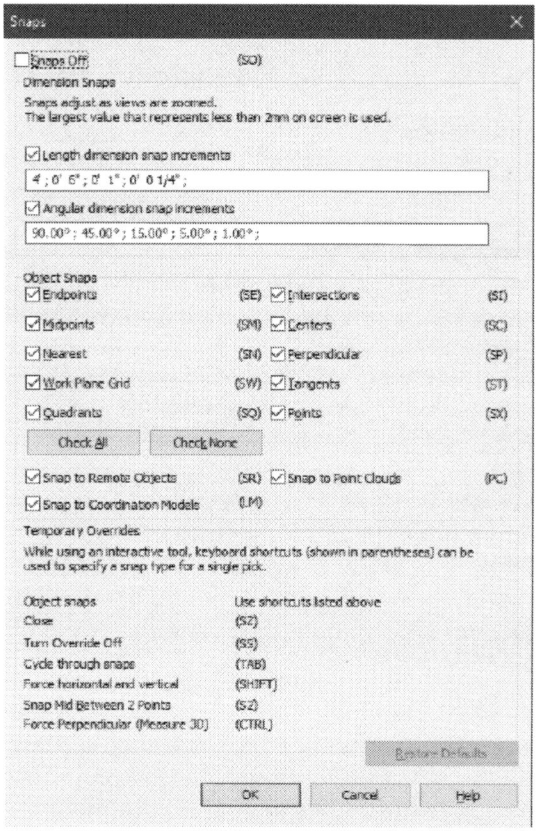

***Figure 2-5** The **Snaps** dialog box*

Dimension Snaps Area

In this area, you can set increments for placing elements or components in a project view. You can set increments for the length and angle dimensions. To set the increment of length dimension, select the **Length dimension snap increments** check box, if it is not selected by default, and then enter the increment values in the edit box below it. The default values entered in this edit box are: **4' ; 0' 6" ; 0' 1" ; 0' 0 1/4"** ; for Imperial (**1000 ; 100 ; 20 ; 5** ; for Metric). Note that every incremental value is separated by a semicolon (;). You can also set increments by typing the values separated by a semicolon. For example, to create an interior layout plan in which the length of the partitions is in 5'(1524 mm) modules, counter top width is 2'(609.6 mm), and the thickness of partitions is 4"(101.6 mm), you can enter the values for the dimension snaps as 5'; 2';4" for Imperial (1524; 609.6;101.6 for Metric). This will enable the cursor to move in these increments and help create the layout with relative ease.

Object Snaps Area

In the **Object Snaps** area, you can specify various object snaps for using them in a project. Object snapping refers to the cursor's ability to snap to geometric points on an element such as endpoints, midpoint, perpendicular, and so on. It is useful for creating and editing elements. The advantage of using object snapping is that you can locate the appropriate point on a drawing object. When enabled, the appropriate object snap is displayed as soon as the cursor is near to an element. For example, it is virtually impossible to pick the exact endpoint to start a wall from an endpoint of an already drawn wall. But when you enable the **Endpoints** object snap, the cursor automatically jumps or snaps to the endpoint of this wall. This helps to start the new wall from the endpoint. This, besides making the drawing accurate, later helps in adding dimensions to the project.

Temporary Overrides Area

The options in the **Temporary Overrides** area provide you the alternative of overriding snaps setting for a single use only. For example, if you have not selected the **Endpoints** object snap in the **Snaps** dialog box and you want to use this option while working with a tool, you need not open the **Snaps** dialog box and set this option. You can instead type the shortcut, **SE** to temporarily activate the endpoint object snap. Once you have used this object snap option, snapping to the endpoint is automatically turned off.

You can toggle between various object snap options available at the same location using the TAB key on the keyboard. Hold down the SHIFT key to create the elements vertically or horizontally. This restricts the movement of the cursor in the orthogonal directions only. Once you release the SHIFT key, the cursor resumes its movement in all directions. You can select the **Snaps Off** check box to disable all types of snapping.

SAVING A PROJECT

You must save your work before closing a project or exiting Revit 2025 session. You have the option of saving the project file in a permanent storage device, which may be a hard disk or a removable disk. Also, you must save your work at regular intervals to avoid loss due to any error in the computer's hardware or software.

Saving the Project File

File menu:	Save As > Project
Shortcut Key:	CTRL+S

To save the project file to a desired location, choose **Save As > Project** from the **File** menu; the **Save As** dialog box will be displayed, as shown in Figure 2-6. Alternatively, you can save the project file by choosing the **Save** button from the **Quick Access Toolbar**. In the **Save As** dialog box, the **Save in** drop-down list displays the current drive and path in which the project file will be saved. The list box below the **Save in** drop-down list shows all the folders available in the current directory. The **File name** edit box can be used to enter the name of the file to be assigned to the project. The **Places List** area on the left of the **Save As** dialog box contains shortcuts for the folders that are frequently used.

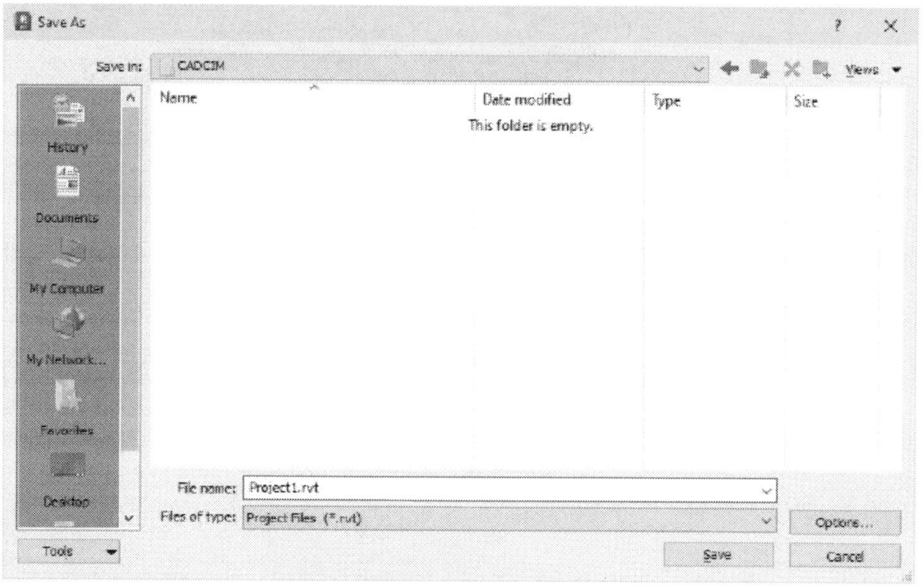

***Figure 2-6** The **Save As** dialog box*

Using the Save Tool

Once the project is saved using the **Save As** tool, you do not need to re-enter the file parameters to save it again. To save a project to a location, choose the **Save** tool from the **File** menu, as shown in Figure 2-7. While saving the project for the first time, the **Save As** dialog box is displayed, even if you invoke the **Save** tool. Alternatively, you can save your project by choosing the **Save** button without prompting you to re-enter the file name and path from the **Quick Access Toolbar**. As you save your project file, Revit 2025 updates it automatically.

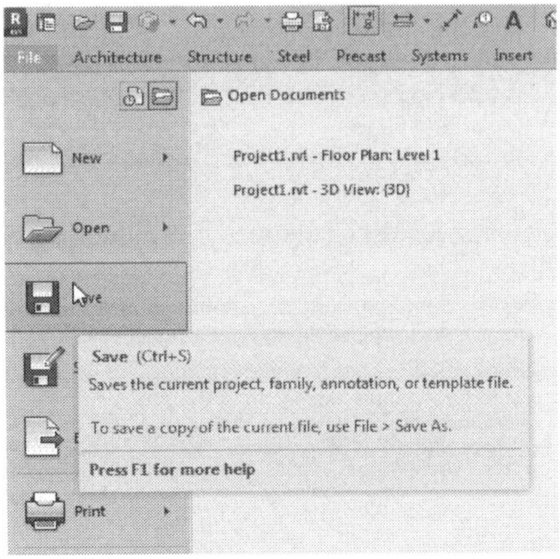

Figure 2-7 Choosing the ***Save*** *tool from the* ***File*** *menu*

CONFIGURING GLOBAL SETTINGS

In Revit, you can configure global settings by using the **Options** dialog box. This dialo box can be invoked by choosing the **Options** button from the **File** menu. The **Options** dialog box, as shown in Figure 2-8, contains nine tabs: **General**, **User Interface**, **Graphics**, **Hardware**, **File Locations**, **Rendering**, **Check Spelling**, **SteeringWheels**, **ViewCube**, and **Macros**. The options in these tabs are discussed next.

General Tab

The **General** tab is chosen by default in the **Options** dialog box. This tab contains the following areas: **Notifications**, **Username**, **Journal file cleanup**, **Worksharing update frequency**, and **View options**, refer to Figure2-8.

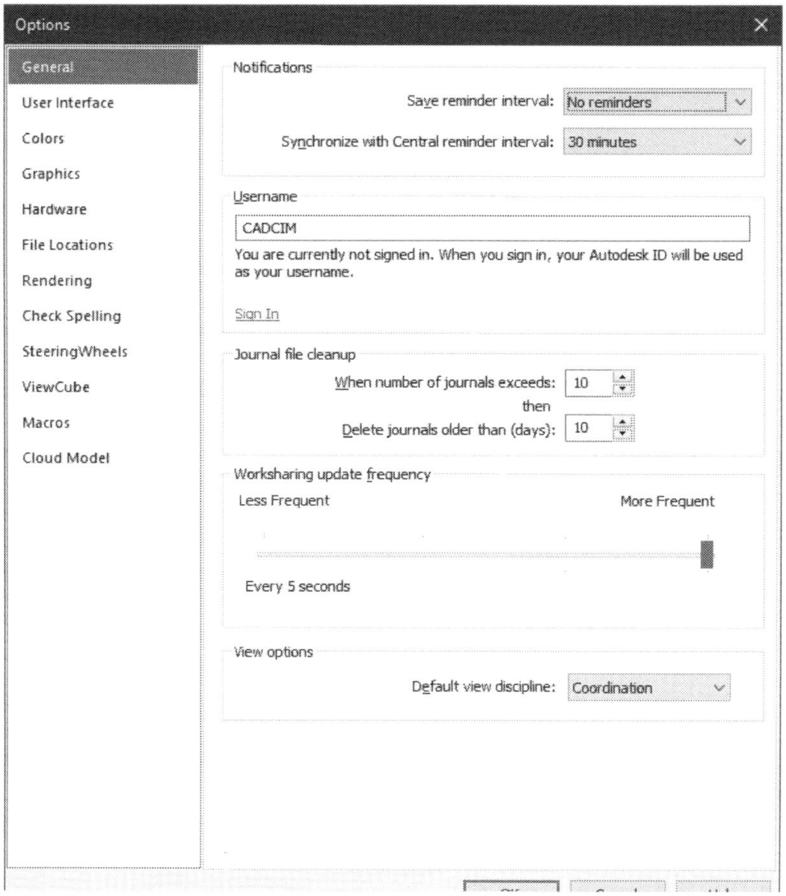

*Figure 2-8 Various areas in the **General** tab*

User Interface Tab

The User Interface tab contains following areas: **Configure**, **Ribbon tab switching**, **View switching**, and **Visual experience**. Using the options in these areas, you can change the display of Revit user interface as per your choice. You can do so by selecting the **Dark** or **Light** option from the **Active theme** drop-down list in the **Visual experience** area. To customize the use of shortcut keys in a project, you can choose the **Customize** button corresponding to the **Keyboard Shortcuts** parameter in the configure area. You can also choose the **Customize** button corresponding to the **Double-click Options parameter**; the **Customize Double-click Settings** dialog box will be displayed, as shown in Figure 2-9.

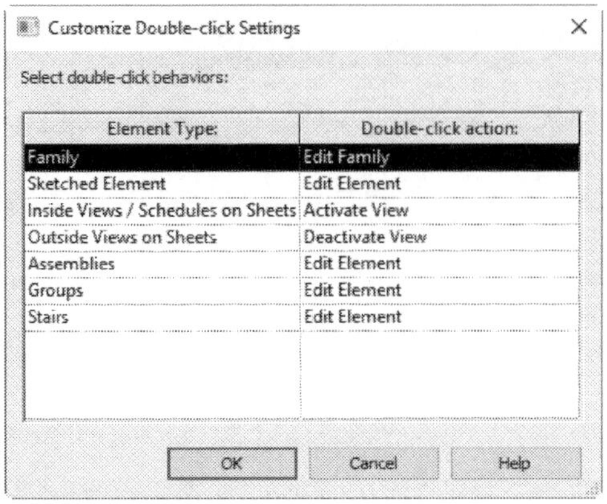

***Figure 2-9** The **Customize Double-click Settings** dialog box*

Graphics Tab

The options in the **Graphics** tab enable you to assign colors to selections, highlights and alerts, and enable anti-aliasing for 3D views. In the **View navigation performance** area of this tab, the **Allow navigation during redraw** check box will be selected by default. As a result, Revit will interrupt the drawing of the model and will allow to navigate the view. In the **View navigation performance** area, the **Simplify display during view navigation** check box will also be selected by default. As a result, Revit will suspend certain graphic effects such as shadows, fill lines, detail lines during the manipulation of the camera view.

In the **Graphics mode** area of this tab, the **Smooth lines with anti-aliasing** check box is selected by default. As a result, the **Allow control for each view in the Graphic Display Options dialog** and **Use for all views (control for each view is disabled)** options will be enabled. The selection of the **Allow control for each view in the Graphic Display Options dialog** check box will allow you to control the aliasing using the **Graphic Display Options** dialog box for each view. The **Graphic Display Options** dialog box can be invoked using the options in the **Visual Style** menu of the Status Bar. The list of features with the barred styles available in the **Visual Style** menu are given in the table below. On selecting the **Use for all view (control for each view is disabled)** option, all sketched lines will become smooth in all views.

Feature	**Visual Styles**			
	Hidden Line	**Shaded**	**Consistent Colors**	**Realistic**
Edges	-----------	Barred	Barred	Barred
Fill Patterns	Barred	Barred	Barred	Barred
Shadows	Barred	Barred	Barred	Barred
Structural Hidden Lines	Barred	Barred	Barred	Barred

Mechanical Hidden Lines	Barred	Barred	Barred	Barred

Hardware Tab

The options in the **Hardware** tab enables you to view information related to hardware setup and change related information for better navigation and rendering view. In the **Hardware setup** area of this tab, the **Use hardware acceleration** check box is selected by default. As a result, the hardware accelerators are enabled. Hardware accelerators help display the larger models faster on refreshing the views. In addition, the hardware accelerators help you speed up the process of switching between the windows of views.

File Locations Tab

The options in the **File Locations** tab can be used to display the link of the default template files present in the project. The options in this tab can also be used to set the path for the template files, user files, family template files that are accessed frequently, and the root path for the point clouds. Figure 2-10 shows various options in this tab.

The path for these files is set at the time of installing Revit. However, you can modify the location by choosing the corresponding **Browse** button and specifying a new path. The default template file location can be changed, in case you wish to use a custom made template file for your projects. Revit uses the default path for user files for saving or opening a project. You can also specify the default path for family template files and for the point clouds by using the corresponding **Browse** button.

In Autodesk Revit, you can add a new template file for the workflows used for system analysis. To do so, choose the **Add Value** button under the **System analysis workflows:** area and specify the desired path for the new template file in the **Browse for System Analysis Workflow** dialog box. To modify the path of the default template file, click on the path given under the **Path** column corresponding to the desired workflow; the **Browse** button will be activated. Now, click on the **Browse** button; the **Browse for System Analysis Workflow** dialog box will be displayed. In this dialog box, specify the desired path for the template file.

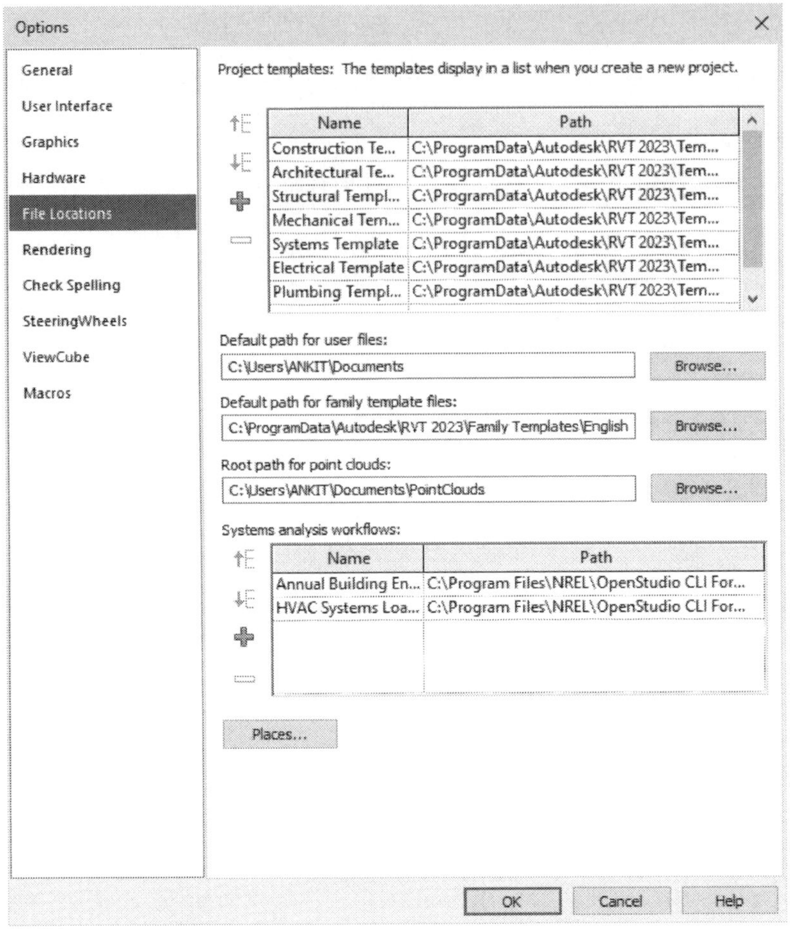

*Figure 2-10 Options in the **File Locations** tab*

Rendering Tab

The options in the **Additional Render Appearance Paths** section of this tab can be used to specify the path of images that are used in textures, bump map, and other procedural maps for rendering. These image files are not present in the software and therefore, you need to specify their paths to use them. To do so, choose the **Add Value** button in the **Additional render appearance paths** area and specify the required path in the displayed field or choose the **Browse** button; the **Browse For Folder** dialog box will be displayed. In this dialog box, select the desired path and choose the **Open** button to add the path to the field. In the **ArchVision content manager location** area of the **Rendering** tab, you can specify the location of the licensed additional RPC content from Archvision. In this area, the **Network** radio button is selected by default. As a result, you can provide the address of the licensed Archvision content in a network so that more than one user can access the content. To provide the network location of the RPC content, you can specify the IP address or the machine name in the **Address** edit box. Also, you can specify the port used by the Archvision Content Manager in the edit box next to the **Address** edit box. By default, the value entered in the **Port** edit box is **14931**. Alternatively, you can select the **Local** radio button in the **ArchVision content manager location** area of the **Rendering** tab, if the additional RPC content is available in the local drive of the computer. On selecting the **Local** radio button, you can specify the location of the local ACM executable file (*rcpACMapp.exe*) in the **Executable Location** edit box. To browse to the local ACM executable file, you can choose the **Browse** button; the **Browse for Template File** dialog box will be displayed. Select the desired executable file and then choose the **Open** button; the path of the file will be added and displayed in the **Executable Location** edit box. In the **Archvision content manager location** area, you can choose the **Get More RPC** button to connect to *https://www.archvision.com/revit* page and get additional RPC required for the project.

Check Spelling Tab

This tab provides you with the option to run spell check in the text and then rectify the errors. You can choose the **Check Spelling** from the **Options** dialog box to display its options. Various self-explanatory settings can be selected from the **Settings** list. You can choose the **Restore Defaults** button available below the **Settings** list to revert back to the default settings. In the **Main dictionary** area of this tab, you can select the type of dictionary to be used as main dictionary for the spelling check from the **Autodesk Revit** drop-down list. Apart from the main dictionary, you can also use additional dictionaries such as the personal and building industry dictionaries. This facilitates the use of various personal and industry related terms in the text matter of the project. There are many words that are not included in the main dictionaries but are frequently used in the building industry. For example, the abbreviation for architecture 'archi' is not available in the main dictionaries. The additional building industry dictionary has many such words and abbreviations that can be used in the text matter of the project without being prompted for errors while checking spellings. You can also add or remove words from your personal and building industry dictionary. Choose any of the **Edit** buttons available in the **Additional dictionaries** area to view the list of words. You can enter or remove any word from these lists using the cursor and keyboard. To run spell check in your drawing, choose the **Check Spelling** tool from the **Text** panel in the **Annotate** tab; the **Check Spelling** dialog box will be displayed, as shown in Figure 2-11, wherein you can rectify the spelling errors in the text by selecting the correct spelling and then choosing the **Change** button in the dialog box, refer to Figure 2-11. Alternatively, you can press the F7 key to display the **Check Spelling** dialog box, refer to Figure 2-11, and then make necessary corrections in the text.

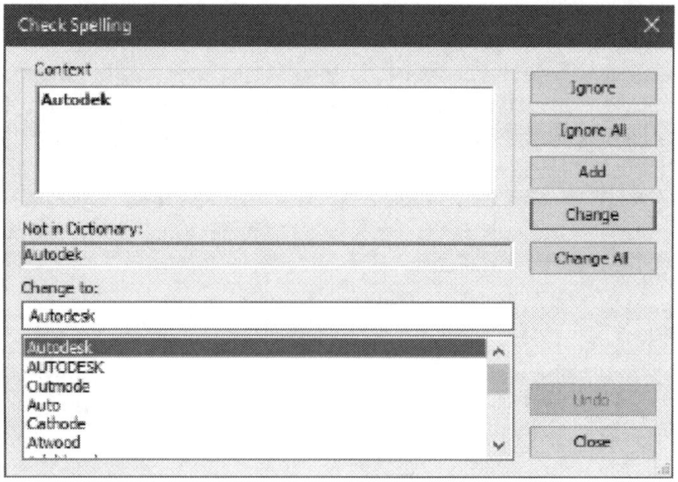

*Figure 2-11 The **Check Spelling** dialog box*

SteeringWheels Tab

The options in the **SteeringWheels** tab in the **Options** dialog box are used to control the visibility, appearance, and operational tools of different types of SteeringWheels.

ViewCube Tab

The options in the **ViewCube** tab in the **Options** dialog box are used to edit various settings of the ViewCube. It has four different areas to modify the ViewCube: **ViewCube appearance**, **When dragging the ViewCube**, **When clicking on the ViewCube**, and **Compass**, as shown in Figure 2-12.

*Figure 2-12 The **ViewCube** tab of the **Options** dialog box*

Macros Tab

The options in the **Macros** tab can be used to set the security level for the **Macros** used in the project. The options in this tab are available in two areas: **Application macro security settings** and **Document macro security settings**.

CLOSING A PROJECT

To close a project, choose the **Close** option from the **File** menu displayed, as shown in Figure 2-13. If you have already saved the latest changes, the project file will be closed. Otherwise, Revit will prompt you to save the changes by using the **Save File** confirmation box. You can save the changes by choosing the **Yes** button or discard them by choosing the **No** button. You can also choose the **Cancel** button to return to the interface and continue working on the project file. You can also use the **Close** button (**X**) in the drawing window to close the project.

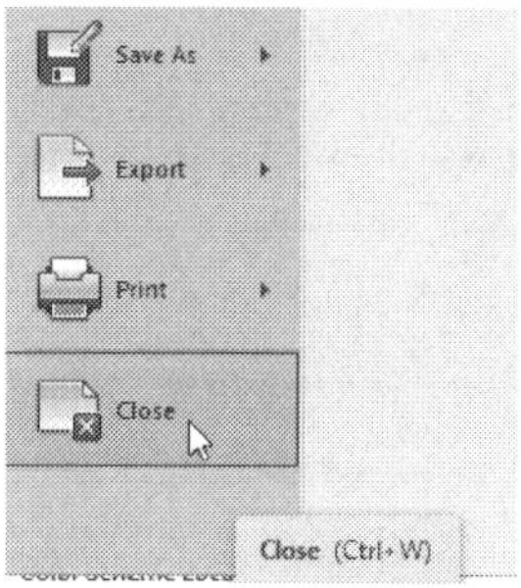

***Figure 2-13** Choosing the **Close** option*

EXITING REVIT

To exit the Autodesk Revit session, choose the **Exit Revit** button from the **File** menu. Even the project is open, you can still choose the **Exit Revit** button to close the file and exit Rev In case, the project has not been saved, it enables you to save the work through the **Save Fi** message box, as shown in Figure 2-14. If you choose the **No** button, all the changes that we not saved will be lost. You can also use the **Close** button (**X**) in the main Revit window to e the Revit session.

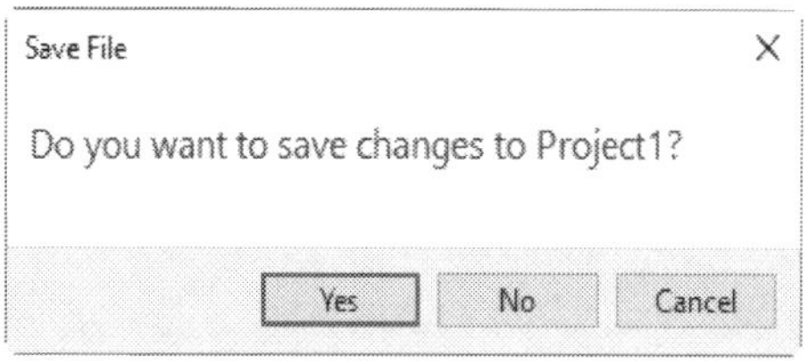

***Figure 2-14** The **Save File** message box*

OPENING AN EXISTING PROJECT

To open an existing project, you can use various options, such as the **Open** tool or the Windows explorer. These options are discussed in detail in the next section.

Opening an Existing Project Using the Open Tool

To open an existing project file, choose **Open > Project** from the **File** menu, as shown in Figure 2-15. Alternatively, you can open a project by choosing the **Open** button from the **Quick Access Toolbar** or by pressing CTRL+O. On doing so, the **Open** dialog box will be displayed, as shown in Figure 2-16. In this dialog box, you can open a particular project file by accessing the appropriate folder using the **Look in** drop-down list.

The **Preview** area of the **Open** dialog box shows the preview of the selected project file. It helps you select a particular file by viewing its contents, even if you are not sure about the file name. The window icons such as the **Views** menu placed along with the **Look in** drop-down list, help you select a project file based on its size, type, and the date when it was last saved. On choosing the **Thumbnails** option from the **Views** menu, you can preview the thumbnails of the project files inside the selected folder in the file list area, refer to Figure 2-17.

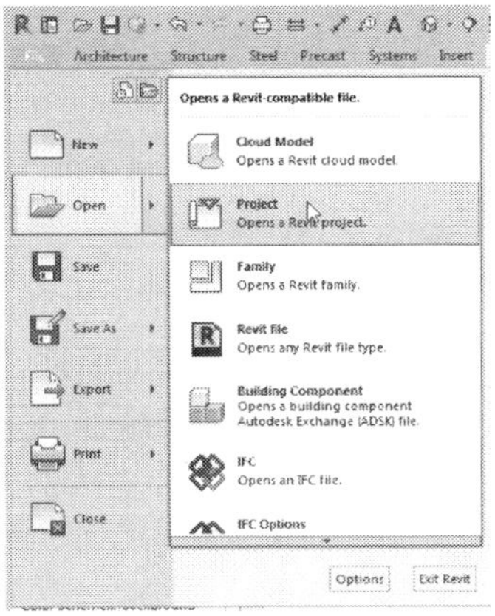

*Figure 2-15 Choosing the **Project** option from the **File** menu*

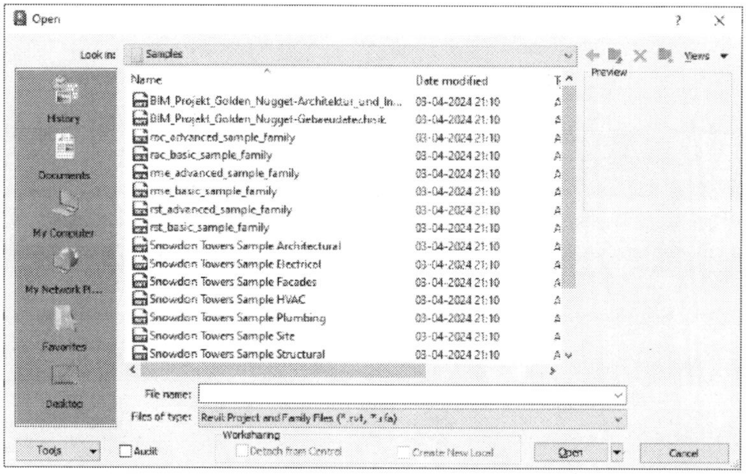

*Figure 2-16 The **Open** dialog box*

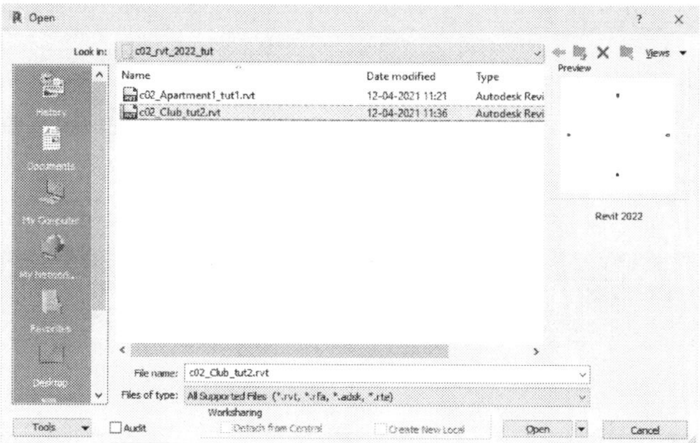

*Figure 2-17 Thumbnails of files in the **Open** dialog box*

The **Places** list is located on the left in the **Open** dialog box. You can add or remove folders from the **Places** list by choosing the **Options** button from the **File** menu. To do so, first close the **Open** dialog box and then choose the **Options** button from the **File** menu. On doing so, the **Options** dialog box will be displayed. Choose the **File Locations** tab from the dialog box, and then choose the **Places** button from it; the **Places** dialog box will be displayed, as shown in Figure 2-18.

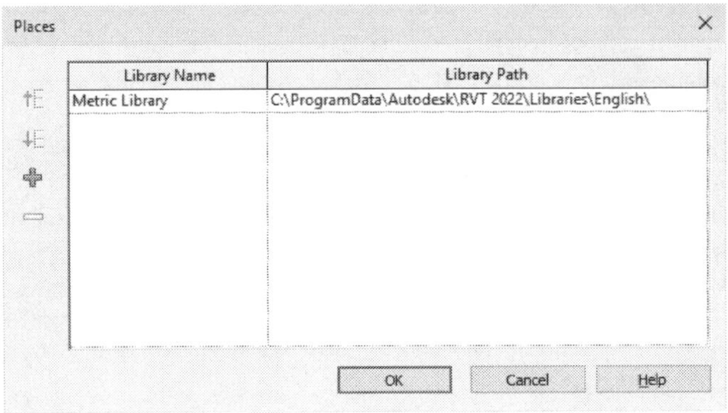

*Figure 2-18 The **Places** dialog box*

MODEL DISPLAY TOOLS

As described earlier, in Revit, you can create the building model using the 3D parametric elements. Various tools are provided to view the building model. Based on the requirement, you can use these tools to navigate and edit elements in the building model.

Using the Zoom Tools

The **Zoom** tools are used to enlarge or reduce a project view in the viewing area. To use these tools, you need to display the **Navigation Bar** in your drawing. Generally, the **Navigation Bar** is displayed by default. If it is not displayed, click on the down arrow on the **User Interface** drop-down in the **Windows** panel of the **View** tab; a drop-down list will be displayed, as shown in Figure 2-19. From the drop-down list, select the **Navigation Bar** check box if it is not selected; the **Navigation Bar** will be displayed in the viewing area. In the **Navigation Bar**, click on the down arrow below the **Zoom All to Fit** tool; a cascading menu will be displayed, as shown in Figure 2-20.

*Figure 2-19 A drop-down list displayed on clicking the **User Interface** drop-down arrow*

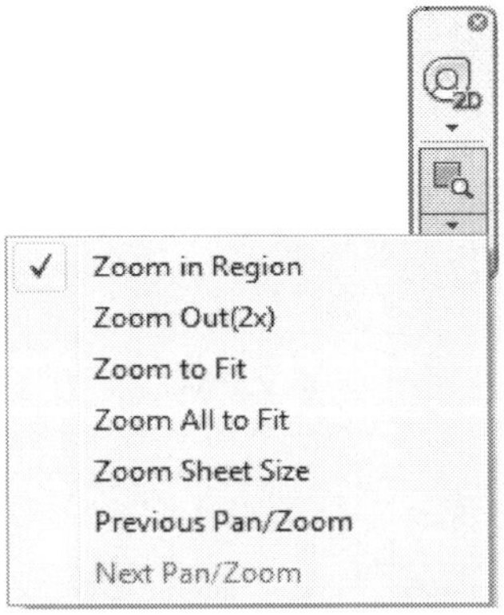

Figure 2-20 Different zooming options in the cascading menu

Navigation Tools

Navigation tools in Revit help you navigate and maneuver into your model in different directions and views.

Other Display Options

Revit provides five options to view the building model in different modes of shading. You can select these options by choosing the **Visual Style** button from the **View Control Bar** and then selecting the desired option from the shortcut menu that will be displayed. There are six options that are available in the shortcut menu and these are discussed next.

Wireframe : Displays the model with all the lines and edges without surfaces.

Hidden Line : Displays the model with lines and edges that are visible.

Shaded : Displays the building model with the surfaces shaded with respect to the material shading and default lighting effect.

Consistent Colors : Displays the building model in different consistent colors, making the model more aesthetic.

Realistic : Displays the realistic view of the model with textures visible in the elements. In this view style, you can specify and visualize the effect of artificial lights and photographic exposure lighting schemes in the project. The use of this style is helpful for quick presentation, where you do not want to render the image using mental ray.

Note

By default, these options affect the current view only. However, you can save the shading effects by saving the shaded view in the project file.

TUTORIAL

In the tutorial of this chapter, you will work on projects: an apartment complex, a club building, and a residential building. However, some portions of these three projects will be completed in the tutorials and the rest will be given as exercises that need to be completed by the students themselves. The tutorials and exercises form a sequence, and therefore, to complete these projects, you need to complete both the tutorials and exercises in the previous chapters.

The following tutorials will familiarize you with the tools and concepts discussed in this chapter such as starting Revit, opening a new project, setting units, setting snaps, saving, and closing a project.

Tutorial 1 Apartment 1

In this tutorial, you will create a new project file for the *Apartment 1* project with the following parameters: **(Expected time: 15 min)**

1. Template file-
   For Imperial *default.rte*
   For Metric *DefaultMetric.rte*

2. Project Units-
   For Imperial **Feet and fractional inches** Rounding- **To the nearest 1/2"**
   For Metric **Millimeters** Rounding- **0 decimal places**

3. Length dimension snap increment-
   For Imperial **5' ; 2'6"; 3"; 0'1/2"**
   For Metric **1500 ; 750 ; 75 ; 10**

4. File name to be assigned-
   For Imperial *c02_Apartment1_tut1.rvt*
   For Metric *M_c02_Apartment1_tut1.rvt*

The following steps are required to complete this tutorial:

a. Start Revit 2025 session.

b. Use the template file for the project.
   For Imperial *default.rte*
   For Metric *DefaultMetric.rte*

c. Set the project units using the **Format** dialog box.
   For Imperial **Feet and fractional inches**
   For Metric **Millimeters**

d. Set the length dimension snap increment using the **Snaps** dialog box.
   For Imperial **5'; 2'6"; 3"; 0'1/2"**
   For Metric **1500 ; 750 ; 75 ; 10**

e. Set **Endpoint**, **Midpoint**, **Nearest**, **Perpendicular**, **Work Plane Grid**, **Snap to Remote Objects**, **Snap to Point Clouds**, and **Intersection** as the object snaps in the **Snaps** dialog box.

f. Save the project using the **Save As** tool.
   For Imperial *c02_Apartment1_tut1.rvt*
   For Metric *M_c02_Apartment1_tut1.rvt*

g. Close the project.

Starting Revit 2025

1. Start Revit 2025 by choosing **All Programs > Autodesk > Revit 2025** from the **Start** menu (for Windows 10 versions). Alternatively, double-click on the Revit 2025 icon on the desktop. As a result, the program is loaded and the user interface screen is displayed.

2. Choose **New > Project** from the **File** menu; the **New Project** dialog box is displayed.

Using the Template File

To use the template file for the project, you need to access the appropriate folder and then select the required template file.

1. In the **New Project** dialog box, choose the **Browse** button from the **Template file** area; the **Choose Template** dialog box is displayed showing a list of the template files available in the **English-Imperial** folder.

2. In the **Choose Template** dialog box, select the **default** template file from the **English-Imperial** folder (**DefaultMetric** template file from the **English** folder), refer to Figure 2-21 and then choose the **Open** button; the **Choose Template** dialog box closes and the selected template file is applied to the new project.

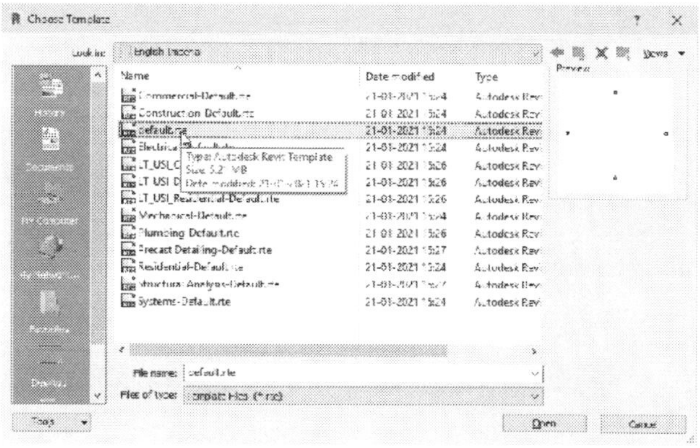

*Figure 2-21 The **Choose Template** dialog box with the **default** file selected*

3. Ensure that the **Project** radio button is selected in the **New Project** dialog box. Next, choose the **OK** button from the dialog box; the *default.rte* (for Metric *DefaultMetric.rte*) template file is loaded. Notice that the **Project Browser** now shows different levels and views that have already been created in the selected template.

Setting Units

You can set units for various measurement parameters using the **Project Units** dialog box.

1. Choose the **Project Units** tool from the **Settings** panel of the **Manage** tab; the **Project Units** dialog box is displayed.

2. Select the **Common** option from the **Discipline** drop-down list and then click on the button displayed in the **Format** column next to the **Length** parameter; the **Format** dialog box is displayed.

3. In the **Format** dialog box, select the required option from the **Units** drop-down list, if not selected by default.

   For Imperial **Feet and fractional inches**
   For Metric **Millimeters**

4. Click on the **Rounding** drop-down list in this dialog box and select the required option, as shown in Figure 2-22.

   For Imperial **To the nearest 1/2"**
   For Metric **1 decimal place**

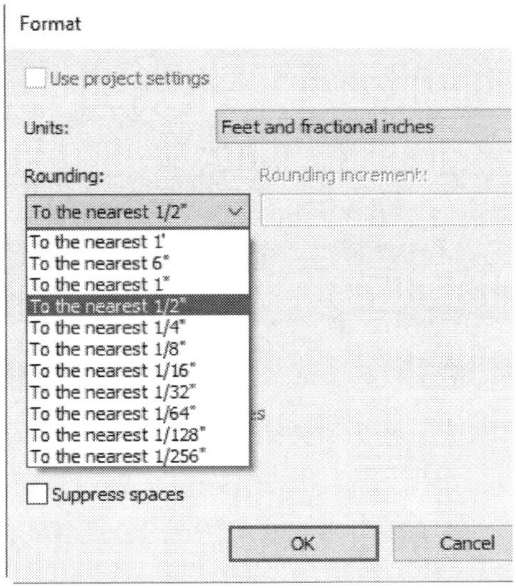

*Figure 2-22 Selecting the **To the nearest 1/2"** option in the **Format** dialog box*

5. Choose the **OK** button; the **Project Units** dialog box is displayed. Choose the **OK** button;

the specified units are applied and the dialog box is closed.

Setting the Dimension and Object Snaps

To set the dimension and object snaps, use the **Snaps** tool. These settings are based on the type of the project and the amount of detailing required.

1. Choose the **Snaps** tool from the **Settings** panel of the **Manage** tab; the **Snaps** dialog box is displayed. In the **Length dimension snap increments** edit box, enter the following values.
   For Imperial **5'; 2'6"; 3"; 0'1/2";**
   For Metric **1500 ; 750 ; 75 ; 10;**

2. In the **Object Snaps** area, clear the **Quadrants**, **Centers**, **Tangents**, and **Points** check boxes. Leave other check boxes selected. Choose the **OK** button; the settings are applied and the **Snaps** dialog box is closed.

3. In the **Manage** tab, select the **Temporary Dimensions** option from **Settings > Additional Settings > Annotations** drop-down list; the **Temporary Dimension Properties** dialog box is displayed. In this dialog box, ensure that the **Centerlines** radio button is selected in the **Walls** and **Doors and Windows** areas.

4. Choose the **OK** button; the **Temporary Dimension Properties** dialog box is closed.

Saving the Project

The project parameters set in the previous steps are an integral part of the project file. To save this project file with these settings, use the **Save** tool.

1. From the **File** menu, choose the **Save** tool from the **File** menu; the **Save As** dialog box is displayed.
2. In this dialog box, browse to the *C* drive and then create a folder with the name **rvt_2025**.

3. Open the *rvt_2025* folder and then create a sub-folder with the name *c02_revit_2025_tut*. Next, open the created folder and save the file with the following name.
   For Imperial *c02_Apartment1_tut1*
   For Metric *M_c02_Apartment1_tut1*

Notice that the **Files of type** drop-down list shows **Project Files (*.rvt)** as the default option refer to Figure 2-23.

4. Choose the **Save** button to save the project with the name *c02_Apartment1_tut1.rvt*. The project is saved at the specified location.

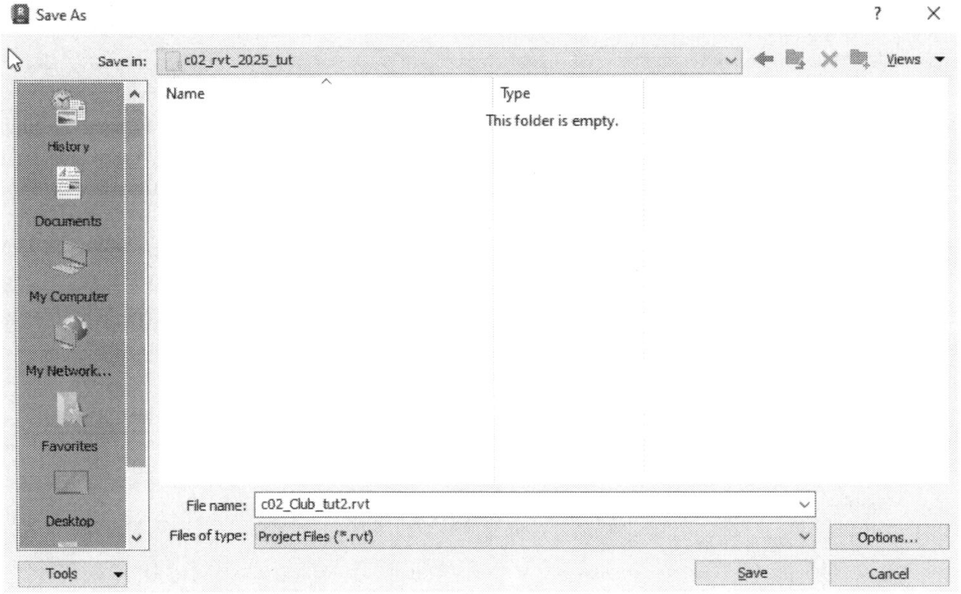

***Figure 2-23** Saving the project using the **Save As** dialog box*

Closing the Project

1. To close the project, choose the **Close** option from the **File** menu.

EXERCISE

Exercise 1 Apartment 2

Create a new project file for the *Apartment 2* project with the following parameters:

(Expected time: 15 min)

1. Template file-
   For Imperial **default.rte**
   For Metric **DefaultMetric.rte**
2. Project Units-
   For Imperial **Feet and fractional inches**
   For Metric **Millimeters**
3. Length dimension snap increment-
   For Imperial **5'; 2'6"; 3"; 0'1/2"**
   For Metric **1500 ; 750 ; 75 ; 10**
4. File name to be assigned-
   For Imperial *c02_Apartment2_ex1.rvt*
   For Metric *M_c02_Apartment2_ex1.rvt*

Chapter 3

Creating Architectural Walls

Learning Objectives

After completing this chapter, you will be able to:

- *Understand the concept of walls*
- *Understand the properties of walls*
- *Create stacked walls*
- *Create exterior and interior walls based on the given parameters*
- *Create wall sweeps and wall reveals*

INTRODUCTION

In the previous chapter, you learned about various tools and options to start, open and save an architectural project in Revit. In this chapter, you will learn about various tools and options to create different types of walls. You will also learn about the functions and parameters of different types of walls. Moreover, you will learn various options to modify the properties of an architectural wall and its type properties will be discussed in detail. Also, you will learn about various tools that are used to add modifiers in a wall, such as sweep and reveal.

CREATING A BUILDING PROJECT

In Revit, the term 'project' comprises not only the physical building model but also its associated documentation such as drawings, views, schedules, areas, and so on. The first step in creating a project is to create the building model. In Autodesk Revit, you can create it using the following two methods:

Method 1: Create a building model using individual building elements such as walls, windows, doors, floors, roofs, and so on.

Method 2: Create a conceptual mass of a building model using the massing tools and conceptualize the overall building shape and volume before working with individual elements.

Tip

You can also use a combination of above two methods. You can generate a building mass using the massing tools and then convert it into a building model with individual building elements using the building maker tools.

The selection of method depends on different project parameters such as project magnitude, building shape, building technology, current documentation stage of a project, industry parameters, and so on. The use of the massing tools to create a building geometry and the usage of individual building components to develop a building model will be described later.

BIM CONCEPTS *While working in a BIM project, Autodesk Revit provides you with several tools to add individual building elements such as wall, floor, roof, and so on for creating a building model. Several predesigned element types have been provided for each building element in Autodesk Revit libraries. You have the flexibility to either use the predesigned element types or create your own element type to create a building model.*

Sequence of Creating a Building Model

The sequence of using building elements for creating a building model depends on various parameters such as building type, building volume, building shape, and so on. For most of the building projects, the sequence given below may be adopted.

Step 1: Start the model by creating the exterior walls of the building at Level 1 (lowest level).

Step 2: Create interior walls at the desired locations.

Step 3: Add doors and windows to the exterior and interior walls at the desired location.

Step 4: Add the floor to the building model.

Step 5: Add the roof to the building model.

Step 6: Add the structural or architectural grid and structural elements.

Step 7: Add stand-alone components such as furniture items and plumbing fixtures.

Step 8: Add text and annotations to different spaces.

Step 9: Create dimensions for different parameters of the project.

Step 10: Create project details and documentation.

Step 11: Create the rendered 3D views and walkthrough.

In Revit, each building element is considered a three-dimensional parametric entity. This means, on adding elements to the model, you also add the related information and specification about the elements. One of the most important elements in a building model is the wall. It defines the basic spatial arrangement of the building that acts as the host for doors and windows.

Understanding Wall Types

Autodesk Revit provides you with several predefined wall types such as **Exterior**, **Interior**, **Retaining**, **Foundation**, **Curtain**, and **Stacked Wall**, based on their usage.

Creating Architectural Walls

Ribbon:	Architecture > Build > Wall drop-down > Wall: Architectural
Shortcut Key:	WA

In this section, you will learn the method of creating and editing architectural walls. In Autodesk Revit, each wall type has specific predefined properties such as its usage, composition, material, characteristics, finish, height, and so on. You can select the wall type based on its specific usage in the project. Walls, like most other model elements, can be created in a plan view or a 3D view.

To create an architectural wall, first you need to invoke the **Wall: Architectural** tool and then select the appropriate wall type and specify various properties. To do so, choose the **Wall: Architectural** tool from the **Build** panel of the **Architecture** tab, refer to Figure 3-1; the **Modify | Place Wall** tab with the **Options Bar** will be displayed. In the **Options Bar**, you can define the direction of the wall by choosing either the **Height** or **Depth** option from the **Height** drop-down list. You can select the desired level upto which the wall is to be created from the **Unconnected** drop-down list. If you select the **Unconnected** option from the drop-down list then the edit box next to it will be activated. Enter the desired height upto which you want the wall to be raised. Select the reference line for sketching the wall from the **Location Line** drop-down list.

Select an exterior or interior wall type from the **Type Selector** drop-down list in the **Properties** palette, as shown in Figure 3-2. Next, from the **Properties** palette, specify and edit various properties of the wall to be created. Various wall properties and the process to specify them are discussed next.

Tip

Besides the ***Properties*** *palette, wall types are also listed under the* ***Families*** *head in the* ***Project Browser****. Left-click on the arrow sign next to the* ***Families*** *head to view the families available in Autodesk Revit. Click the arrow symbol next to the* ***Walls*** *subhead and then the one next to the* ***Basic Wall*** *subhead to display various basic wall types. Select and right-click on any wall type to display a shortcut menu with the options that can be used for editing.*

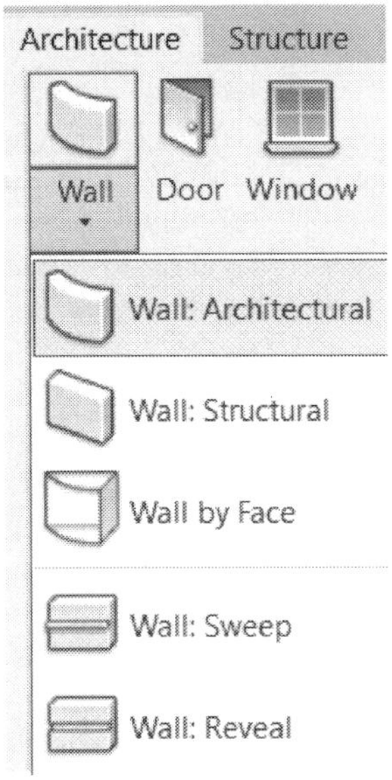

Figure 3-1 *Invoking the* ***Wall: Architectural*** *tool from the* ***Build*** *panel*

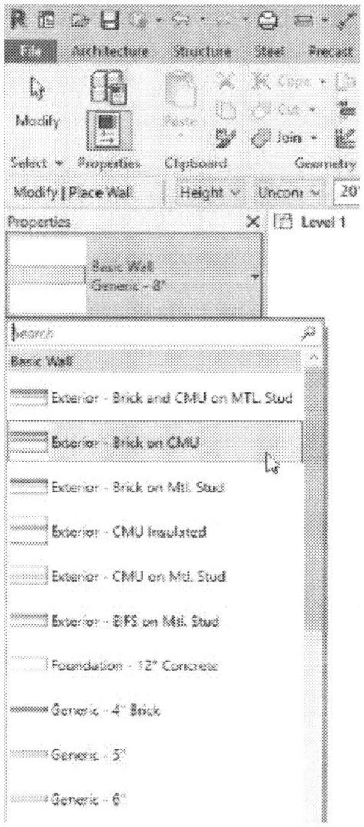

***Figure 3-2** Selecting the wall type from the **Type Selector** drop-down list*

CREATING SLANTED WALLS

In Revit 2025, you can create slanted walls. To create a slanted wall, sketch the desired wall profile and select the **Slanted** option from the drop-down list corresponding to the **Cross-Section** parameter in the **Properties** palette. Now, specify a value for the angle of inclination of the wall from the vertical in the **Angle From Vertical** option.

CREATING TAPERED WALLS

You can create a tapered wall or taper an existing wall. To taper an existing wall, make sure that the wall has a variable thickness. Select the wall from the model and then from the **Properties** palette, select the **Tapered** option from the drop-down list corresponding to the **Cross-Section** parameter. Next, select the check box corresponding to the **Override Type Properties** parameter; the **Exterior Angle** and **Interior Angle** parameters will be enabled. Enter a desired value corresponding to these parameters to taper the selected wall. Figure 3-3 displays a tapered wall in a project.

Tip

*Before you taper a wall, make sure that the layer of the wall that you want to taper is set to variable thickness. To set the layer of the wall to variable thickness, invoke the **Edit Assembly** dialog box from the **Type Properties** dialog box of the wall and then select the **Variable** check box corresponding to the layer for which you want to have a variable thickness.*

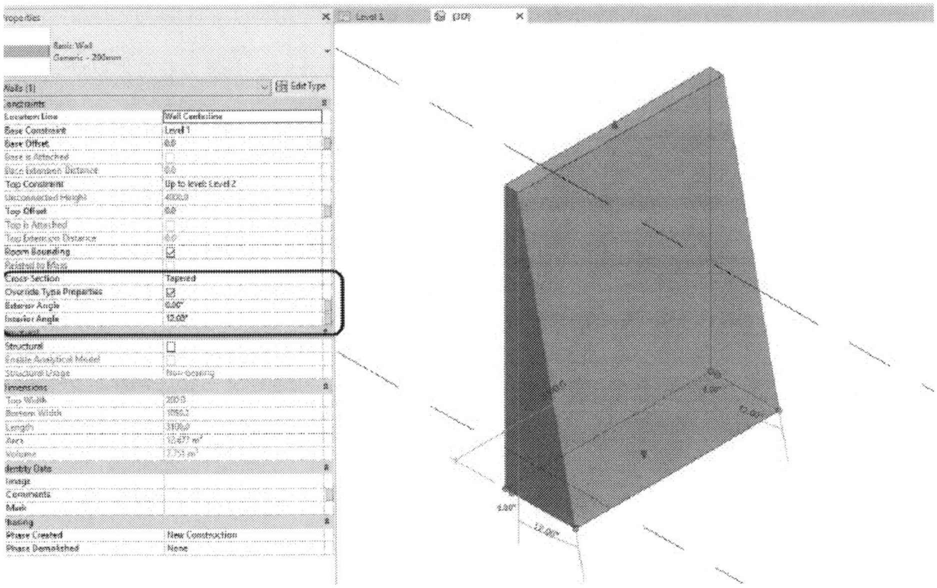

Figure 3-3 A tapered wall in a project

WORKING WITH STACKED WALLS

In Autodesk Revit, besides basic wall and curtain wall there is a special kind of wall called a stacked wall. A stacked wall is a wall created using two or more basic walls that are stacked on top of another to form a complex design. These walls are assembled in such a way that the entire facade acts as a single wall. In the next section, you will learn about various options that are used to create a stacked wall and then add it to the project.

Creating a Stacked Wall

To create a stacked wall, choose the **Wall: Architectural** tool from **Architecture > Build > Wall** drop-down; the **Modify | Place Wall** tab with the instance parameters in the **Properties** palette will be displayed. In the **Properties** palette, select the stacked wall type from the **Type-Selector** drop-down list. After selecting the stacked wall type, choose the **Edit Type** button from the **Properties** palette; the **Type Properties** dialog box will be displayed.

ADDING INTERIOR WALLS

In Autodesk Revit, interior walls are of a different wall types. They differ from the exterior wall types based on their usage, material specifications, and non load-bearing character. Several predefined interior wall types are provided in the Autodesk Revit libraries.

To view the interior wall types, choose the **Wall: Architectural** tool from **Architecture > Build > Wall** drop-down; the instance parameters of the wall will be displayed in the **Properties** palette. From this palette, you can select different types of walls available in Revit. The wall types that may be used as interior walls have been assigned the prefix **Interior**. Based on the project requirement, you can select the appropriate interior wall type from the drop-down list. For example, **Interior - 5" Partition (2-hr)** is a type of interior wall for Imperial (**Interior - 135mm**

Partition (2-hr) for Metric) with a 2-hour fire rating and can be selected from the **Properties** palette.

To view and specify the properties of an interior wall type, you can use the **Properties** palette, as shown in Figure 3-4. This palette shows various parameters of the selected wall such as **Unconnected Height**, **Location Line**, **Structural Usage**, **Top Constraint**, and so on. You can modify these parameters by entering a new value or selecting the required value from the corresponding value field in the **Properties** palette.

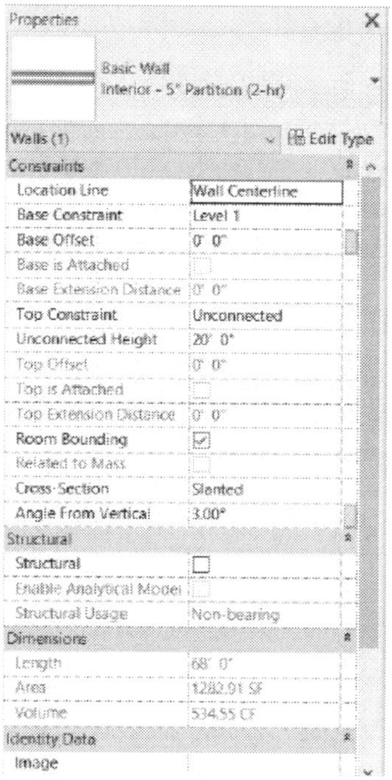

*Figure 3-4 The **Properties** palette for the **Interior - 5" Partition (2-hr)** wall type*

> **Tip**
> *After creating a wall, right-click on any end of the wall; a shortcut menu will be displayed. Choose the **Disallow Join** option from the shortcut menu to prevent the wall end from joining with other walls.*

The sketching tools explained earlier (for exterior walls) can also be used for sketching interior walls. When you sketch the interior walls, the top constraint of the interior walls is automatically set to a level above the current level.

> **Tip**
> *The **Location Line** parameter in the **Properties** palette of the selected walls is useful for creating interior walls. You can create interior walls at some specified distance from another existing wall using the **Finish Face: Interior** option.*

Note

The interior walls do not create a neat junction with the exterior walls as they have different composition and characteristics.

Walls are the host elements for other building elements such as doors and windows. Therefore, deleting walls will delete all their dependent elements as well.

The basic composition of an interior wall type can be viewed by choosing the **Type Properties** tool from the **Properties** panel in the **Modify | Place Wall** tab. On doing so, the **Type Properties** dialog box will be displayed, showing different type parameters of the selected wall type.

To view the structural composition of a wall type, choose the **Edit** button in the **Structure** type parameter; the **Edit Assembly** dialog box will be displayed, showing layers for the corresponding wall type. For example, the **Interior - 5" Partition (2-hr)** for Imperial or (**Interior-135 mm Partition (2-hr)** for Metric) wall type comprises of seven layers with different materials and thickness. You can use the **Insert** and **Delete** buttons to modify the wall type. You can choose the **Preview** button to display the graphic view of the **Interior - 5" Partition (2-hr**) for Imperial (**Interior - 135mm Partition (2-hr**) for Metric) wall type in the preview area, as shown in Figure 3-5.

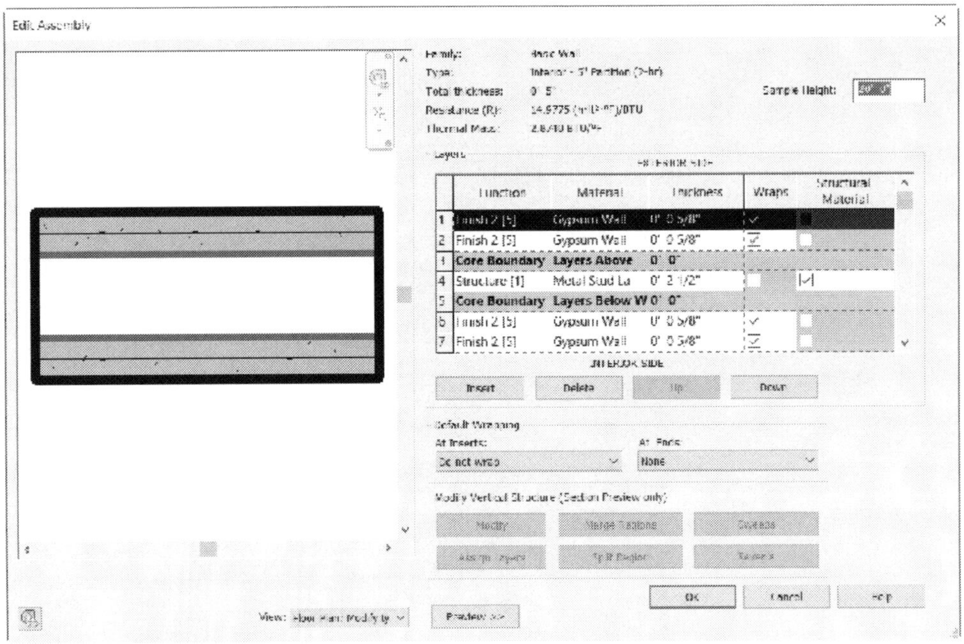

***Figure 3-5** The **Edit Assembly** dialog box for the **Interior - 5" Partition (2-hr)** wall type*

ADDING WALL SWEEPS AND REVEALS

In an architectural project, wall sweep and wall reveal act as modifiers in a wall. Each type of wall sweep and wall reveal has predefined properties. You can select and use any type of wall sweep and wall reveal based on its usage in project. Wall sweep is used to add a baseboard, crown molding or other type of decorative horizontal or vertical projections to a wall. Wall reveal is used to add a decorative horizontal or vertical cutout to a wall. Wall sweep and wall reveal provide advancement to the building model by adding skirtings, projections and cutout to a building.

Note
Wall sweep and wall reveal can be added to a wall only in 3D view or in elevation view.

Wall Sweeps

Ribbon: Architecture > Build > Wall drop-down > Wall: Sweep

To add details to a wall, you can add a wall sweep by using the **Wall Sweep** tool. Wall Sweep can be selected either in an elevation view or in 3D view. Choose the **Default 3D View** tool from the **Quick Access Toolbar**; the user interface will be displayed. Next, invoke the **Wall: Sweep** tool from the **Build** panel of the **Architecture** tab, as shown in Figure 3-6. On doing so, the **Modify | Place Wall Sweep** tab will be displayed. In the **Placement** panel of this contextual tab, the **Horizontal** tool is selected by default. As a result, the wall sweep will be added horizontally to the face of the wall. Select the required type of the wall sweep from the **Type Selector** drop-down list. Now, move the cursor over the face of the wall; a preview of the wall sweep along with the temporary dimensions will be displayed on the wall face. Click at the desired level of elevation; a wall sweep will be created, as shown in Figure 3-7. After adding the wall sweeps at desired faces of the wall, press ESC, or choose the **Modify** button from the **Select** panel to exit from the contextual tab. Similarly, you can add a vertical wall sweep at the face of the wall. To create a vertical wall sweep, choose the **Vertical** tool from the **Placement** panel.

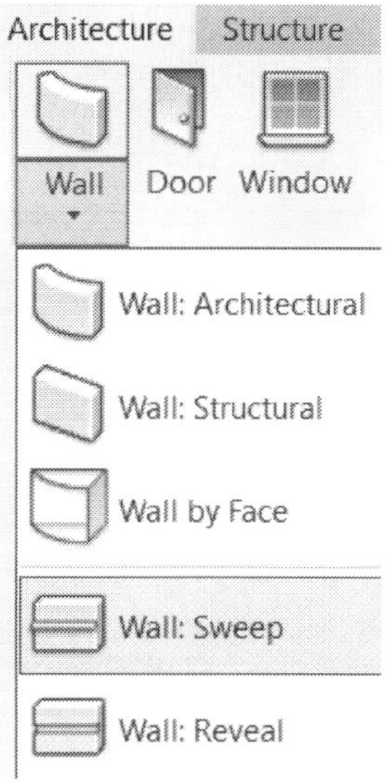

*Figure 3-6 Invoking the **Wall: Sweep** tool from the **Build** panel*

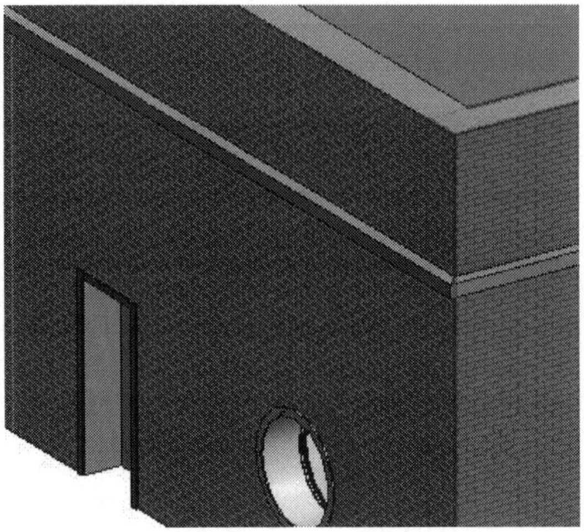

Figure 3-7 Wall sweep added to the wall

Wall Reveals

Ribbon: Architecture > Build > Wall drop-down > Wall: Reveal

To complete the design of our wall, you need to add decorative cutout in the wall. To do so, you can use the wall reveal. To add a wall reveal, you need to be in an elevation view or in 3D view and then invoke the **Wall: Reveal** tool from the **Build** panel of the **Architecture** tab, refer to Figure 3-8; the **Modify | Place Reveal** tab will be displayed. In this contextual tab, select the desired orientation of the wall reveal by choosing either the **Horizontal** or **Vertical** tool from the **Placement** panel. Next, select the type of the wall reveal from the **Type Selector** drop-down list and then move the cursor over the face of the wall; a preview of the wall reveal with the temporary dimension will be displayed along the wall face. Click at desired level of elevation; a wall reveal will be added, as shown in Figure 3-9. You can add the wall reveal to other walls by clicking on their faces. After adding the wall reveal at desired faces of the wall, press ESC or choose the **Modify** button from the **Select** panel to exit the contextual tab. After adding the wall reveal to the wall, you can change the properties of the reveal such as its profile. To change the profile of the reveal, invoke the **Edit Type** button from the **Properties** palette; the **Type Properties** dialog box will be displayed. Choose the **Duplicate** button from the upper left corner of this dialog box; the **Name** dialog box will be displayed. In this dialog box, enter the desired name in the **Name** edit box and choose the **OK** button to return to the **Type properties** dialog box. In this dialog box, you can select the desired **Wall Reveal** type from the **Profile** drop-down list by clicking on the value field of this dialog box.

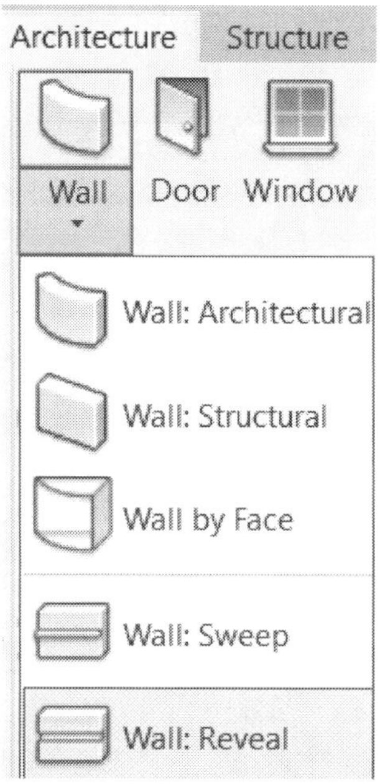

***Figure 3-8** Invoking the **Wall: Reveal** tool from the **Build** panel*

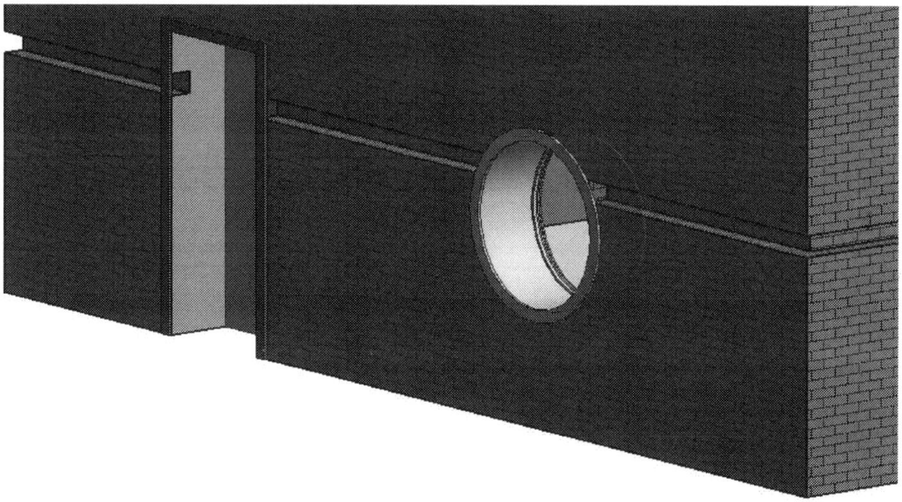

***Figure 3-9** Wall reveal added to the wall*

TUTORIAL

The files used in the tutorials can be downloaded from the CADSOFT website. These files are compressed in zip file format and are required to be extracted before using them in the tutorials. The path of the files is as follows: *Textbooks > Civil/GIS > Revit Architecture > Exploring Autodesk Revit 2025 for Architecture*. For example, the tutorial file, *c02_Apartment_tut1.rvt* that is used in Tutorial 1 of Chapter 3 is compressed in the *c02_rvt_2025_tut.zip file*.

Tutorial 1 Apartment 1 - Exterior Walls

In this tutorial, you will create the exterior walls of a two-room apartment based on the sketch plan shown in Figure 3-10. The dimensions have been given only for reference and are not to be used in this tutorial. The project file and the parameters to be used for creating the exterior walls are given next. **(Expected time: 30 min)**

1. Project file-
   For Imperial *c02_Apartment_tut1.rvt*
   For Metric *M_c02_Apartment_tut1.rvt*
2. Exterior wall type- **Basic Wall: Exterior - Brick on Mtl. Stud**.
3. Location line parameter- **Wall Centerline**; Top Constraint- **Up to Level 2.**

The following steps are required to complete this tutorial:

a. Open the *Apartment 1* project file created in Tutorial 1 of Chapter 2.
b. Invoke the **Wall**: **Architectural** tool from the ribbon.
c. Select the exterior wall type **Exterior - Brick on Mtl. Stud** from the **Properties** dialog box.
d. Modify **Top Constraint- Up to level: Level 2** and **Location Line - Wall Centerline** as wall properties using the **Properties** palette.
e. Invoke the **Line** sketching tool and then sketch the exterior walls based on the given parameters.
f. Save the project using the **Save As** tool.
   For Imperial *c03_Apartment1_tut1.rvt*
   For Metric *M_c03_Apartment1_tut1.rvt*
g. Close the project.

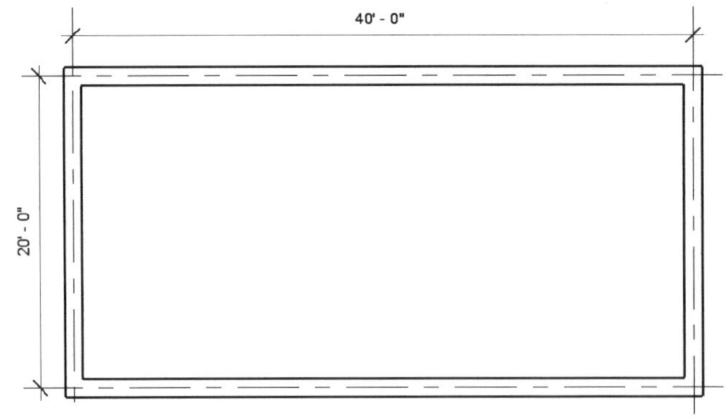

Figure 3-10 *Sketch plan for creating exterior walls for the Apartment 1 project*

Opening the Project File

Choose **Open > Project** from the **File** menu and open the *c02_Apartment1_tut1.rvt (M_c02_Apartment1_tut1.rvt for Metric)* project file. You can also download this file from *https://www.cadsoft.com*. The path of the file is as follows: *Textbooks > Civil/GIS > Revit Architecture > Exploring Autodesk Revit 2025 for Architecture*.

Invoking the Wall: Architectural Tool and Selecting the Wall Type

To start sketching the wall, you must invoke the **Wall: Architectural** tool from the ribbon and select the wall type to be used **(Exterior - Brick on Mtl. Stud** in this case).

1. Invoke the **Wall: Architectural** tool from **Architecture > Build > Wall** drop-down; the **Modify | Place Wall** tab is displayed.

2. In the **Type Selector** drop-down list of the **Properties** palette, select the **Exterior - Brick on Mtl. Stud** wall type.

Modifying Properties of the Exterior Wall

After selecting the wall type, you need to modify the instance properties of the wall type using the **Properties** palette.

Note

The default.rte or (DefaultMetric.rte for Metric) template file used for this project has two predefined levels: ***Level 1*** *and* ***Level 2***.

1. In the **Properties** palette, ensure that the **Location Line** parameter has **Wall Centerline** as the default value. Click on the column adjacent to the **Top Constraint** instance parameter; a drop-down list is displayed. Select **Up to level: Level 2** from this drop-down list, and choose the **Apply** button, as shown in Figure 3-11.

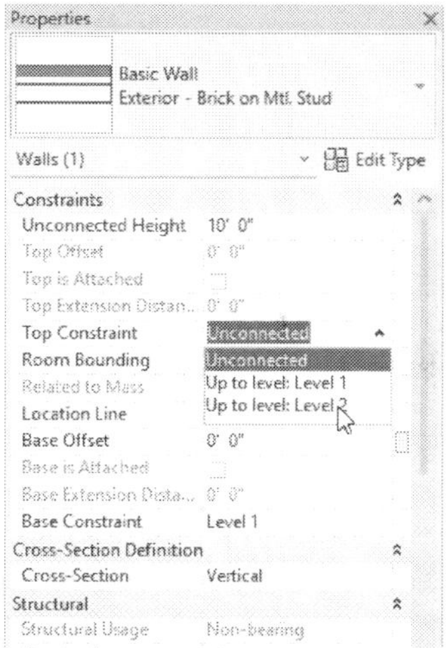

Figure 3-11 *Setting the* ***Top Constraint*** *parameter using the* ***Properties*** *palette*

Tip

If the ***Properties*** *palette is not displayed in the User interface screen by default, select the* ***Properties*** *check box from* ***View > Windows> User-Interface*** *drop-down.*

Sketching the First Exterior Wall Segment

To sketch a wall, you need to choose an appropriate sketching tool from the **Draw** panel in the **Modify | Place Wall** tab. The exterior walls of the given sketch of the *Apartment 1* project can be created using the **Rectangle** tool. You will however use the **Line** tool to learn and understand the usage of this tool for sketching the straight walls.

1. Ensure that the **Line** tool is chosen in the **Draw** panel in the **Modify | Place Wall** tab and the **Chain** check box is cleared in the **Options Bar**.

2. Ensure that the **Allow** option is selected in the **Join Status** drop-down list.

3. Click between the four inward arrow keys to specify the start point of the first wall segment. Next, move the cursor toward the right hand side. On doing so, a wall segment starts from the specified point with temporary dimension appearing on it. The dimension changes dynamically as you move the cursor. This shows the length of the wall segment at any given location of the cursor.

4. Right-click in the drawing area; a shortcut menu is displayed. Choose the **Zoom In Region** option from the shortcut menu and zoom into the area to get a closer view of the sketched wall segment, as shown in Figure 3-12 (for zooming techniques, refer to Chapter 2- Starting an Architectural Project).

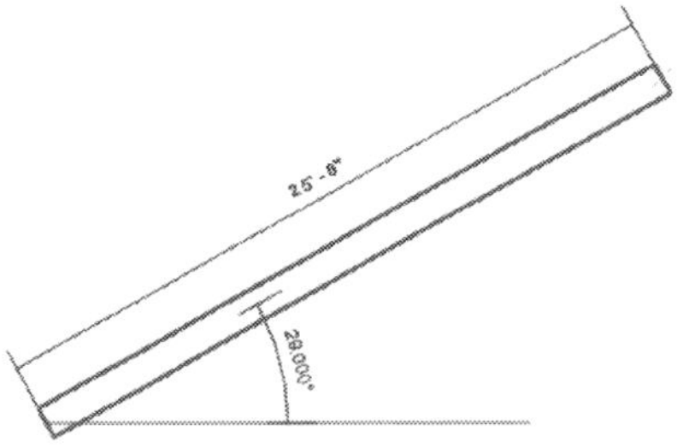

Figure 3-12 *The temporary dimensions displayed on the wall*

5. Move the cursor on the horizontal axis such that a dashed line appears at the central axis of the wall segment, as shown in Figure 3-13. Notice the two-sided arrow attached to the endpoint of the wall. This indicates that the wall segment being sketched is horizontal. A tooltip indicating the horizontal alignment is also displayed with two-sided arrow.

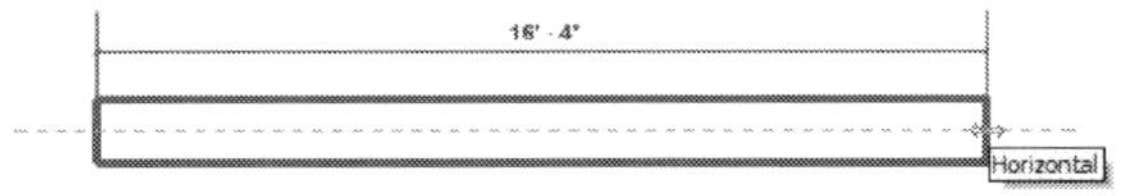

Figure 3-13 *Sketching a horizontal wall segment*

6. Move the cursor to the right until the temporary dimension shows a value more than 40'- 0"(12192 mm), as shown in Figure 3-14. Click at this location as the endpoint of the wall segment and press ESC twice; the wall is created. Note that, if the dimension is not displayed, you need to click on the created wall to display the dimension. Press ESC.

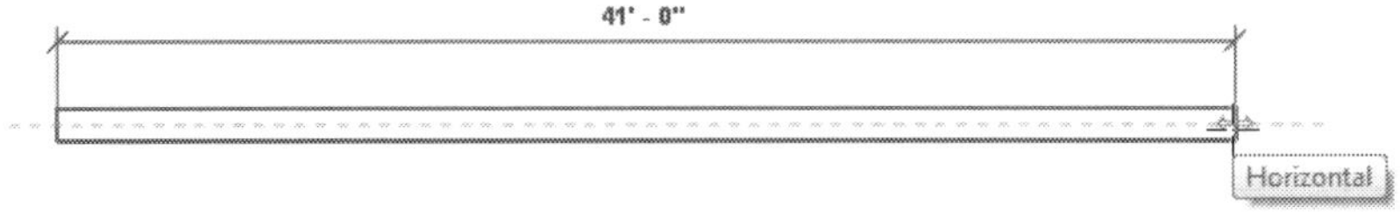

Figure 3-14 *The sketched horizontal wall with its controls*

Note

You can create a wall of exactly 40'0"(12192 mm) length using the dimension snaps set in the ***Snaps*** *dialog box. The only purpose of creating a wall of length more than the desired length is to explain how to modify the length of the sketched wall to the exact value.*

Modifying the Length of the Sketched Exterior Wall

You will now modify the length of the sketched wall to the actual dimension, as given in the *Apartment 1* sketch.

1. Select the created wall and click on its temporary dimension; an edit box appears showing the current dimension of the wall segment.

2. Enter **40' (12192 mm)** in the edit box, as shown in Figure 3-15 and then press ENTER; the length of the wall is modified to 40'0"(12192 mm). Press ESC to exit the **Modify | Walls** tab.

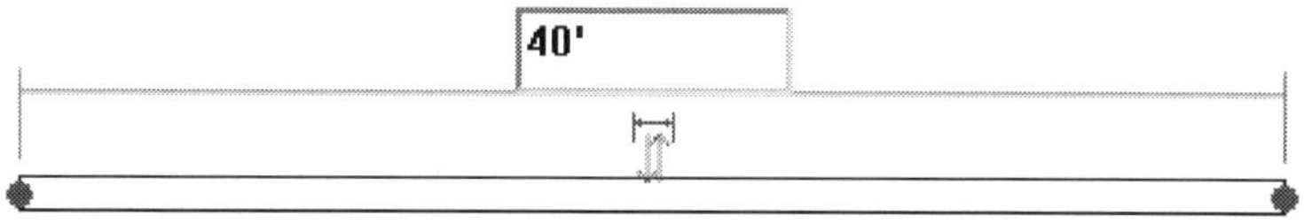

Figure 3-15 *Modifying the length of a wall segment using temporary dimensions*

Tip

To exit the current tool, you can also right-click and choose ***Cancel*** *from the shortcut menu displayed.*

Note

By default, the exterior faces of the walls drawn from the left to right are on the upper face, and those drawn from top to bottom are on the right face. Similarly, the exterior faces of the walls drawn from right to left are on the lower face, and those drawn from the bottom to top are on the left side. Hence, you can minimize the use of the ***flip*** *tool by sketching the walls in appropriate direction.*

Sketching Other Exterior Wall Segments

In this section, you need to create other exterior wall segments using the **Endpoint** object snap tool.

1. Choose the **Wall: Architectural** tool from **Architecture > Build > Wall** drop-down. Now, bring the cursor close to the right endpoint of the first wall segment. When the cursor shows a square box at the endpoint (indicating the **Endpoint** object snap), as shown in Figure 3-16, click to specify the start point of the second wall segment.

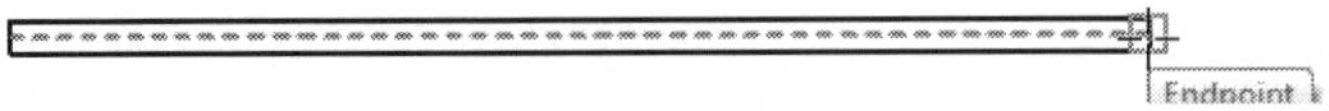

Figure 3-16 *Starting a second exterior wall segment using the* ***Endpoint*** *object snap option*

2. Move the cursor; a wall is dynamically created with one end attached to the specified point and the other end attached to the cursor. Move the cursor vertically downward. A dashed vertical line is visible inside the wall segment. Now, enter **20'0"(6096 mm)** as the value of the length; an edit box is displayed with the dimension that you have entered, as shown in Figure 3-17. Press ENTER; the second wall segment is sketched exactly to **20'0"(6096 mm)** length.

   Notice that the intersection of the first and second wall segments has been intuitively filled or completed.

Figure 3-17 *Creating the second exterior wall segment*

3. To create the third wall segment, select the **Chain** check box from the **Options Bar**.

4. Move the cursor close to the endpoint of the second wall segment and click when the endpoint object snap is displayed. On doing so, the third wall segment starts from this specified point. Move the cursor horizontally toward the left. Now, hold the SHIFT key while moving the cursor.

   Notice that the cursor can now move only in orthogonal directions (in horizontal and vertical direction). When the length of the wall segment is around **40'0"(12192 mm)**, a vertical dashed line originates from the start point of the first wall segment. The alignment line shows the point on the third wall segment that is in plumb with the first point. An intersection snap symbol indicated by an X appears at this point, as shown in Figure 3-18.

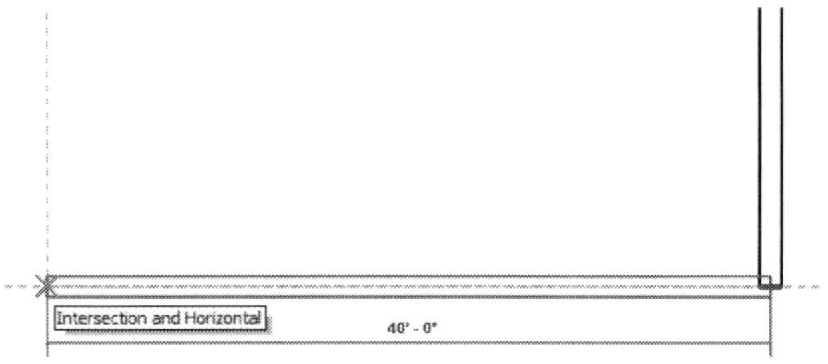

Figure 3-18 Creating the third wall segment

5. Click to specify the location of the endpoint of the third wall segment when the intersection snap symbol is displayed.

6. As you have enabled the **Chain** option, the next wall segment automatically starts from the last specified point. Move the cursor vertically upward and enter **25'(7620 mm)**, as shown in Figure 3-19, and then press ENTER. The fourth wall segment is created.

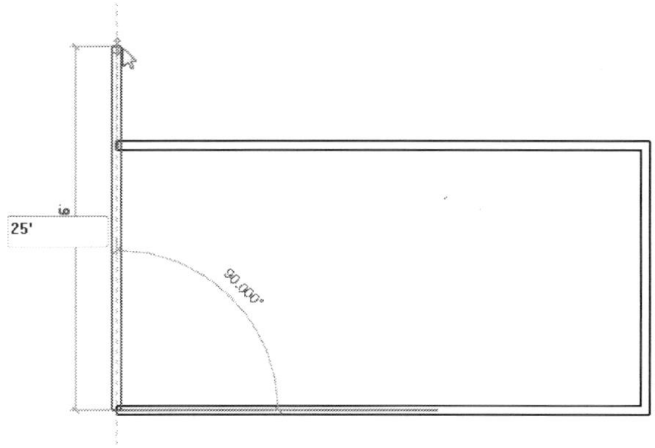

Figure 3-19 Sketching the fourth exterior wall segment

7. Press ESC twice to exit the **Wall: Architectural** tool.

Note
The purpose of creating a wall more than the desired length is only to explain how to stretch the wall to the exact length.

Stretching the Wall Segment

You will now stretch the wall segment to change its length to the desired dimension by using the drag controls.

1. Select the fourth wall segment to display its controls. The two blue dots at its two endpoints are the drag controls. Move the cursor near the upper dot; the color of the drag control symbol changes, as shown in Figure 3-20. Press and hold the left mouse button at this point and drag the cursor vertically downward and bring it close to the start point of the first wall segment. On doing so, the endpoint object snap is displayed at the intersection of two walls and the tooltip shows **Endpoint and Vertical**.

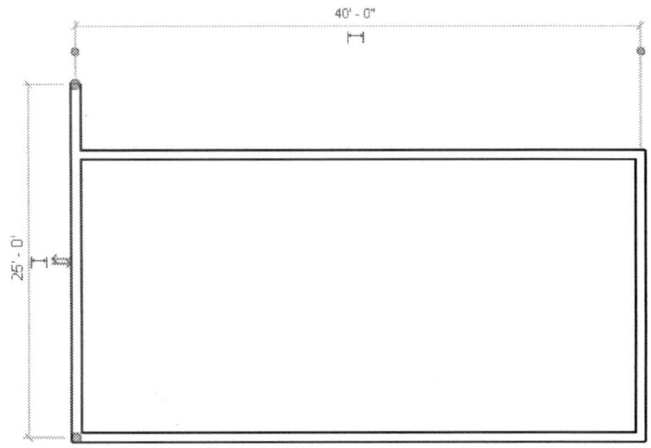

Figure 3-20 *Using the drag control to modify the length of wall*

2. Release the left mouse button at the intersection point; the first and fourth wall segments are joined at the corner with their ends completed.
   Note that if these wall segments do not join at the corner on releasing the left mouse button, you need to click at their intersection.

3. Press ESC to remove the wall segment from the selection set. The external wall profile is completed for this tutorial, as shown in Figure 3-21.

Figure 3-21 *The completed exterior wall profile*

Saving the Project

In this section, you will save the project file using the **Save As** tool.

1. Choose **Save As > Project** from the **File** menu; the **Save As** dialog box is displayed. Create a folder with the name **c03_revit_2025_tut** and then save the file with the name **c03_Apartment1_tut1** (**M_c03_Apartment1_tut1**).

2. Choose the **Close** option from the **File** menu; the file is closed.

Tutorial 2 — Club

Create the exterior walls of the Club building. The sketch plan is shown in Figure 3-22. The dimensions have been given only for reference and are not to be used in this tutorial. The parameters to be used for creating the exterior walls of the club building are given next.

(Expected time: 30 min)

1. Project file-
   For Imperial *c02_Club_tut2.rvt*
   For Metric *M_c02_Club_tut2.rvt*
2. Exterior wall type- **Exterior - Split Face and CMU on Mtl. Stud**.
3. Unconnected height of walls- **15'0" (4572 mm)**.
4. All inclined walls are at 45-degree to the horizontal axis.

The following steps are required to complete this tutorial:

a. Open the project file.
   For Imperial *c02_Club_tut2.rvt*
   For Metric *M_c02_Club_tut2.rvt*
b. Invoke the **Wall: Architectural** tool.
c. Select the exterior wall type:
   For Imperial **Exterior - Split Face and CMU on Mtl. Stud**
   For Metric **Exterior - Brick on Mtl. Stud**

d. Set the unconnected height to **15'0"** or **4572mm**.
e. Change the **Location Line** parameter to **Finish Face: Exterior**, refer to Figure 3-23.
f. Select the **Line** sketching option and sketch the inclined wall profile using the **Chain** option, refer to Figures 3-24 through 3-30.
g. Use the **Center-ends Arc** sketching option to create the curved wall, refer to Figures 3-31 and 3-33.
h. Save the project using the **Save As** tool.
   For Imperial *c03_Club_tut2.rvt*
   For Metric *M_c03_Club_tut2.rvt*
i. Close the project.

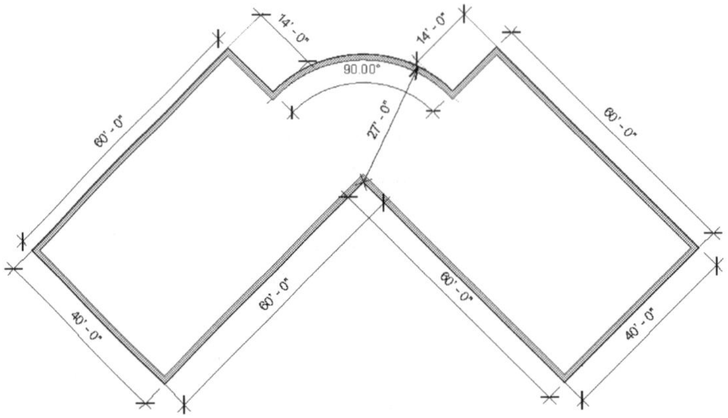

Figure 3-22 *Sketch plan for the Club project*

Opening an Existing Project

Choose **Open > Project** from the **File** menu and open the *c02_Club_tut2.rvt (M_c02_Club_tut2.rvt* for Metric*)* project file created in Tutorial 2 of Chapter 2. You can also download this file from *https://www.Cadsoft.com*. The path of the file is as follows: *Textbooks > Civil/GIS > Revit Architecture> Exploring Autodesk Revit 2025 for Architecture*.

Invoking the Wall: Architectural Tool and Selecting the Wall Type

First, you will invoke the **Wall: Architectural** tool from the ribbon and then select the specified exterior wall type, as given in the project parameters.

1. Choose the **Wall: Architectural** tool from the **Architecture > Build > Wall** drop-down; the **Modify | Place Wall** tab is displayed.

2. In the **Properties** palette, select the required option from the **Type Selector** drop-down list.
   For Imperial **Exterior - Split Face and CMU on Mtl. Stud**
   For Metric **Exterior - Brick on Mtl. Stud**

Modifying Properties of the Exterior Wall

Next, you will use the **Properties** palette to modify the unconnected height to 15'0" or 4572mm. The dimensions given in the sketch are exterior wall face dimensions. Therefore, you need to set the **Location Line** parameter to **Finish Face: Exterior**.

1. In the **Properties** palette, click in the value field corresponding to the **Unconnected Height** parameter and replace the current value with **15'0" (4572mm)**.

2. Click in the value field of the **Location Line** parameter and select the **Finish Face: Exterior** option from the drop-down list displayed, as shown in Figure 3-23. Choose the **Apply** button to accept the specified values.

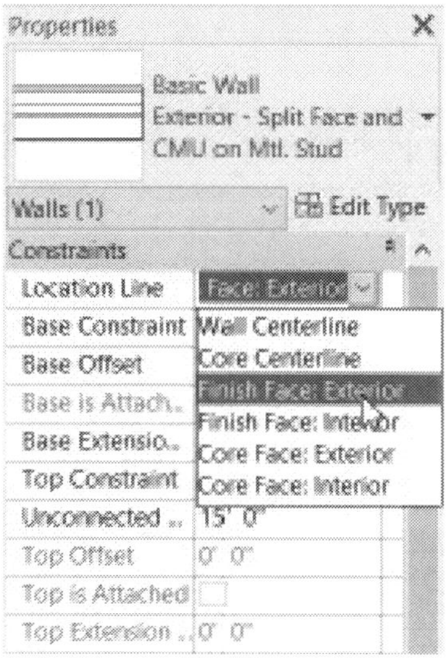

*Figure 3-23 Selecting the **Finish Face : Exterior** option for the **Location Line** parameter*

Sketching the Inclined Exterior Walls

Start creating the exterior wall profile by first sketching the inclined walls in a sequence such that the exterior face of the wall is on the external side. You need to select the **Chain** check box to sketch the continuous wall profile. Once you have created the first inclined wall, the other parallel and perpendicular walls can easily be created using the alignment lines and different object snaps options.

1. To create the straight wall, Choose the **Line** tool in the **Draw** panel of the **Modify | Place Wall** tab, if not chosen by default.
2. In the **Options Bar**, select the **Chain** check box, if not selected. Also, select the **Allow** option from the **Join Status** drop-down list.
3. To start sketching the first inclined wall segment, click inside the four arrow keys in the drawing window and move the cursor upward toward the right and then move it to an inclination such that the angle subtended at the horizontal axis is 45 degrees.
4. Enter the value **14' (4267 mm)** to specify the length of the first wall segment; the value is displayed in the edit box, as shown in Figure 3-24. Press ENTER to create the first wall segment of the specified length.
5. As the **Chain** check box is selected in the **Options Bar**, the second wall segment will start from the last specified point. Now, move the cursor downward toward the right and right-click to invoke the shortcut menu. Next, choose **Snap Overrides > Perpendicular** from the shortcut menu displayed; the perpendicular snap symbol appears at the end of the wall. Enter **60'(18288 mm)**, see Figure 3-25; a wall is created perpendicular to the first inclined wall.

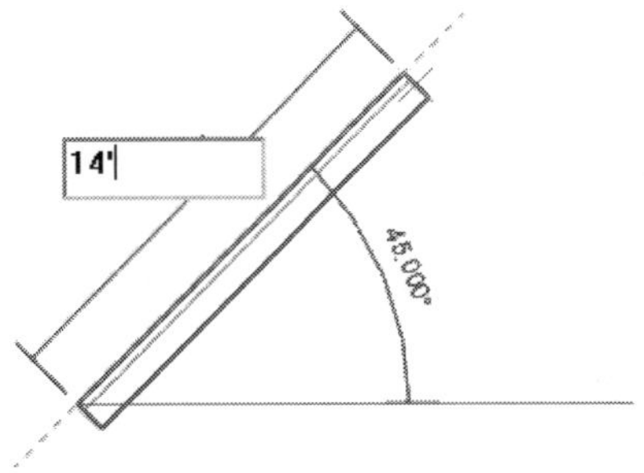

Figure 3-24 Sketching the first inclined wall segment

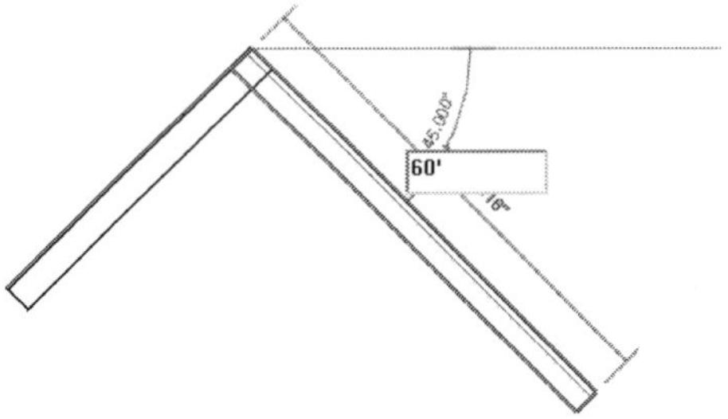

Figure 3-25 Sketching the second inclined wall segment

6. Similarly, for creating the next wall, move the cursor downward to the left and choose the perpendicular snap override as in step 5. Enter **40'(12192 mm)** as length, as shown in Figure 3-26. Now, press ENTER to create the third inclined wall.

7. To create the next wall, move the cursor upward toward the left and invoke the perpendicular snap override as in step 5. Enter **60' (18288 mm)** as the length of the wall segment and press ENTER to create the fourth inclined wall segment, as shown in Figure 3-27.

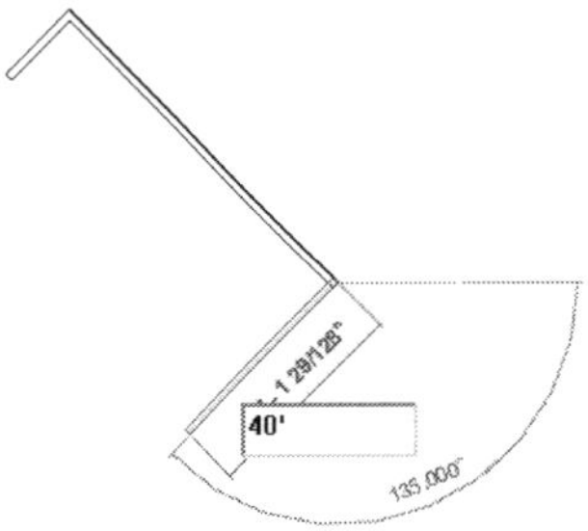

Figure 3-26 Sketching the third inclined wall segment

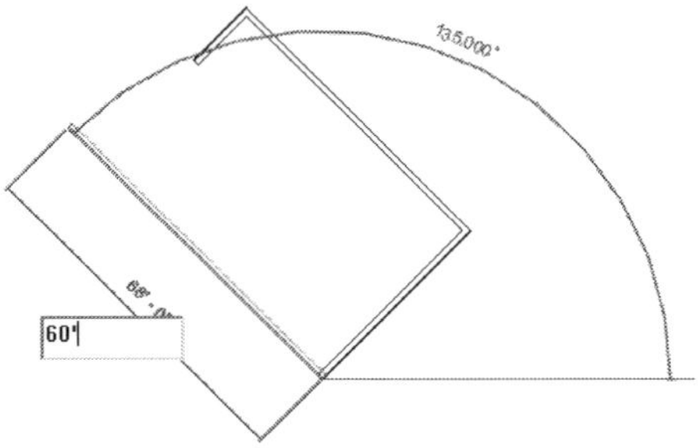

Figure 3-27 Sketching the fourth inclined wall segment

8. Similarly, create the fifth inclined wall segment of **60'(18288 mm)** length, as shown in Figure 3-28.

9. Next, create the connected wall segment of **40'(12192 mm)** length, as shown in Figure 3-29.

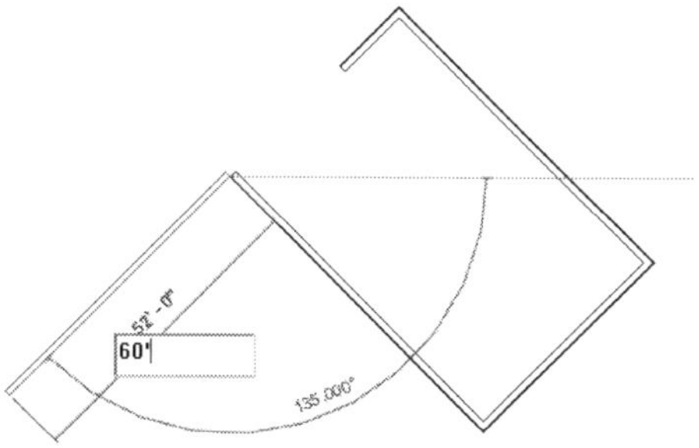

Figure 3-28 Sketching the fifth inclined wall segment

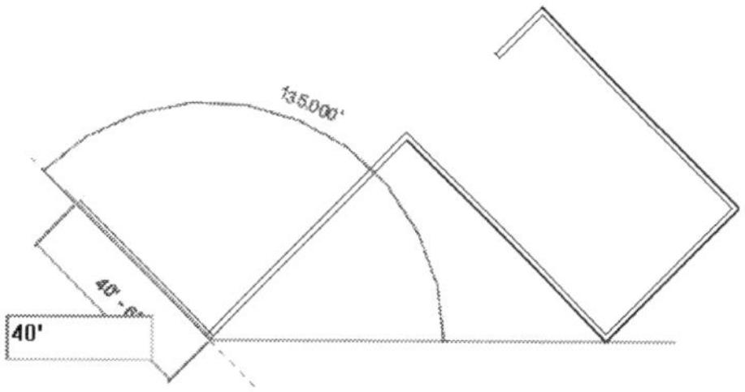

Figure 3-29 Sketching the sixth inclined wall segment

10. Now, create the next two wall segments of lengths **60'(18288 mm)** and **14'(4267 mm)** to complete the inclined wall exterior profile, as shown in Figure 3-30.

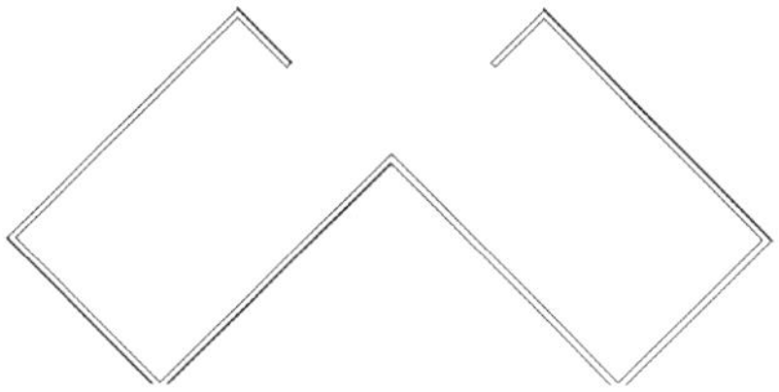

Figure 3-30 The sketched inclined wall exterior profile

11. Press the ESC key to discontinue the wall at this point and finish sketching the inclined walls.

Sketching the Curved Exterior Wall

Next, you will sketch the curved exterior wall profile based on the given parameters. You need to use the **Center-ends Arc** tool to create the curved wall segment.

1. Choose the **Center-ends Arc** tool in the **Draw** panel of the **Modify | Place Wall** tab. Move the cursor in the drawing window and click on the outer intersection of the inclined walls to specify the center of the curved wall, refer to Figure 3-31.

2. Move the cursor near the endpoint of the last sketched inclined wall segment. When 135-degree is displayed as the angular dimension, enter the value **27' (8100 mm)** as the radius of the curved wall and press ENTER, as shown in Figure 3-31. On doing so, the curved wall with the specified radius starts from the specified point.

Note

*While tracking the wall, if the perpendicular or other snapping symbol does not appear, you can right-click and choose **Snap Overrides**; a cascading menu will be displayed.*

Choose any snapping option from the cascading menu. On doing so, the desired snapping symbol will appear for the specified point or action.

3. Move the cursor toward the right and click when the cursor snaps to the endpoint of the inclined wall, as shown in Figure 3-32. Now, press ESC twice to complete the exterior wall profile, as shown in Figure 3-33.

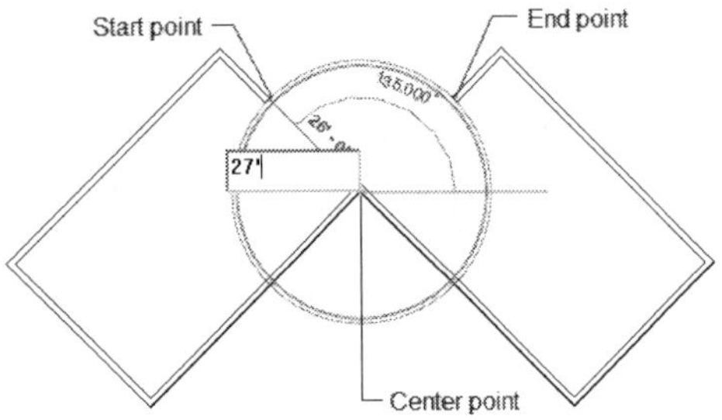

Figure 3-31 *Starting the curved wall segment*

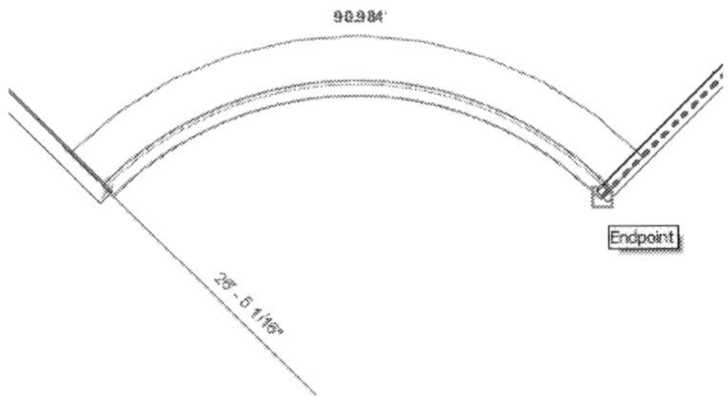

Figure 3-32 *Completing the curved wall segment*

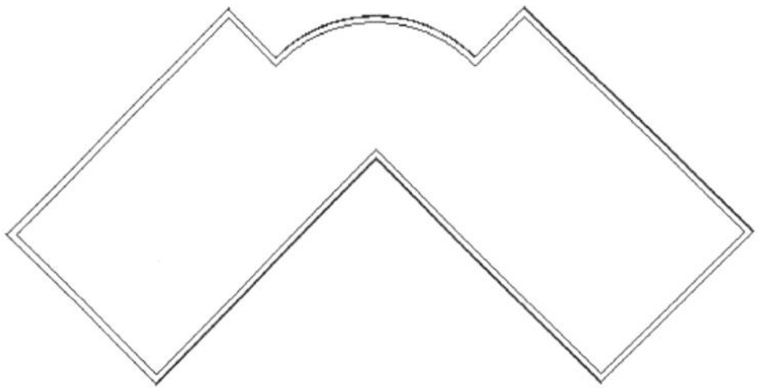

Figure 3-33 *Completed layout of the exterior walls for the Club project*

4. Choose the **Default 3D View** tool from **View > Create > 3D View** drop-down; a 3D view of the building model is displayed, as shown in Figure 3-34.

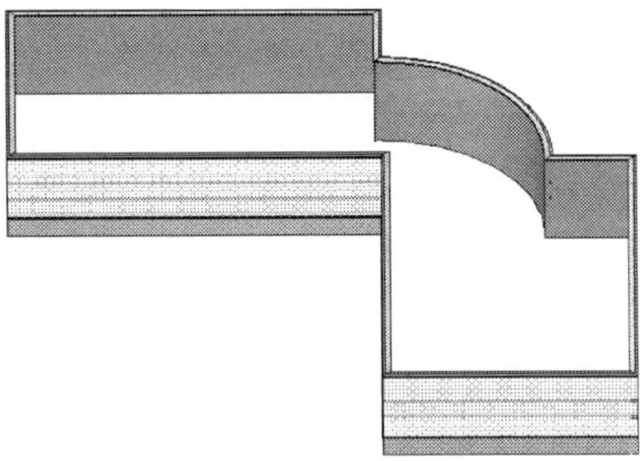

Figure 3-34 *3D view of the completed exterior wall profile*

5. From **View Control Bar**, choose the **Visual Style** button; a flyout is displayed. Choose the **Shaded** option from the flyout.

6. Under the **Floor Plans** head in the **Project Browser**, double-click on **Level 1** to return to the plan view.

Saving the Project

In this section, you will save the project file using the **Save As** tool.

1. Choose **Save As > Project** from **File** menu; the **Save As** dialog box is displayed. Now, browse to the folder **c03_revit_2025_tut** and enter **c03_Club_tut2** (**M_c03_Club_tut2** for Metric) and choose **Save**.

2. Choose the **Close** option from the **File** menu; the file is closed.

EXERCISES

Exercise 1 Apartment 2

Create the exterior and interior walls of the *Apartment 2*, refer to Figure 3-35. The thick walls are the exterior walls and the thin walls are the interior walls. The dimensions and texts are not to be added. The project parameters for this exercise are given next.

(Expected time: 30 min)

1. Project file -
   For Imperial *c02_Apartment2_ex1.rvt.*
   For Metric *M_c02_Apartment2_ex1.rvt.*
2. Exterior wall type- **Exterior - Brick on Mtl. Stud**.
3. Interior wall type-
   For Imperial **Basic Wall: Interior - 5" Partition (2-hr)**.
   For Metric **Basic Wall: Interior - 135mm Partition (2-hr)**.
4. Height of the wall- **Top Constraint** - **Up to Level 2**.
5. Location line parameter for the exterior walls- **Wall Centerline**.
6. Name of the file to be saved-
   For Imperial **c03_Apartment2_ex1**
   For Metric **M_c03_Apartm**

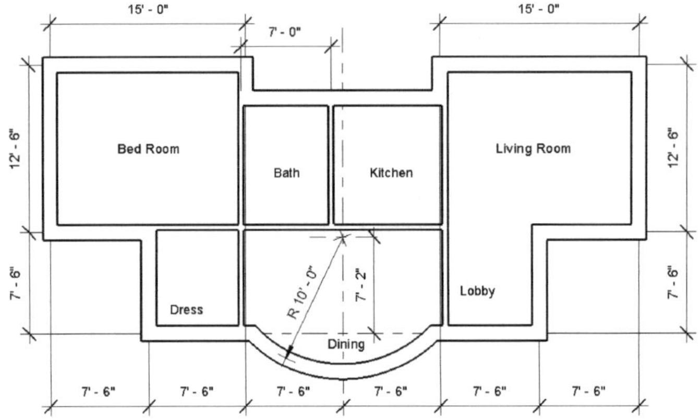

Figure 3-35 The sketch plan for creating the exterior and interior walls for the Apartment 2 project

Exercise 2 Elevator and Stair Lobby

Create the exterior walls of the *Elevator and Stair Lobby* project, refer to Figure 3-36. Do not add dimensions or texts as they are given only for reference. The project parameters for this exercise are given next. **(Expected time: 30 min)**

1. Project file -
   For Imperial *c02_ElevatorandStairLobby_ex2.rvt*
   For Metric *M_c02_ElevatorandStairLobby_ex2.rvt*
2. Exterior wall type - **Basic Wall Exterior Brick on Mtl. Stud**.
3. Height of the wall - **Top Constraint- Up to Level 2**.
4. Location line parameter - **Wall Centerline**.
5. Name of the file to be saved -
   For Imperial **c03_ElevatorandStairLobby_ex2**.
   For Metric **M_c03_ElevatorandStairLobby_ex2**.

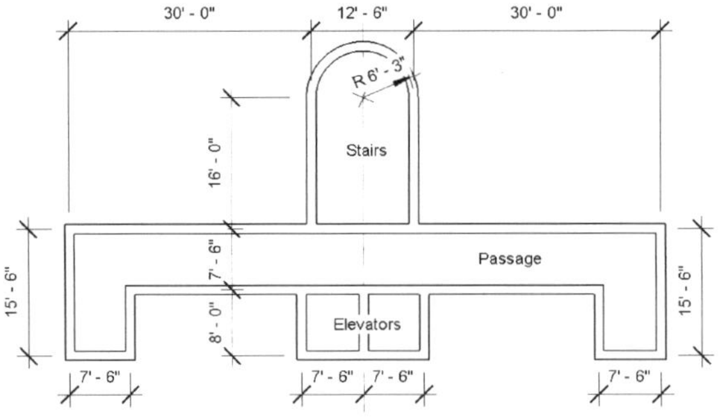

Figure 3-36 Sketch plan for creating the exterior walls for the Elevator and Stair Lobby project

Chapter 4

Using Basic Building Components-I

Learning Objectives

After completing this chapter, you will be able to:

- *Understand the concept of doors*
- *Understand various properties of doors*
- *Add doors to the exterior and interior walls*
- *Understand the concept of windows*
- *Understand various properties of windows*
- *Add windows to a building model*
- *Create door and window openings in walls*
- *Add openings to the walls*

INTRODUCTION

In the previous chapter, you learned to create different types of exterior and interior walls in a building profile. Similar to walls, the positioning of doors and windows also plays an important role in designing a building profile. In this chapter, you will learn the process of selecting and adding doors and windows to a building model. Autodesk Revit has several in-built families of doors and windows which help in adding different types of doors and windows in a building model. In Revit, doors and windows act as hosted components that can be added to any type of wall. In this chapter, you will also learn to use these components and modify their properties.

ADDING DOORS IN A BUILDING MODEL

A door is one of the most frequently used components in a building model. It helps in accessing various exterior and interior spaces in a project. Autodesk Revit provides a variety of predefined door types. You can access these door types by using the options from the **Type Selector** drop-down list of the **Properties** palette. You can also load other door types from the **English-Imperial** (for imperial) or **English** (for metric) folder. A wall acts as a host element for doors. This means that a door can be placed only if a wall exists there. When you add a door to a wall, Autodesk Revit intuitively creates an opening in it. The procedure for adding doors to a building model is described next.

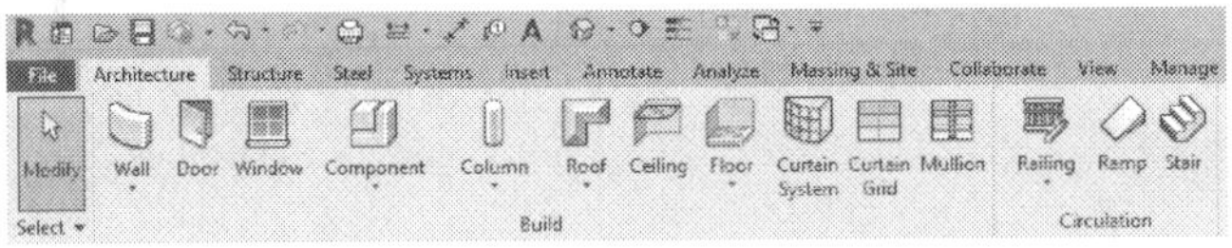

***Figure 4-1** Invoking the **Door** tool from the **Build** panel*

Adding Doors

In a project, you can add doors to a wall by using the **Door** tool. You can invoke this tool from the **Build** panel, as shown in Figure 4-1. Alternatively, you can type **DR** to invoke this tool.

To add a door, invoke the **Door** tool; the **Modify | Place Door** tab will be displayed in the ribbon. Select the desired door type from the **Type Selector** drop-down list in the **Properties** palette; the instance properties of the door will be displayed, as shown in Figure 4-2. This drop-down list displays the loaded door types. Next, after selecting a door type, click on the desired location of a wall to add the door. After adding the door to the building model, you can view and modify its properties. To do so, select the door; the instance properties of the door will be displayed in the **Properties** palette, as shown in Figure 4-3. You will learn more about adding doors to a wall further in this chapter.

Now, to modify the type properties of the selected door type, choose the **Edit Type** button from the **Properties** palette; the **Type Properties** dialog box for the selected door type will be displayed, as shown in Figure 4-4. In this dialog box, change the parameters as per your requirement and then choose the **Preview** button to view the graphical image of the selected door type.

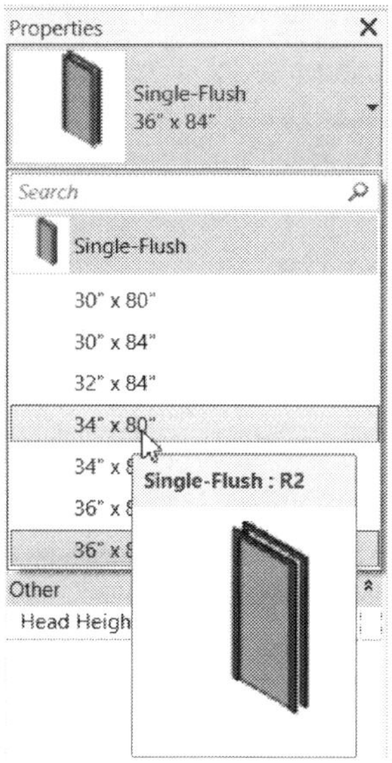

Figure 4-2 Selecting the door type from the **Properties** palette

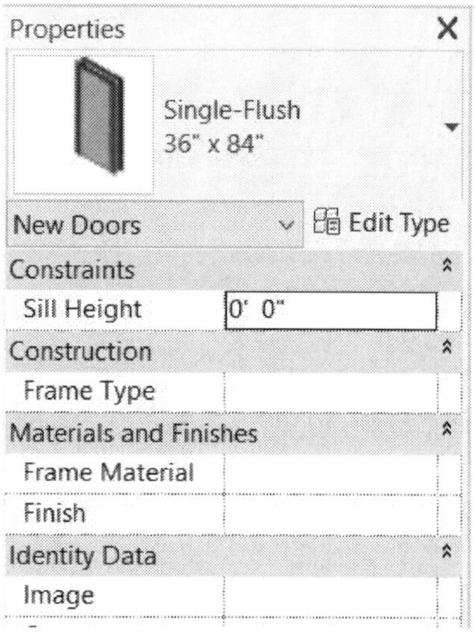

Figure 4-3 The instance properties displayed for the ***Single-Flush: 34" x 84"*** door type

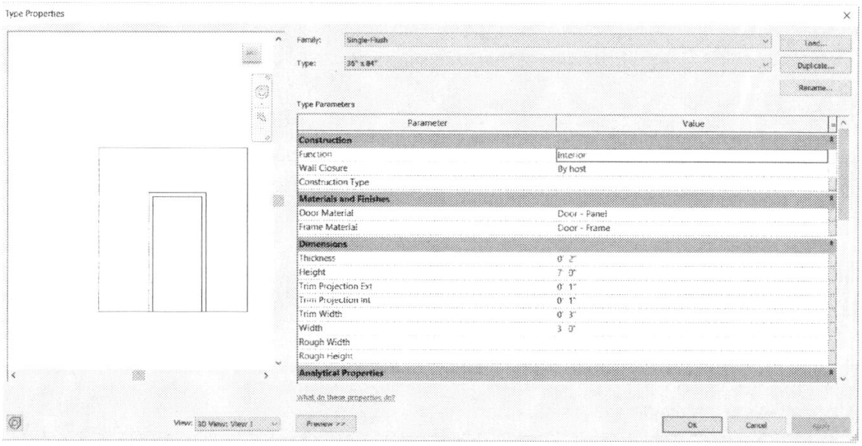

Figure 4-4 The **Type Properties** dialog box

Understanding Door Properties

The **Type Properties** dialog box displays the values of different parameters of a door. The type parameters are arranged under various heads based on their properties such as **Construction**, **Analytical Properties**, **Materials and Finishes**, **Dimensions**, **Identity Data**, and **IFC Parameters**. The parameters associated with each property are listed under each head. For example, the **Dimensions** head includes the dimensional type parameters associated with the door type such as **Height**, **Width**, **Trim Width**, and so on. You have the option of either using the default values for these parameters or modifying them to the desired values. However, before modifying the type properties of a predefined door type, it is recommended to create a copy of the original door type and modify the desired properties of the new door type. The **Duplicate** button available on the top right corner in this dialog box is used to create a copy of the door type. When you choose this button, the **Name** edit box will be displayed, prompting you to assign a name to the new door type. Enter a name in this edit box and choose the **OK** button; a new door type is created which inherits the properties of the parent door type. Now, its parameters can be modified as desired. Use the **Rename** button to give a new name to the selected door type. You can use the **Load** button provided on the top right corner of the dialog box to load a door type of the required family from the additional libraries.

Note

*Renaming a door type does not alter its properties. For example, if you rename **34"x 84"** to **36"x 84"**, its width will still remain 34", irrespective of its name.*

Adding a Door to a Wall

Doors can be added to a wall in the plan, section, elevation or a 3D view by clicking at the desired location on the wall. After invoking the **Door** tool and selecting the door type, move the cursor over the wall on which it is to be added. As you move the cursor over the wall, you will notice the preview of the door type. When the cursor comes close to the upper or the lower face of the wall, the door symbol appears on it, as shown in Figure 4-5. Also, a temporary dimension appears depicting the distance of the center of the door opening from the adjacent walls or wall edges.

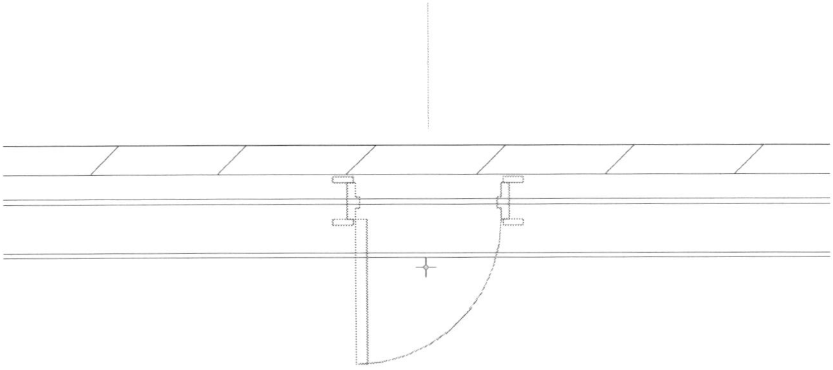

Figure 4-5 *Adding a door to a wall*

Using the temporary dimensions displayed, you can add the door at the desired location. Alternatively, you can first add the door at an approximate location on the wall and then modify its location to the exact dimension. When a door is added, an opening is automatically created in the wall and the edges are also completed, as shown in Figure 4-6. After selecting the door, you will notice that the horizontal and vertical twin arrow symbols appear in the door along with a tag number. To flip the swing-side of the door, click on the twin arrow key that appears parallel to the wall edge. To shift the door to the other side of the wall, click on the twin arrow key that appears perpendicular to the wall edge (vertical, in this case).

To move the added door to the exact dimension, click on the particular temporary dimension text and enter the exact value, as shown in Figure 4-7. The door shifts to the desired location.

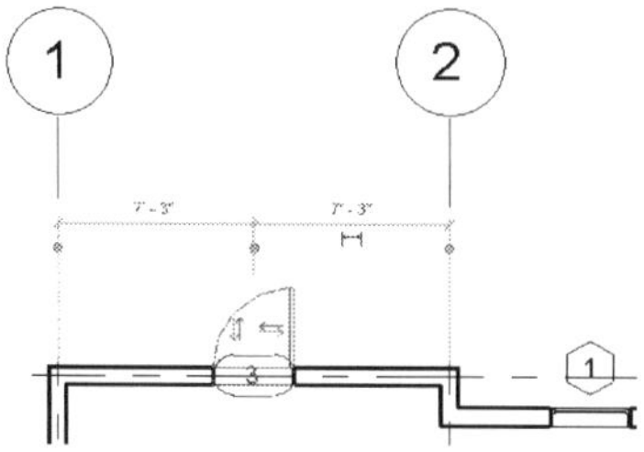

Figure 4-6 *A door added along with its controls*

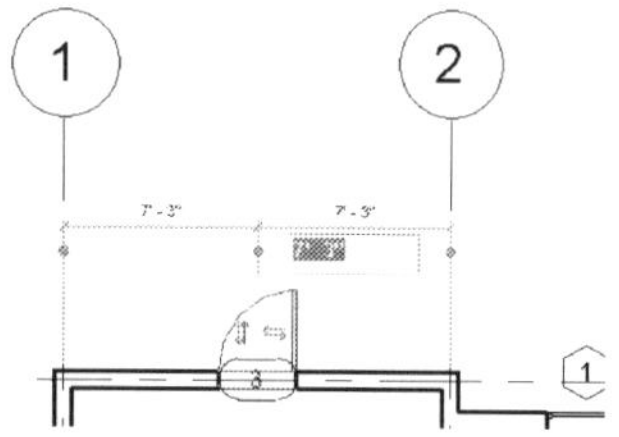

Figure 4-7 *Specifying the exact location of the door*

Tip

To flip the swing of the door, select the door and press the SPACEBAR. Each time you press the SPACEBAR, the door will cycle through all the four possibilities.

Note

A door can be placed in any type of wall regardless of the height of the door. Autodesk Revit 2025 displays alert messages if the door is not placed appropriately.

In Autodesk Revit, when doors are inserted, the door tags are inserted if you choose the **Tag on Placement** button from the **Tag** panel of the **Modify | Place Door** contextual tab. The value associated with the door tag depends on the type of family associated with the tag.The door tags assists in marking the doors and later arranging them in the form of a schedule. The door type tag number increases sequentially when you place a door by using the **Door** tool or when you copy and paste it into the building model. You can, however, give a specific mark or tag to doors individually. The visibility of door tags can be controlled by using the **Visibility/Graphics Overrides for Floor Plan** dialog box. This dialog box can be invoked by typing **VG**, or by selecting the desired door from the drawing and choosing the **Visibility/ Graphics** tool in the **Graphics** panel in the **View** tab. Next, if you want to hide the door tags, choose the **Annotation Categories** tab from the **Visibility/Graphics Overrides for Floor Plan** dialog box and clear the **Door Tags** check box. When you invoke the **Door** tool, the option for adding a horizontal or vertical door tag becomes available in the **Options Bar**. You can set the orientation of the door tags to horizontal or vertical by selecting the **Horizontal** or **Vertical** option from the drop-down list displayed in the **Options Bar**.

To change a door type, select the particular instance; the properties and the type of the selected door will be displayed in the **Properties** palette. In the **Properties** palette, select a new door type from the **Type Selector** drop-down list; the selected door will be replaced with the new door type.

In Autodesk Revit, you can create a new custom door type by using various tools. To create a new custom door type, choose the **Model In-place** tool from the **Mode** panel of the **Modify | Place Door** tab; the **Name** dialog box will be displayed. In this dialog box, enter a name in the **Name** edit box and choose the **OK** button; the dialog box will be closed and the **Modify** tab will be displayed. Using the tools from this tab, you can create the doors that are specifically designed for that instance.

Note

*On invoking the **Model In-place** tool, the family editor mode will be activated. In this mode, you can use various massing tools from the **Architecture** tab to create an in-place door assembly.*

Apart from the door types available in the **Type Selector** drop-down list, you can use other door types from additional libraries. To access them, choose the **Load Family** tool in the **Mode** panel of the **Modify | Place Door**; the **Load Family** dialog box will be displayed. Additional door types are available in the **Doors** sub folder of the **English-Imperial** folder. Select a door type to view its image in the **Preview** area, as shown in Figure 4-8. After selecting the door type, choose the **Open** button. The selected family of doors will be added to the **Type Selector** drop-down list in the **Properties** palette for that project.

For a door type, you can specify swing angle of a door displayed in a Floor Plan. To do so, select the desired door type from the **Type Selector** drop-down list in the **Properties** palette, specify the swing angle in the edit box corresponding to the **Swing Angle** option and click on the **Apply** button in the **Properties** palette. You will notice that the swing angle has changed.

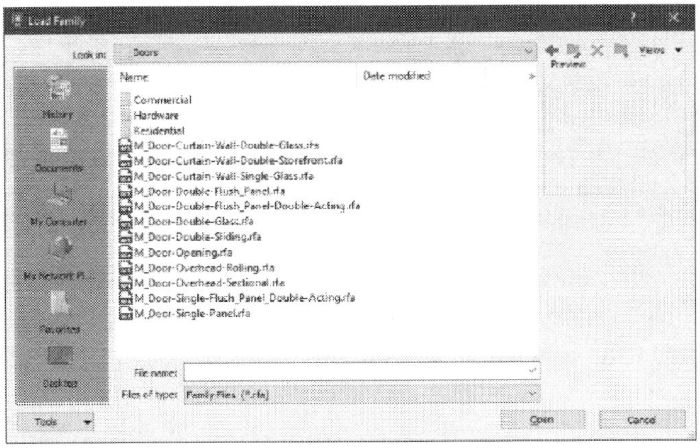

*Figure 4-8 Using the **Load Family** dialog box to load additional door types*

Adding a Door to a Slanted Wall

In Autodesk Revit 2025, you can add a door to a wall and then orient the door with the wall slant. To do so, after adding the desired door family to the slanted wall, double click the **Esc** button. Now select the door; **Modify|Door** tab will be displayed. Now choose the **Slanted** option from the drop-down list corresponding to the **Orientation** option from the **Properties** palette; the door will orient itself with the wall angle.

Tip
*Using the **Options Bar**, you can set certain parameters before placing the components. For example, before placing the door, you can set the horizontal or vertical position of door tags and select the leader as well.*

ADDING WINDOWS TO A BUILDING MODEL

Windows form an integral part of any building project. Autodesk Revit provides several in-built window types that can be easily used and added to a building model. Like doors, windows are also dependent on the walls that act as their host element.

Adding Windows

Ribbon: Architecture > Build > Window
Shortcut Key: WN

In Autodesk Revit, you can add windows to a building model by using the **Window** tool. To do so, invoke the **Window** tool from the **Build** panel of the **Architecture** tab, as shown in Figure 4-9; the **Modify | Place Window** tab will be displayed. In this tab, select the window type from the **Type Selector** drop-down list in the **Properties** palette, as shown in Figure 4-10. To add a window in the building model, move the cursor over the wall and click to place it either in the upper face or in the lower face. After adding the window, you can view or change its instance parameters. To do so, select the window from the drawing; the **Properties** palette will be displayed, as shown in Figure 4-11. You can use this palette to view and modify various instance parameters of the selected window such as its Level, Head Height, Sill Height, and others. To modify and view the type parameters of the window, choose the **Edit Type** button in the **Properties** palette; the **Type Properties** dialog box will be displayed. In this dialog box, choose the **Preview** button to view the graphical image of the selected window type, as shown in Figure 4-12.

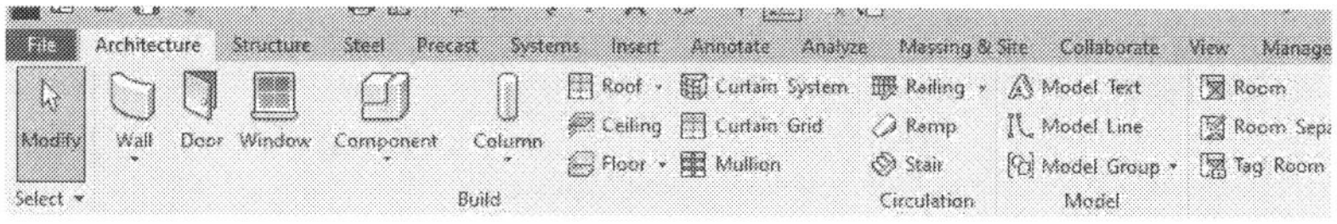

*Figure 4-9 Invoking the **Window** tool from the **Build** panel*

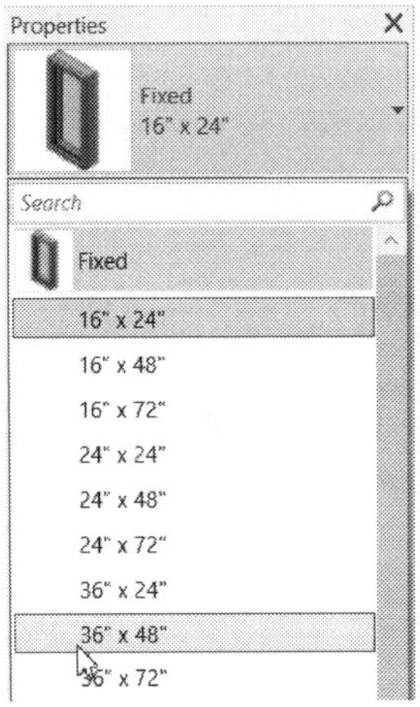

Figure 4-10 Selecting the window type from the ***Type Selector*** *drop-down list*

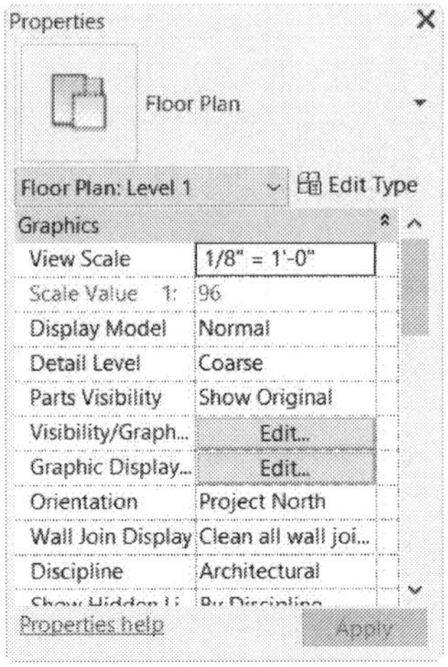

Figure 4-11 The ***Properties*** *palette for the* ***Fixed: 36" x 48"*** *window type*

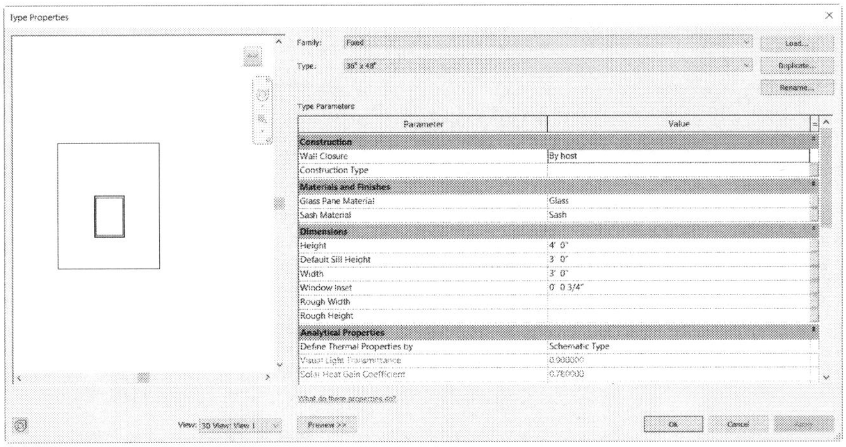

Figure 4-12 The ***Type Properties*** *dialog box for the* ***Fixed: 36" x 48"*** *window type with the* ***Preview*** *button chosen*

Understanding Window Properties

The **Type Properties** dialog box displays the type parameters for different components of a window such as **Glass Pane Material**, **Sash Material**, **Default Sill Height**, **Width**, **Height**, and so on. You can click in the value column of each parameter and select from the available options or use the **Load** button provided on the top right corner of the dialog box to load a window family type from the additional libraries. The **Duplicate** button is used to create a copy of the window type with a different name. The new window inherits the properties of the parent window type and its parameters can be modified. By using the **Rename** button, you can give a new name to the selected window type.

Adding a Window to a Wall

To add a window to a wall, invoke the **Window** tool from the **Architecture** tab of the ribbon and select the window type from the **Type Selector** drop-down list in the **Properties** palette. Next, move the cursor close to the wall on which it needs to be added. The window symbol appears when the cursor is moved over the wall. When the cursor comes close to the lower face of the wall, the window appears on that face. The window appears on the upper face when the cursor comes near to it, as shown in Figure 4-13. Also, notice a temporary dimension, which depicts the distance of the center of the window opening from the adjacent walls or wall edges.

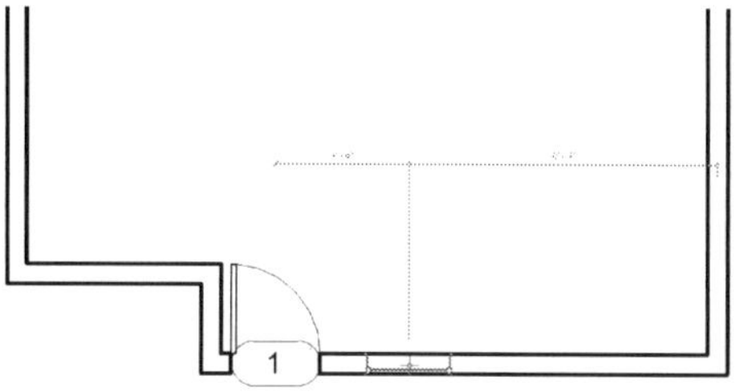

Figure 4-13 *Adding a window to an existing wall*

You can use the temporary dimensions to add the window by clicking at the appropriate location on the wall. Autodesk Revit automatically creates an opening in the wall and the edges also get completed, as shown in Figure 4-14.

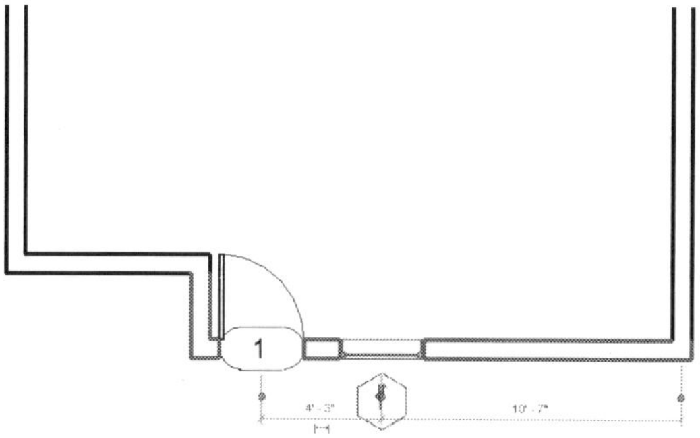

Figure 4-14 *Window added to an existing wall*

You will also notice a twin arrow symbol on the window, along with a tag number. Click on the symbol to flip the orientation of the window. The flip arrow symbol will appear on the side that is interpreted as the inner side of the window. You can specify the exact position of the window with respect to the adjacent walls or edges after placing it at the approximate location. To specify the dimensional location of a window, click on the particular temporary dimension text and type the exact dimension, as shown in Figure 4-15. The window will shift to the desired location.

Window tags are automatically generated in Autodesk Revit. Unlike doors, however, each window of the same type bears the same tag. Different window types have preassigned marks, which can be changed using the **Type Properties** dialog box. Window tags are used for creating the window schedule.

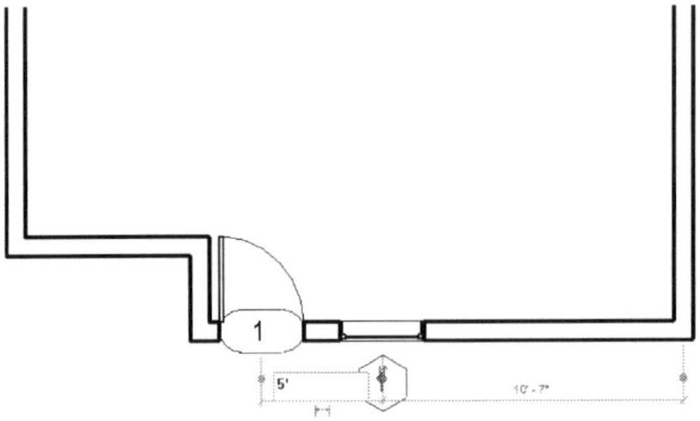

Figure 4-15 *Changing the location of the added window*

Tip
The depth of the window is automatically created based on the thickness of the host wall. If you change the type of the host wall of a window, the depth of the associated window will change accordingly.

Note
As you place windows in a project, the tag number assigned to them increase by one. You can also give them a specific tag number individually.

Similar to the door tags, the visibility of the window tags can also be controlled in the drawing. To hide the window tags, invoke the **Visibility/Graphic for Floor Plan** dialog box and choose the **Annotation Categories** tab from it. Next, clear the **Window Tags** check box.

Note
On clearing the ***Window Tags*** *check box in the* ***Annotation Categories*** *of the* ***Visibility/Graphic for Floor Plan*** *dialog box, the tags will get hidden but will still remain a part of the project.*

In Autodesk Revit, you can create the windows that are specifically designed for a location or an instance. To do so, choose the **Model In-place** tool from the **Mode** panel of the **Modify |Place Window** tab; the **Name** dialog box will be displayed. In this dialog box, enter the name of the window type in the **Name** edit box and choose the **OK** button; the dialog box will be closed and the **Modify** tab will be displayed. Now, the family editor mode will be activated, and in this mode, you can use various tools to create the in-place window. Apart from the window types displayed in the **Type Selector** drop-down list, you can access other window types as well. To add other window types in the **Type Selector** drop-down list, choose the **Load Family** tool from the **Mode** panel of the **Modify | Place Window** tab; the **Load Family** dialog box will be displayed. In this dialog box, add additional window types from the **Library > English-Imperial > Windows** folder. Alternatively, you can use the **Revit Web Content Library** to download more window types to your project.

Adding a Window to a Slanted Wall

In Autodesk Revit 2025, you can add a window to a wall and then orient it to the slanted wall. To do so, after adding the desired window family to the slanted wall, double-click the **Esc** button. Next, select the window; the **Modify|Windows** tab will be displayed. Now choose the **Slanted** option from the drop-down list corresponding to the **Orientation** option from the **Properties** palette; the window will orient itself with the wall angle.

DOORS AND WINDOWS AS WALL OPENINGS

Autodesk Revit also provides in-built opening types for door and windows. These types are available in the **Openings** subfolder of the **Libraries > English-Imperial** folder for Imperial (**Libraries > English** folder for Metric) that can be accessed from the **Load Family** dialog box. To invoke this dialog box, choose the **Load Family** tool from the **Load from Library** panel of the **Insert** tab.

To add an opening for the door, invoke the **Load Family** dialog box and then from the **Openings** subfolder, select any of the following family types such as: For Imperial- **Opening with Trim**, **Opening**, **Opening-Door**, **Passage Opening-Cased**, **Opening-Window-Round, Passage Opening-Elliptical Arch, and Opening-Window-Square** or for Metric **M_Opening with Trim**, **M_Opening**, **M_Opening-Door**, **M_Passage Opening-Cased**, **M_Opening-Window-Round, M_Passage Opening-Elliptical Arch, and M_Opening-Window-Square**. After selecting any of the family type, choose the **Open** button in the **Load Family** dialog box; the selected family will be loaded in the project. Now, to place the selected opening type in the wall, choose the **Place a Component** tool from **Architecture > Build > Component** drop-down; the **Modify | Place Component** tab will be displayed. In the **Properties** palette, ensure that the desired opening type is selected in the **Type Selector** drop-down list. Next, click in a desired location in the wall to insert the door opening. In the **Properties** palette, you can also edit the width and height of the door opening.

Similar to adding a door opening, you can insert a window opening. To do so, invoke the **Load Family** dialog box and then select any of the two family types from the **Openings** sub folder: **Opening with Trim** or **Opening**. To place the window opening in the wall, you will follow the same method as used for inserting the door opening.

OPENINGS IN THE WALL

Ribbon: Architecture > Opening > Wall

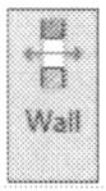

You can create rectangular openings in a curved or straight wall. To cut an opening in a wall, you can use a plan, elevation, or sectional view. Generally, the sectional or elevation view is preferred as locating and placing such views are easy. To cut a rectangular opening in a wall, open a preferred elevation or a section view in which the host wall of the opening is visible. (You will learn about creating the elevation and section views in Chapter 6 of this textbook). Next, choose the **Wall** tool from the **Opening** panel of the **Architecture** tab and then select the wall that will host the opening. Next, click in the desired area in the wall to mark the start point or the first corner point of the rectangular opening. As you move the cursor, a rectangle with its temporary dimensions appears. Click at the desired point in the wall to mark the other corner point of the rectangular opening. Now, you can use the temporary dimensions displayed on the opening to modify the placement of the opening. Next, choose the **Modify** button from the **Select** panel to exit the tool. After exiting the tool, you can select the created opening to modify its properties such as its height, top offset, base offset, and so on from the **Properties** palette.

> **Tip**
> *Press SM (shortcut keys) to snap the inserts to the midpoint of the wall. This is known as jump snap.*

TUTORIALS

Tutorial 1 Apartment 1

In this tutorial, you will add doors and windows to the apartment project file created in Tutorial 3 of Chapter 3. Refer to Figure 4-16 for adding these elements. The dimensions and the text have been given for reference and are not to be added. The project file name and parameters to be used are given next. **(Expected time: 30 min)**

1. Project file-
   For Imperial *c03_Apartment1_tut3.rvt*
   For Metric *M_c03_Apartment1_tut3.rvt*

2. Door types to be used
   For Imperial 1- **Single - Flush: 30"x 84"**
   2 and 3- **Door - Interior - Single - 4 - Panel - Wood - 36"x 84"**
   4- **Door-Exterior-Single-Two_lite - 48" x 96"**
   Door openings- **Passage Opening-Cased**
   For Metric 1- **M_Single - Flush: 0762 x 2134 mm**
   2 and 3- **M_Door -Interior-Single 4-Panel-Wood-0762 x 2134 mm**
   4- **M_Door-Exterior-Single-Two_lite - 1200 x 2400 mm**
   Door openings- **Passage Opening-Cased** resize to **2134 mm**

3. For Imperial Window types to be used
   1- **Fixed: 24" x 48"**
   2- **Fixed: 48" x 48"** (with modified width)
   For Metric Window types to be used
   1- **Fixed: 0610 x 1220 mm**
   2- **Fixed: 1220 x 1220 mm** (with modified width)

The following steps are required to complete this tutorial:

a. Open the file created in Chapter 3.
   For Imperial *c03_Apartment1_tut3.rvt*
   For Metric *M_c03_Apartment1_tut3.rvt*
b. Invoke the **Door** tool and add doors at approximate locations.
c. Load and add door openings.
d. Place the door openings to the exact location based on the given parameters.
e. Invoke the **Window** tool. Select the window type by using the **Type Selector** drop-down list.
f. Place the windows at approximate locations.
g. Place the windows at the exact location as per the given dimensions.
h. Save the project using the **Save As** tool.
   For Imperial *c04_Apartment1_tut1.rvt*

For Metric **M**_*c04_Apartment1_tut1.rvt*

i. Close the project.

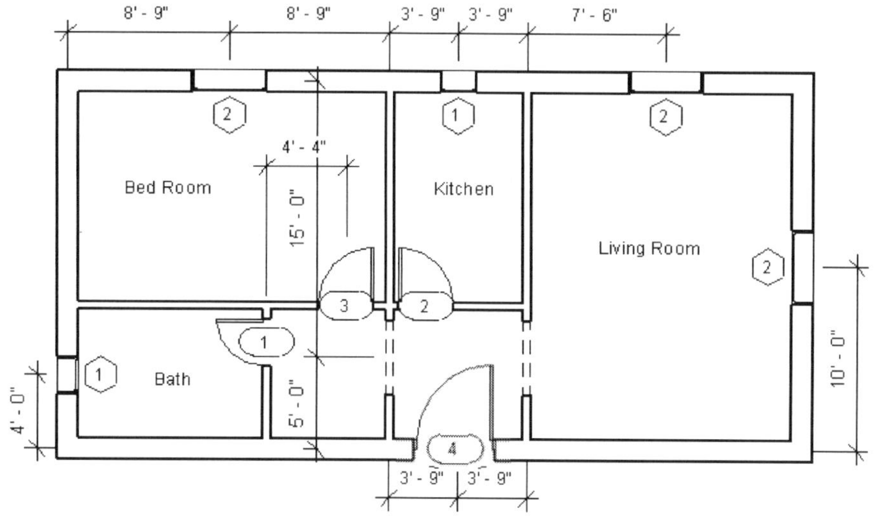

Figure 4-16 *Sketch for adding doors and windows to the Apartment 1 project*

Opening the Existing Project and Invoking the Door Tool

First, you need to open the specified project and invoke the **Door** tool from the **Architecture** tab.

1. Choose **Open > Project** from the **File** menu and open the *c03_Apartment1_tut3.rvt* for Imperial or *M_c03_Apartment1_tut3.rvt* for Metric project file created in Tutorial 3 of Chapter 3.

   You can also download this file from *https://www.Cadsoft.com*. The path of the file is as follows: *Textbooks > Civil/GIS > Revit Architecture > Exploring Autodesk Revit 2025 for Architecture*

2. Double-click on **Level 1** under the **Floor Plans** head in the **Project Browser**; the floor plan view is displayed.

3. Invoke the **Door** tool from the **Build** panel of the **Architecture** tab; the **Modify | Place Door** contextual tab is displayed.

4. On invoking the **Door** tool, the properties of the door to be added are displayed in the **Properties** palette. In the palette, select the type of door from the **Type Selector** drop-down list.

   For Imperial **Single-Flush: 30" x 84"**
   For Metric **M_Single - Flush: 0762 x 2134 mm**

5. Ensure that the **Tag on Placement** tool is chosen from the **Tag** panel of the **Modify | Place Door** contextual tab.

Adding Doors

In this section, you need to add doors at the approximate location and then modify its location by specifying the exact dimension. You will also use the **Load Family** tool to load the desired door type in the project.

1. Move the cursor close to the interior wall of the **Bath** area to display the door symbol, as shown in Figure 4-17. Notice that as you move the cursor, the side of the door is changed. Click on the interior wall; the door is created at the specified location.

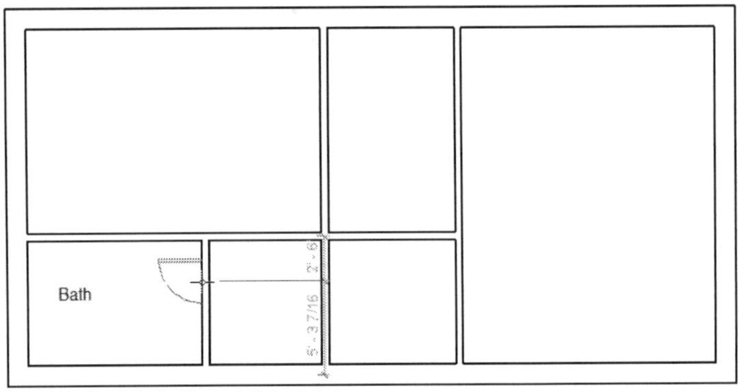

Figure 4-17 *Specifying the location of the bath door*

Note

*You need to select the **Centerline** radio button in the **Temporary Dimension** dialog box for measuring center to center temporary dimensions.*

2. To move the door to the exact location, choose the **Modify** button in the **Select** panel of the **Modify | Doors** tab. Next, select the door added to the drawing; the selected door gets highlighted in blue with the controls and the related temporary dimensions displayed in it.

3. Since the location of the door is given with reference to the upper interior wall, click on the upper temporary dimension, and then enter **2'0"(610 mm)**, as shown in Figure 4-18. Next, press ENTER; the door moves to the specified location.

4. To place the door type 2, invoke the **Door** tool from the **Build** panel of the **Architecture** tab; the **Modify | Place Door** contextual tab is displayed.

5. Choose the **Load Family** tool in the **Mode** panel of this tab; the **Load Family** dialog box is displayed.

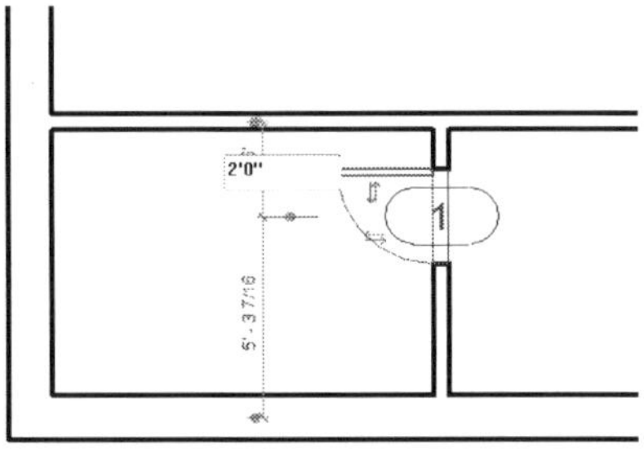

Figure 4-18 *Moving the door to the exact location*

6. In this dialog box, choose the **Doors > Residential** subfolder from the **English-Imperial** folder. Next, select the required door family.
   For Imperial **Door-Interior-Single-4-Panel-Wood**
   For Metric **M_Door-Interior-Single-4-Panel-Wood**

   Next, choose the **Open** button; the **Specify Types** dialog box is displayed. In this dialog box, select the door type **36" x 84"(900 x 2100 mm)** from the **Types** area and choose the **OK** button.

7. Next, ensure that the following door type is selected in the **Type Selector** drop-down list.
   For Imperial **Door-Interior-Single-4-Panel-Wood 36" x 84"**
   For Metric **M_Door-Interior-Single-4-Panel-Wood-900 x 2100 mm**

8. Move the cursor near the interior wall common to the Lobby and the Kitchen areas, as shown in Figure 4-19, and click to add the kitchen door close to this location.

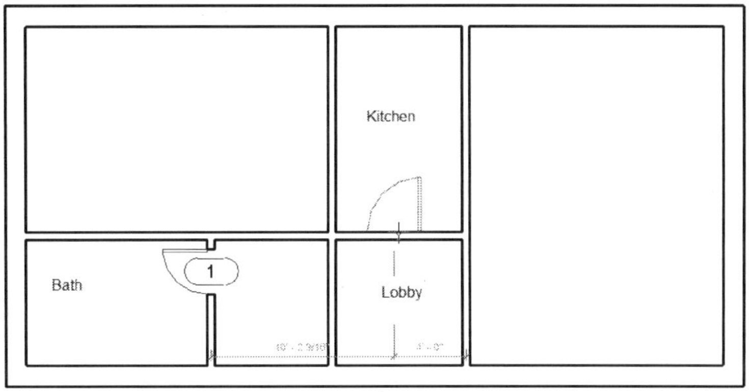

Figure 4-19 *Adding the door to the kitchen area*

9. Choose the **Modify** button from the **Select** panel. Next, select the door inserted in the drawing; the door gets highlighted and its controls are displayed.

   Notice that the door placed has the swing on the right side whereas the sketch shows a left side opening door, so you need to flip the swing side.

10. Click on the horizontal arrow key to flip the swing side. The swing side of the door is changed, as shown in Figure 4-20.

11. Click on the right side dimension to set it to **5'6" (1677 mm)**, as shown in Figure 4-20.

12. Add the door of the same type in the bedroom area at a distance of **4'4" (1320 mm)** from the internal wall, as shown in Figure 4-21.

13. To create the entrance door of the apartment, choose the **Load Family** tool from the **Mode** panel of the **Modify | Place Door** tab; the **Load Family** dialog box is displayed. In this dialog box, add the **Door-Exterior-Single-Two Lite** for Imperial (**M_Door-Exterior-Single-Two Lite** for Metric) family type from the **Doors > Commercial** subfolder in the **English-Imperial** folder.

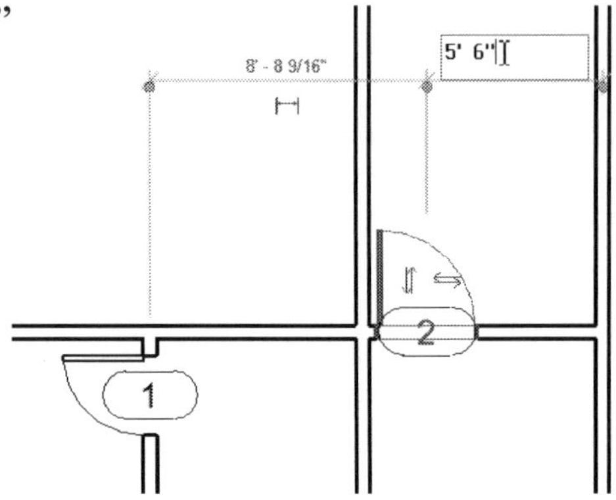

Figure 4-20 *Using the flip control to change the swing side of the door*

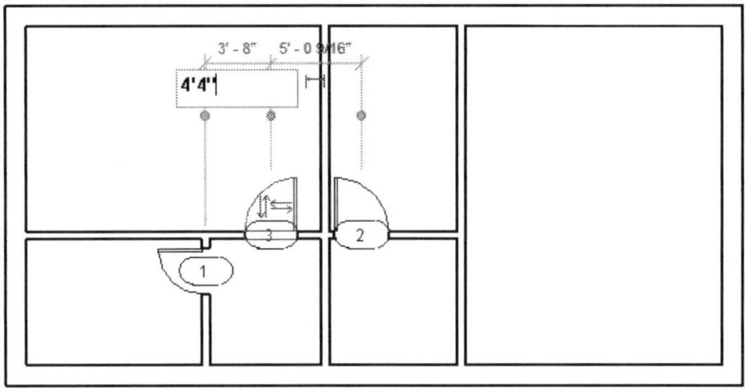

Figure 4-21 *Specifying the location of the door in the bedroom area*

14. Ensure that the **Door-Exterior-Single-Two_lite - 48" x 96"** option for Imperial or **M_Door-Exterior-Single-Two_lite - 1200 x 2400 mm** option for Metric is selected in the **Type Selector** drop-down list, and then add the door to the lobby area, as per the sketch plan, refer to Figure 4-22. While adding the door, you can use the spacebar to flip the door swing side.

15. After adding the door, choose the **Modify** button and select the recently created door; the temporary dimension is displayed. Click on the left of the temporary dimension and then enter the value **3'9"(1143 mm)**, as shown in Figure 4-22, to move the door to the exact location as given in the sketch plan. Now, press ESC to exit the **Modify | Doors** tab.

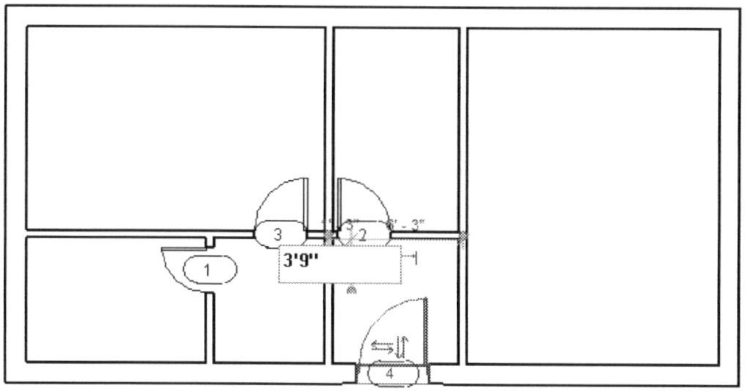

Figure 4-22 Specifying the location of the entrance door

Adding Door Openings

Next, you need to create two door openings in the lobby walls as given in the sketch plan. The openings can be loaded as door families but are added to the model as components.

1. Choose the **Insert** tab and then choose the **Load Family** tool from the **Load from Library** panel; the **Load Family** dialog box is displayed.

2. In the **Load Family** dialog box, select the **Passage Opening-Cased** family from the **Openings** sub-folder of the **English-Imperial** folder or select the **M_Passage Opening-Cased** family from the **Openings** sub-folder of the **English** folder. Next, choose the **Open** button to load this family; the **Passage Opening - Cased** family is added as a component type.

3. Invoke the **Place a Component** tool from **Architecture > Build > Component** drop-down; the **Modify | Place Component** tab is displayed.

4. Select the **Passage Opening - Cased 36" x 84"** option for Imperial or **M_Passage Opening-Cased 915 x 2134 mm** for Metric from the **Type Selector** drop-down list in the **Properties** palette, if it is not selected by default.

5. Move the cursor near the location shown in Figure 4-23 and then click to specify the location for the center of the opening. The opening is created.

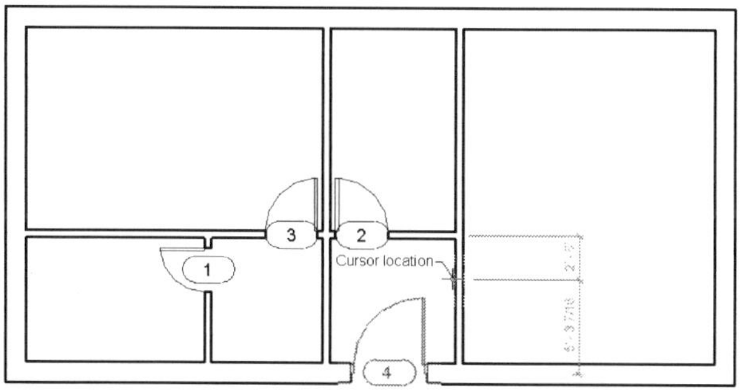

Figure 4-23 Adding openings

6. Now to modify the width of the opening, choose the **Modify** button from the **Select** panel. Select the recently created opening from the drawing; the **Modify | Generic Models** tab is displayed.

7. In this tab, choose the **Edit Type** tool from the **Properties** panel; the **Type Properties** dialog box is displayed.

8. Choose the **Duplicate** button; the **Name** dialog box is displayed. In this dialog box, enter **48" x 84"** for Imperial or **1220 x 2134 mm** for Metric in the **Name** edit box. Next, choose the **OK** button to close the **Name** dialog box.

9. Click in the **Value** field of the **Width** parameter and enter **4'0" (1220 mm)** as the new value.

10. Choose the **OK** button in the **Type Properties** dialog box to close it.

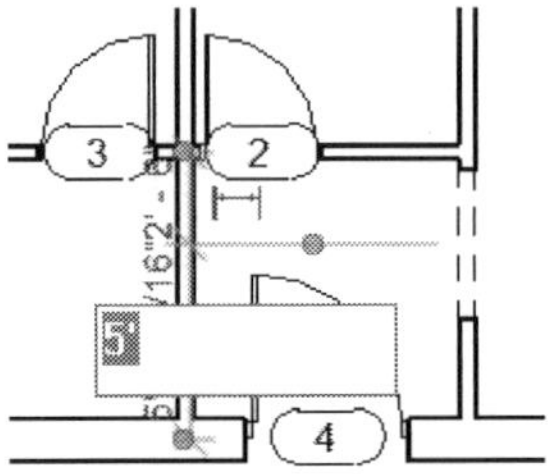

Figure 4-24 *Specifying the exact location of the opening*

11. To move the opening to the exact location, click on lower temporary dimension, and enter **5'0"(1524 mm)**, as shown in Figure 4-24. Next, press ENTER and ESC to accept the value entered and exit the editing mode.

12. Similarly, create the opening of the same size and at the same distance on the opposite wall to complete the creation of doors and opening for the *Apartment 1* project. After adding them, the layout plan of the project will look similar to the plan shown in Figure 4-25.

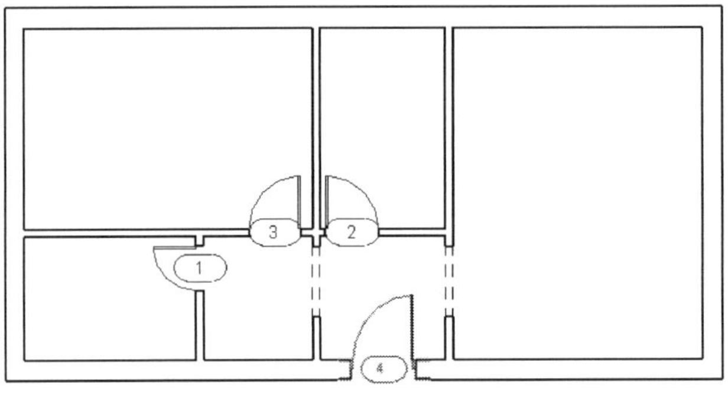

Figure 4-25 *Layout plan with all doors and openings added*

Tip

Doors and door openings can also be added to 3D view, sections, and elevations. You should, however, choose appropriate view to place the door. You can use the temporary dimensions to place the door at the exact location. In case you add a door incorrectly (for example, if the door is not entirely placed on the wall), Autodesk Revit displays a message, alerting you about the conflict and prompting you to take an appropriate action.

Adding Windows

The procedure of adding windows is quite similar to that of adding doors. Invoke the **Window** tool, select the window type, and add the window at an approximate location. You can then modify its location based on the sketch provided next in this tutorial.

1. Invoke the **Window** tool from the **Build** panel of the **Architecture** tab; the **Modify | Place Window** tab is displayed.

2. Ensure that the **Tag on Placement** tool is chosen from the **Tag** panel of the **Modify| Place Window** tab.

3. Click on the **Type Selector** drop-down list to view the in-built window types. To create the window number 1, select **Fixed 24" x 48"** for Imperial or **M_Fixed: 0610 x 1220 mm** for Metric from the drop-down list.

4. Move the cursor close to the exterior wall of the kitchen to display the window symbol, as shown in Figure 4-26. Add the window by clicking on the inner face of the exterior wall. The window is created at the specified location.

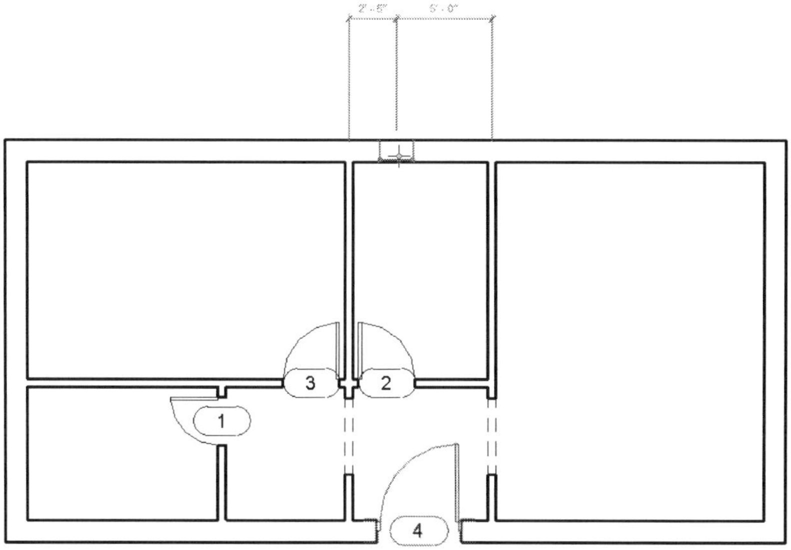

Figure 4-26 *Adding the kitchen window to the exterior wall*

5. To move the window to the exact location, choose the **Modify** button from the **Select** panel and then select the window from the drawing; it gets highlighted in blue and its controls are displayed.

6. Click on the left temporary dimension and enter **3'9" (1143 mm)**, as shown in Figure 4-27. Press ENTER; the window is moved to the specified location. Similarly, add another window of the same type near the external wall of the bath by invoking the **Window** tool from the **Build** panel.

7. After adding the window in the external wall of the bath, you need to move it to the exact location. To do so, select the window; it is highlighted in blue. Click on the lower temporary dimension; an edit box is displayed. Enter **4'0" (1220 mm)** in the edit box, refer to Figure 4-28, and then press ENTER. The window is moved to the desired location. Press the ESC key and exit.

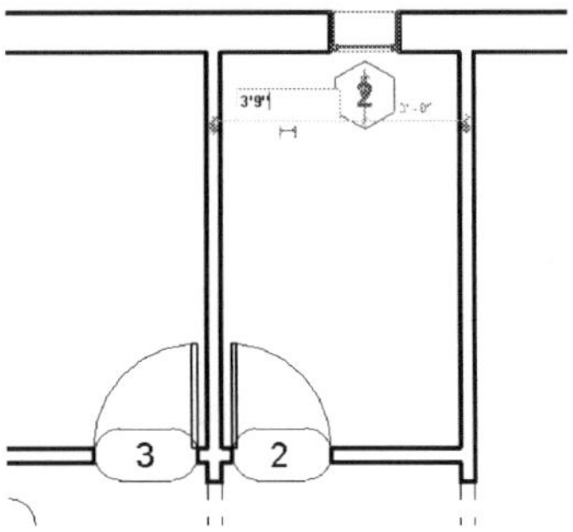

Figure 4-27 *Specifying the exact location of the window*

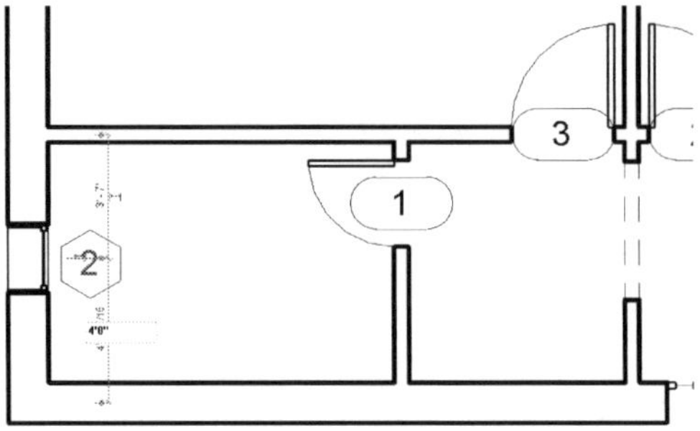

Figure 4-28 *Specifying the location of the bath window*

Now, you need to add the windows of type 2 to the drawing. These windows have a modified width of **4'0"**(**1220 mm**). To add these windows, select the **Fixed: 24" x 48"** for Imperial or **M_Fixed: 0610 x 1220 mm** for Metric window type and create a window type duplicate. Modify window width to 4'0"(1220 mm) by using the **Type Properties** dialog box and add the window at the desired location.

8. Invoke the **Window** tool from the **Build** panel of the **Architecture** tab; the **Modify | Place Window** tab is displayed.

9. In the **Type Selector** drop-down list, select the **Fixed: 24" x 48"** (**M_Fixed: 610 x 1220 mm** for Metric) window type. Next, choose the **Edit Type** button in the **Properties** palette; the **Type Properties** dialog box is displayed.

10. In this dialog box, choose the **Duplicate** button; the **Name** dialog box is displayed. In the **Name** edit box, enter **48" x 48" (1220 x 1220 mm)** and then choose the **OK** button.

11. In the **Value** field for the **Width** type parameter, enter **4'0" (1220 mm)** and then choose the **Apply** and **OK** to close the **Type Properties** dialog box and return to the drawing window.

    Notice that the **Fixed: 48" x 48" (M_Fixed: 1220 x 1220 mm** for Metric) window type is added to the **Type Selector** drop-down list.

12. Move the cursor near the exterior wall of the bedroom and click to place the window. Press ESC twice to exit.

13. Now, select the added bedroom window; the window gets highlighted in blue color and its controls are displayed. Click on the temporary dimension displayed at the left and enter **8'9" (2667 mm)** to specify the location of the bedroom window, as shown in Figure 4-29.

The window moves at the desired location. Similarly, add the windows of the same type to the walls of the living room and specify their respective locations based on the sketch plan given, refer to Figures 4-30 and 4-31.

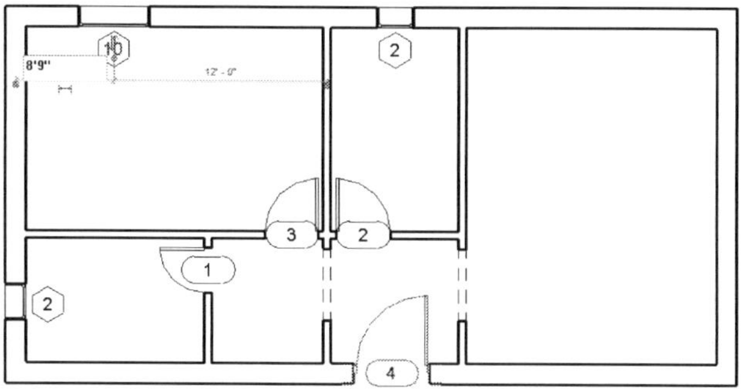

Figure 4-29 *Specifying the location of the bedroom window*

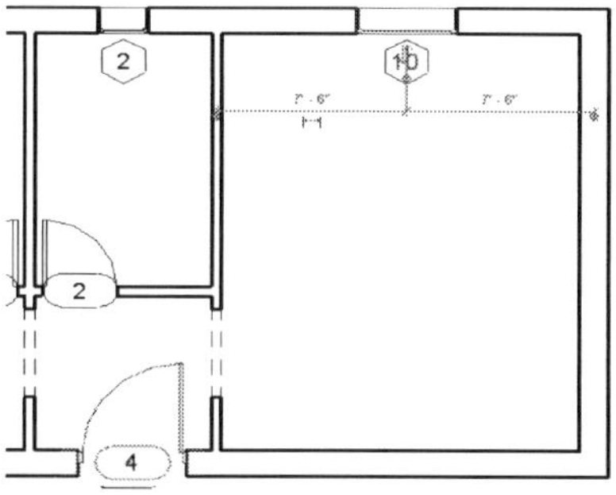

Figure 4-30 *Specifying the location of the first living room window*

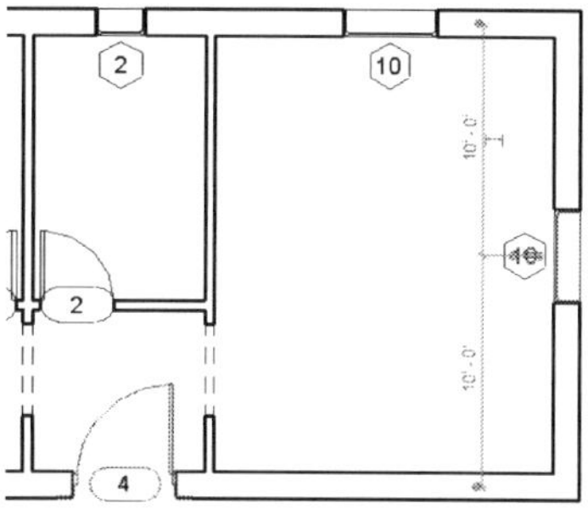

Figure 4-31 *Adding the second living room window using temporary dimensions*

14. To change the window tag numbers, move the cursor over the number in any of the window tags marked 13 (tag number may vary) and double-click; an edit box is displayed. Enter **1** in the edit box and press ENTER; the **Revit** message box is displayed. Choose **Yes** in the message box; the window tag is renamed. Similarly, replace the other window tags marked 13 with the tag mark 1 and those marked 10 with the tag mark 2, refer to Figure 4-32. You will notice that the tag marks of other instances of the same window type are immediately modified.

The Level 1 plan should look similar to the plan shown in Figure 4-32.

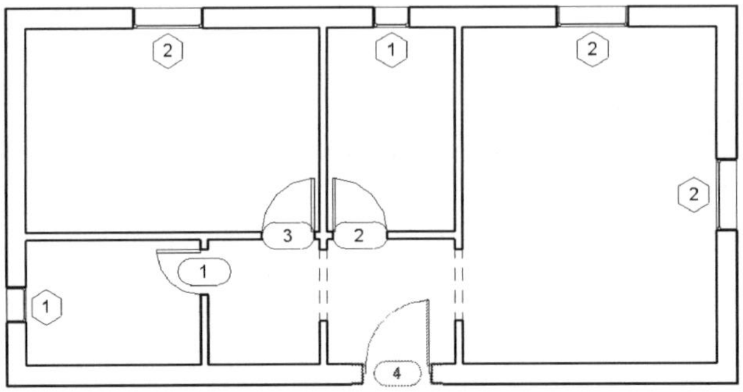

Figure 4-32 *Completed project plan with renamed window tags*

15. To view the plan in 3D, choose the **Default 3D View** tool from the **View > Create > 3D View** drop-down; the 3D view of the project is displayed, as shown in Figure 4-33.

Saving the Project

In this section, you will save the project file using the **Save As** tool.

1. Choose **Save As > Project** from **File** menu; the **Save As** dialog box is displayed. Create a folder with the name **c04_revit_2025_tut** and enter the desired name in the **File name** edit box.

   For Imperial **c04_Apartment1_tut1**
   For Metric **M_c04_Apartment1_tut1**

Choose **Save**; the file is saved.

2. Choose **Close** from the **File** menu to close the project file.

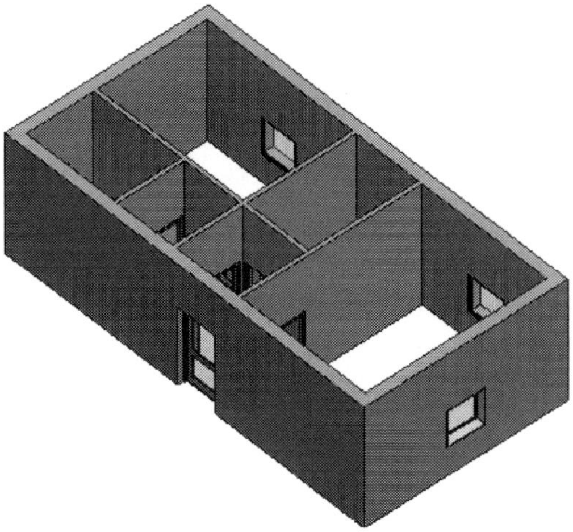

Figure 4-33 *3D view of the Apartment 1 project*

EXERCISE

Exercise 1 Apartment 2

Add doors, windows, and openings to the *Apartment 2* project created in Exercise 1 of Chapter 3, refer to Figure 4-34. Do not mention dimension and text as they are given for reference only. Use the following project parameters for completing this exercise.

(Expected time: 30 min)

1. Project file-
   For Imperial *c03_Apartment2_ex1.rvt*
   For Metric *M_c03_Apartment2_ex1.rvt*

2. Door types to be used:
   For Imperial Bath and kitchen doors- **Door- Single - Panel: 36" x 84"**
   Bedroom and Lobby door- **Door-Interior-Double-Full Glass Wood - 36" x 84"**
   Door openings in dining (dashed lines)- **4'0" x 7'0"** using **Passage-Opening Cased: 36"x 84"**
   Distance of the door centerlines from wall centerlines- **2'0"**
   For Metric Bath and kitchen doors- **M_Door- Single- Panel: 0915 x 2134 mm**
   Bedroom door- **M_Door-Interior-Double-Full Glass Wood- 1500 x 2000 mm**
   Lobby door- **M_Double-Interior-Double-Full Glass Wood - 900 x 2100 mm**

Door openings in dining (dashed lines)- **1220 x 2134 mm**
using **M_Passage-Opening Cased : 0915 x 2134 mm**
Distance of the door centerlines from wall centerlines- **610 mm**

3. Window types to be used
   For Imperial — Bedroom and living room windows - **Fixed: 36" x 48"**
   Bath and kitchen windows- **Fixed: 24" X 48"**
   For Metric — Bedroom and living room windows - **M_Fixed: 0915 x 1220 mm**
   Bath and kitchen windows- **M_Fixed: 0610 X 1220 mm**

4. File name to be saved-
   For Imperial **c04_Apartment2_ex1**
   For Metric **M_c04_Apartment2_ex1**

For the distance of window centerlines from wall centerlines refer to Figure 4-34.

Chapter 5

Using the Editing Tools

Learning Objectives

After completing this chapter, you will be able to:

- *Create a selection set of elements*
- *Use the tools in the Status Bar*
- *Move and copy elements*
- *Use the Trim and Extend tools*
- *Use the Cut and Paste tools*
- *Use the Rotate, Mirror, and Offset tools*
- *Create an array of elements using the Array tool*
- *Use the Match, Align, Delete, Lock, and Group tools*
- *Split walls using the Split Elements and Split with Gap tools*

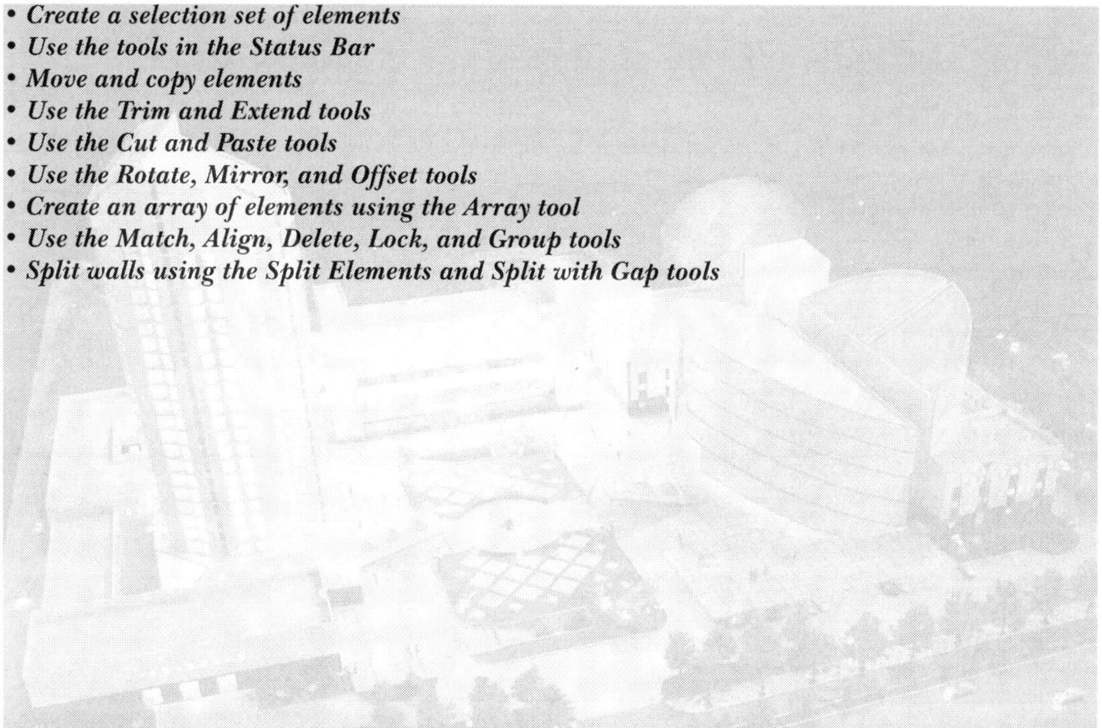

INTRODUCTION

In the previous chapter, you learned to create building elements such as walls, doors, and windows. In this chapter, you will learn to use various editing tools provided in Autodesk Revit. You can use these tools to add specific design requirements to the elements that you have created. Primarily, you will learn to select and edit elements using object controls and editing tools such as **Move**, **Copy**, **Trim**, **Extend**, **Cut**, **Delete**, **Rotate**, **Mirror**, **Array**, **Align**, **Match**, and so on. You will also learn to use different tools located in the **Status Bar**.

CREATING A SELECTION SET

In order to edit an element or a group of elements, you must first select them. To create a selection set, first you need to exit the currently invoked tool. To do so, choose the **Modify** button from the **Select** panel of the contextual tab or type **MD** to exit the current tool. After exiting the current tool, move the cursor near the element that you wish to select. On doing so, you will notice that it gets highlighted and its description is displayed in the **Status Bar**. In case, there are elements in proximity of the element to be selected, press the TAB key to cycle between them. Click to select the element when it is highlighted. By default, a selected element will appear semi-transparent and blue in color.

MOVING AND COPYING ELEMENTS

Moving and copying elements are frequent tasks that you need to perform while working in a project. In Revit platform, there are different options or methods by which you can perform these tasks. You can perform both copying and moving tasks at the same time. The methods used for moving and copying elements are described next.

Moving the Elements by Changing the Temporary Dimensions

When you create an element, its temporary dimension from a nearby object is also displayed. As explained in Chapters 3 and 4, you can move the element by changing this dimension. Click the temporary dimension and enter the new value to move the element to the desired location. For example, a door can be moved by specifying its new value from the adjacent wall or door (see Chapter 4).

Moving the Elements by Dragging

Dragging is a method used to move elements to the desired location. To drag an element, choose the **Modify** button from the **Select** panel and then select the element from the drawing. After making selection, move the cursor over the element, click and hold the left mouse button. Next, move the cursor toward the desired location. You will notice that the selected element moves along with the cursor, as shown in Figure 5-1. When the element reaches the desired location, release the left mouse button; the element is moved to the new location, as shown in Figure 5-2.

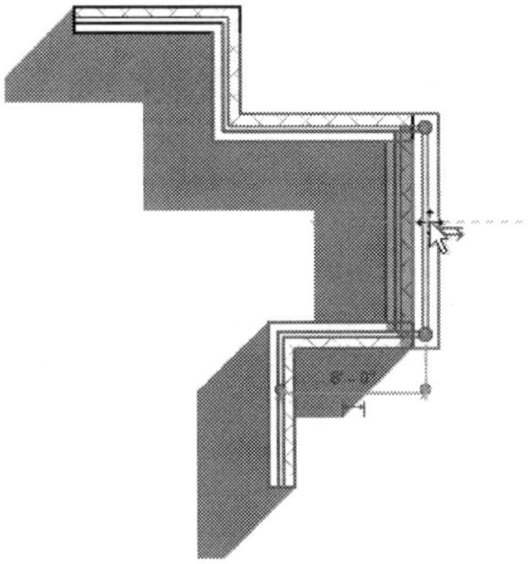

Figure 5-1 *Moving the walls by dragging*

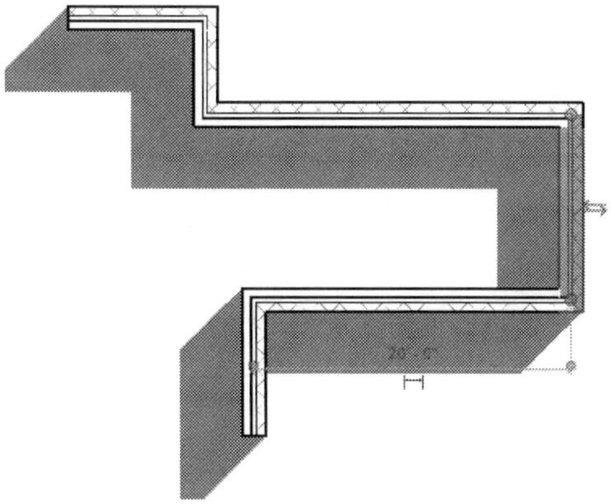

Figure 5-2 *Moved wall with completed edges*

Note

In case of walls, the joined walls adjust and get completed automatically, refer to Figure 5-2.

Moving the Elements by Dragging the End-Joint Components

Using the drag control symbol, you can drag and move more than two components simultaneously provided they have a common joint. This feature is applicable for the line based components such as walls, beams, or braces. To drag the common joined elements, select one of the members of the common joint from the drawing and right-click; a shortcut menu will be displayed. Choose the **Select Joined Elements** option from the shortcut menu. All the members joined at the common end will be highlighted displaying the circular shaped drag controls, as shown in Figure 5-3. Bring the cursor at the common joint; the **Drag Wall End** tooltip will be displayed. Drag the common joint to a new location, as shown in Figure 5-4. You will notice that all the components joined to that common joint have been dragged.

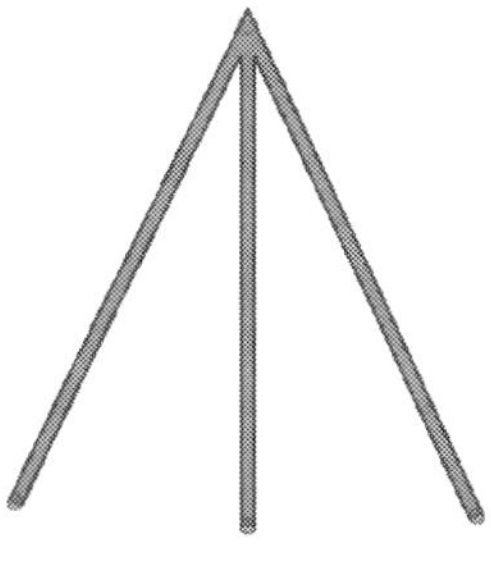

Figure 5-3 *Three walls joined at a common end*

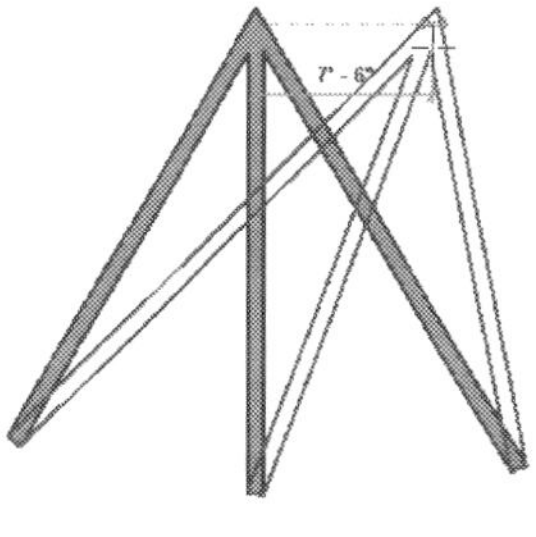

Figure 5-4 *Dragging the end-joined walls simultaneously*

Moving the Elements by Selecting and Dragging

You can modify the location of an element by pressing and dragging the cursor over it. To do so, choose the **Drag Elements On Selection** tool from the **Status Bar** and place the cursor on the desired element; the element gets highlighted. Now, press and drag the element and the element will get relocated.

Note

It is recommended to keep the ***Drag Elements On Selection*** *tool disabled to avoid accidental displacement of the elements.*

Using the Move Tool

Ribbon: Modify | (Elements / Components) > Modify > Move
Shortcut Key: MV

The function of the **Move** tool is similar to that of dragging with the only difference that the preview of the elements is not displayed. It is more useful for moving elements by a specific distance, thereby increasing the accuracy of the building model. To move elements, select them; a contextual tab will be displayed. Note that the name of the contextual tab will depend on the element type selected from the drawing. Invoke the **Move** tool from the **Modify** panel of the contextual tab, or type **MV**; the selected elements will be highlighted inside a dashed rectangle and the **Constrain**, **Disjoin**, and **Multiple** check boxes will be displayed in the **Options Bar**.

Note

The ***Move*** *tool, like other editing tools such as* ***Copy****,* ***Rotate****,* ***Mirror****,* ***Array****, and* ***Scale****, is only available after a selection is made. You can access these tools from the contextual tabs such as* ***Modify / Walls****,* ***Modify / Windows****, and others. The name of the contextual tab will depend upon the type of element or component you select.*

Using the Copy Tool

Ribbon: Modify | (Elements / Components) > Modify > Copy
Shortcut Key: CO

The usage of the **Copy** tool is similar to that of the **Move** tool. After invoking the **Copy** tool, click in the drawing window to specify the start point and then move the cursor to the desired distance. To specify the endpoint, click at the appropriate location. You can also specify the distance between the original and copied elements by entering the value.

Tip

While copying elements, you can click on either side of the original element to create copies or you can enter negative or positive values in the edit box.

TRIMMING AND EXTENDING ELEMENTS

You can trim or extend one or more elements to a boundary defined by the element of the same type. You can also extend or trim (if the elements intersect) the elements that are not-parallel to create a corner. You can trim or extend a reference line or a wall to another reference line or wall. In Autodesk Revit, you can trim and extend elements by using three tools namely: **Trim/ Extend to Corner**, **Trim/ Extend Single Element**, and **Trim/Extend Multiple Elements**. You can use these tools with walls, lines, beams, or braces. These tools are discussed next.

Using the Trim/Extend to Corner Tool

Ribbon: Modify > Modify > Trim/Extend to Corner

This tool is used to extend a wall to another wall or to trim any extended portion of the building element at the apparent intersection to form a corner.

> **Tip**
> *The **Trim/Extend to Corner** tool can be used with walls as well as with lines (for example, while creating 2D details for your project).*

Using the Trim/Extend Single Element Tool

Ribbon: Modify > Modify > Trim/Extend Single Element

This tool enables you to extend or trim an element by defining a reference line or boundary. Using this tool, you can trim a wall or extend a wall to another wall without creating a corner at the intersection. You can also select the desired side of the wall to be retained.

Using the Trim/Extend Multiple Elements Tool

Ribbon: Modify > Modify > Trim/Extend Multiple Elements

The **Trim/Extend Multiple Elements** tool enables you to extend or trim a number of elements using a common boundary or edge which can be defined by clicking on an element. After specifying the boundary, select the elements to be extended or trimmed. Autodesk Revit intuitively extends or trims them at the defined boundary.

> **Tip**
> *While selecting walls or lines for trimming, ensure that the portion of the element to be retained is clicked. In trimming, the portion of the element on the other side of the intersection is removed.*

CUTTING AND PASTING ELEMENTS

Using various cut and paste tools, you can cut some elements and then paste them at another location. You can use these tools efficiently to speed up your work and reduce the time taken in creating new elements. Various cutting and pasting tools are discussed next.

Cutting Elements

Ribbon: Modify | (Elements / Components) > Clipboard > Cut to Clipboard
Shortcut Keys: CTRL+X

The **Cut to Clipboard** tool is used to remove one or more elements from the project. The selected elements that are cut using this tool are copied on the clipboard where they stay until another selection is cut. When you cut an element and paste it to the clipboard, the existing selection is overwritten by the new one. These elements can then be pasted into the same or another project. The **Cut to Clipboard** tool becomes available only after a selection is made in the drawing. To create a selection set, you can use any selection tool described earlier in this chapter. Select the element(s) that you want to cut from the drawing area and choose the **Cut to Clipboard** tool from the **Clipboard** panel of the contextual tab. Alternatively, press the CTRL+X keys. On doing so, the selected elements will be removed from the project and pasted on the clipboard.

Copying Elements to the Clipboard

Ribbon:	Modify | (Elements / Components) > Clipboard > Copy to Clipboard
Shortcut Keys:	CTRL+C

This tool is used to create a copy of the selected elements on the clipboard and retain the original ones in the current project. Select the elements or components that you want to copy from the drawing area and invoke the **Copy to Clipboard** tool from the **Clipboard** panel of the contextual tab. Alternatively, press the CTRL+C keys. A copy of the selected elements will be created on the clipboard.

Pasting Elements from the Clipboard

Ribbon:	Modify | (Elements / Components) > Clipboard > Paste drop-down > Paste from Clipboard
Shortcut Key:	CTRL+V

The Paste from Clipboard tool, which is used to add one or more elements from the clipboard to the project, becomes available only after the element(s) have been cut or copied to it earlier. The selection can be pasted on the current view or on a different view. To paste the elements from the clipboard to the drawing, invoke the Paste from Clipboard tool from the Clipboard panel. Alternatively, press the CTRL+V keys to paste the selected elements on the drawing view. You can then paste the element(s) at the desired location. To paste a single element, such as a door or a window on the clipboard, move the cursor to the desired location on the wall and click; the copied door or window will be placed in the host wall. If there are a number of elements in the pasted selection, they will appear in the preview box. If they are pasted into the same project, a temporary dimension will be displayed. This indicates the distance and direction of the cursor from the original selection. You can enter the value of the dimension and the angle to paste the elements. Alternatively, to paste the elements graphically, move the box to the desired location and click. Once these elements are pasted, they remain selected as a group in a dashed rectangular box and the **Modify | Model Groups** tab is displayed. Using this tab, you can use other editing tools such as **Move**, **Copy**, **Rotate**, **Align**, and so on to modify them. In the **Options Bar**, choose the **Activate Dimensions** button to view the dimensions of the pasted elements from the nearby element(s). After completing the pasting procedure, choose the **Finish** button from the **Edit Pasted** panel to return to the **Architecture** tab (default mode). To abort the pasting operation after the entities have been pasted and while the group of elements is selected, choose the **Cancel** button in the **Modify | Model Groups** tab. On doing so, the recently pasted group of elements will be removed from the project and you will exit from the **Paste from Clipboard** tool.

ROTATING ELEMENTS

Ribbon: Modify (Elements / Components) > Modify > Rotate
Shortcut Keys: RO

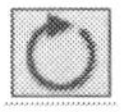

The **Rotate** tool is used to rotate elements about a specified axis or point. You can rotate an element about an axis perpendicular to its plan, elevation, and section views. In 3D view, elements rotate about an axis perpendicular to the current work plane. This tool is available only after the elements have been selected.

Select the element(s) or component(s) from the drawing that you need to rotate and invoke the **Rotate** tool from the **Modify** panel. Alternatively, you can type **RO** to invoke the **Rotate** tool. On invoking the **Rotate** tool, a rotation symbol is displayed at the center of the selected element(s). It represents the perpendicular axis or the center about which the elements will be rotated. You can snap to the symbol and drag it to the desired center of rotation. Once the center of rotation has been specified, a ray is generated from this point, which moves along the direction of the cursor. Click at the desired start point of the rotation to display a line indicating its direction. You can use various object snap options to specify the start point of the rotation. Once the starting point has been specified, another line rotates with the cursor. A preview box, indicating the selected elements, also rotates with it. Now, move the cursor to the required angle and click in the drawing area to specify the second point of rotation. On doing so, the element will rotate toward a new direction. On rotating the entities, you will notice that a temporary angular dimension is displayed. You can enter the exact value of the angle of rotation, if required. This rotates the element to the specified angle. When you invoke the **Rotate** tool, five options are displayed in the **Options Bar**, as shown in Figure 5-5.

Figure 5-5 *Rotate options available in the* ***Options Bar***

MIRRORING ELEMENTS

In Autodesk Revit, you can create a mirror image of the elements selected from the drawing about an axis defined by an imaginary line passing through the two specified points. You can use various tools to mirror the selected elements. Using these tools, you can create a symmetrical profile and also create a copy of the elements which are equidistant from the specified axis. You can mirror elements using any of the two tools, **Mirror - Pick Axis** or **Mirror- Draw Axis**. These tools are available only after the elements to be mirrored have been selected. To mirror the element(s) in the drawing, select them and then choose **Mirror - Pick Axis** tool or **Mirror - Draw Axis** tool from the **Modify** panel of the contextual tab. Alternatively, you can type **MM** or **DM** to mirror elements. Note that when you type **MM**, the **Mirror - Pick Axis** tool is invoked and on typing **DM**, the **Mirror - Draw Axis** tool is invoked. On invoking these tools, you will be prompted to select the axis of reflection from the drawing area. Also, you will notice that the **Options Bar** displays the **Copy** option. By default, this option is selected, and therefore you will be able to create a mirrored copy of the elements. Clear the **Copy** check box if you do not require a copy.

CREATING AN OFFSET

Ribbon: Modify > Modify > Offset
Shortcut Key: OF

The **Offset** tool is used to create a copy of the selected element at a specified distance from the original element. You can select a single element or multiple elements. The selection can be made even after invoking this tool. By default, Autodesk Revit creates a copy of the selected element while creating an offset. To offset an element, invoke the **Offset** tool from the **Modify** panel of the **Modify** tab. Alternatively, you can type **OF** to invoke the **Offset** tool. On invoking the **Offset** tool, you will notice that the **Options Bar** displays the **Copy** check box. By default, this check box is selected. As a result, you will be able to create a copy of the element at the offset distance. Clear the **Copy** check box, if you do not require a copy. In this case, the object selected will be shifted to the offset distance and no copy will be created. The **Options Bar** also shows two radio buttons, **Graphical** and **Numerical**, which can be used to specify the mirror axis.

CREATING AN ARRAY OF ELEMENTS

Ribbon: Modify | (Elements / Components) > Modify > Array

The **Array** tool creates copies of the selected element in a rectangular or radial pattern. It is available only after the elements are selected for an array. You can use any of the methods discussed earlier in this chapter to create a selection. Autodesk Revit provides you the option of arraying multiple elements that are associated as a group. Select the element(s) that you want to array from the drawing and then invoke the **Array** tool from the **Modify** panel. Alternatively, you can type **AR** to invoke the array tool. On invoking the **Array** tool, you will notice that the **Options Bar** displays certain array options. From the **Options Bar**, select the **Group and Associate** check box to create a group of the selected elements, which are then copied as a group rather than individual elements. Clear this check box, if you want to array elements individually. The two options available for creating an array are discussed next.

Linear

Linear, the default array option, can be used to create copies of the selected elements in a rectangular or linear pattern. In the **Options Bar**, additional options are available for creating an array. In the **Number** edit box, enter the number of copies you want to create. You can select the **2nd** radio button to specify the distance between items in the array. The **Last** option is used to specify the entire distance of the array. The items specified are then placed equidistantly between the first and the last specified points. You can use the **Activate Dimension** button to view various dimensions of the elements.

Tip

Autodesk Revit displays error messages if the elements cannot be arrayed. For example, if the specified angle or the number of items do not fit into the length of the wall, or overlap each other, a suitable warning message is displayed to help you modify the settings accordingly.

Radial

You can create radial arrays using the **Radial** option available in the **Options Bar**. The other setting options available in the **Options Bar** such as **Group and Associate**, **Number**, **Move to: 2nd** and **Last** are the same as those given for the **Linear** option. The **Angle** edit box is used to specify the angle about which the array is to be created. If the **Move To: 2nd** option is selected, the angle specified becomes the angle between each item, whereas if the **Move To: Last** option is selected, the number of items are created by equally dividing the specified angle. The procedure for creating a radial array is similar to that of using the **Rotate** tool. After selecting the element(s) of array and choosing the **Radial** option, you will notice that a rotational symbol is displayed at the center of the selection, which you can drag to the location to be used as the center of the array.

MATCHING ELEMENTS

Ribbon: Modify > Clipboard > Match Type Properties
Shortcut Keys: MA

In Autodesk Revit, you can use the **Match Type Properties** tool to copy the properties of a source element and transfer them to another element belonging to the same family. For example, you can select a wall and then another wall and match its properties with the first wall.

To match the property of a source element to another element of the same family, invoke the **Match Type Properties** tool from the **Clipboard** panel. After invoking this tool, you will notice that as you move the cursor in the drawing window, it changes to a paint brush. To select a type of element to match, move the paint brush near the element and click. The empty paint brush changes into a filled one indicating that the element type has been selected. To match another element of the same family type as the first selection, move the dropper over it and click. The dropper empties, indicating that the element type has been changed. In case of walls, the matching wall type does not change the instance parameters. Autodesk Revit has the improved feature of matching multiple family elements to a selected element of the same family.

ALIGNING ELEMENTS AND WORKING WITH CONSTRAINTS

Ribbon: Modify > Modify > Align
Shortcut Keys: AL

The **Align** tool is used to arrange elements in a collinear manner. You can choose an element that acts as a reference line or a plane along which the other elements can be aligned. When you align them, a padlock key will be displayed which represents the alignment constraint. Using this key, you can lock or unlock the alignment of the walls. Once you lock the alignment, the walls move together. The **Align** tool can be used only in plan or elevation views. In Autodesk Revit, you can use the **Align** tool to align multiple elements in your project at once. Invoke the **Align** tool from the **Modify** panel. In the **Options Bar**, the **Multiple Alignment** check box can be selected to align a number of elements to a common line. While aligning walls, you can choose to align the wall centerlines or wall faces by selecting the appropriate option from the **Prefer** drop-down list in the **Options Bar**. To select the reference line, click when the appropriate line is displayed.

DELETING ELEMENTS

Using the **Delete** tool, you can remove the selected elements from the project. The deleted elements are permanently removed and are not pasted on the clipboard. The tool is available only after the elements have been selected. To invoke the **Delete** tool, select the element you want to delete, the contextual tab will be displayed. From the **Modify** panel of the contextual tab, choose the **Delete** tool; the selected elements will be deleted from the drawing. Alternatively, you can press the DELETE key or type **DE** to delete the selected elements.

SPLITTING ELEMENTS

Ribbon:	Modify > Modify > Split Element
Shortcut Keys:	SL

The **Split Element** tool is used to break a wall or a line into two parts at a specified point. Once a wall or a line is split, the parts can be edited independently. Invoke the **Split Element** tool from the **Modify** panel and select the element that you want to split from the drawing. The **Options Bar** has the **Delete Inner Segment** check box, which you can select to delete the inner segment created between the two consecutive splits. On moving the cursor in the drawing window, you will notice that it changes to a cutting tool. Now, move the cursor over the wall or a line until the temporary dimension is displayed. You can move the cursor to the desired distance and then click to specify the point of the split. A temporary dimension, indicating the location of the split, will be displayed. Click on it and enter the value of the distance. As a result, the split moves to the specified location.

SPLITTING WITH GAP

Ribbon:	Modify > Modify > Split with Gap

The **Split with Gap** tool is used to split a wall into two separate walls with a defined gap defined between them. To split a wall, invoke the **Split with Gap** tool from the **Modify** panel; the cursor will change to a cutting tool. In the **Options Bar**, enter a value in the **Joint Gap** edit box to set the spacing between two consecutive splits. Note that you can enter a value between 0 1/16" and 1' in the **Joint Gap** edit box. Next, move and place the cursor over the wall that you want to split. Click at the desired point in the wall; the wall will split into two separate walls with a gap value specified in the **Joint Gap** edit box.

SPLITTING FACES

Ribbon:	Modify > Geometry > Split Face

The **Split Face** tool is used to divide a face of a wall, architectural column, floor, roof, or any other element into different regions for applying different materials. To split the face of an element, choose the **Split Face** tool from the **Geometry** panel; the cursor with a selected symbol will be displayed. Also, prompt you to select a face. Move the cursor toward the face of wall or column and click when the face is highlighted. On doing so, the **Modify | Split Face > Create Boundary** tab will be displayed. From the **Draw** panel of this tab, choose a desired sketching tool and sketch the boundary line that will split the face. After sketching the boundary, choose the **Finish Edit Mode** button from the **Mode** panel; the face of the wall or column will be divided into two different regions and you can apply materials to them.

CREATING PARTS

Ribbon: Modify > Create > Create Parts

Revit allows you to create a part of a layered element in both 2D and in 3D view. These parts can then be selected, scheduled, tagged, and edited independently. Parts can be generated from elements with layered structures, such as Walls, Floors, Roofs, Ceilings and Structural slab foundations.

Note

1. Parts created are dependent on parent element. Parts are automatically updated and regenerated when the original element from which they are originated is modified. Modifications may include adding or removing layers, changing wall type, layer thickness, wall orientation, geometry, materials, or openings.

2. Deleting a part will also delete all other parts created from the original element.

Resizing Created Parts

The created parts can be resized according to your project needs. To do so, select the part you want to resize and select the **Show Shape Handles** option from the **Properties** palette of the **Modify|Parts** contextual tab. As you check the **Show Shape Handles** box, you will notice the selected part is now provided with drag points which can now be used to adjust the dimensions of each created Part.

Changing the Material of Created Parts

To change the material of a part, select the desired part and make sure the **Material By Original** check box is cleared in the **Properties** palette of the **Modify|Parts** contextual tab. As you clear the **Material By Original** check box; an editable **Material** option is appears in the **Properties** palette. The **Material** option can be used to change the material of the part entirely different from the parent material.

Dividing Parts

You can also divide the created parts to create new sub-parts and represent them as skirting, glass pannel, plywood and so on in the parent element. To create a sub-part, select the created part and choose the **Divide Parts** option from the **Part** panel of the **Modify|Part** contextual tab; the **Modify|Division** contextual tab will be displayed, as shown in Figure 5-6. Now select the **Edit Sketch** option from the **Draw** panel; the **Modify|Division > Edit Sketch** contextual tab will be displayed, as shown in Figure 5-7. To create the sub-part, you need to set the work plane at the face of the required part. Now using the tools provided in the **Draw** panel, sketch the profile of the required sub part; temporary dimensions will be activated which can be used to edit the dimensions of the profile exactly as required, as shown in Figure 5-8. Click on **Finish Edit Mode** from the **Mode** panel of the **Modify|Division > Edit Sketch** contextual tab. Now click on **Finish Edit Mode** from the **Mode** panel of the **Modify|Division** contextual tab to finish the profile. The material and shape of the created profile can be further edited using the tools discussed above.

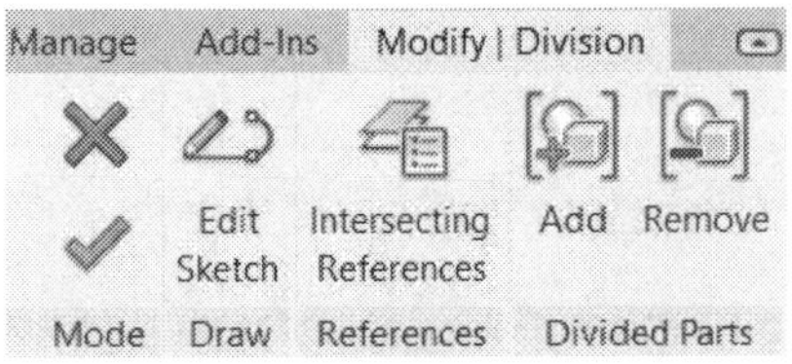

***Figure 5-6** The **Modify|Division** contextual tab*

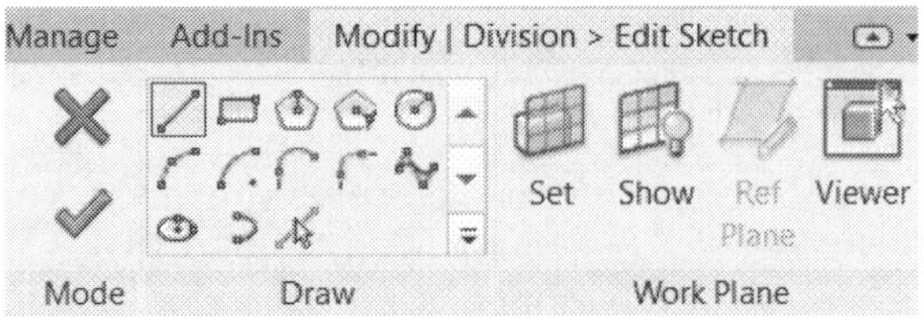

***Figure 5-7** The **Modify|Division > Edit Sketch** contextual tab*

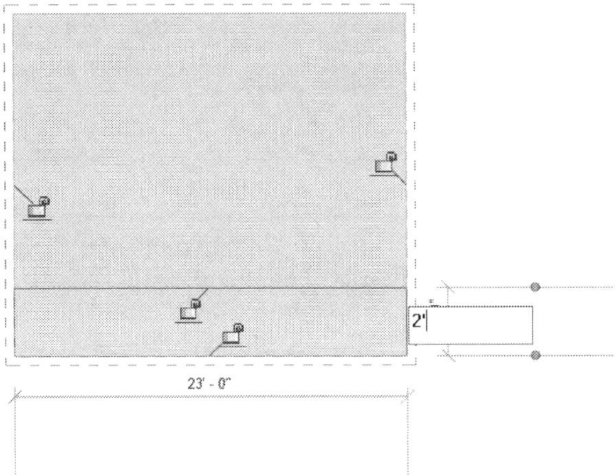

***Figure 5-8** Editing temporary dimensions of the profile*

Note

*If the parts are not visible, choose the **Show Parts** option from drop-down corresponding to the **Parts Visibilty** tab in the **Properties** palette.*

GROUPING ELEMENTS

Ribbon:	Architecture > Model > Model Group drop-down > Create Group

Grouping elements enables you to associate a single or a number of elements as a single entity or group. Once grouped, the elements act as a single entity. They can be selected as a single element so that you can move or copy a number of elements as a single unit. Each instance of the group acts as a single unit and any modifications made in the group are propagated in all its instances. For example, you may create a group with a cluster of walls, windows, and doors, and then create their multiple instances. When you modify the door type in one group, the change will reflect in all instances of the group. You can create three types of groups, model groups for the model elements, detail groups for the view-specific elements, and attached detail groups for view-specific elements associated with a specific model group. You can create groups either by selecting the elements first and then grouping them, or by creating groups first and then later adding the elements to the groups by selecting them from the project views. To create groups, invoke the **Create Group** tool from the **Model Group** drop-down of the **Model** panel of the **Architecture** tab; the **Create Group** dialog box will be displayed in which you can enter the name of the group.

Note

*1. If you invoke the **Create Group** tool after selecting the elements, the **Create Model Group** dialog box will be displayed.*

*2. You can invoke the **Create Group** tool from the **Create** panel of the **Modify| (Elements/ Components)** tab. Alternatively, type GP.*

Creating Groups by Selecting Elements in Project Views

You can create a group by selecting only the model elements from the project views or using the **Group Editor**. To create groups of elements, select the elements to be grouped and then choose the **Create Group** tool from the **Create** panel of the **Modify | Multi-Select** tab; the **Create Model Group** dialog box will be displayed, as shown in Figure 5-9. Alternatively, after selecting the elements to be grouped, choose the **Create Group** tool from **Architecture > Model > Model Group** drop-down; the **Create Model Group** dialog box will be displayed.

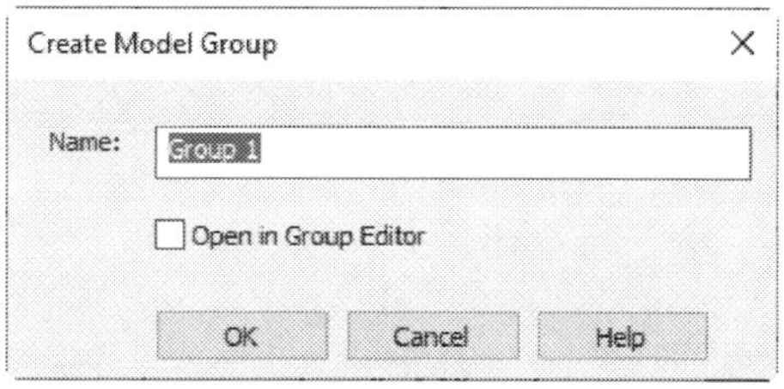

***Figure 5-9** The **Create Model Group** dialog box*

Note

*While creating a group, if you have selected one element type, the corresponding **Modify / (Elements / Components)** tab will be displayed instead of the **Multi-Select** tab.*

In the **Create Model Group** dialog box, specify the name of the group in the **Name** edit box and choose the **OK** button; the name of the created group will be automatically displayed under the **Group** heading in the **Project Browser**. The default name given to a model group is **Group 1** and the number in it increments as more groups are created. You can rename a group by selecting and right-clicking its name and choosing **Rename** from the shortcut menu.

Creating Groups Using the Group Editor

The Group Editor allows you to create groups prior to the selection of the elements. You can create the **Model** group or the **Annotation** group based on the requirement and later on select the elements from the project views and add them to the groups. To do so, choose the **Create Group** tool from **Architecture > Model > Model Group** drop-down; the **Create Group** dialog box will be displayed, as shown in Figure 5-10. In this dialog box, specify the name of the group in the **Name** edit box. The **Group Type** area in this dialog box displays two radio buttons for creating group, **Model** and **Detail**. Specify the type of group by selecting the **Model** or **Detail** radio button and then choose the **OK** button; the screen will display edit mode with a yellow background and the **Edit Group** panel will be displayed in the **Architecture** tab. Figure 5-11 shows the options in the **Edit Group** panel. In the **Edit Group** panel, you can choose the **Add** button to add elements into a group. On choosing the **Add** button, the add (+) symbol will appear with the cursor. You can select the elements from the project view to group them. If there are no elements in the project view, you can create the elements using different tools from the **Build** panel in the **Architecture** tab. The elements will be automatically grouped. You can remove the elements from the existing group, attach details, and even view properties of the group using the respective buttons from the toolbar. Choose the **Finish** button after adding elements to the group.

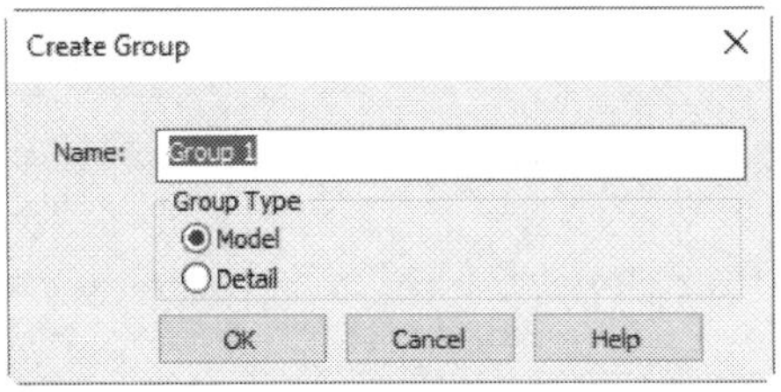

*Figure 5-10 The **Create Group** dialog box*

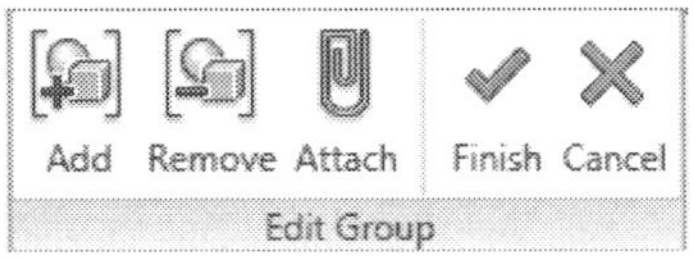

*Figure 5-11 The options in the **Edit Group** panel*

Creating a Detail Group

Revit allows you to group all the detail lines, fill regions, and the detail components separately. You will learn about them in Chapter 12 of this textbook. To group detail components, select them in the drawing and choose the **Create Group** tool from **Annotate > Detail > Detail Group** drop-down; the **Create Model Group** dialog box will be displayed. Specify the name of the detail group in the **Name** edit box and choose the **OK** button; the name of the detail group will be displayed in the **Groups > Detail** head in the **Project Browser**.

Note

If you select incorrect components to add to a particular type of group, Revit will display a warning about the components that cannot be grouped together. For example, you cannot add tags, dimensions, and spot dimensions to a group of detail components.

Creating Model and Attached Detail Groups

Autodesk Revit allows you to group all the dimensions, tags, and the detail components as an attachment to the model group. This attached group consists of dimensions, tags, detail lines, or detail components that refer to the model group. To create an attached detail group, select the model elements and detail elements simultaneously using a crossing window; the **Modify | Multi-Select** tab will be displayed. Choose the **Create Group** tool from the **Create** panel; the **Create Model Group and Attached Detail Group** dialog box will be displayed, as shown in Figure 5-12. Enter the name of the model group in the **Name** edit box of the **Model Group** area. Next, enter a name of the detail group in the **Name** edit box of the **Attached Detail Group** area and choose the **OK** button; the attached detail group will be created consisting of the detail components, dimensions, and detail elements. The details will be attached automatically to the required model group and displayed in the **Project Browser** under the model group that it belongs to.

You can create a new group by duplicating an existing group. You can also modify the group without affecting the parent group. To create a duplicate group, select the group of which you want to create a duplicate in the **Project Browser** and right-click; a shortcut menu will be displayed. Choose **Duplicate** from the shortcut menu; a new group will be added to the **Project Browser** under the parent group. You can rename the group, if required.

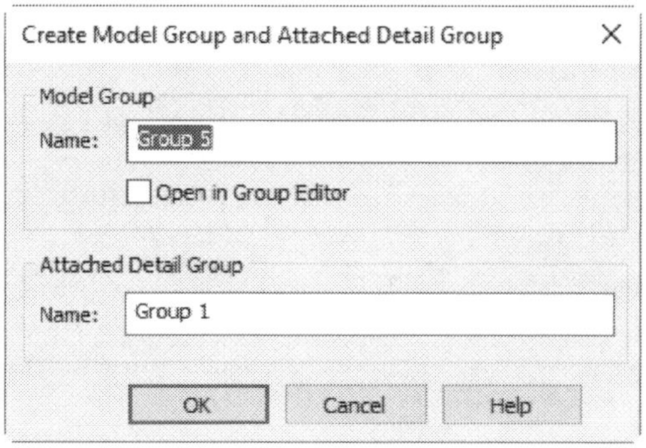

Figure 5-12 *The* ***Create Model Group and Attached Detail Group*** *dialog box*

Tip
You can group elements belonging to different families such as walls, doors, windows, furniture, into a single group. If dependent elements are grouped without their respective hosts, Autodesk Revit will display relevant error messages.

Note
In Revit, the detail group behaves as clipped element. On selecting the detail group, a clipped symbol will appear and the group will get selected even if the ***Select Pinned Element*** *option is disabled in the* ***Status Bar****. But on selecting the Model group and detail group together and making them a pinned element, the detail group will act as a pinned element and will not get selected on disabling the* ***Select Pinned Elements*** *option.*

Placing Groups

Once a group is created, you can easily place its instances in the project by dragging the group from the **Project Browser**. Select the required **Detail** or the **Model** group from the **Project Browser**, drag and drop it at the desired location in the drawing area, and then click to place it. Alternatively, to place a model group, choose the **Place Model Group** tool from **Architecture > Model > Model Group** drop-down; the **Modify | Place Group** tab will be displayed. Select the required group from the **Type Selector** drop-down list and click in the drawing area to place the model group.

Tip
The origin and rotation axis controls of a group are useful for saving time in positioning and swapping the group.

Alternatively, to place a detail group, choose the **Place Detail Group** tool from **Annotate > Detail > Detail Group** drop-down; the **Modify | Place Group** tab will be displayed. Now, select the desired detail group from the **Type Selector** drop-down list and click in the drawing area to place the detail group.

While placing a model group that has an attached detail group, you can control the display of the attached detail group. To do so, you need to place an instance of a model group that has a detail group attached with it in the drawing, and then select it; the **Modify | Model Groups** tab will be displayed. Now, choose the **Attached Detail Groups** tool from the **Group** panel; the **Attached Detail Group Placement** dialog box will be displayed. In the **Attached detail groups** area, a list of attached detail groups along with their corresponding check boxes will be displayed. You can select or clear the desired check box(es) to display or hide the detail group(s) along with the model group(s), when placed. Now, choose **OK**; the **Attached Detail Group Placement** dialog box will close and the selected model group will display or hide the attached detail group depending on whether the corresponding check box of the detail is selected or cleared in the **Attached Detail Group Placement** dialog box.

Once a group is placed in a drawing, you can position it at an appropriate location by moving or dragging it. Autodesk Revit provides the origin and rotation controls to set the position of the group. The origin control defines the base point of the grouped elements, whereas the rotation axis defines the rotation of the group. You can drag the two controls to the required location. By doing so, the orientation or position of the group will not change.

Swapping Groups

You can easily replace one group with another group. To do so, select a group from the drawing area; the **Modify | Model Groups** tab will be displayed. In the **Type Selector** drop-down list, the name of the available groups will be displayed. You can click on the **Model Group** drop-down list and select a different group to replace the selected group.

In case of swapping an attached detail group, Revit has the ability to update and replace the attached detail group of the old group with the attached detail group of the swapped group having the same name. Revit also displays a warning indicating the elements for which the reference could not be found out.

Modifying Groups

You can modify a group by ungrouping it. To modify a group, select it from the drawing; the **Modify | Model Groups** tab will be displayed. The **Group** panel of this tab displays the **Edit Group**, **Ungroup**, and **Link** buttons. On choosing the **Ungroup** button, the elements in the selected group will get disassociated and ungrouped. Using the **Edit Group** button, you can add or remove elements from the group and also attach details to it. When you choose this button, the screen will enter into the edit mode displaying the **Edit Group** panel. The **Add** and **Remove** buttons can be chosen to add and remove elements from the group, respectively. After editing, choose the **Finish** button to complete the modifications. The **Cancel** button provides the option of discarding the modifications and returning to the **Modify | Model Groups** tab. You can use different editing tools on the grouped elements such as **Move**, **Copy**, **Mirror**, **Cut**, **Paste**, **Rotate**, and so on.

Excluding Elements from a Group

You can exclude elements from the group if some elements of a group overlap with the elements of the other group while placing them together. If the elements host some other elements such as doors and windows, Revit will automatically rehost the elements in the adjacent walls. To exclude an element from a group, place the cursor over the element that you want to exclude. Next, press TAB to highlight it, and click; a blue color icon will be displayed. When you place the cursor on the icon, a tooltip will be displayed prompting you to click on the icon to exclude the selected element. Click on the icon and press ESC; the element will be excluded from the group, and will not be displayed in the drawing. Alternatively, to exclude an element from the group, select the element in the drawing as explained above and right-click; a shortcut menu will be displayed. Choose **Exclude** from the shortcut menu; the element will disappear from the drawing and will be excluded from the group. The elements that are excluded will not be visible in the drawing and will not be included in the schedule. To restore the excluded elements in the group, select the group and right-click; a shortcut menu will be displayed. Choose **Restore All Excluded** from the shortcut menu; the elements will be restored in the group and will be displayed in the drawing.

Tip

You can use the shortcuts to exclude an element from a group. For example, select the element in the drawing and press EX, DELETE, or CTRL+X to exclude the selected element.

Saving and Loading Groups

In Revit, you can save the group as the Revit file with *.rvt* extension. You can edit the groups independent of the project in which they are loaded. To save groups, choose **Save As > Library > Group** from the **File** menu; the **Save Group** dialog box will be displayed, as shown in Figure 5-13. Select the group to be saved from the **Group To Save** drop-down list and then save it using either the same name as the group name or enter a new name in the **File name** edit box. The group(s) will be saved with the assigned name with *.rvt* extension in case of a project file, or *.rfa* extension in case of a family file. The saved groups can also be loaded into another Autodesk Revit project that is open in the current session by choosing the **Load as Group into Open Projects** tool from **Architecture > Model > Model Group** drop-down; the **Load into Projects** dialog box will be displayed. In this dialog box, the list box displays the name of the project files, along with their check boxes, that are opened in the current session. To load the group(s) to a specified file, you can select the check box that is displayed along with the file name, from the list box. In the **Include** area of the **Load into Projects** dialog box, you can select the **Attached Details**, **Levels**, or **Grids** check boxes, if you want to load the details, grids, and levels in the project with the group. The detail elements will be loaded in the project as attached detail groups. Now, choose **OK**; the **Load into Projects** dialog box will close and the window of the selected project file(s) will be displayed. Notice that the group(s) that you have loaded is displayed under the **Groups** head in **Project Browser**.

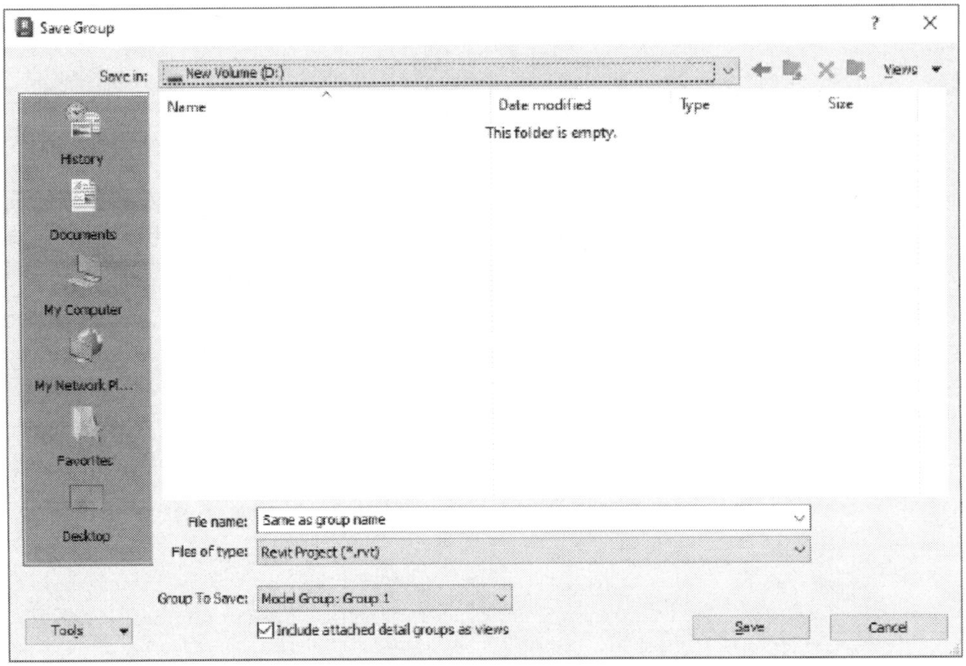

*Figure 5-13 The **Save Group** dialog box*

Converting Groups into Linked Models

To convert the groups into linked models, select the group to be converted into a linked model from the drawing; the **Modify | Model Groups** tab will be displayed. In this tab, choose the **Link** tool from the **Group** panel; the **Convert to Link** dialog box will be displayed, as shown in Figure 5-14.

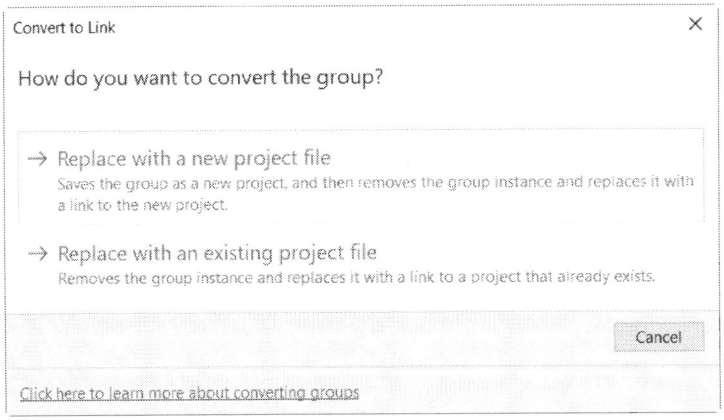

*Figure 5-14 The **Convert to Link** dialog box*

To create a new linked file, choose the **Replace with a new project file** button from the dialog box; the **Save Group** dialog box will be displayed. Specify the name in the **File name** edit box, browse to the required location, and choose the **Save** button to save the file. Choose the **Replace with an existing project file** button to replace the group with an already existing linked Revit model. On choosing this button, the **Open** dialog box will be displayed. Browse to the required location, select the existing file, and choose the **Open** button; the model group will be replaced by the existing linked file. You can choose the **Cancel** button to cancel the conversion.

If the name assigned to the linked group file already exists in the project then Revit will display a message box, as shown in Figure 5-15. Choose the **Yes** button to replace the group file. On choosing the **No** button, the **Save As** dialog box will be displayed. Enter a new name for the linked group file to save the file with a different file name.

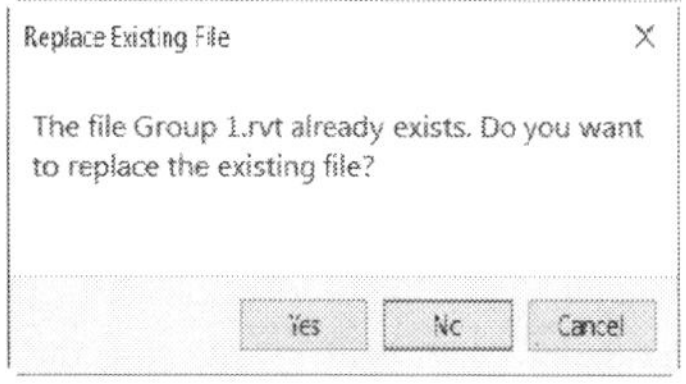

Figure 5-15 *The message box asking to replace existing file*

Deleting Groups

The created groups can be deleted only after all the instances of the group are deleted from the project. To delete all the instances of a group, right-click on the group name in the **Project Browser** and choose **Select All Instances > In Entire Project** from the shortcut menu. Then, choose the **Delete** button from the **Modify** panel of the contextual tab or press the DELETE key. The name of the group will be displayed even after all the instances are deleted. To delete the group from **Project Browser**, right-click its name and choose **Delete** from the shortcut menu. The **Delete** option is available only when all the instances of the group have been deleted from the project.

Tip

You can also create nested groups with sub-groups within groups. When you ungroup a group with sub-groups, the association between the sub-groups is terminated. The ***Ungroup*** *button can be used again to edit individual elements in the sub-group. To delete or edit an element from an instance of a group, you need to ungroup it first and then modify the desired element. After the necessary changes are made, the elements can be regrouped.*

CREATING SIMILAR ELEMENTS

Ribbon: Modify (Elements / Components) > Create > Create Similar
Shortcut Key: CS

Autodesk Revit provides the **Create Similar** tool to create and place copies of an element or a group at specified points. This tool is available only after a selection has been made. To create an element similar to an existing element in the drawing, select it; the **Modify | (Elements / Components)** tab will be displayed. Now, choose the **Create Similar** tool from the **Create** panel; the creation tool which was used to create the selected element will be invoked. Now, create the element of similar instance parameter as that of the selected element in your drawing. For example, if you select a window from the drawing and then invoke the **Create Similar** tool, the **Modify | Place Window** tab will be displayed with the window type similar to that of the selected window, which is selected in the **Type Selector** drop-down list. Now, you can place windows with similar instance parameter as that of the selected window in the drawing. Alternatively, to invoke the **Create Similar** tool, select the element and then right-click; a shortcut menu will be displayed. Next, choose **Create Similar** from the shortcut menu; the **Create Similar** tool will be invoked.

PINNING AND UNPINNING ELEMENTS

Ribbon: Modify | (Elements / Components) > Modify > Pin / Unpin
Shortcut Key: PN, UP

Elements can be pinned to their position using the **Pin** tool. To do so, select an element from the drawing; the **Modify | (Elements / Components)** tab will be displayed. Invoke the **Pin** tool from the **Modify** panel. Alternatively, you can type **PN** to invoke the **Pin** tool. Once pinned, a graphical pushpin symbol appears with the element, as shown in Figure 5-16, indicating that it is pinned and cannot be moved from its position. You can also pin the position with respect to other elements and groups. Click on the pin symbol to pin or unpin an element. For example, you can pin the position of columns with the intersecting grid lines. Once pinned, the column and the grid lines cannot be moved without first unpinning their position. You can use the **Unpin** tool to unpin the position constraint for the pinned elements. Note that the **Unpin** tool will replace the **Pin** tool after an element is pinned.

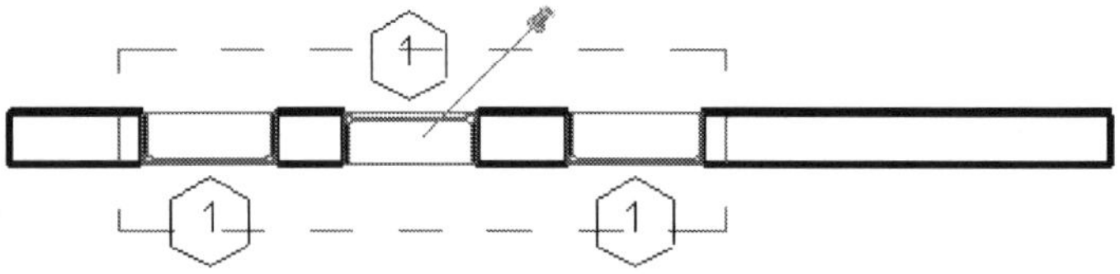

Figure 5-16 *Group of windows pinned at their position*

Note

The pin symbol is not visible after invoking other tools. The element or group must be selected to display the pin symbol. In Revit, you cannot delete the pinned elements. In case of multiple selection, if pinned elements get selected for deletion, then Revit will warn you that the elements are pinned and cannot be deleted. Also, it will instruct you to unpin the element.

SCALING ELEMENTS

Ribbon: Modify (Elements / Components) > Modify > Scale
Shortcut Key: RE

The **Scale** tool is used to modify the scale of a single or a selection of elements and is available only for walls, lines, images, *.dwg* and *.dxf* imports, reference planes and position of dimensions. After making a selection, invoke the **Scale** tool from the **Modify** panel. The **Options Bar** displays two radio buttons that can be selected for scaling, **Graphical** and **Numerical**.

You can select the **Graphical** radio button from the **Options Bar** to resize an element graphically. On selecting this radio button, you will be prompted to specify three points. First to specify the origin of scaling and the other two to define the scale vectors. Once the origin is specified, move the cursor and click to specify the length of the first vector. Move the cursor and click again to define the length of the second vector. Autodesk Revit calculates the proportion of the two vectors and resizes the selected element in the same proportion using the specified origin. The temporary dimensions displayed can also be used to resize the element to the desired scaling. For example, if the length of the first vector is 10' and the second vector is 15', the selected element is resized 1.5 times.

You can select the **Numerical** radio button in the **Options Bar** and specify the numeric value of the scaling in the **Scale** edit box. You then need to specify the origin of the scaling by clicking it. Autodesk Revit instantly resizes the element by the specified scaling proportion. For example, when you resize a 10' long wall by a numerical factor of 2.5, its length is resized to 25'.

Tip

Scaling moves the position of location line of the wall but does not change height or thickness of the wall. You can use either the location line or the face of the wall for specifying the origin point.

USING DIAGNOSTIC TOOLS

Autodesk Revit provides tools to assist you in seeking information about the elements in the project. You can use these tools to measure distances and angles, get information of the element identification, and more.

Measuring Distance Between References and Along an Element

Ribbon: Modify > Measure > Measure drop-down > Measure Between Two References
Measure Along an Element

You can use the measure tools to quickly measure and temporarily display the length (and angle from the horizontal, if applicable) of individual walls or lines that you select in the project view (plan and elevation). The dimensions generated by this tool will remain on the screen until you exit from the current tool or start the next measurement.

In Revit, you have two measuring tools: **Measure Between Two References** and **Measure Along an Element**. You can invoke the **Measure Between Two References** tool to find out the exact distance between any two points and also the angle of the distance vector with the horizontal. Invoke this tool from the **Measure** panel; you will be prompted to select the start point. Now, move the cursor to the desired point and click; the start point of the line to be measured will be specified. Now, move the cursor towards the second point; a temporary dimension appears between the specified point and the current location of the cursor. Now, click to specify the location of the second point and the end of the line to be measured. A temporary dimension, along with the angular dimension, is created between the two specified points. Notice that in the **Options Bar**, the **Total Length** text box displays the running total for the length of the chain, if the **Chain** check box is selected.

You can invoke the **Measure Along an Element** tool to display the length and angle of an element along its axis line. Invoke this tool from the **Measure** panel and then select an element from the drawing; the temporary dimensions will be displayed along the axis of the element.

Tip
While using the measure tools, you can use the object snaps to measure the distance accurately. The location line snaps can also be used to get this information.

Selecting Elements Using the Element ID

In Autodesk Revit, each created element, group or annotation symbol is referred by a unique identification number. Error messages that are displayed refer to an element by its identification number. To display the identification of an element, select it from the drawing and then choose the **IDs of Selection** tool from the **Inquiry** panel of the **Manage** tab. On doing so, the **Element IDs of Selection** dialog box will be displayed, as shown in Figure 5-17. The **Id(s)** text box in this dialog box displays the identification number of the selected element.

You can invoke the **Select by ID** tool from the **Inquiry** panel of the **Manage** tab to identify and select an element by entering its identification number. On invoking this tool, the **Select Elements by ID** dialog box will be displayed, as shown in Figure 5-18. You can enter more than one Id using a ',' (comma) between them. Choose the **Show** button to preview the elements with the entered Ids in the drawing. Choose the **OK** button to create a selection of these elements.

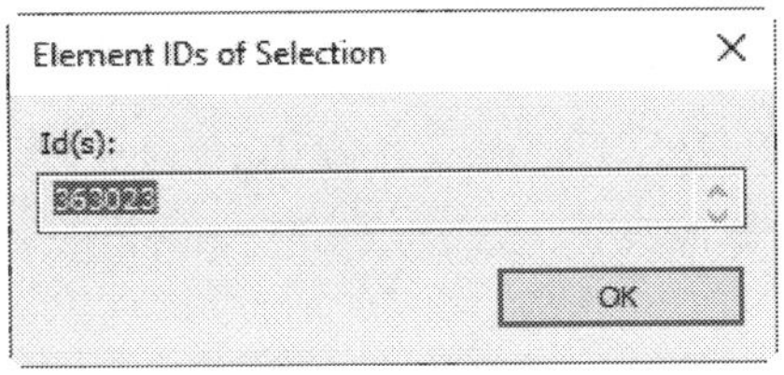

***Figure 5-17** The **Id(s)** text box showing the Id of the selected elements*

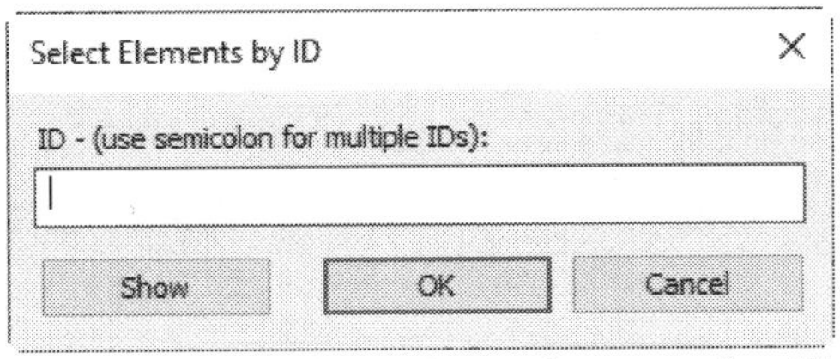

***Figure 5-18** The **Select Elements by ID** dialog box*

Note
When you select a chain of entities, the controls are displayed at joints and vertices. Select the control at the joint and drag it by pressing and holding the left mouse button. The joint remains intact.

ASSEMBLIES

In Revit, you can combine multiple elements to form a single entity called an assembly. The assembly include elements like model elements, annotations, detail items, groups, images, and so on. These assemblies help in identifying, classifying, and quantifying the unique element combinations in the model and are used for scheduling, tagging, and filtering.

The assemblies are listed as a type in the Project Browser. You can select the assembly type from the Project Browser and generate one or more types of isolated views of the assembly as well as parts lists, material takeoffs, and sheets. Assemblies can be placed in the project or assembly sheet views.

Creating Assemblies

To create an assembly, first select the elements that you want to include in the assembly; the **Modify | Multi- Select** tab will be displayed. Choose the **Create Assembly** tool from the **Create** panel of the **Modify|Multi-Select** tab; the **New Assembly** dialog box will be displayed. If the assembly includes elements of different categories like walls, doors and windows then the type name will depend on the option selected from the **Naming Category** drop-down list. If the assembly includes elements of same categories, the **Naming Category** drop-down list will be disabled and you can edit the type name in the **Type Name** edit box. Now, choose the **OK** button; the **New Assembly** dialog will be closed and the created assembly will be listed in the **Project Browser**.

Editing Assemblies

In Revit, you can add elements, remove elements, or can perform edits on the assembly. To edit an assembly, select it from the drawing area; the **Modify|Assemblies** tab will be displayed. In this tab, choose the **Edit Assembly** tool from the **Assembly** panel, you will enter into the editing mode and the floating **Edit Assembly** panel and origin with its planes in the drawing area will be displayed.

The origin determines the orientation of the assembly in relation to the project orientation. The planes of the origin helps in determining the orientation of assembly in the corresponding elevations while creating assembly views. On changing the type of an assembly instance, the origin of the new type is applied to that instance. When you select the assembly origin, the drag controls are displayed. You can use the drag controls to modify the origin.

This floating **Edit Assembly** panel will allow you to add or remove the elements from the drawing area. After adding or removing the desired elements from the assembly, choose the **Finish** button from the floating **Edit Assembly** panel, the created assembly will be modified.

You can remove the created assembly at any time from the project by choosing the **Disassemble** tool from the **Assembly** panel of the **Modify|Assemblies** tab.

Creating Assembly Views and Sheets

You can create the assembly views and sheets for an assembly type. To create the assembly view, select the assembly; the **Modify|Assemblies** contextual tab will be displayed. Now, choose the **Create Views** tool from the **Assembly** panel of the **Modify|Assemblies** tab; the **Create Assembly Views** dialog box will be displayed. In this dialog box, you can set the view scale from the **Scale** drop-down list. You can also set your own scale value. To do so, choose the **Custom** option from the **Scale** drop-down list; the **Scale Value** edit box will be enabled. In the **Scale Value** edit box, you can set the scale value for the view. In the **Views to Create** area of the dialog box, all the check boxes are selected by default. As a result, all the views will be created for the assembly. For the assembly, you can create the schedules as well. To create a schedule, select the check box(es) from the **Schedule** drop-down list. You can include the sheets as well by selecting the required sheet from the **Sheet** drop-down list. You can also load the sheets, if required, by selecting the **Load** option from the **Sheet** drop-down list. Now, choose the **OK** button; the assembly views are created under the **Assembly** head in the **Project Browser**.

TUTORIAL

Tutorial 1 Apartment 1

In this tutorial, you will modify the *c04_Apartment1_tut1.rvt* project file created in Tutorial 1 of Chapter 4. You will copy, move, and mirror the windows to create the project, refer to Figure 5-19. Do not add the dimensions and text as they are given only for reference. Use the following project parameters for completing this tutorial: **(Expected time: 30 min)**

The following steps are required to complete this tutorial:

a. Open the required file created in Chapter 4.
   For Imperial *c04_Apartment1_tut1.rvt*
   For Metric *M_c04_Apartment1_tut1.rvt*
b. Create the window type W5 using the **Window** tool.
c. Use the **Copy** tool to create the windows at the desired locations.
d. Use the **Group** tool to group the twin windows W3.
e. Use the **Mirror** tool to mirror windows.
f. Save the project using the **Save As** tool.
   For Imperial *c05_Apartment1_tut1.rvt*
   For Metric *M_c05_Apartment1_tut1.rvt*
g. Close the project.

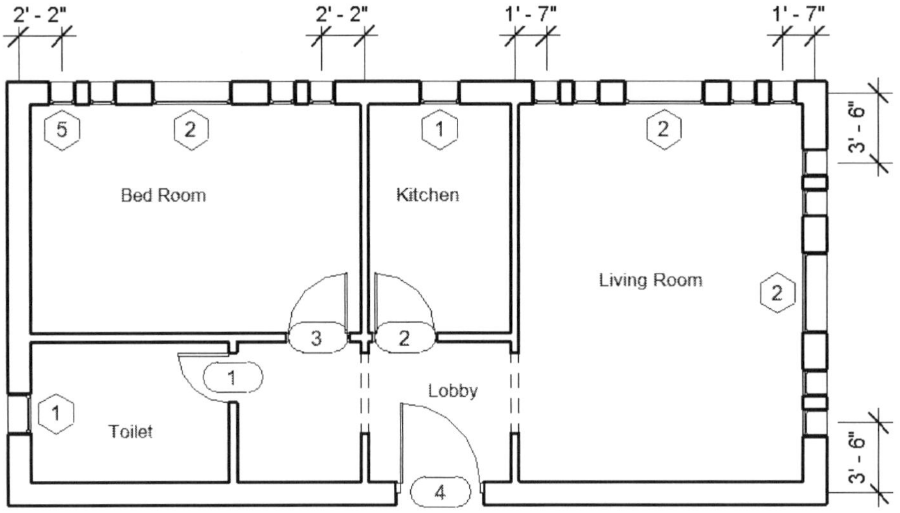

Figure 5-19 *Sketch plan of the Apartment1 project*

Opening the Project and Creating the Window

You need to open the specified project and add the left side bedroom window using the **Window** tool (see Adding Windows section in Chapter 4).

1. Choose **Open > Project** from the **File** menu and open the file created in Tutorial 1 of Chapter 4.

   For Imperial *c04_Apartment1_tut1.rvt*
   For Metric *M_c04_Apartment1_tut1.rvt*

   You can also download this file from *https://www.Cadsoft.com*. The path of the file is as follows: *Textbooks > Civil/GIS > Revit Architecture > Exploring Autodesk Revit 2025 for Architecture.*

2. In the Project Browser, double-click on **Level 1** under the **Floor Plans** head.

3. Invoke the **Window** tool from the **Build** panel of the **Architecture** tab. Next, in the **Properties** palette, select the required option from the **Type Selector** drop-down list.
   For Imperial **Fixed 16"x48"**
   For Metric **Fixed 406 x 1220mm**

4. Click near the top exterior wall of the bed room to create a window and then modify the temporary dimensions, as shown in Figure 5-20. Press ESC twice to exit the **Window** tool.

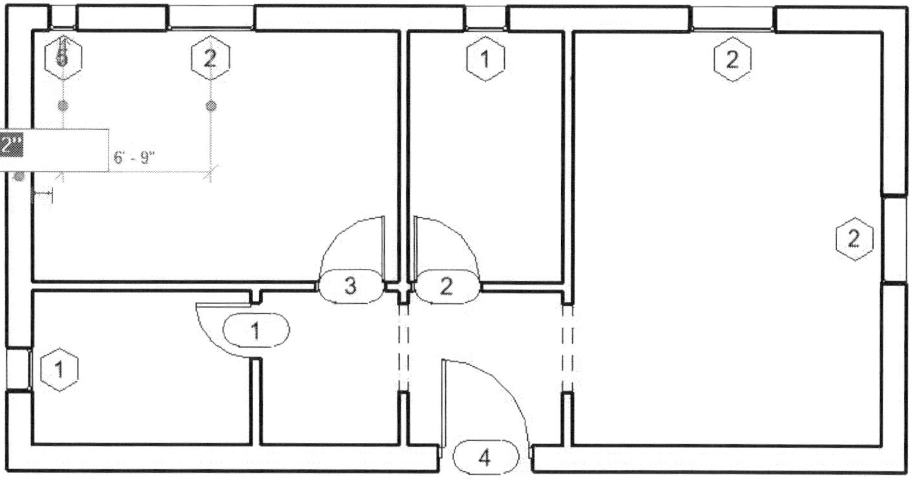

***Figure 5-20** Modifying dimension to add window to the Apartment 1 project*

Creating a Copy of the Window

Next, you need to copy the window type created in the previous section and then create another window at the specified distance.

1. Select the recently created window from the drawing area by clicking on it. On doing so, the **Modify | Windows** tab is displayed.

2. Invoke the **Copy** tool from the **Modify** panel; the window enclosed in the preview box is displayed indicating that it has been selected for copying. Ensure that the **Constrain** check box is selected in the **Options Bar**.

3. Move the cursor near the midpoint of the window you just created and click to specify the first point of the copy distance when the mid-point snap is displayed, as shown in Figure 5-21.

4. Move the cursor toward the right and type **2'0"(610 mm)** in the edit box displayed. Next, press ENTER; the window is copied at the specified distance, as shown in Figure 5-22. Now, press ESC to exit the **Modify | Windows** tab.

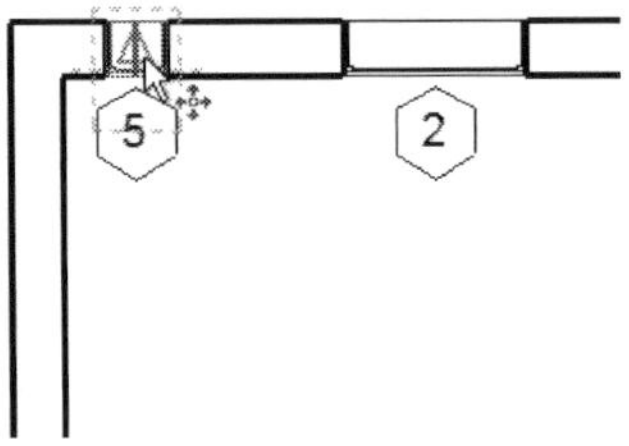

***Figure 5-21** Specifying the first point of the copy distance*

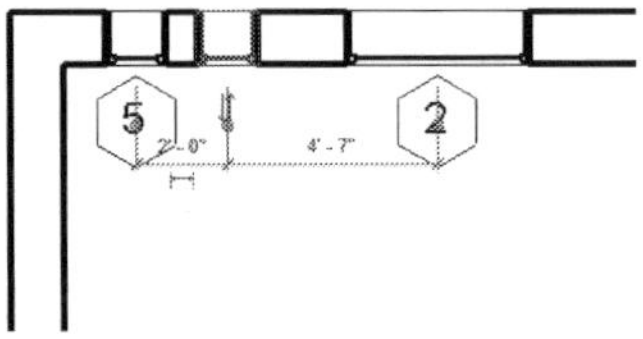

***Figure 5-22** Specifying the distance at which the copy of the window will be placed*

Grouping Windows

In this section, you will create a group using the **Create Group** tool. Since these windows appear at the same distance at all occurrences in the project, you can group them as a single entity and then create the copies. Use the pick box option for selecting the windows and then use the **Create Group** tool to group them.

1. Using the pick box selection method, select the two windows; the two windows are highlighted, refer to Figure 5-23.

2. Invoke the **Create Group** tool from the **Create** panel of the **Modify | Windows** tab; the **Create Model Group and Attached Detail Group** dialog box is displayed.

3. Specify **2XW3** as the name of the group in the **Name** edit box and choose the **OK** button.

   The windows are grouped and a preview box is displayed. Also, the group origin controls are displayed, as shown in Figure 5-24.

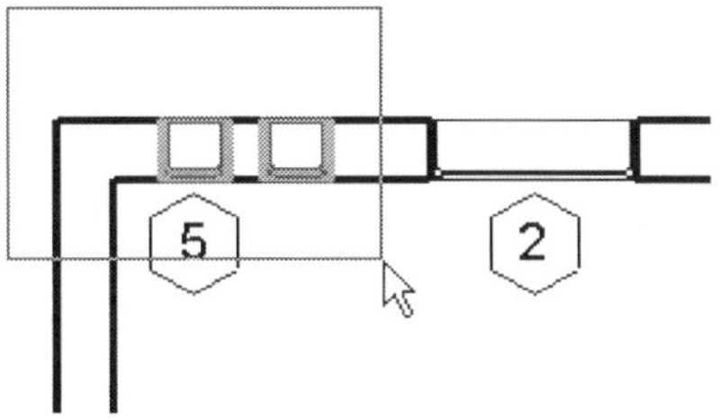

Figure 5-23 Selecting the windows by using the pick box selection method

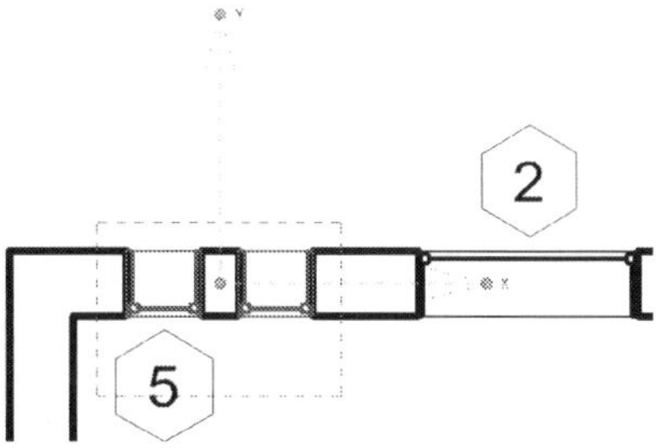

Figure 5-24 Grouped windows with the group origin controls displayed

Mirroring and Copying Windows

Now, the twin windows need to be mirrored and copied using the **Mirror - Pick Axis** tool. The sketch plan shows that the twin windows of the bedroom are equidistant from the center of the window tagged 2(W2). You will use this point to create a mirror axis to mirror the windows.

1. Invoke the **Mirror - Pick Axis** tool from the **Modify** panel in the **Modify | Model Groups** tab. The mirror tools are available because the created group is already selected and highlighted.

2. Move the cursor near the midpoint of the window tagged 2 (W2) until a vertical line is displayed, as shown in Figure 5-25. This line indicates the midpoint and the mirror axis. Click at this location; the windows are mirrored on the other side of the mirror axis, as shown in Figure 5-26.

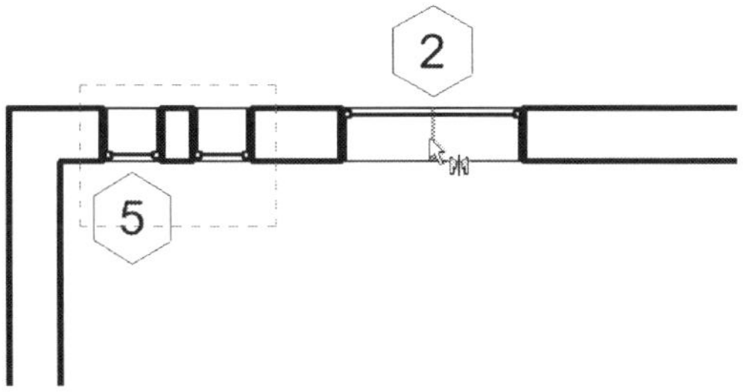

Figure 5-25 Specifying the mirror axis to mirror the twin windows

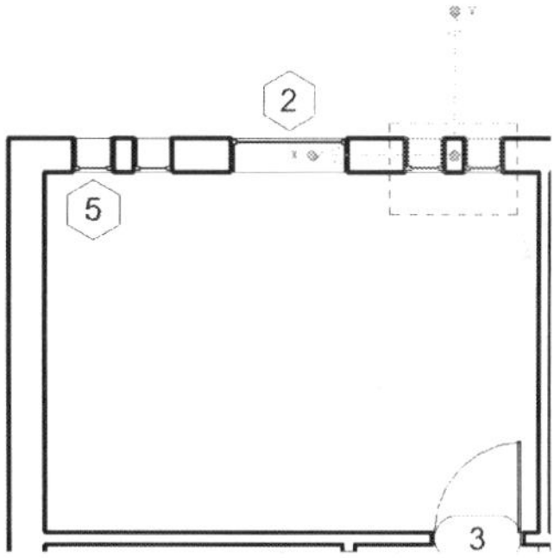

Figure 5-26 Windows created after mirroring

Placing Grouped Windows

Next, you need to place the grouped windows on the exterior wall of the living room by dragging and dropping the group in the drawing.

1. In the **Project Browser**, click on the '+' sign in the **Groups** node of the **Project Browser** to display the created groups. Click on the '+' sign corresponding to the **Model** sub-node; it displays **2XW3** as the group in the project.

2. To place this group in the project, click on the group name in the **Project Browser** and drag and drop it on the exterior wall of the living room, as shown in Figure 5-27. Click to place the group.

3. Select the placed group, as shown in Figure 5-28, and choose the **Activate Dimensions** button from the **Options Bar**. Click on the left dimension and enter **1'7" (483 mm)**; the group moves to the desired location.

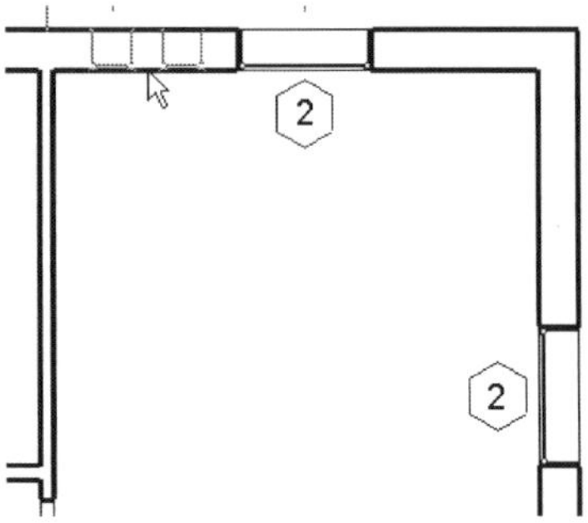

*Figure 5-27 Dragging and dropping a group from the **Project Browser***

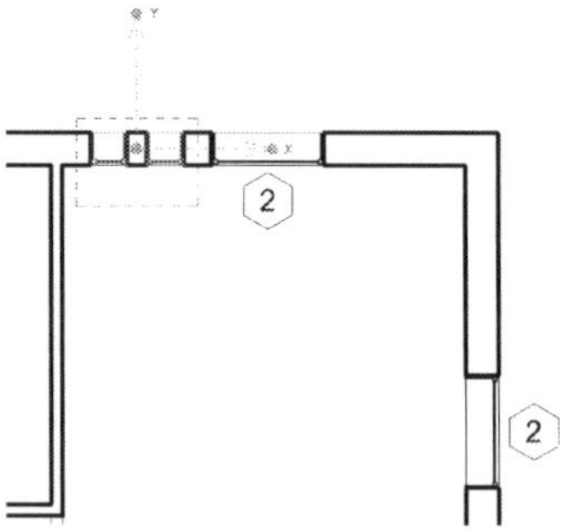

Figure 5-28 Selecting the group to move to the desired location

4. Create a mirror image of the twin windows using the midpoint of the central window as the mirror axis. The windows are mirrored, as shown in Figure 5-29.

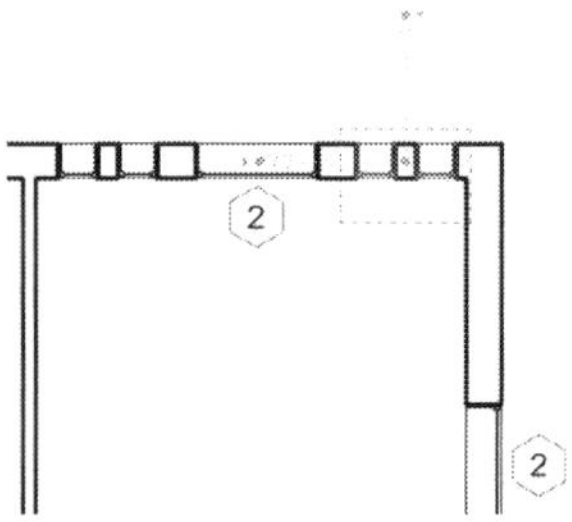

Figure 5-29 Mirrored windows

5. Repeat steps 1, 2, and 3 to place the window group on the upper side of the right exterior wall of the living room. Use the same distance parameter as specified in the sketch plan, refer to Figure 5-19. After placing the window group, create a mirror image of the group using the midpoint of the central window. Use the dimensions shown in Figure 5-19 to specify the exact location of the group. Press ESC to exit the **Modify | Model Groups** tab.

Saving the Project

In this section, you will save the project file using the **Save As** tool.

1. Choose **Save As > Project** from the **File** menu; the **Save As** dialog box is displayed. Browse to the folder **c05_revit_2025_tut** and choose the **Open** button; the **File Name** edit box is displayed.

Enter the required file name in the **File name** edit box.
For Imperial **c05_Apartment1_tut1**
For Metric **M_c05_Apartment1_tut1**

Choose the **Save** button; the file gets saved and the **Save As** dialog box is closed.

2. Choose **Close** from the **File** menu to close the project file.

The completed project will look similar to the one shown in Figure 5-30.

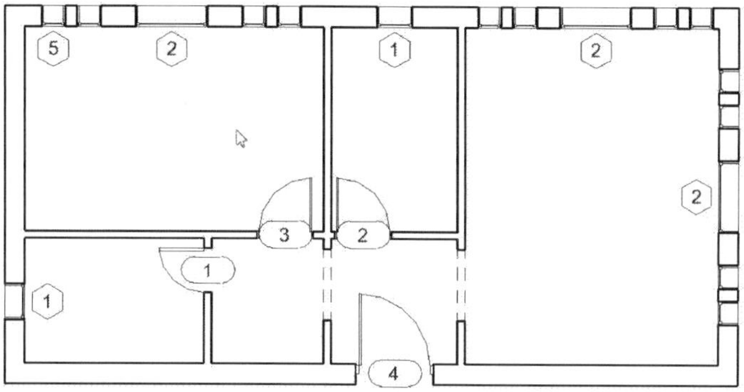

***Figure 5-30** The completed Apartment 1 project*

Tutorial 2 Club

In this tutorial, you will modify Hall 1 and Lounge of the *Club* project created in Tutorial 2 of Chapter 4. You will move and copy the windows to their desired locations and modify their size and type, refer to Figure 5-31. Do not create the dimensions and text as they are given only for reference. Use the following project parameters to complete this tutorial:

(Expected time: 30 min)

1. Project File-
For Imperial *c04_Club_tut2.rvt*
For Metric *M_c04_Club_tut2.rvt*

2. Window types to be used:
For Imperial W1- **Fixed 36"x48"**
For Metric W1- **Fixed 0915 x 1220 mm**

The following steps are required to complete this tutorial:

a. Open the required file created in Chapter 4.
For Imperial *c04_Club_tut2.rvt*
For Metric *M_c04_Club_tut2.rvt*

b. Create an array of four windows of Hall 1 by selecting the **Linear** option of the **Array** tool.

c. Create an array of windows of the Lounge by selecting the **Radial** option of the **Array** tool.

d. Use the **Create Group** tool to group the four windows W1 as a single entity.

e. Use the **Mirror - Pick Axis** and **Mirror - Draw Axis** tools to mirror the grouped windows.

f. Use the **Copy** tool to copy the grouped windows.

g. Save the project using the **Save As** tool.
For Imperial *c05_Club_tut2.rvt*
For Metric *M_c05_Club_tut2.rvt*

h. Close the project.

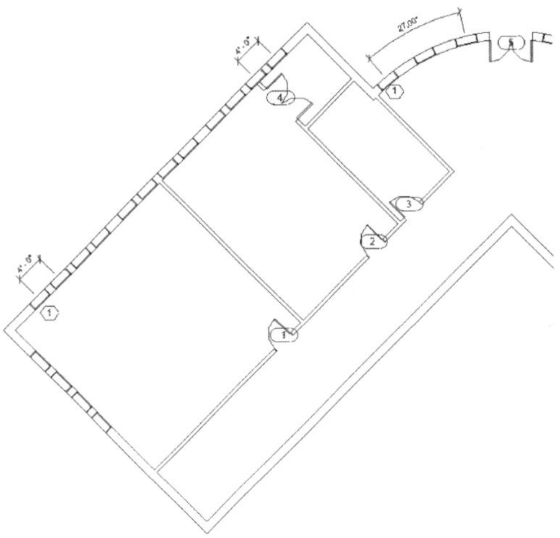

***Figure 5-31** Sketch plan for moving and copying the windows in the Club project*

Opening an Existing Project and Creating an Array of Windows

In Figure 5-25, notice that the windows of Hall 1 appear in three clusters of four windows each in the sketch plan. The four windows are equidistant, 4'0" (1220 mm) apart. You can create them using the **Linear** option of the **Array** tool. Similarly, to create the four windows of the Lounge, use the **Radial** option.

1. Choose **Open > Project** from the **File** menu and open the desired file.
For Imperial *c04_Club_tut2.rvt*
For Metric *M_c04_Club_tut2.rvt*

You can also download this file from *https://www.cadsoft.com*. The path of the file is as follows: *Textbooks > Civil/GIS > Revit Achitecture> Exploring Autodesk Revit 2025 for Architecture*

2. Move the cursor on the lower left window tagged 1(W1) of Hall 1 and click to select the window.

3. Invoke the **Array** tool from the **Modify** panel of the **Modify | Windows** tab; the window is enclosed in a dashed rectangular box indicating that it has been selected for creating an array. 

4. In the **Options Bar**, ensure that the **Linear** button is chosen and the **2nd** radio button is selected in the **Move To** area. Also, ensure that the **Constrain** check box is cleared.

5. In the **Number** edit box of the **Options Bar**, enter **4**.

6. Now, move the cursor near the midpoint of the selected window and click when the midpoint object snap is displayed, as shown in Figure 5-32; the start point of the linear array is specified. Ensure that the **Group and Associate** check box is selected.

7. Move the cursor upward along the wall incline and enter the value **4'0" (1220 mm)**, as shown in Figure 5-33; a text box with the number of array elements is displayed.

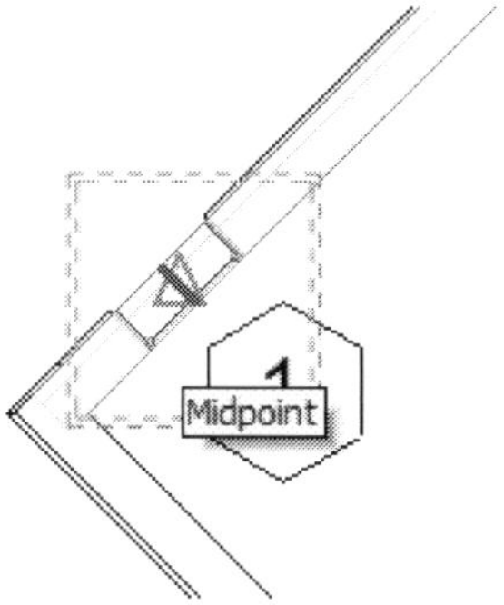

Figure 5-32 *Specifying the first point of the array distance*

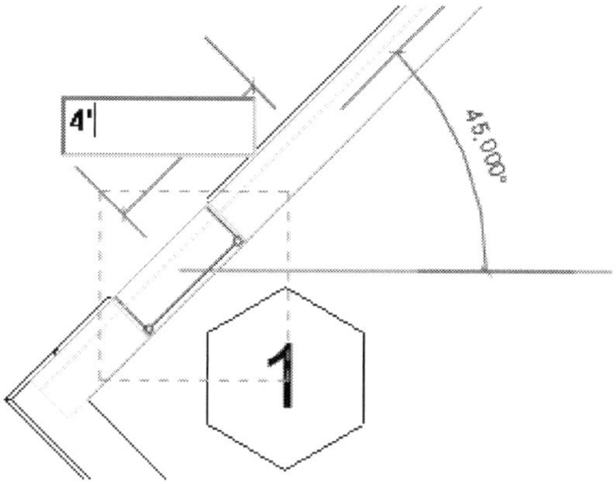

Figure 5-33 *Entering the value of the array distance*

8. Press ENTER to accept the value; an array is created. Notice that as you move the cursor, a preview box showing the location of the selected elements is displayed.

   Similarly, the windows of the Lounge can be created using the **Radial** option of the **Array**.

9. Move the cursor over the exterior window at the Lounge area tagged 1(W1), and click when it is highlighted; the window is selected.

10. Next, invoke the **Array** tool from the **Modify** panel of the **Modify | Windows** tab.

11. In the **Options Bar**, choose the **Radial** button. Also, ensure the **Group and Associate** check box is selected.

12. Select the **Last** radio button in the **Move To** area of the **Options Bar**. To create three windows,

enter the value **3** in the **Number** edit box if entered by default.

As the center of the required radial array is the center of the circular wall, you need to move the rotation point to it and then create the radial array.

13. In the **Options Bar**, choose the **Place** button and place the cursor at the center of the curved wall, as shown in Figure 5-34. Alternatively, you can move the cursor near the rotation key of the selected window and drag it to the center of the curved wall.

    Notice that the radial array line now originates from the new center point.

14. Move the cursor near the midpoint of the Lounge window and click when the **Midpoint** object snap is displayed, as shown in Figure 5-35; the start point of the radial array is specified.

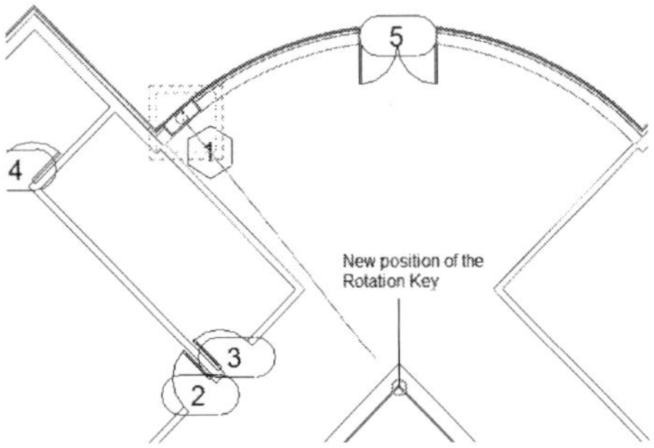

Figure 5-34 *Rotation key moved to the new rotation point*

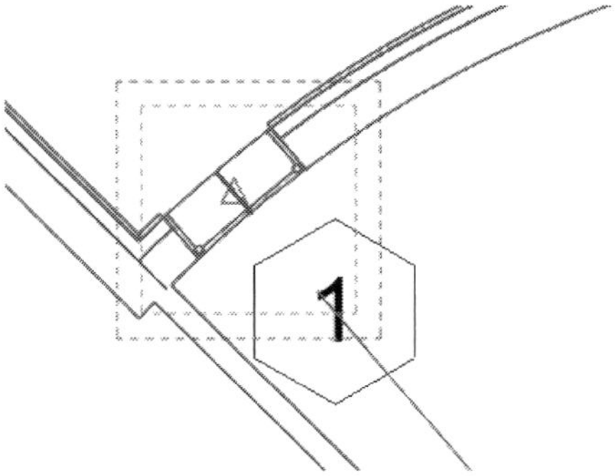

Figure 5-35 *Specifying first point of the array*

15. Move the cursor clockwise along the curved wall and enter **27** as the temporary angular dimension, as shown in Figure 5-36.

16. Next, press ENTER twice to accept the value specified for the angular dimension and the number of array elements; radial array is created, as shown in Figure 5-37.

17. Choose the **Modify** button from the **Select** panel to exit the current selection.

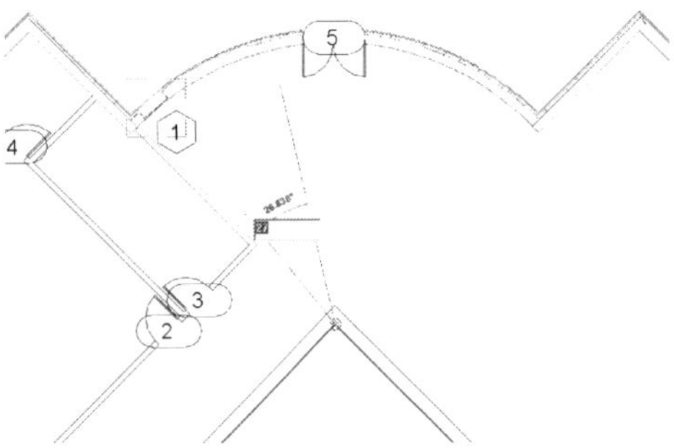

Figure 5-36 *Specifying the array angle*

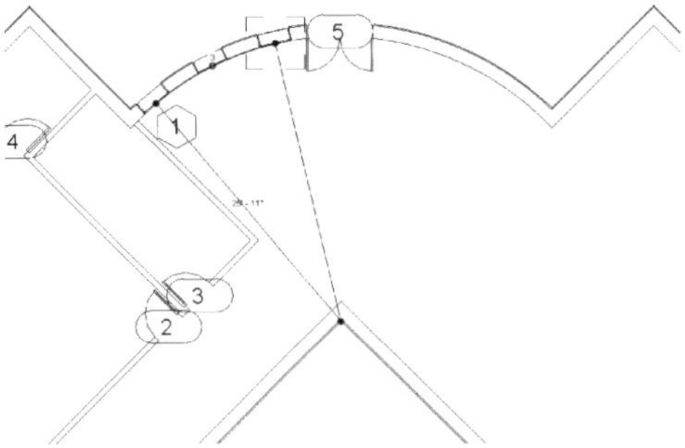

Figure 5-37 *Radial array of the selected window*

Grouping the Cluster of Windows

The sketch plan shows the cluster of four Hall 1 windows appearing together. Group them for the ease of copying. You can use the **Create Group** tool to group the array of windows.

1. Move the cursor near the four windows of the Hall 1 and select them using the CTRL key.

2. Invoke the **Create Group** tool from the **Create** panel of the **Modify | Model Groups** tab; the **Create Model Group** dialog box is displayed. The default name of the group of windows in the dialog box is Group 1.

3. Specify **4XW1- Linear Windows** as the group name in the **Name** edit box.

4. Choose the **OK** button in the dialog box; the windows are grouped and the group name is displayed in the **Groups** node of the **Project Browser** in the **Model** subhead.

Mirroring the Grouped Windows

You need to mirror the grouped windows of Hall 1 and the radial windows of the Lounge using the **Mirror - Pick Axis** and **Mirror - Draw Axis** tools.

1. Move the cursor near the grouped windows of Hall 1 and click; the grouped windows are selected and highlighted, as shown in Figure 5-38.

2. Choose the **Mirror - Pick Axis** tool from the **Modify** panel of the **Modify | Model Groups** tab. Also, ensure that the **Copy** check box is selected in the **Options Bar**.

3. Move the cursor near the interior wall and click when the cursor snaps to the wall centerline; the wall centerline is selected as the mirror axis, as shown in Figure 5-39, and the grouped windows are mirrored to the other end of the exterior wall. Ignore the warning message that is displayed due to the conflicts between the interior wall and the window. Now, choose the **Modify** button from the **Select** panel or press ESC to exit the current selection.

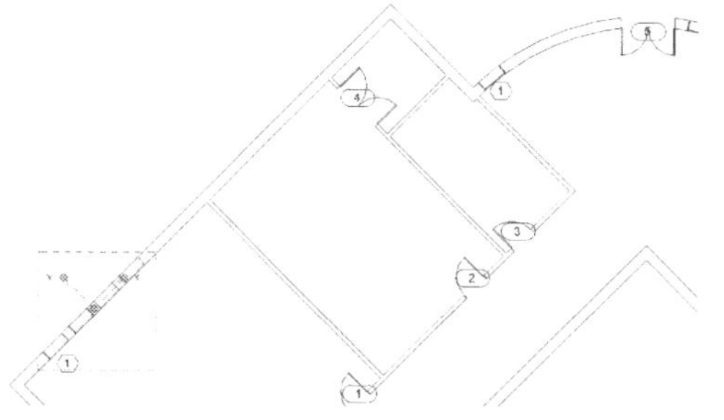

Figure 5-38 Grouped windows selected and highlighted

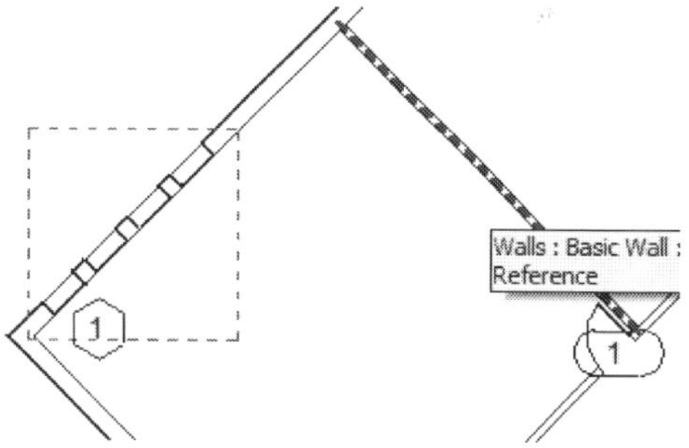

Figure 5-39 Using the wall centerline as the mirror axis

Notice that the same window type is also placed on the lower left perpendicular wall. You can use the **Mirror - Draw Axis** tool to mirror the grouped windows to the perpendicular wall.

4. Click on the original grouped windows to select them.

5. Invoke the **Mirror - Draw Axis** tool from the **Modify | Model Groups** tab.

6. In the **Options Bar**, ensure that the **Copy** check box is selected.

7. Move the cursor to the intersection of the two perpendicular walls and click when the endpoint object snap is displayed; the first point of the mirror line is specified.

8. Move the cursor horizontally and click when the dashed line is displayed, as shown in Figure 5-40; the second point of the mirror axis is specified. Now, choose the **Modify** button from the **Select** panel to exit the current selection.

9. The grouped windows are mirrored to the perpendicular wall, as shown in Figure 5-41. Similarly, mirror the three radial windows of the Lounge to the other side of the door using the **Mirror - Draw Axis** tool. Draw the vertical axis from the center of the curved wall as the mirror axis.

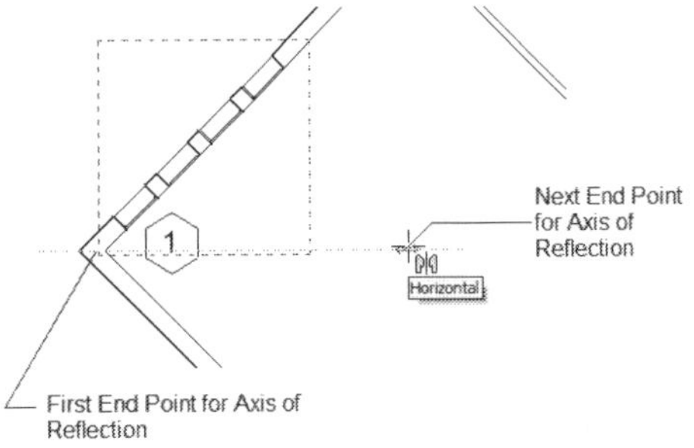

Figure 5-40 Using the horizontal axis as the mirror reference line

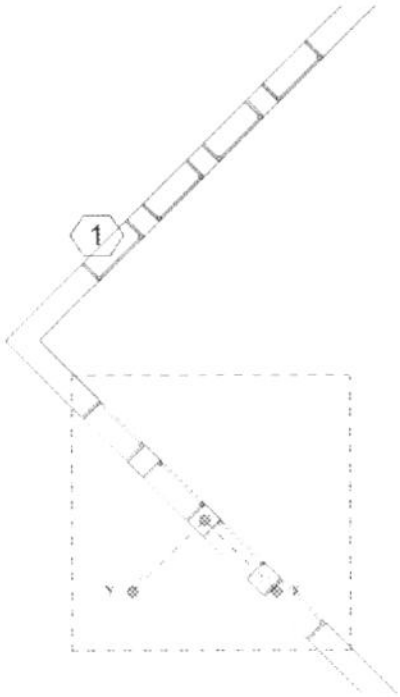

Figure 5-41 Mirrored windows on the perpendicular wall

Copying the Grouped Windows

You can use the **Copy** tool to copy the grouped windows of Hall 1.

1. Select the original grouped window of Hall 1.

2. Invoke the **Copy** tool from the **Modify | Model Groups** tab.

3. Select the group origin or the midpoint of the grouped windows as the start point of the copy distance.

4. Move the cursor upward along the external wall and choose the endpoint of the central internal wall as the second point of the copy distance, as shown in Figure 5-42. On doing

so, the grouped window is copied at the desired location, as shown in Figure 5-43. Ignore the warning message, if any, displayed.

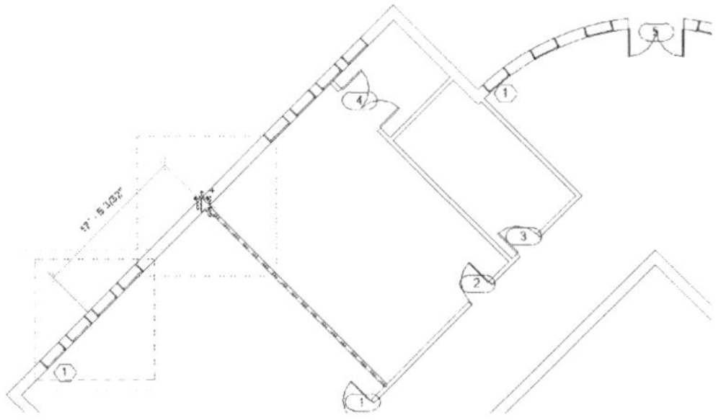

Figure 5-42 *Specifying the start and end points of the copy distance*

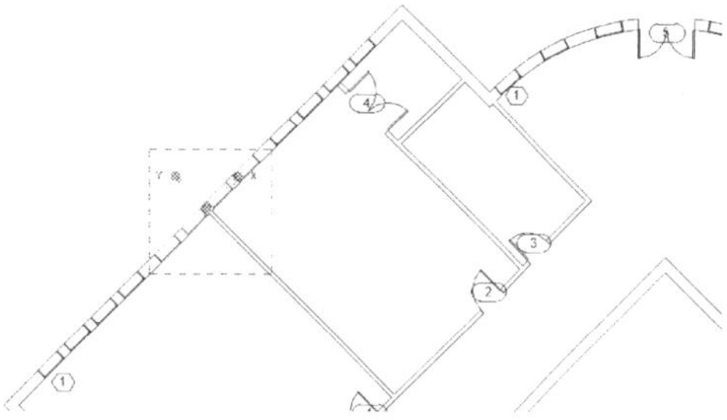

Figure 5-43 *Grouped windows copied*

Saving the Project

In this section, you will save the project file using the **Save As** tool.

1. Choose **Save As > Project** from the **File** menu; the **Save As** dialog box is displayed. Browse to the folder **c05_revit_2025_tut** and choose the **Open** button; the **File name** edit box is displayed.

   Enter the required file in the **File name** edit box
   For Imperial **c05_Club_tut2**
   For Metric **M_c05_Club_tut2**
   Choose **Save**; the file is saved and the dialog box gets closed.

2. Choose **Close** from the **File** menu to close the project file.

EXERCISES

Exercise 1 Apartment 2

Add windows to the exterior wall of the dining area of the *Apartment 2* project created in Exercise 1 of Chapter 4 based on Figure 5-44. The dimensions and text are given only for reference and are not to be added. The parameters to be used for the project are given next.

(Expected time: 30 min)

1. Project-
   For Imperial *c04_Apartment2_ex1.rvt*
   For Metric *M_c04_Apartment2_ex1.rvt*
2. Window type to be added in the dining area-
   For Imperial **Fixed: 24"X48"**
   For Metric **Fixed: 0610 x 1220 mm**
3. Name of the file to be saved-
   For Imperial **c05_Apartment2_ex1**

For Metric **M_c05_Apartment2_ex1**

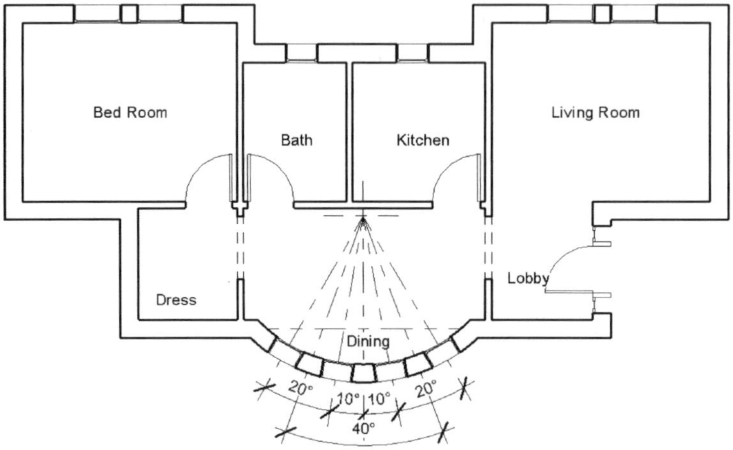

Figure 5-44 *Sketch plan for adding windows to the Apartment 2 project*

Exercise 2 Club

Add windows, as shown in Figure 5-45, to the Hall 2 of the *Club* project created in Tutorial 2 of this chapter. The dimensions and text have been given only for reference and need not be added.

(Expected time: 30 min)

1. Project File-
   For Imperial *c05_Club_tut2.rvt*
   For Metric *M_ c05_Club_tut2.rvt*
2. Window types to be used -
   For Imperial W1- **Fixed: 36" X 48"**
   For Metric W1- **Fixed: 0910 X 1220 mm**
3. Name of the file to be saved -
   For Imperial **c05_Club_ex2**
   For Metric **M_c05_Club_ex2**

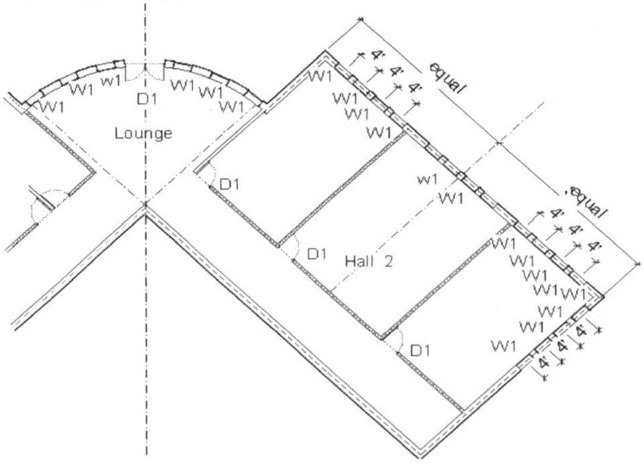

Figure 5-45 *Sketch plan for adding windows to Hall 2 of the Club project*

Chapter 6

Working with Datum Plane and Creating Standard Views

Learning Objectives

After completing this chapter, you will be able to:

- *Understand the concept of levels*
- *Create multiple levels in a project*
- *Understand the concept of using grids in a project*
- *Create rectangular and circular grids in a project file*
- *Use the Work Plane and Reference Plane tools in a project*
- *Create a plan view in a project*
- *Create and understand the usage of elevation view*
- *Create section views and modify their properties*
- *Use the Scope Box tool*

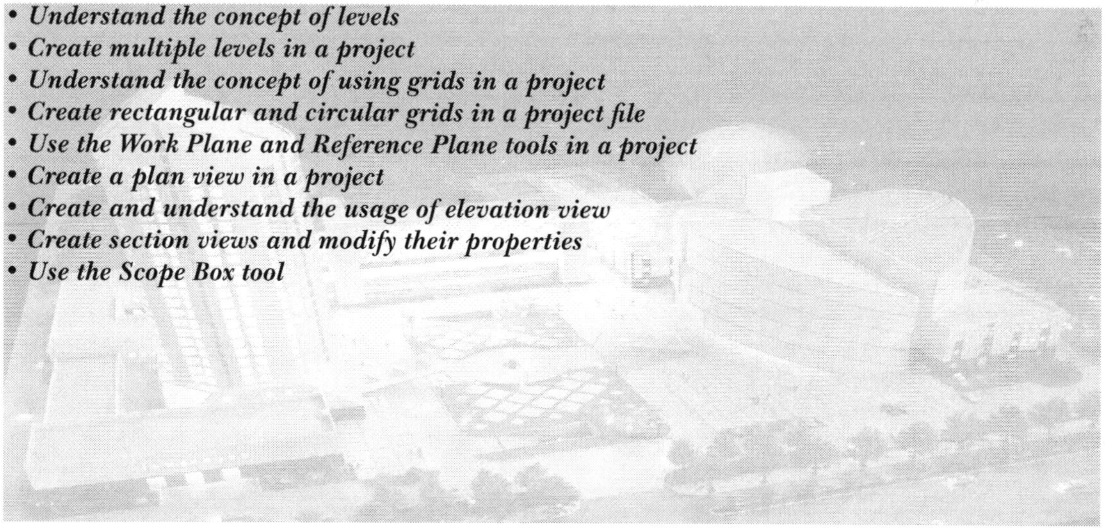

INTRODUCTION

In the previous chapters, you learned to use editing tools to create groups and modify their properties in a project. In this chapter, you will learn to add datum elements and the procedure to create standard views in an architectural project. Datum elements and the non-physical elements form the basic framework in a project. These datum elements include levels, reference planes, and grids. Depending on the complexity of a project, you can use some or all of these components to systematize and provide a reference to a building model. These elements are also used for editing a project, creating schedules, and also creating other reference documentation for a project.

In Autodesk Revit, a standard view can be created and displayed with relative ease. Creating standard views in a project not only help in its presentation but also assist in creating and editing various building and stand-alone elements of a project.

BIM
CONCEPTS

Levels and grids represent datum objects in a BIM project. While working on a BIM project, it is important to define standard levels and grids for placement and coordination of object. Every levels and grids are represented by a unique IfcGUID, which helps in identifying them in the project.

WORKING WITH LEVELS

For a multistory building project, you need to specify floor levels within the building volume. Each floor level or story can contain a different set of building components. Levels, in a typical multistory building, may be understood as infinite horizontal planes that define each story. Revit uses levels as references for level-hosted elements such as floor, roof, and ceiling. The distance between levels can be used to define the story-height of a building model, as shown in Figure 6-1. Revit also provides flexibility to create a non-story level or a reference level such as sill level, parapet level, and so on. For example, a multistory office building may have different story heights for each floor. Building components such as exterior walls, windows, doors, and furniture may also differ on each floor. You can create levels based on the story height of a building. You can then create various building elements on each level such as an entrance door on the first floor level, bay windows on the second floor level, an elevator room on the roof level, and so on.

***Figure 6-1** Levels displayed in a multistory building project*

Understanding Level Properties

A typical level is represented by a level line, level bubble, level name, level elevation, and so on, refer to Figure 6-2. Using these parameters and controls, you can modify the appearance of a level. The level name is a modifiable parameter that is used to refer to each level. The level elevation is the distance of the level from the base level. The visibility of the level bubble on either side of the level line can be controlled by using the bubble display control. The length alignment control can be used to align levels lines. Autodesk Revit provides the 2D or 3D extents control for datums when they are selected. This enables you to change their extents in one or multiple views in which they are visible. When a datum is in the 3D mode, any modifications made in the 3D view will be propagated in all views of a building model. On the other hand, the 2D mode can be used to modify a datum in a specific view, thereby making it view-specific.

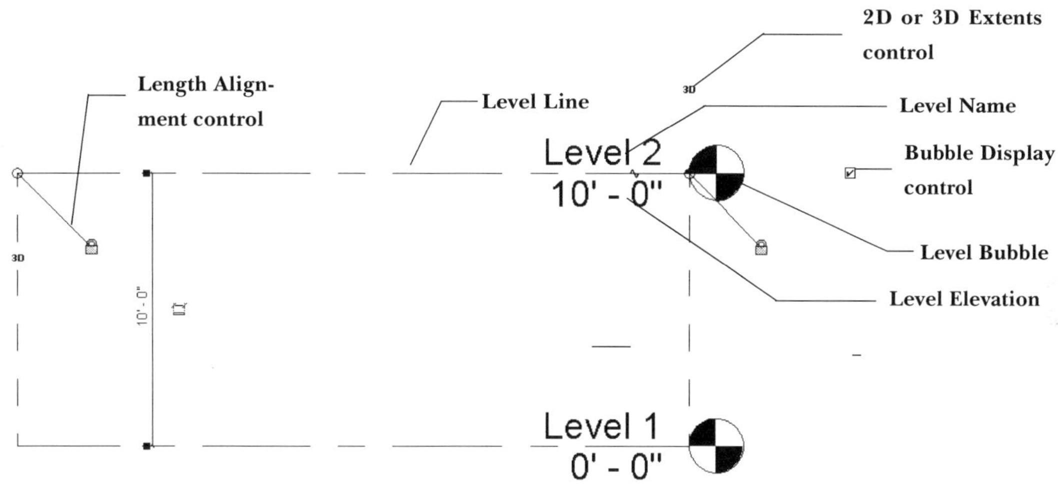

Figure 6-2 *Various levels defined in the default template file*

Adding Levels

Ribbon: Architecture > Datum > Level
Shortcut Key: LL

In Revit, you can create multiple levels based on your project requirements. Note that the **Level** tool remains inactive in the **Datum** panel for all the plan views. The **Level** tool will only be active in an elevation or a section view. To create a level, invoke the **Level** tool from the **Datum** panel, as shown in Figure 6-3; the **Modify | Place Level** tab will be displayed. In this tab, choose any of the sketching options displayed in the **Draw** panel to create levels in your project. You can also invoke the **Level** tool by typing **LL**. In the **Modify | Place Level** tab, you can select a type of level from the **Type Selector** drop-down list to modify an existing level. This drop-down list has two types of levels, **Level: 1/4" Head** and **Level: No Head**. To make the level head visible, select the **Level: 1/4" Head** option. Else, select the **Level: No Head** option.

***Figure 6-3** The **Level** tool displayed in the **Datum** panel of the **Architecture** tab*

Modifying Level Parameters

You can change a level type by selecting the level from the drawing and then selecting a different level type from the **Type Selector** drop-down list in the **Properties** palette. On doing so, the current level will be modified into the selected level. You can specify the properties of a level using the **Properties** palette before adding it. Some of the parameters of the level can also be modified by clicking on the level in the drawing window and entering the new value.

For example, after selecting a level, you can click on its name and assign a new name to it. As you start typing the new name, an edit box appears with the new value, as shown in Figure 6-4. Also, Autodesk Revit prompts you to specify whether you want to rename all the associated views such as the floor plan, ceiling plan, and so on. If you choose to rename the views, the name of the associated views will change in the **Project Browser**. Similarly, you can modify the elevation of a level by selecting it and entering the new temporary dimension value, as shown in Figure 6-5. When you enter the new value, the level automatically moves to the specified elevation.

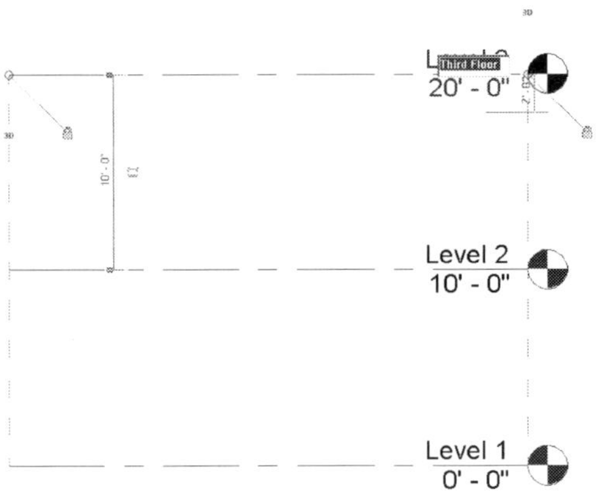

***Figure 6-4** Editing the name of the level*

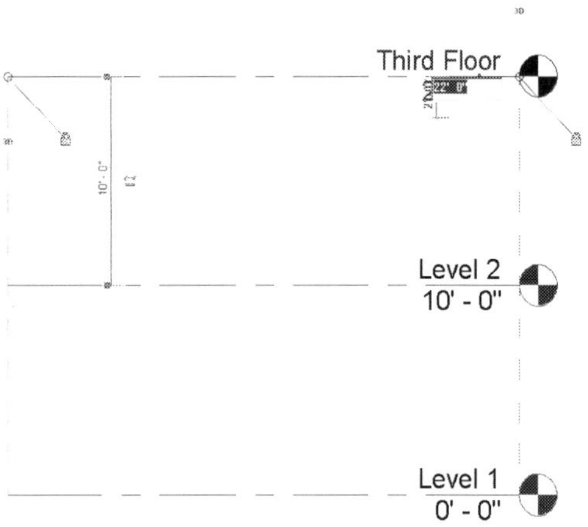

***Figure 6-5** Modifying the level elevation*

HIDING ELEMENTS IN A VIEW

Revit allows you to hide elements and their categories, including annotations. You can also restore the visibility of hidden elements.

To hide an element in a view, select the view in which you want to hide a particular element or a category, select the element in the drawing, and then right-click to display a shortcut menu.

In the shortcut menu, choose **Hide in View > Elements** / **Category** to hide an element or a category in the current view.

To restore the visibility of a hidden element or a category, choose the **Reveal Hidden Elements** button from the **View Control Bar**; the screen will enter the **Reveal Hidden Elements** mode with a magenta boundary and also a new panel **Reveal Hidden Elements** will be displayed in the ribbon. All hidden elements inside the magenta boundary will be displayed in magenta color. Select a hidden element and then choose the **Unhide Element** button from the **Reveal Hidden Elements** panel to restore the visibility of hidden element. Now, choose the **Toggle Reveal Hidden Elements Mode** button in the **Reveal Hidden Elements** panel to exit the **Reveal Hidden Elements** mode. Another method to restore the visibility of the hidden element(s) is to select the element(s) in the **Reveal Hidden Elements** mode and then right-click; a shortcut menu will be displayed. Choose **Unhide in View > Elements** / **Category** from the shortcut menu to restore the visibility of the hidden elements.

> **Tip**
> *After selecting the element(s) or a category in the **Reveal Hidden Elements** mode, you can type EU or VU to unhide the element(s) or the category in the current view.*

Controlling the Visibility of Levels

You can control the visibility of one level or all levels in any of the project views. On right-clicking a selected level in an elevation or 3D view, a shortcut menu will be displayed, as shown in Figure 6-6. Choose **Hide in View > Category** from the shortcut menu; the level category will be hidden in the current view. To hide a particular level, choose **Hide in View > Elements** from the shortcut menu; only the selected level will be hidden in the current view. The level will, however, be displayed in all other views.

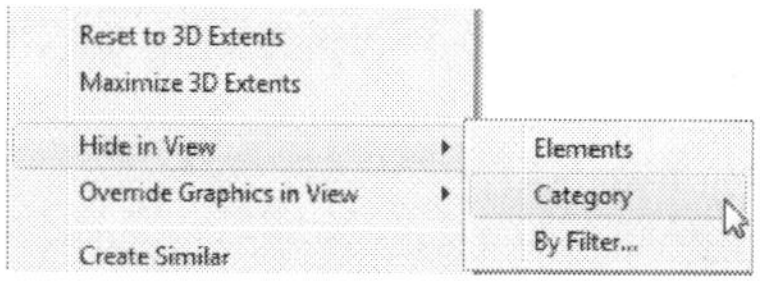

Figure 6-6 *Partial view of the shortcut menu displayed on selecting the level*

To hide all the levels in a current view, select any of the levels in the project view. Next, choose the **Override by Category** tool from the **Modify | Levels > View > Override Graphics in View** drop-down; the **View-Specific Category Graphics** dialog box will be displayed. Clear the **Visible** check box in this dialog box and choose **OK**; all levels will be hidden from the current view. You can use the **Scope Boxes** feature to control the visibility of levels. You will learn about this feature later in this chapter.

> **Tip**
> *After selecting the element, you can type EH or VH to hide an element or a category in the current view.*

WORKING WITH GRIDS

Ribbon: Architecture > Datum > Grid
Shortcut Key: GR

Autodesk Revit provides you the option of creating rectangular or circular grids for your projects. Using these grids, you cannot only create building profiles easily but also place the structural elements at the specific locations and intersections.

Creating Grids

You can create the grid patterns based on the project requirements. Grid patterns can be rectangular or radial, depending on the project geometry. A rectangular grid pattern can be created using straight grid lines, whereas a radial grid pattern can be formed using arc grids. The created grids are visible in all plan, elevation, section views, and in a 3D View.

In Revit 2025 you can see the grids in 3D views. Also, you can control the visibility of Grids as 3D Planes in 3D Views. The presentation of the grids in the 3D view is similar to that of highlighted reference planes in previous Revit versions

To create a grid line, invoke the **Grid** tool from the **Datum** panel of the **Architecture** tab, as shown in Figure 6-7; the **Modify | Place Grid** tab and the **Properties** palette with the properties of grids will be displayed. The **Modify | Place Grid** tab, as shown in Figure 6-8, displays various options to draw and modify grids in a drawing. The **Draw** panel in this tab displays various tools to draw grids as lines and curves or to convert existing model lines into grids. In this **Properties** palette, you can modify the type and instance properties of the grid, before or after creating it. You can also change the type of grid by selecting a grid type from the **Type Selector** drop-down list.

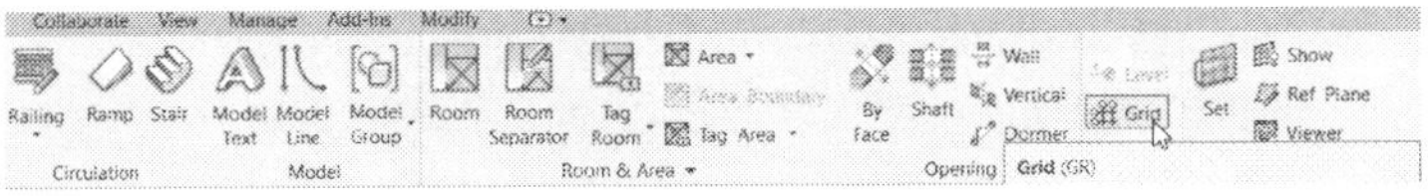

*Figure 6-7 Invoking the **Grid** tool*

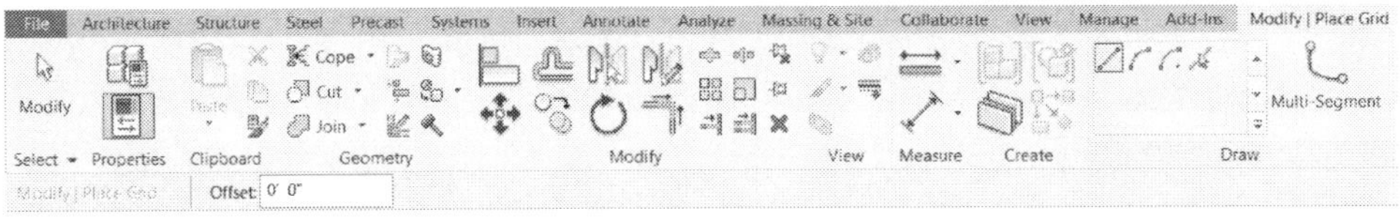

*Figure 6-8 Various options in the **Modify / Place Grid** tab*

Modifying Grids

As mentioned earlier, parameters of grids can be modified before or after grids are created. To modify a grid after it is created, you need to select the grid and modify its properties from the **Properties** palette.

For example, after selecting a grid, you can click on its name and assign it a new name. Similarly, you can modify the distance between grids by selecting the temporary dimension and entering a new value, as shown in Figure 6-9. When you enter a new value, the grid automatically moves to the specified distance.

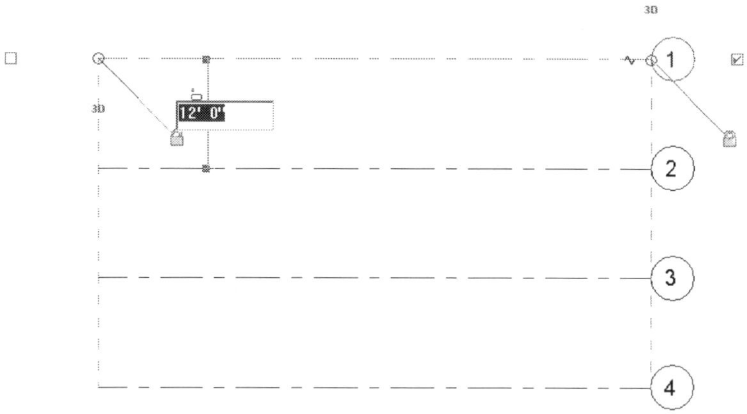

Figure 6-9 *Modifying the distance between grids*

Grid Properties

Like levels, grids too have associated properties. A typical grid consists of a grid line, grid bubble, grid name, and other controls, as shown in Figure 6-10. The usage of the controls such as grid bubble visibility, 2D or 3D extents, and grid bubble break is similar to the usage described for levels.

To view and modify the properties of a grid, select it; the instance parameter of the selected grid will be displayed in the **Properties** palette. In this palette, choose the **Edit Type** button; the **Type Properties** dialog box will be displayed, as shown in Figure 6-11. You can use this dialog box to view and modify the type properties of the selected grid.

> **Tip**
> *It is recommended that you draw the grid lines with the endpoints aligned to each other. You can use the alignment line that is displayed when the cursor is moved to its proximity. The* **Lock** *constraints can be used to drag the extents of the single or multiple grid lines.*

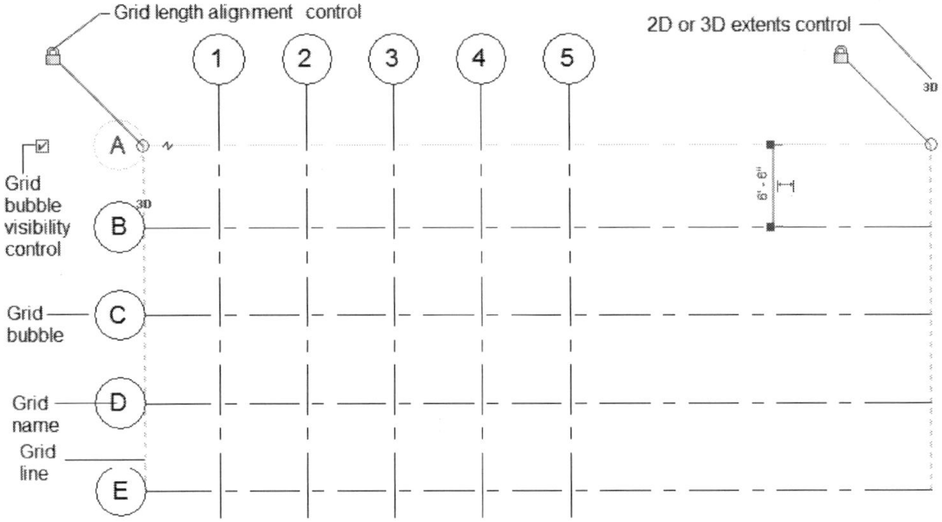

Figure 6-10 *A typical grid line pattern with its controls*

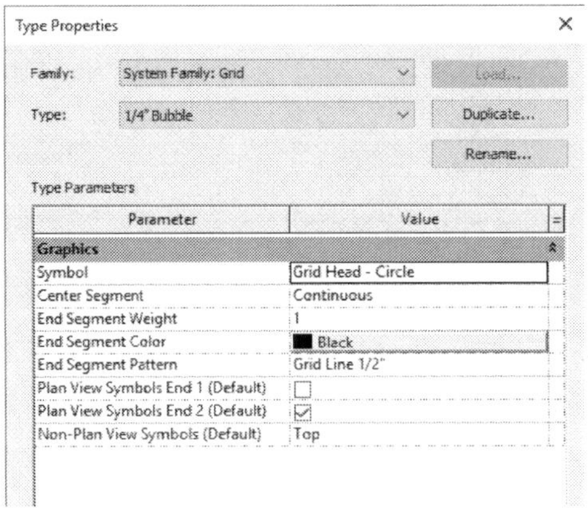

Figure 6-11 *Partial view of the* ***Type Properties*** *dialog box of a grid*

Customizing the Grid Display

In Revit, you can customize the grid display according to your need. You can change the color, line weight, and line type of the entire grid line or part of the grid line such as center segment and end segment. At the same time, you can change the display of symbols at the end of the grid line.

Controlling the Visibility of Grids

You can control the visibility of a grid or all grids in any of the project views. To control the visibility of a grid, select it and right-click; a shortcut menu will be displayed. Choose **Hide in View > Category** from the shortcut menu; the grid category will be hidden in the current view. Similarly, you can also hide one particular grid by choosing **Hide in View > Elements** from the shortcut menu. On doing so, the selected grid will be hidden in the current view. The grid will, however, be displayed in all other views.

REFERENCE PLANES

Ribbon:	Architecture > Work Plane > Ref Plane
Shortcut Key:	RP

Reference planes are useful while sketching and adding building elements to a design. They can be used as datum planes to act as a guideline for creating elements. They can also be used effectively for creating new family elements. To create a reference plane, choose the **Ref Plane** tool from the **Work Plane** panel. On invoking the **Ref Plane** tool, the **Modify | Place Reference Plane** tab will be displayed. Invoke the tools from the **Draw** panel and start drawing the reference plane in the drawing. Alternatively, you can type **RP** to invoke the **Ref Plane** tool.

After invoking the tool, click at the desired location in the drawing window to start a line that defines the reference plane. Now, move the cursor to a new location and click to specify the endpoint of the reference line; the reference plane will be created. To assign the reference plane a name, select it from the drawing. Next in the **Properties** palette, enter the name of the selected reference plane in the value field of the **Name** instance parameter.

WORK PLANES

As the name suggests, the work plane is a plane that can be used for sketching elements. In Autodesk Revit, you can create and edit only those elements that are in the current work plane. The work plane can be horizontal, vertical, or inclined at any specified angle. Each generated view has an associated work plane. This plane is automatically defined for some standard views such as floor plans. For others such as sections, elevations, and 3D views, you can set the work plane based on the location of the elements that are to be created or edited. The concept of work planes is especially useful for creating elements in elevations, sections, or inclined planes.

Setting a Work Plane

You can set a work plane based on your project requirement. To set a work plane, choose the **Set** tool from the **Work Plane** panel in the **Architecture** tab; the **Work Plane** dialog box will be displayed, as shown in Figure 6-12. It shows the current work plane and assists you in specifying the parameters to set a new work plane. The **Name** radio button has a drop-down list of available views. Select an available work plane from the drop-down list, which contains the names of levels, grids, and named reference planes. You can select the **Pick a plane** radio button in the **Work Plane** dialog box to set a work plane along an existing plane. On selecting this radio button and choosing the **OK** button in this dialog box, you will be prompted to select an existing plane in the drawing to which you want to align the work plane. The existing plane can be a face of a wall, floor, or roof. You can also select the **Pick a line and use the work plane**

it was sketched in radio button from this dialog box to create a work plane that is coplanar with the plane on which the selected line was created.

Note

In the project environment of Autodesk Revit, you can choose the ***Viewer*** *tool from the* ***Work Plane*** *panel of the* ***Architecture*** *tab to display the* ***Workplane Viewer*** *window. In this window, you can modify the workplane dependent element easily.*

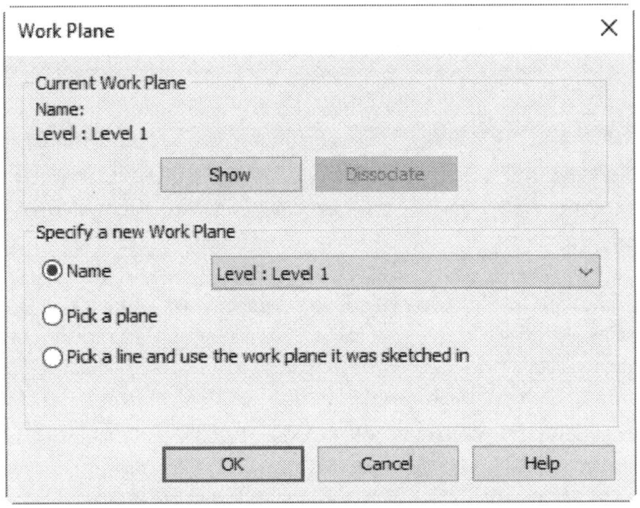

Figure 6-12 *The* ***Work Plane*** *dialog box*

Controlling the Visibility of Work Planes

You can control the visibility of the current work plane by choosing the **Show** button from the **Work Plane** panel of the **Architecture** tab. As a result, the work plane appears as a grid in the current view, as shown in Figure 6-13. To hide it, choose the **Show** button again.

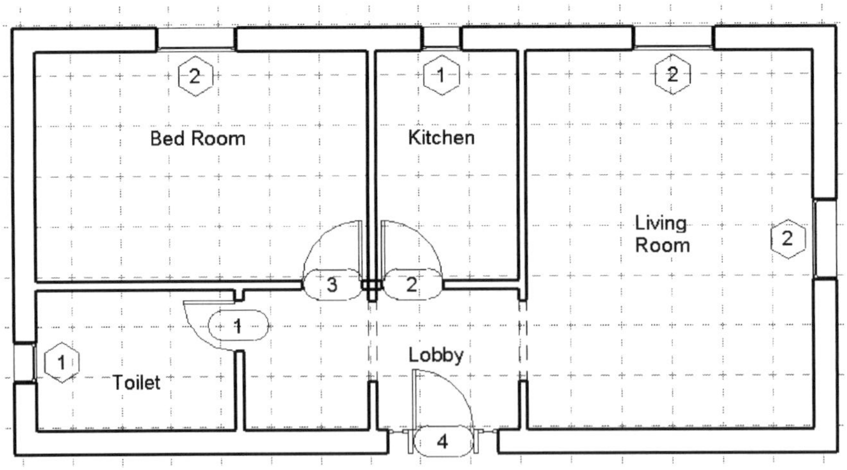

Figure 6-13 *The display of the work plane as a grid*

You can also set the grid spacing for a work plane by selecting it from the drawing. On doing so, the work plane gets highlighted. Specify the spacing by entering the new value in the **Spacing** edit box in **Options Bar**. You can snap to the work plane grid using the object snap tools. For example, these tools can be used to create wall profiles based on a square grid or a flooring pattern for an inclined ramp.

> **Tip**
> *The drop-down list, next to the **Name** radio button in the **Work Plane** dialog box, shows all levels, grids, and reference planes that have been created in the project. If you select a plane that is perpendicular to the current view, the **Go to View** dialog box will be displayed and you can select the appropriate view from it.*

WORKING WITH PROJECT VIEWS

While working on the building model, you may need to view its different exterior and interior portions in order to add or edit the elements in the design. Revit provides various features and techniques that can be used to view the building model. In this section, you will recapitulate some concepts introduced earlier and also learn about the tools that will help you in working with the views.

Viewing a Building Model

The default template file has certain predefined standard project views which are displayed in the **Views** head of the **Project Browser**, as shown in Figure 6-14. These include floor plans, ceiling plans, and elevations. To open any of these views, double-click on the name; the corresponding view will be displayed in the viewing area. You can control the visibility of the **Project browser** by selecting the **Project browser** check box from **View > Windows > User Interface** drop-down.

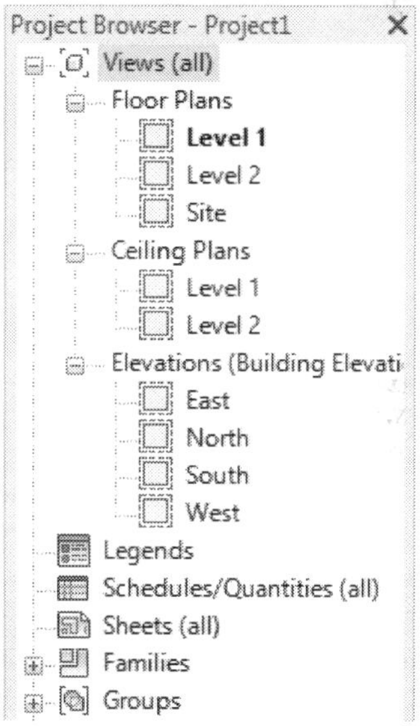

***Figure 6-14** The **Project Browser** showing the standard building views*

When you open a new project, the viewing area displays four inward arrow symbols in the floor plan view, which indicate the four side elevations: North, East, South, and West. You can use these symbols to view the appropriate building elevation by double-clicking on them.

Apart from these standard building views, you can use different viewing tools to view the building model from various angles. As described earlier (see Chapter 2), you can use the **Zoom** tool by choosing it from the navigation bar to zoom-in or zoom-out of the current view. You can also use the **Orient** tool to view the model using the preset standard viewpoints. The **SteeringWheels** and **ViewCube** tools assist you in navigating and viewing the model with various options. You can pan, zoom, walk, and change viewpoints using these tools.

To restrict the visibility of certain categories of elements, choose the **Visibility/ Graphics** tool from the **Graphics** panel of the **View** tab; the **Visibility/ Graphic Overrides** dialog box for the specific view will be displayed. The **Model Categories** tab of this dialog box contains a list of model elements such as walls, doors, windows, and so on, whereas the **Annotation Categories** tab contains annotations such as dimensions, door tags, and so on. You can clear the category of elements that you want to hide from the current view using this dialog box.

Visibility/ Graphic Overrides of an Element

You can override the visibility and graphics of any element in a view. To do so, select an element and right-click; a shortcut menu will be displayed. Choose **Override Graphics in View > By Element** from the shortcut menu; the **View-Specific Element Graphics** dialog box will be displayed, as shown in Figure 6-15.

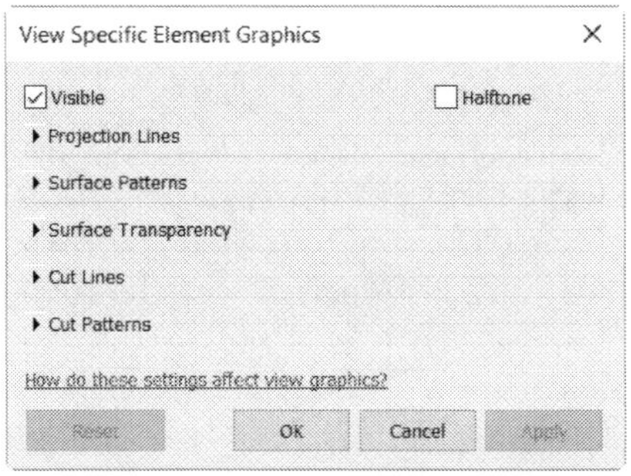

***Figure 6-15** The **View-Specific Element Graphics** dialog box*

> **Tip**
> *You can also use the keyboard shortcuts **VV** and **VG** to open the **Visibility/ Graphics Overrides** dialog box.*

Visibility/ Graphic Overrides of an Element Category

To edit the visibility and graphics of an element category, open the required project view. Next, select an element of the required category and right-click; a shortcut menu will be displayed. Choose **Override Graphics in View > By Category** from the shortcut menu; the **View-Specific Category Graphics** dialog box for the loaded view will be displayed. In this dialog box, choose the **Open the Visibility Graphics dialog** button; the **Visibility/ Graphic Overrides for Floor Plan** dialog box will be displayed, refer to Figure 6-16.

The dialog box contains the tabs to edit the visibility and graphic display of the models, annotations, and imported elements. Choose the required tab to edit the visibility and graphics of the selected category. In the **Visibility** column, clear or select the visibility check box available on the left of the category name to edit the visibility in the view. To specify overrides to a category, click on the name of the category in the **Visibility** column of the **Visibility/ Graphics Overrides for Floor Plan** dialog box; the **Overrides** button will be displayed for the respective columns in the specified tab. To override patterns for the selected category, choose the **Override** button displayed in the **Patterns** column corresponding to the selected category; the **Fill Pattern Graphics** dialog box will be displayed. Set the visibility, color, and pattern for the **Visible**, **Color**, and **Pattern** options in the dialog box and choose the **OK** button. Similarly, to override the transparency of the selected object, choose the **Override** button displayed in the **Transparency** column; the **Surfaces** dialog box will be displayed. In this dialog box, you can use the **Transparency** slider to control the transparency of the selected element.

Similarly, click in the required columns in the **Visibility/ Graphics Overrides for Floor Plan** dialog box to edit the graphic display of the category. Select the check boxes in the **Halftone** column if required. You can set the detail level of the selected category in the **Detail level** column. To do so, click in the **Detail Level** column and choose the down arrow displayed on the left. Select the detail level from the **Detail Level** drop-down list. Choose the **Apply** button to view the changes in the visibility and graphic display of the selected category. Choose the **OK** button to retain the changes and close the **Visibility/ Graphics Overrides for Floor Plan** dialog box.

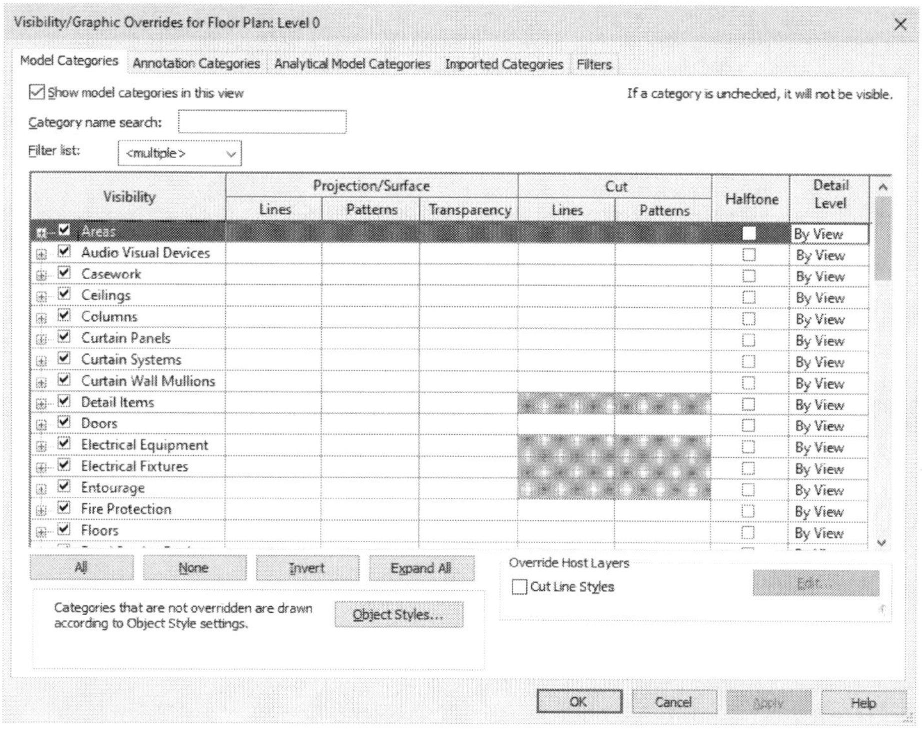

Figure 6-16 *The* ***Visibility/ Graphic Overrides for Floor Plan*** *dialog box*

Making Elements Transparent

Revit provides you with a tool to make elements transparent so that you can see through an object. This tool can be used to view the interior of a building from the top, even after adding roofs or ceilings.

To make an element transparent, select the element and right-click; a shortcut menu will be displayed. Choose **Override Graphics in View > By Element** from the shortcut menu; the **View Specific Element Graphics** dialog box will be displayed. In this dialog box, choose the arrow on the left of the **Surface Transparency** area; the options in this area will be displayed. In this area, you can use the **Transparency** slider to control the transparency of the selected object. You can also increase the transparency by moving the slider toward right. Alternatively, you can enter a desired value in the edit box, next to the **Transparency** slider, to control the transparency of the selected element. After setting the transparency in the **Surface Transparency** area, choose the **OK** button; the selected element will become transparent and you will be able to see through it. Although the element becomes transparent, the edges and surface pattern of the element will still be visible. You can view the change before and after using the Transparency slider, refer to Figures 6-17 and 6-18.

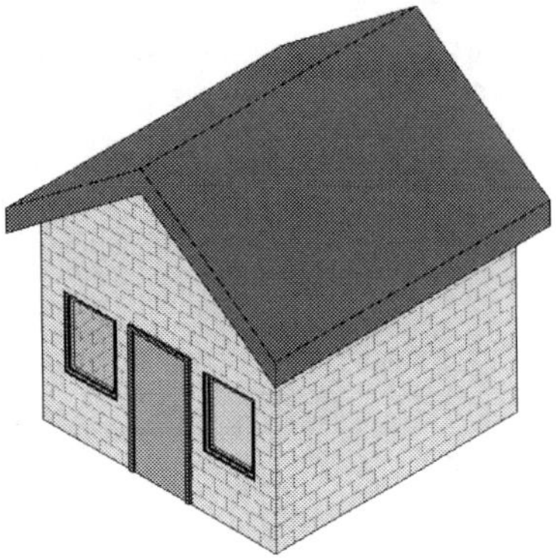

***Figure 6-17** Elements before selecting the **Transparent** check box*

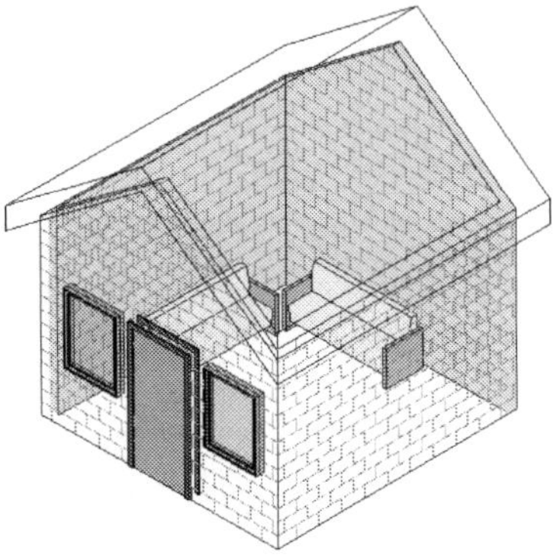

***Figure 6-18** Elements after selecting the **Transparent** check box*

Using the Temporary Hide/Isolate Tool

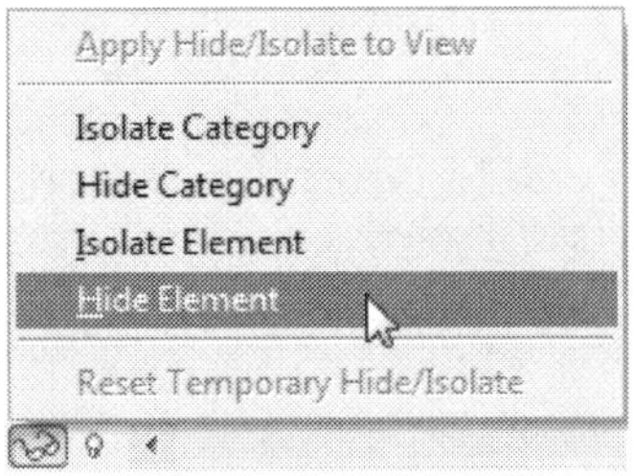

***Figure 6-19** Choosing the **Hide Element** tool*

The **Temporary Hide/Isolate** tool can be used to hide or isolate elements temporarily from a project view. This tool will be available only after a selection has been made. Select the element or elements that you need to hide or isolate in the project view and then choose the **Temporary Hide/Isolate** tool from the **View Control Bar**; a cascading menu will be displayed. You can hide or isolate elements or their categories using the tools given in the cascading menu. The cascading menu contains six tools. On choosing any of the tools from it, a cyan color border will be displayed in the drawing area, indicating that the elements in the drawing are in the **Temporary Hide/Isolate** mode. You can choose the **Hide Element** tool from the cascading menu, as shown in Figure 6-19, to hide the selected element in a view. On choosing the **Isolate Element** tool, only the selected elements will be displayed while the rest will be hidden in the view. The **Hide Category** tool is used to hide all the elements of the category of the selected element. The **Isolate Category** is used to display only the elements belonging to the category of the selected element in the view.

The **Reset Temporary Hide/Isolate** tool is used to revert to the original view without saving the temporary hide or isolate changes in the view. The **Apply Hide/Isolate to view** option is used to hide or isolate the temporary hidden or isolated elements permanently in the view. After you choose this tool, the blue boundary becomes invisible and the temporarily hidden or isolated elements and categories are permanently hidden or isolated. On hiding or isolating an element or category, the sunglasses symbol of the **Temporary Hide/Isolate** tool in the **View Control Bar** is highlighted in the same color as that of the boundary, indicating that certain elements have been hidden temporarily. When you choose the **Reset Temporary Hide/Isolate** option, the sunglasses symbol is no longer highlighted.

Plan Views

In Autodesk Revit, you can use the floor and ceiling views to view the building model plan. By default, Autodesk Revit opens the floor plan view when you create a new project. The default template file contains two floor plans and two corresponding ceiling plans. These views are enlisted under the **Floor Plans** and **Ceiling Plans** heads in the **Project Browser**.

Elevation Views

An elevation view refers to the view of the building model from four sides, North, East, South, and West. The four side elevation views are automatically created by Autodesk Revit when the default template file is used. Using the elevation view, you cannot only visualize the building from its exterior but also create the views of the interior walls of various internal spaces.

For example, for a residential building, you can easily view the four side elevations and also create the elevation views of the walls of the interior spaces such as kitchen and bathroom. In the plan view, the elevation is represented by the elevation symbol pointing in the direction of the elevation view side.

Section Views

Section views are generated by cutting sections through the building model. These views are created to display various wall elevations, floor heights, and special vertical features of the project. They are also useful in creating and editing elements added to the interior spaces of the building model.

For example, in an office building, to emphasize the salient features of the central atrium, you may need to show a section through the central atrium. Autodesk Revit enables you to create it with relative ease. You can also modify the sectional view to create a section that displays the interior spaces.

Using the Scope Box Tool

With the **Scope Box** tool, Autodesk Revit provides you the option of controlling the visibility of the datum elements in the project views. As described earlier in this chapter, the datum planes have an infinite scope and extend throughout the project. Using the **Scope Box** tool, you can limit the boundary for the visibility of the datum planes. You can also specify the views in which these datum planes become visible.

Note *The **Scope Box** tool is activated only in the plan view and it is view specific.*

TUTORIALS

Tutorial 1 Apartment 1

In this tutorial, you will add levels and grids to the *c05_Apartment1_tut1.rvt (M_c05_Apartment1_tut1.rvt* for Metric*)* project. Also, you will create wall elevation for the kitchen walls and sections using the Figure 6-20 as reference. Do not create the dimensions and text as they have been given only for your reference. Use the following project parameters:

(Expected time: 30 min)

1. Rename Level 1 as **First Floor** and Level 2 as **Second Floor**

2. Levels to be added:

   For Imperial **Third Floor- Elevation 20'0"**
   Fourth Floor- Elevation 30'0"
   Roof Floor- Elevation 40'0"

   For Metric **Third Floor- Elevation 6000 mm**
   Fourth Floor- Elevation 9000 mm
   Roof Floor- Elevation 12000 mm

   Note that in the metric template level 2 is at 4000 mm. Change that level to 3000 mm and then add further levels.

3. Grids and sections to be created using Figure 6-20 as reference. The horizontal section to be renamed as **Section X** and the vertical section as **Section Y**.

The following steps are required to complete this tutorial:

a. Open the file created in Chapter 5.
   For Imperial *c05_Apartment1_tut1.rvt*
   For Metric *M_c05_Apartment1_tut1.rvt*
b. Add levels using the **Level** tool.
c. Add grids using the **Grid** tool.
d. Create section views using the **Section** tool.
e. Create and view the four interior wall elevations of the kitchen walls.
f. Save the project using the **Save As** tool.
   For Imperial *c06_Apartment1_tut1.rvt*
   For Metric *M_c06_Apartment1_tut1.rvt*
g. Close the project.

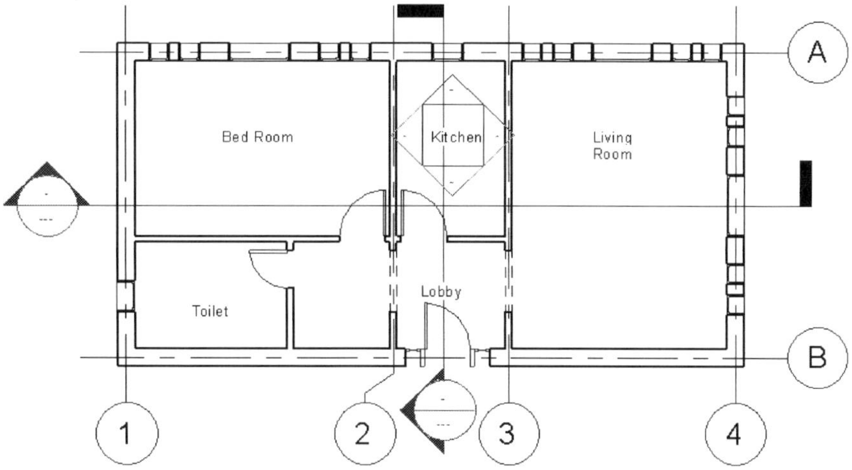

Figure 6-20 Sketch plan for adding grids, sections, and elevations to the Apartment 1 project

Opening an Existing Project and Hiding Tags

Before starting this tutorial, you need to open the specified project, and then you need to hide the door and window tags to simplify the project view using the **Visibility/ Graphics** tool.

1. Choose **Open** > **Project** from the **File** menu and then open the *c05_Apartment1_tut1.rvt* file (for *Metric M_c05_Apartment1_tut1.rvt*), created in Tutorial 1 of Chapter 5. You can also download this file from *https://www.Cadsoft.com*. The path of the file is as follows: *Textbooks > Civil/GIS > Revit Architecture >Exploring Autodesk Revit 2025 for Architecture*

2. Ensure that the **Floor Plans** view for Level 1 is opened. Next, to hide the doors and windows tags, choose the **Visibility/ Graphics** tool from the **Graphics** panel of the **View** tab; the **Visibility/ Graphic Overrides for Floor Plan: Level 1** dialog box is displayed.

3. Choose the **Annotation Categories** tab from the dialog box and clear the check boxes for **Door Tags** and **Window Tags**.

4. Choose the **Apply** button to apply the changes and then the **OK** button to close the dialog box and return to the drawing window.

Adding Levels

In this section, you will create levels in the elevation view.

1. Move the cursor to the **Project Browser** and double-click on **North** under the **Elevations (Building Elevation)** head. The north elevation with the existing levels is displayed in the drawing window. Make sure that the **Hidden Line** option is selected in the **Visual Style** menu of the **View Control Bar**. Select the levels and drag the endpoints of the level lines near the building profile; the view is adjusted, as shown in Figure 6-21.

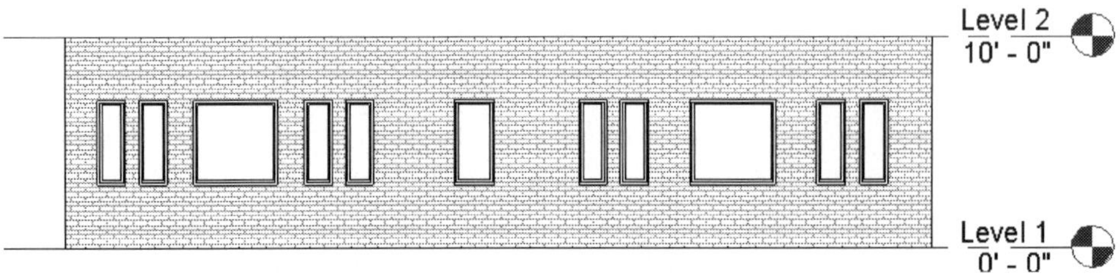

Figure 6-21 *The north elevation view of the Apartment 1 project*

2. Invoke the **Level** tool from the **Datum** panel of the **Architecture** tab.

Note

*In the metric unit system, by default, the Level 2 is at 4000 mm. Shift the Level 2 to 3000 mm by clicking on the value and entering the value as **3000**.*

3. Next, move the cursor near the left endpoint of the level line of Level 2 and then move it upward and when the alignment line appears, enter **10'0"(3000 mm)** as the distance of the new le

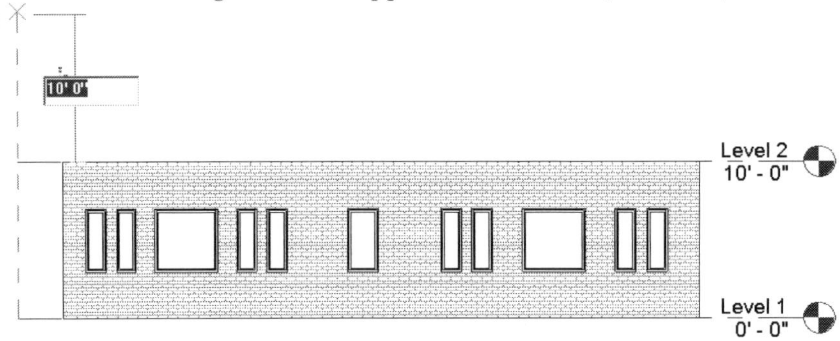

Figure 6-22 *Specifying the distance of the new level*

4. Move the cursor toward right until the alignment line is displayed above the Level 2 bubble, as shown in Figure 6-23, and click to complete the level line. The new level line is created showing the level name as **Level 3**, and the elevation as **20'0"(6000 mm)**. Press ESC to exit.

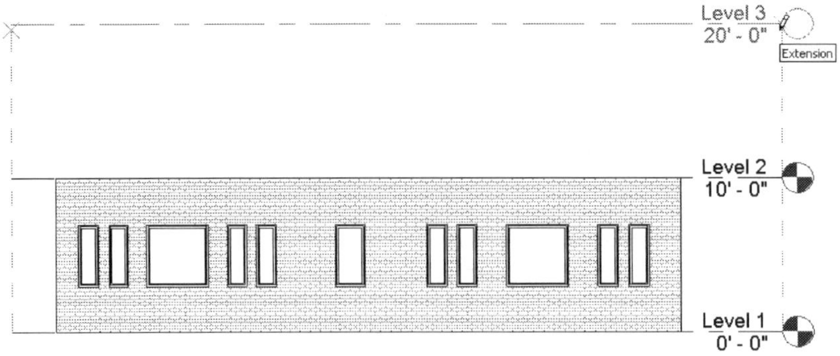

Figure 6-23 *Creating the new level line using the alignment line*

5. Repeat the procedure followed in steps 3 and 4 to create two more levels, Level 4 and Level 5 above Level 3 at a distance of **10'0"**(**3000 mm**) each. Press ESC twice to exit the **Level** tool.

   Notice that the **Project Browser** now shows the floor and ceiling plans for Level 3, Level 4, and Level 5.

6. To rename the levels, move the cursor over Level 1 under the **Floor Plans** head in the **Project Browser** and then right-click to display a shortcut menu.
7. Choose the **Rename** option from the shortcut menu; the edit box is displayed.
8. Enter **First Floor** in the edit box and choose the **OK** button. Autodesk Revit asks whether you want to rename the corresponding levels and views.
9. Choose the **Yes** button to rename the levels and views. Level 1 is renamed to **First Floor** in the elevation view.

10. Repeat the procedure followed in steps 6 through 9 to rename Level 2 as **Second Floor**, Level 3 as **Third Floor**, Level 4 as **Fourth Floor**, and Level 5 as **Roof**. Figure 6-24 shows all levels renamed in the north elevation.

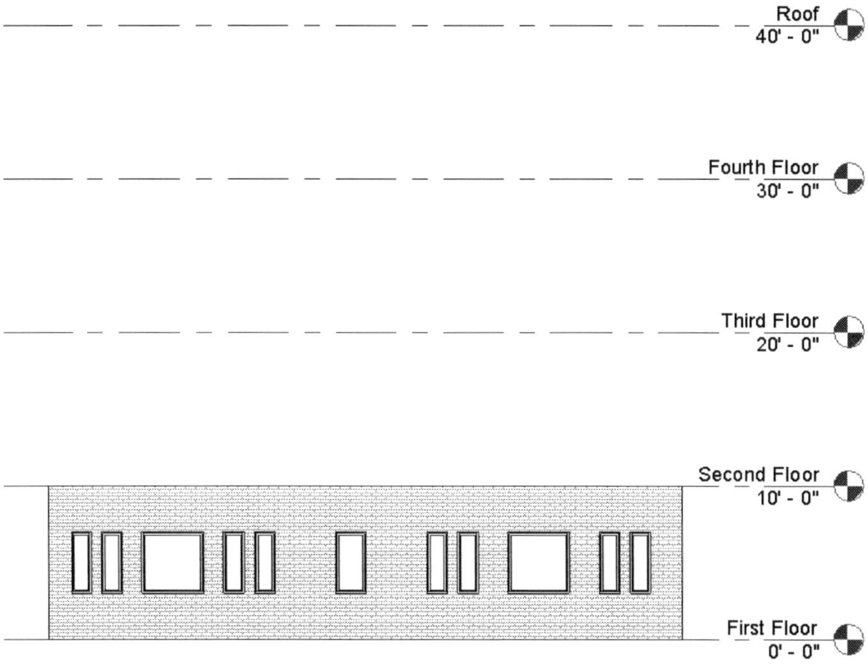

Figure 6-24 *North elevation showing the renamed levels*

Adding Grid Lines

You can use the plan view to add grids to the project using the **Grid** tool. Grids are automatically numbered as they are created. You will create grids in the sequence, as shown in the sketch plan.

1. Double-click on **First Floor** from the **Floor Plans** head in the **Project Browser** to display the ground floor plan in the drawing window.

2. Next, choose the **Grid** tool from the **Datum** panel of the **Architecture** tab; the **Modify | Place Grid** tab is displayed.

   Notice that the sketch plan shows the grid lines as the centerlines of walls. You can use the **Line** tool (default) from the **Draw** panel to create them.

3. Now, ensure that the **Line** tool is chosen in the **Draw** panel.

4. Move the cursor near the top left corner of the exterior wall profile and move the cursor up till a vertical extension line is displayed. Click to specify the start point of the grid line when the temporary dimension of **3'0"(900 mm)** is displayed from the centerline of the exterior wall, as shown in Figure 6-25.

5. Move the cursor vertically downward and click outside the south wall to specify the endpoint of the grid line, as shown in Figure 6-26.

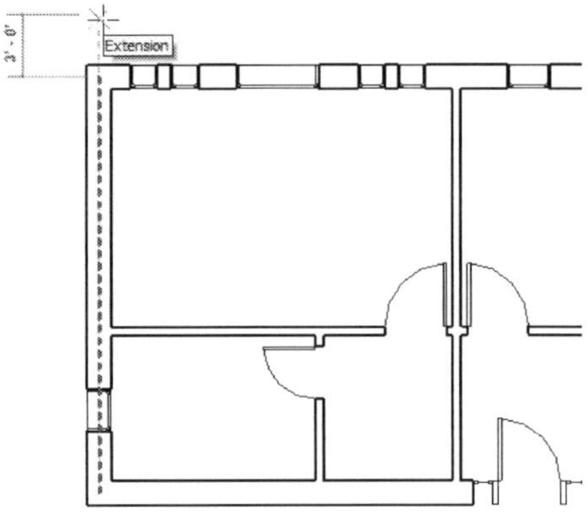

Figure 6-25 *Specifying the start point of the grid line*

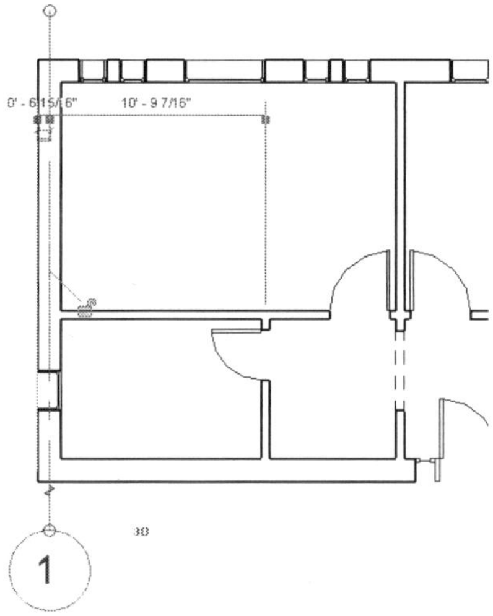

Figure 6-26 *Specifying the end point of the first vertical grid line*

Note

*If grid lines are not visible in the drawing, choose the **Visibility/ Graphics** tool from the **Graphics** panel of the **View** tab to display the **Visibility/ Graphic Overrides for Floor Plan** dialog box. In the **Annotation Categories** tab of this dialog box, select the **Grids** check box.*

The same procedure can be followed to draw grid lines for the interior walls. As the thickness of the interior wall is **5"(127 mm)**, you can specify **2 1/2"(64 mm)** as the offset distance to draw grid lines for the interior walls.

6. Repeat the procedure followed in steps 3, 4, and 5 to create other vertical and horizontal grid lines in the sequence of their numbers using the alignment line feature. After adding grid lines, press ESC twice to exit. Figure 6-27 shows the floor plan after adding grid lines.

7. To rename the grid 5, select it and then click on its name. The edit box with the current number is displayed.

8. Enter the value **A** in the edit box and press ENTER to rename the grid.

9. Repeat steps 7 and 8 to rename the grid 6 as **B**.

Note

You can rename a grid by selecting it from the drawing and then entering a value corresponding to the value column for the ***Name*** *parameter in the* ***Properties*** *palette. The entered value will be the new name of the grid.*

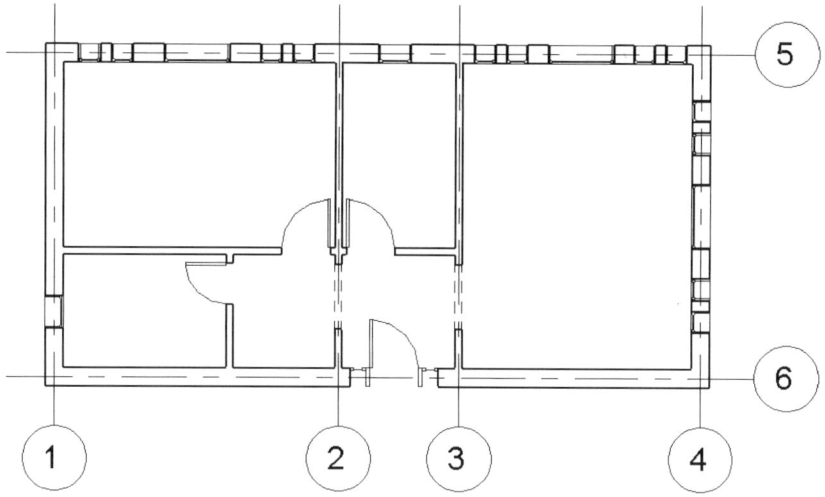

Figure 6-27 *The horizontal and vertical grids created for the wall centerlines*

Creating Section Views

The Section views can be created by using the **Section** tool.

1. Choose the **Section** tool from the **Create** panel of the **View** tab.

2. Move the cursor near the midpoint of the west wall. Click to specify the start point of the section line, refer to Figure 6-28.

3. Move the cursor horizontally across the building plan until you cross the exterior face of the east wall, as shown in Figure 6-28. Click to specify the endpoint of the section line. The section view is created and the name **Section 1** is displayed in the **Sections (Building Section)** head in the **Project Browser**.

4. Repeat the procedure followed in steps 2 and 3 to create a transverse section through the apartment building, as shown in Figure 6-29. Choose a point outside the main entrance as the start point and move the cursor vertically upward beyond the exterior wall of the kitchen to specify the endpoint of the section line. As the bubble of the grid line 2 touches the section line, you can move the grid bubble using the **Modify the grid by dragging its model end** tool.

5. Select the grid line 2 and click on the **Add elbow** symbol displayed near the grid bubble.

6. Drag the two blue dots to a location outside the exterior wall, refer to Figure 6-29. Press ESC to exit.

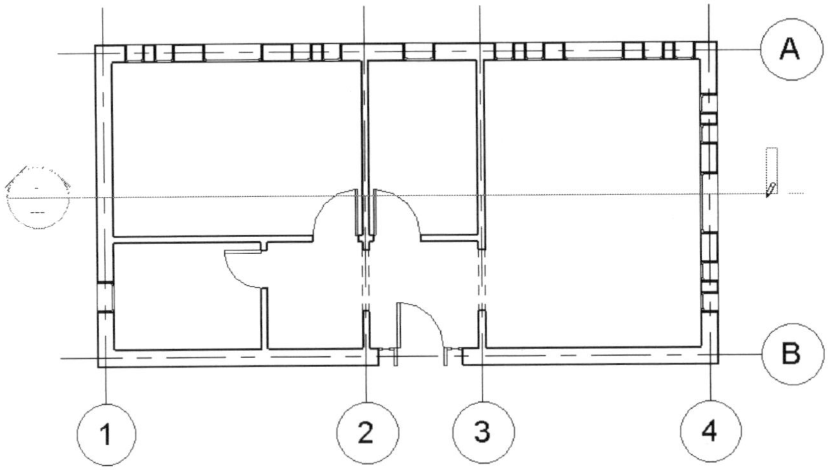

Figure 6-28 *Creating a section line for the section view*

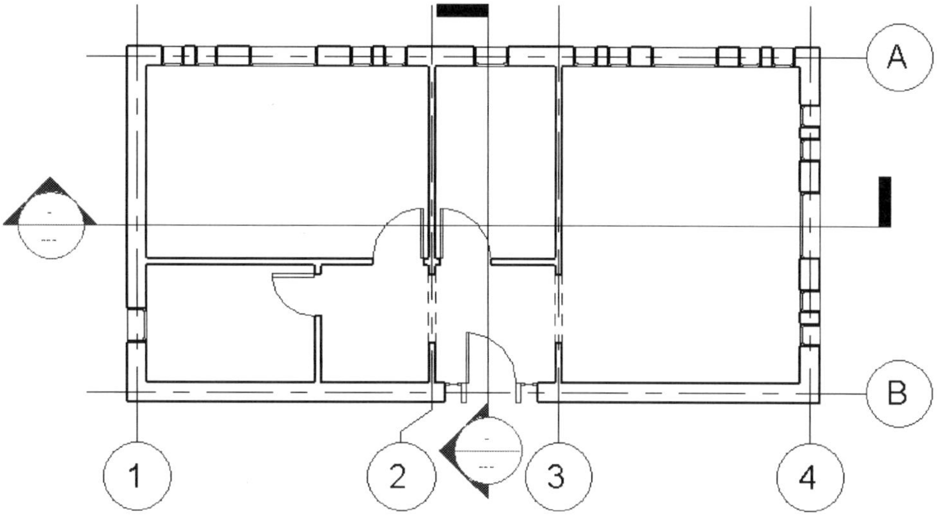

Figure 6-29 *The section lines created for the section views*

7. Click on the '+' symbol on the left of the **Sections (Building Section)** head in the **Project Browser** to display the section name created. Right-click on **Section 1** and then choose **Rename** from the shortcut menu; the edit box is displayed.

8. Enter **Section X** in the edit box and then choose **OK**; the section view is renamed.

9. Similarly, rename **Section 2** as **Section Y**.

10. Double-click on **Section X** to display the corresponding section in the drawing window, as shown in Figure 6-30.

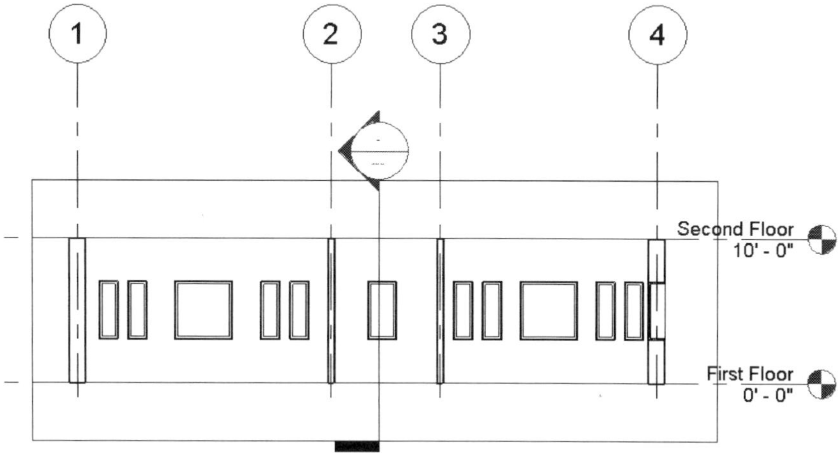

Figure 6-30 *Section view of* ***Section X***

Creating Elevation Views

To create the interior wall elevation of the kitchen walls, use the **Elevation** tool. You can use the first floor plan view to create new elevations.

1. Double-click on the name **First Floor** in the **Floor Plans** head of the **Project Browser** to display the first floor plan in the drawing window.
2. Choose the **Elevation** tool from **View > Create > Elevation** drop-down.
3. Move the cursor and place on a wall near the door in the kitchen area until the arrowhead points upward, toward the kitchen window and click to create the elevation.
4. Press the ESC key twice to exit the **Elevation** tool.
5. Select the elevation symbol to display its controls.
6. Click to select all the three cleared check boxes to create interior elevations. The new elevations are added under the **Elevations (Building Elevation)** head in the **Project Browser**.
7. Right-click the elevation views 1-a , 1-b , 1-c , and 1-d and using the corresponding shortcut menus, rename them as Kitchen-North Wall, Kitchen-East Wall, Kitchen-South Wall, and Kitchen-West Wall, respectively .
8. Double-click on the **Kitchen-North Wall** option in the **Project Browser** to display the corresponding elevation view. The crop-region can be picked by selecting the box. On selecting the box, drag controls appear. Drag the view crop region using the drag controls and grids of the kitchen area, as shown in Figure 6-31. Similarly, create the **Kitchen- South Wall** elevation, as shown in Figure 6-32.

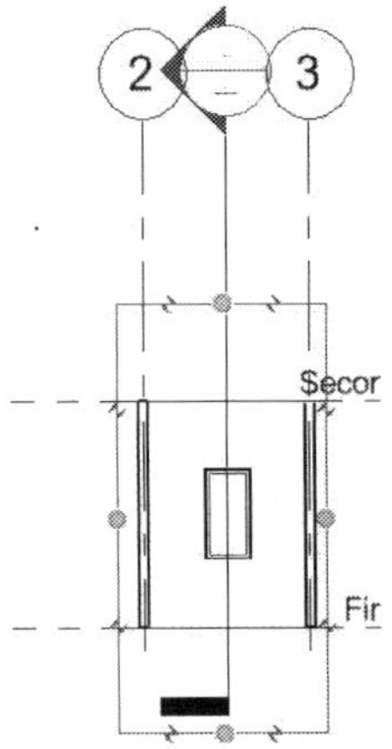

Figure 6-31 *The Kitchen-North Wall elevation*

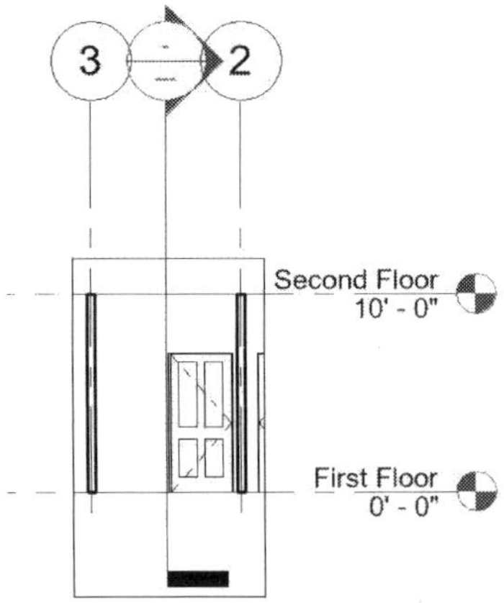

Figure 6-32 *The Kitchen-South Wall elevation*

Note

*If the crop-region boundary is not visible in the drawing, then choose the **Show Crop Region** button from the **View Control Bar**.*

9. Double-click on **First Floor** under the **Floor Plans** head in the **Project Browser** to return to the floor plan view.

Saving the Project

In this section, you will save the project file using the **Save As** tool.

1. Choose **Save As > Project** from the **File** menu; the **Save As** dialog box is displayed. Enter **c06_Apartment1_tut1** in the **File name** edit box and then choose **Save** to save the project.

2. Choose the **Close** option from the **File** menu.

EXERCISE

Exercise 1 Apartment 2

Add levels and grids to the *c05_Apartment2_ex1.rvt (M_c05_Apartment2_ex1.rvt* for Metric) project created in Exercise 1 of Chapter 5. Create the wall elevation for the dining room walls and also sections in the sketch plan shown in Figure 6-33. Do not create dimensions and texts as they are given only for your reference. Use the parameters given below.

(Expected time: 30 min)

1. Rename Level 1 as the **First Floor** and Level 2 as the **Second Floor**.
   In Metric unit system, shift the Level 2 to **3000** mm.
2. Levels to be added:

   For Imperial **Third Floor- Elevation 20'0"**
   Fourth Floor- Elevation 30'0"
   Roof- Elevation 40'0"

   For Metric **Third Floor- Elevation 6000 mm**
   Fourth Floor- Elevation 9000 mm
   Roof- Elevation 12000 mm
3. Grids and Sections to be created in the sketch plan shown in Figure 6-33. Name the horizontal section as **Section X1** and the vertical section as **Section Y1.**
4. Name of the file to be saved-

   For Imperial **c06_Apartment2_ex1**
   For Metric **M_c06_Apartment2_ex1**

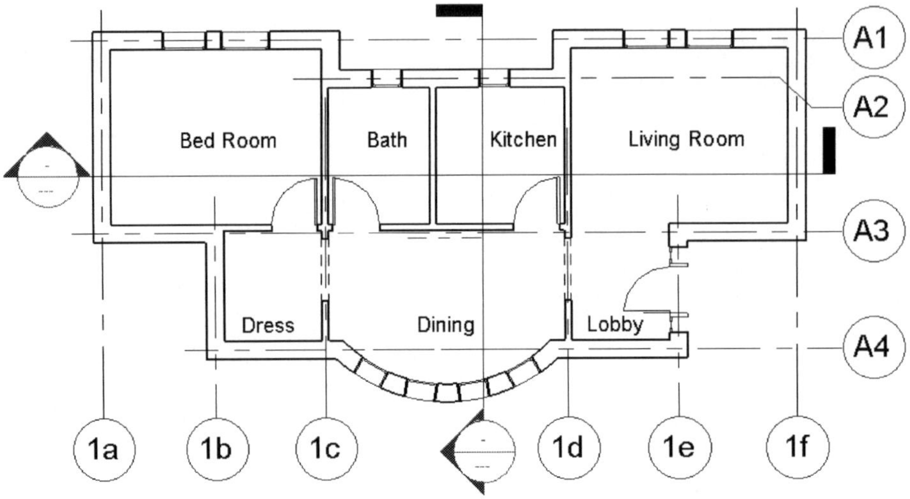

Figure 6-33 *Sketch for adding grids and sections to the Apartment 2 project*

Chapter 7

Using Basic Building Components-II

Learning Objectives

After completing this chapter, you will be able to:

- *Understand the concept of floor*
- *Create floors using the Floor tool*
- *Create roofs using the Roof tool*
- *Use the shape editing tools for floor, roof, and slabs*
- *Create ceilings using the Ceiling tool*
- *Create rooms using the Room tool*
- *Create openings in the floor, roof, and ceiling*
- *Join walls with the other walls and roofs*

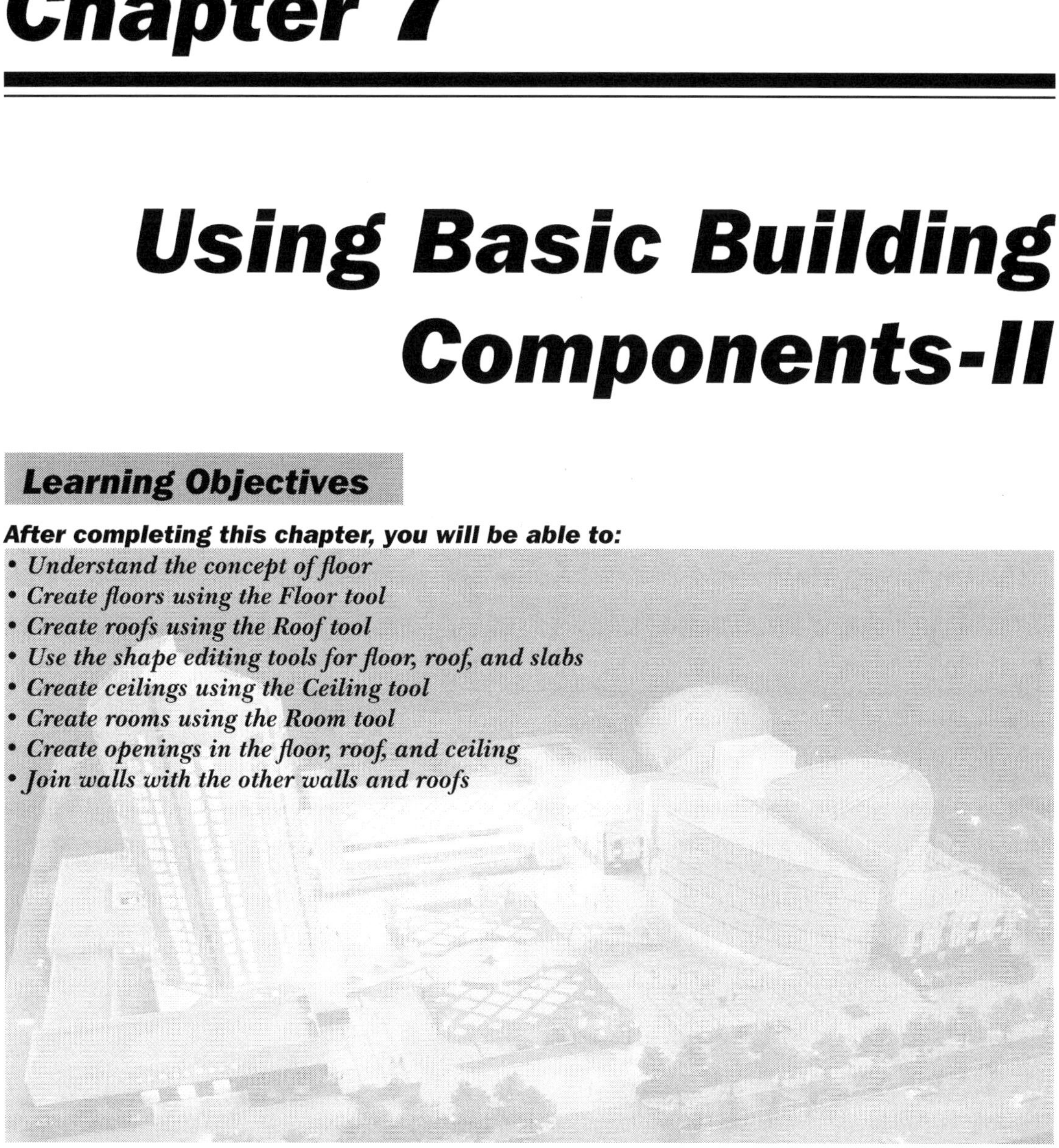

INTRODUCTION

In the earlier chapters, you learned to use levels, grids, and datums in a project. Also, you have learned that elements such as walls, doors, windows, and so on can be added from the library of in-built elements provided in Autodesk Revit. This chapter describes creation of sketched elements such as floors, roofs, and ceilings in a project. For creating these sketched elements, you need to draw sketch.

BIM
CONCEPTS

For a BIM project, floors and roofs are created as per the Level of Development specified in the EIR. Also, the rooms plays an essential part of the project to keep track of various BIM objects and also to assist in various spatial planning and analysis.

CREATING ARCHITECTURAL FLOORS

Ribbon: Architecture > Build > Floor drop-down > Floor: Architectural

You can add architectural and structural floors to your project. You can add an architectural floor to the current level of a building model using the **Floor: Architectural** tool. You can invoke this tool from the **Build** panel of the **Architecture** tab. On invoking this tool, the **Modify | Create Floor Boundary** contextual tab will be displayed, as shown in Figure 7-1.

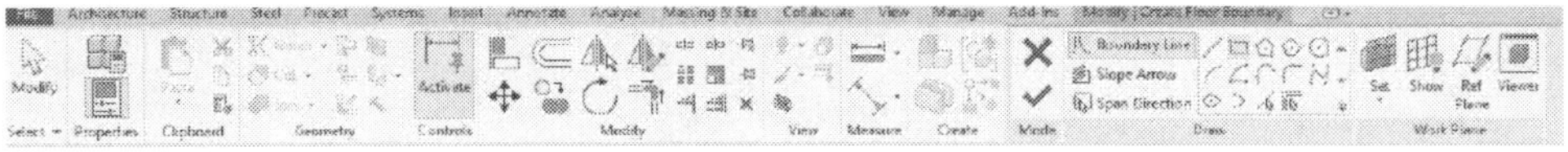

Figure 7-1 *The* ***Modify / Create Floor Boundary*** *tab*

You can use this tab to draw, annotate, and edit a floor boundary for your building model as well as to assign properties to it. The **Draw** panel of the **Modify | Create Floor Boundary** tab consists of various tools that are used to draw the floor sketches. These sketches define the boundary of the floor. To define the boundary of the floor, you can either pick the existing walls or sketch the boundary in the plan view by using lines or sketch the boundary in the 3D view, provided that the work plane is set to the plan view.

Similar to the other model components, the floor too has associated type and instance properties. Once you sketch the floor, the **Properties** palette will display the instance parameters of the floor, as shown in Figure 7-2. The **Type Selector** drop-down list in this palette displays the floor types available in Autodesk Revit's template file, refer to Figure 7-2. This palette can be used to modify the instance properties of the floor such as the level at which the floor is to be created, the height offset of the floor from the specified level, and so on. To modify the type properties of a floor, choose the **Edit Type** button from the **Properties** palette; the **Type Properties** dialog box will be displayed, as shown in Figure 7-3. You can set the type parameters for the floor in it. You can also create a new floor type by using the **Duplicate** button. To edit the structural elements of the floor, choose the **Edit** button in the **Value** column for the **Structure** type parameter; the **Edit Assembly** dialog box will be displayed, showing the structure of the floor type with its different layers. In the **Edit Assembly** dialog box, the **Insert** or **Delete** button can be used to customize the new floor type based on the specific project requirement.

Note

The type and instance properties vary depending upon the floor type selected. Autodesk Revit Help provides a detailed explanation of all the properties associated with the floor types.

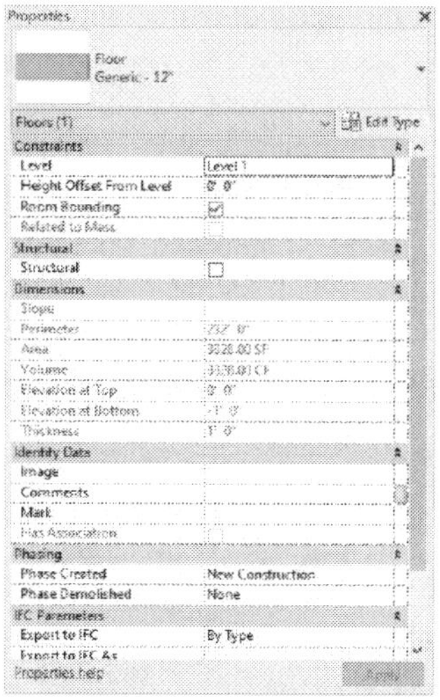

Figure 7-2 *Partial view of the* ***Properties*** *palette displaying the instance parameters*

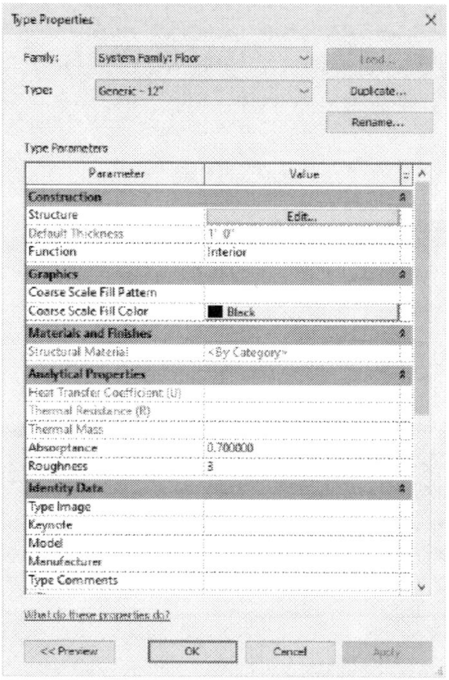

Figure 7-3 *The* ***Type Properties*** *dialog box*

Tip
*The **Preview** button in the **Type Properties** dialog box can be used to display a graphical image of the floor type which changes when you add or remove a layer to the floor structure.*

Sketching the Floor Boundary

To create a floor, you need to sketch its boundary. There are two methods to sketch a boundary. The first method is to pick the already created walls using the **Pick Walls** tool to define the floor boundary. The other method to define the floor boundary is to draw the floor profile using the draw tools such as line, rectangle, polygon, arc, and others from the **Draw** panel of the **Modify|Create Floor Boundary** tab.

By default, the **Pick Walls** tool is chosen in the **Draw** panel of the **Modify | Create Floor Boundary** tab. This tool can be used to sketch the floor for the spaces bound by the connected walls. The **Option Bar** displays the **Offset** edit box; which can be used to specify the offset distance of the floor sketch line from an existing wall. The **Extend into wall (to core)** check box is used to extend the floor to the wall core and assists in creating a joint between the floor and the wall core. If the **Pick Walls** tool is not chosen by default, then invoke this tool and move the cursor near the wall. You will notice that the wall is highlighted as the cursor is brought near its face. You can choose either of the faces of the wall to sketch the floor. Click when the wall will be highlighted at appropriate location; a magenta line with flip and drag controls as well as two parallel lines will be displayed on the wall, refer to Figure 7-4. The flip control can be used to flip the line between the two faces of the wall.

Similarly, you can select other walls to define the floor boundary. The sketched boundary must form a closed profile with all edges connected to each other.

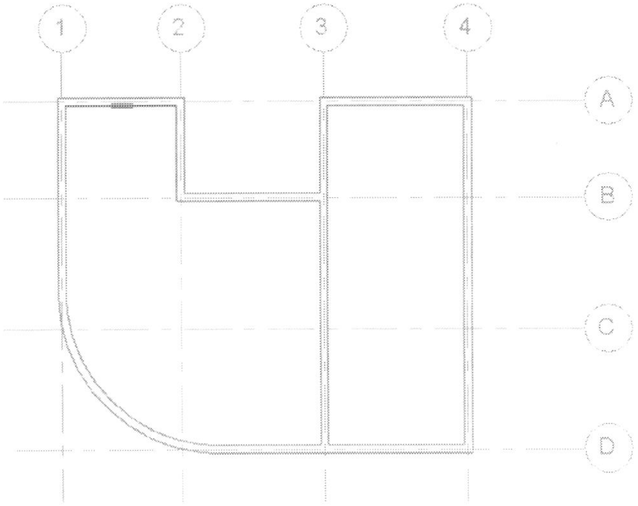

Figure 7-4 *A boundary line created from one of the wall faces while sketching the floor boundary*

CREATING ROOFS

A roof is a structure that covers the uppermost part of a building. The primary function of a roof is to protect the building and its contents from rain. The other functions of a roof depend upon the nature of the structure that it is protecting. For dwellings, a roof can protect against heat, cold, sunlight, and wind.

In Autodesk Revit, you can create a roof structure using various tools displayed in the **Roof** drop-down in the **Build** panel of the **Architecture** tab, as shown in Figure 7-5. You can use the extrusions or mass instances present in the drawing to create a roof.

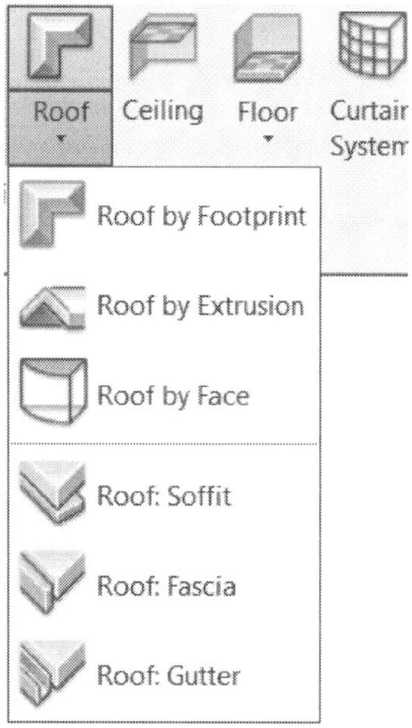

Figure 7-5 *Various tools displayed in the* ***Roof*** *drop-down*

Creating Roofs by Footprint

Ribbon: Architecture > Build > Roof drop-down > Roof by Footprint

The roof footprint is a two dimensional sketch of the perimeter of a roof. In Autodesk Revit, you can create a roof footprint by drawing lines or picking walls that will define the perimeter of the roof to be created.

The sketch of the footprint which is to be created, will be placed at the same level of the plan view where it was sketched. The sketch of the footprint should be a closed loop geometry that may also contain other closed loops, which will define the opening in the roof.

Creating Roofs By Extrusion

Ribbon: Architecture > Build > Roof drop-down > Roof by Extrusion

The second method of creating a roof is by using the **Roof by Extrusion** tool. You can sketch the roof profile and then extrude it to create the roof by using this tool. The profile can be created in the elevation view by using a specific work plane. Also, the profile created must have a series of connected lines that do not form a closed boundary.

To create a roof by extrusion, invoke the **Roof by Extrusion** tool from the **Build** panel; the **Work Plane** dialog box will be displayed, as shown in Figure 7-6. In this dialog box, select the work plane for sketching the roof profile.

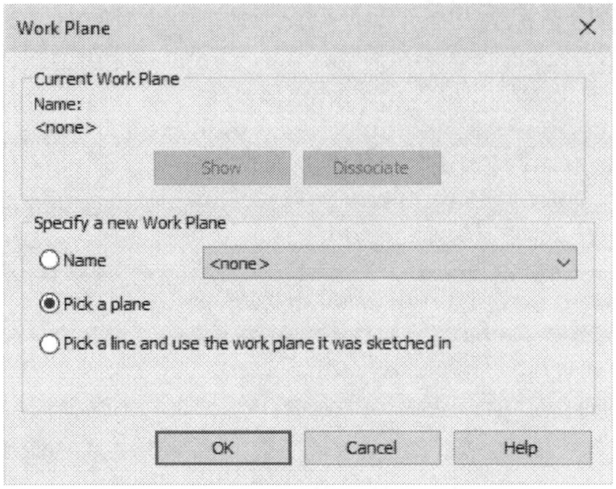

Figure 7-6 *Specifying the work plane for sketching the roof profile in the* ***Work Plane*** *dialog box*

Modifying Roof Properties and Editing Shapes

In Autodesk Revit, you can modify the properties and edit the shape of the roof that you have created for the building model. The properties of a roof can be classified into the following three categories:

a) **Roof Boundary Line Properties** - properties of the lines used in sketching the roof footprint such as length, slope angle, offset from the base, and so on.

b) **Roof Instance Properties** - instance properties of the selected roof such as maximum ridge height, base offset from level, and so on.

c) **Roof Type Properties** - properties of the roof, composition, and cost such as structure, thickness, material, and so on.

Tip

You can combine different roof shapes and join or unjoin them by selecting the corresponding roof and then choosing the ***Join Geometry /Unjoin Geometry*** *tool from the* ***Join*** *drop-down in the* ***Geometry*** *panel of the* ***Modify*** *tab.*

You can modify the roof instance and type properties by using the **Properties** palette. To modify these properties from the palette, select the desired roof; the properties related to it will be displayed in the **Properties** palette. You can select a roof type from the **Type Selector** drop-down list in this palette, as shown in Figure 7-7.

The **Properties** palette, as shown in Figure 7-8, lists various instance parameters under the following five main headings: **Constraints**, **Construction**, **Dimensions**, **Identity Data**, and **Phasing**.

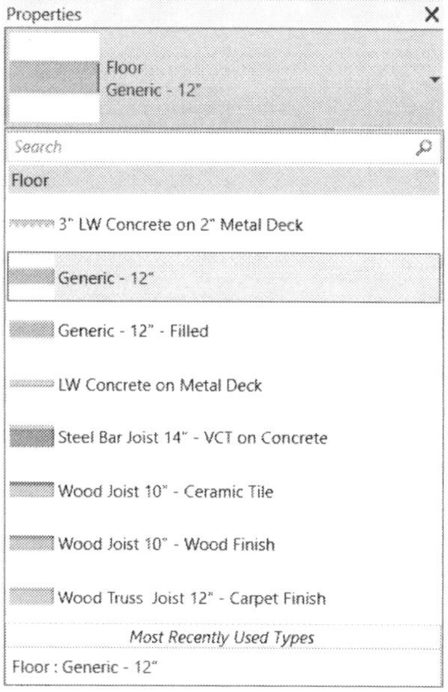

Figure 7-7 *Selecting an option from the* ***Type Selector*** *drop-down list*

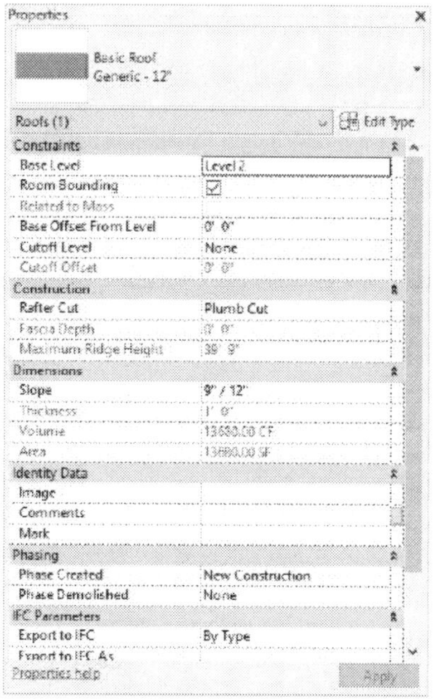

Figure 7-8 *Partial view of the* ***Properties*** *palette displaying various instance properties of a roof*

SHAPE EDITING TOOLS FOR FLOORS

Autodesk Revit provides you with a set of shape editing tools. Using these tools, you can modify the horizontal surface of floors, roofs (without slope), and structural floors by adding slopes to the surfaces for draining. You can add slopes to the horizontal surfaces without defining the sloping edges. To do so, you need to add low or high elevation points to the horizontal surface of the selected floor, roof, or structural floor, thereby splitting the surface into subregions, which in turn slope into different directions. These tools basically belong to the Revit Structure program, but you can also access these tools in Revit Architecture.

To access the shape editing tools, select the floor, roof, or structural floor created in the drawing; the shape editing tools will be displayed in the **Shape Editing** panel of the **Modify | Floors** tab of the ribbon, as shown in Figure 7-9. The shape editing tools are **Modify Sub Elements**, **Add Point**, **Add Split Line**, **Pick Supports**, and **Reset Shape**. These tools are discussed next.

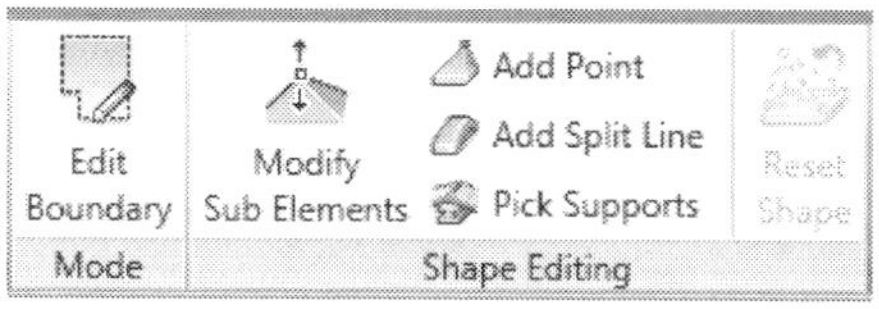

Figure 7-9 *The shape editing tools displayed in the* ***Shape Editing*** *panel*

Modify Sub Elements

This tool helps you select and modify the points or edges on a selected slab, floor, or roof surface. On invoking the **Modify Sub Elements** tool from the **Shape Editing** panel, a green colored boundary line and green colored corner grips will be displayed at the corners, as shown in Figure 7-10.

On selecting an edge, it will be highlighted in blue color and a drag control symbol consisting of two arrows will be displayed. Drag the arrows to move the selected point or edge vertically to different elevations. Figure 7-11 shows the drag control symbol displayed after selecting the left corner point and Figure 7-12 shows the floor after dragging the corner point vertically upward. You can also specify elevation for points or edges by typing the elevation value in the **Elevation** edit box in the **Options Bar**. If you select a boundary line, the drag control symbol will be displayed at the center of the boundary which can be dragged for specifying different elevations but the other two points of the floor which are at the ends of edges will remain at the same elevation.

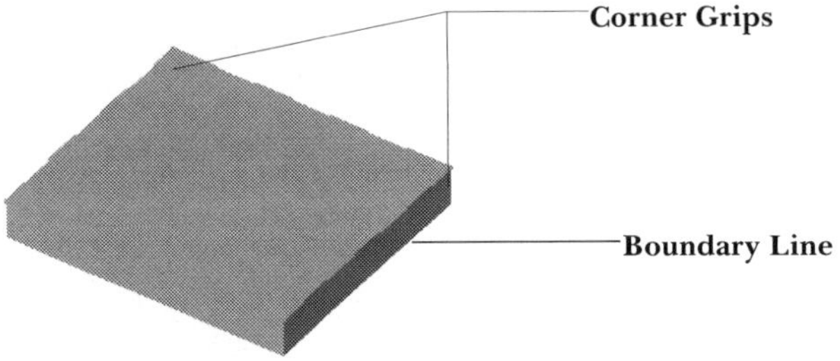

Figure 7-10 *The boundary line and the corner grips displayed after selecting the floor*

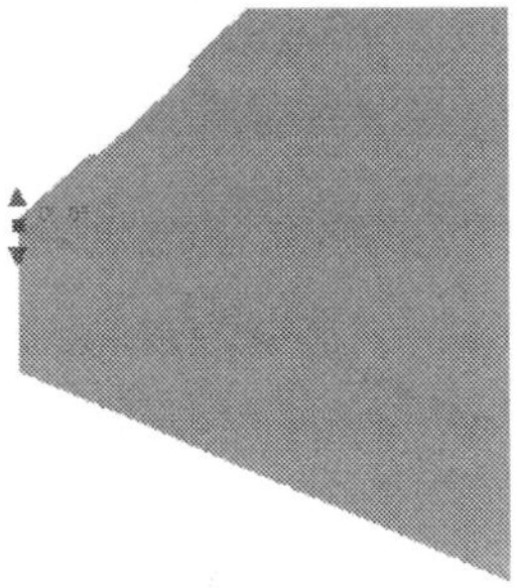

Figure 7-11 *Drag control symbol displayed after selecting the left corner point*

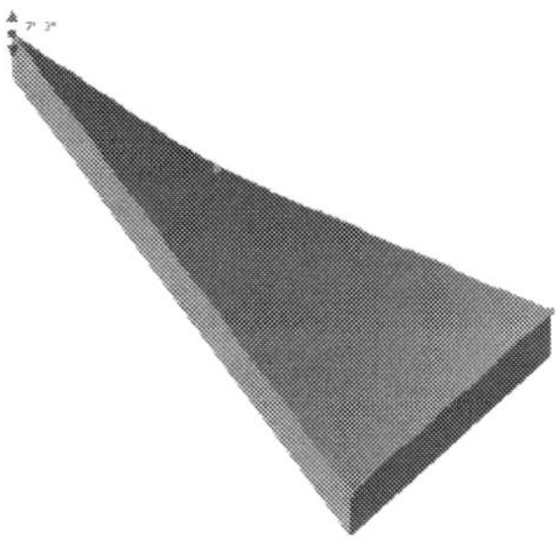

Figure 7-12 *The floor after dragging the left corner point vertically upward*

Note

The shape editing tools are available only after selecting the existing space floor, roof (without slope), or slab.

Add Point

This tool is used to assign a spot elevation by adding single or multiple points to the surface. These points are then used to modify the geometry of the selected slab, floor or roof. On invoking this tool, the **Elevation** edit box and the **Relative** check box will be displayed in the **Options Bar**. The **Relative** check box is selected by default, which means that the point will be added at a specified elevation relative to the surface to which it is added. You can specify the elevation value in the **Elevation** edit box. By default, the elevation value is **0'0"** which means that the points will be added on the surface on which they are created. On clearing the **Relative** check box, the points will be added on the surface at the elevation value displayed in the **Elevation** edit box. This value represents the project elevation. Figure 7-13 shows the floor after adding a point at the center and dragging the point vertically downward.

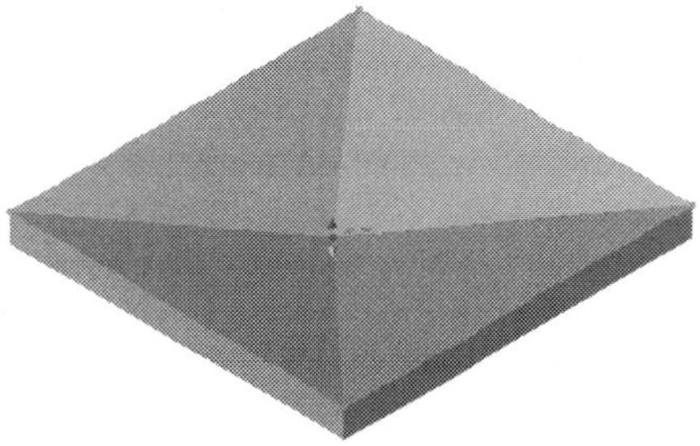

Figure 7-13 *Adding a point at the center of the surface and then dragging the point*

Add Split Line

This tool is used to split or divide the slab, floor, or roof surface into different sub regions by adding the split lines to the surface. These split lines act as edges and split the surface into different regions. You can then drag these split edges to different elevations, thereby providing slopes to these subregions. Each subregion created will slope independently in different directions. To add split lines, invoke this tool and select a vertex, edge, point, or face of the surface to start the split line. Select another vertex, point, or edge to complete the split line between the points, vertices, or edges. Figure 7-14 shows the floor shape after adding three split edges, one joining the two corners and the other two joining the centers of two edges.

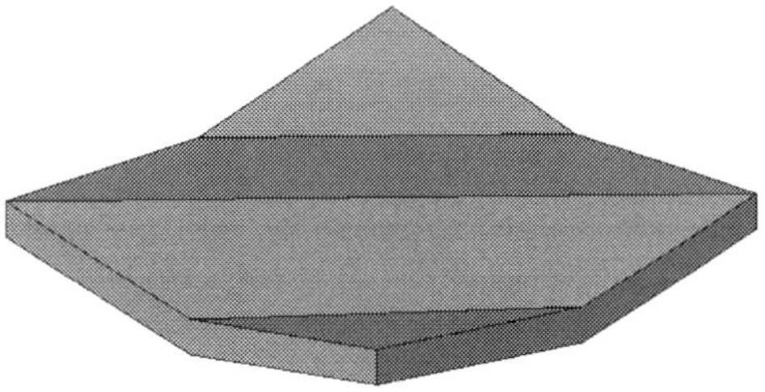

Figure 7-14 *The floor after adding three split lines and dragging them to different elevations*

Pick Supports

This tool is used to select the beams to define the split lines. Invoke this tool and select the beams from the drawing for reference; a new split edge will be created using the elevations of the endpoints of the picked reference. The elevations will shift from the bottom face to the top face of the selected slab or roof according to the slab or roof thickness.

Reset Shape

This tool allows you to reset the geometry and shape of the slab, floor, or roof. After invoking this tool, the selected floor, roof, or slab reverts to the original shape.

CREATING CEILINGS

Ribbon: Architecture > Build > Ceiling

You can add a ceiling to a building model by using the **Ceiling** tool. To do so, invoke this tool from the **Build** panel; the **Modify | Place Ceiling** tab with various tools for creating the ceiling will be displayed. Since the ceilings are not visible in the floor plan, they are created in the ceiling plan. You can add a ceiling to a project using three different methods: by automatic ceiling, by sketching the ceiling, and by using the pick walls method.

Creating an Automatic Ceiling

The first method to add a ceiling to a project is by creating an automatic ceiling. To do so, choose the **Ceiling** tool from the **Build** panel; the **Modify | Place Ceiling** tab will be displayed. In this tab, the **Automatic Ceiling** tool is selected by default in the **Ceiling** panel. In the **Properties** palette, select the ceiling type from the **Type Selector** drop-down list. You can select different types of built-in ceiling that can be used from this list. Now, to add the ceiling to the entire room, open the ceiling plan, move the cursor inside the room, and click when the ceiling boundary is displayed. Autodesk Revit will automatically create ceiling from the center of the room. For example, when you move the cursor inside the room for creating an automatic ceiling, its boundary is highlighted in red, as shown in Figure 7-15. Click to create the ceiling automatically. If you have created the ceiling in a floor plan, Autodesk Revit will display a warning indicating that the ceiling created will not be visible in it. You can then open the ceiling plan of the corresponding level to view the created ceiling, as shown in Figure 7-16. You can also open the section view through the room to view the cross-section of the ceiling.

Figure 7-15 *Selecting the room to create an automatic ceiling*

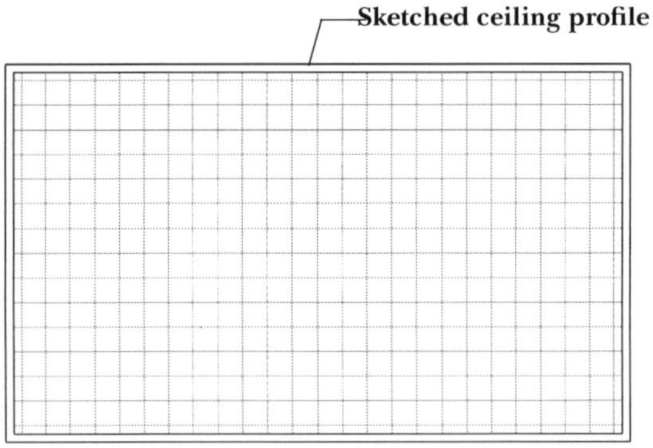

Figure 7-16 *Automatic ceiling created*

Sketching a Ceiling

You can sketch a ceiling boundary and create a ceiling. To do so, choose the **Ceiling** tool from the **Build** panel; the **Modify | Place Ceiling** tab will be displayed. In this tab, choose the **Sketch Ceiling** tool from the **Ceiling** panel; the **Modify | Create Ceiling Boundary** tab will be displayed. This tab contains various tools for sketching and modifying the ceiling boundary. To start the sketch of a ceiling boundary, choose the **Boundary Line** tool from the **Draw** panel; a list box containing various drawing tools will be displayed on the right in this panel, as shown in Figure 7-17. The **Line** tool is chosen by default in the **Draw** panel. You can use this tool and other drawing and modification tools displayed in the **Modify | Create Ceiling Boundary** tab to complete the sketch. Figure 7-18 shows a ceiling boundary that is sketched using various tools displayed in the **Modify| Create Ceiling Boundary** tab. After completing the sketch, choose the **Finish Edit Mode** button from the **Mode** panel of the **Modify | Create Ceiling Boundary** tab; the ceiling will be created within the sketched profile, as shown in Figure 7-19. Display the section view to view the ceiling in it. The ceiling will appear at a certain height from the floor level, as shown in Figure 7-20.

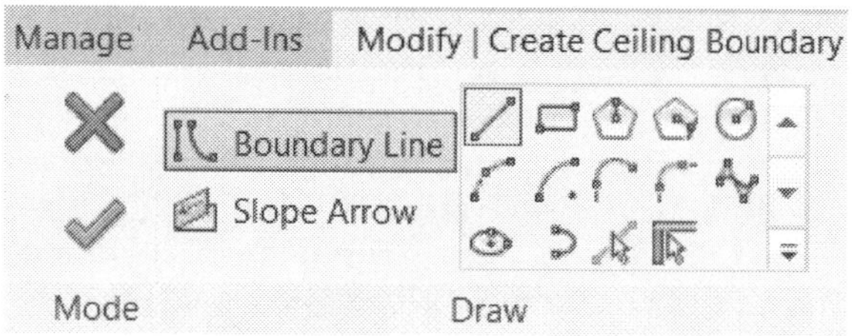

Figure 7-17 *The list box containing the tools to sketch the boundary line*

Figure 7-18 *Sketching the ceiling boundary*

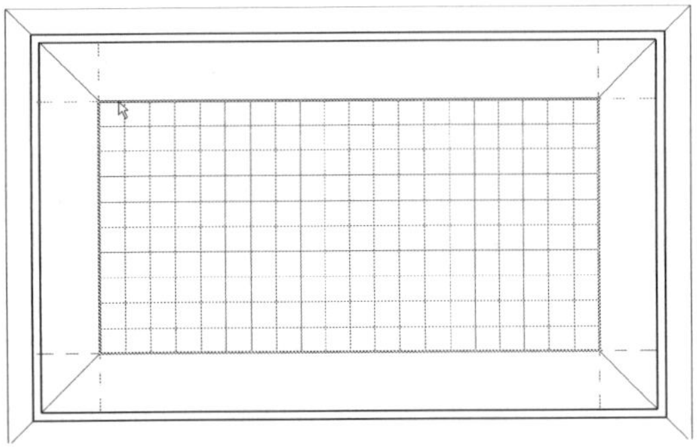

Figure 7-19 Ceiling created within the sketched boundary

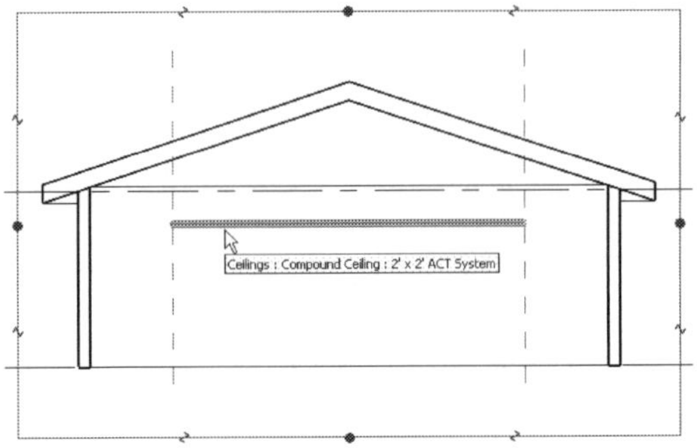

Figure 7-20 Ceiling displayed in the section view

Using the Pick Walls Method

The third method to create the ceiling is by picking the wall faces. This is a common method for creating ceilings. To create a ceiling by picking the wall faces, choose the **Ceiling** tool from the **Architecture** tab; the **Modify | Place Ceiling** tab will be displayed. In this tab, choose the **Sketch Ceiling** tool from the **Ceiling** panel; the **Modify | Create Ceiling Boundary** tab will be displayed. In this tab, choose the **Pick Walls** tool from the list box of the **Draw** panel; you will be prompted to pick the walls to define the boundary of the ceiling. Specify the offset value for the ceiling in the **Offset** edit box in the **Options Bar**. Select the **Extend into Wall (to core)** check box, if you want the offset to be measured from the core layer of the walls. Next, pick the walls to create the sketch for the ceiling. In the **Properties** palette, select the ceiling type from the **Type Selector** drop-down list. Next, choose the **Finish Edit Mode** button from the **Mode** panel of the **Modify | Create Ceiling Boundary** tab; the ceiling will be created.

Modifying a Ceiling

You can modify a ceiling after it has been created. To modify the ceiling, select it; the **Modify|Ceiling** tab will be displayed. From this tab, you can use various modification and editing tools to make changes on the selected ceiling. For editing the selected ceiling, you can use the editing tools such as **Copy**, **Rotate**, **Move**, and **Mirror**. You can also modify the properties of the selected ceiling based on its sketch. To do so, choose the **Edit Boundary** tool from the **Mode** panel; the **Modify | Ceiling > Edit Boundary** tab will be displayed and the sketch mode will be activated for the selected ceiling in the drawing. Now, you can modify the sketch profile by using the drag controls and other editing tools. Select the edge(s) of the ceiling boundary; the instance properties for the selected edges(s) of the ceiling profile will be displayed in the **Properties** palette. In this palette, you can enter the required values of various instance parameters for the selected edge(s) of the ceiling boundary.

You can view and edit the instance properties of a ceiling. To do so, select the ceiling from the drawing; the instance properties of the ceiling will be displayed in the **Properties** palette. This palette displays the instance parameters of the selected ceiling such as the level, height offset from level, and so on, as shown in Figure 7-21.

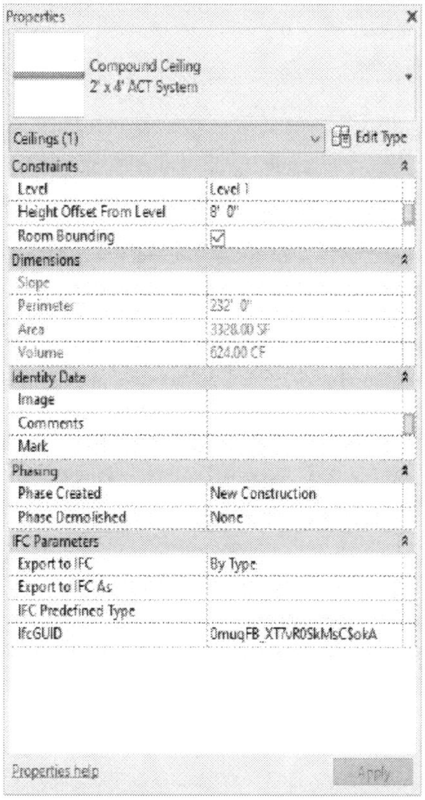

***Figure 7-21** The **Properties** palette for selected ceiling with various instance properties*

Being a level-based element, ceilings are created at a specified distance from the base level. This distance is specified in the **Height Offset From Level** instance parameter in the **Properties** palette. For example, to create a ceiling at 10' height from Level 2, you can select it as the value for the **Level** instance parameter and enter the value **10'** in the **Height Offset From Level** instance parameter.

ROOMS

A room in Revit is considered as spatial division of a project based on elements such as walls, floors, roofs, and ceilings. Revit can compute areas, volume and perimeter of the rooms and can schedule them as well. Autodesk Revit provides you the flexibility of creating rooms independent of room tags. Rooms can be created only in the plan view. You can also add rooms from the room schedules. Rooms and areas have the same graphical representation.

Adding Rooms

Ribbon: Architecture > Room & Area > Room

Figure 7-22 Invoking the ***Room*** *tool*

You can add rooms to the plan view by using the **Room** tool. To do so, invoke this tool from the **Room & Area** panel, as shown in Figure 7-22; the **Modify | Place Room** tab will be displayed. Now, as you move the cursor inside a closed boundary that you need to define as a room in the drawing, a symbol with a cross hair graphics attached to the cursor will appear. Notice that the size of this cross hair graphics will change according to the area of the room space of the closed boundary. Before you insert the room, you can select the type of room from the **Type Selector** drop-down list in the **Properties** palette. You can select any of the three options namely, **Room Tag**, **Room Tag With Area**, and **Room Tag With Volume** from the **Type-Selector** drop-down list. You can select the **Room Tag** option to display the inserted room with the tag only. Similarly, you can select the **Room Tag With Area** or **Room Tag With Volume** option from the **Type Selector** drop-down list to attach the room tag to the room to display the information regarding the area or volume of the room. After selecting the required option from the **Type Selector** drop-down list, choose the **Highlight Boundaries** tool from the **Room** panel of the **Modify | Place Room** tab; the **Autodesk Revit 2025** message box will be displayed, informing that the room bounding elements are highlighted in the drawing. Choose the **Close** button to close the message box.

Calculating Room Volumes

To enable the volume computations and calculate room volumes, choose the **Area and Volume Computations** tool from the **Room & Area** drop-down; the **Area and Volume Computations** dialog box will be displayed. In the **Volume Computations** area of the **Computations** tab, select the **Areas and Volumes** radio button. By default, the **Areas only** radio button is selected in the **Volume Computations** area. Next, choose the **OK** button; the volume computation of room will be enabled and displayed in its instance properties. To view the calculated volume of a room, select it from the drawing; the **Properties** palette will be displayed. In the **Properties** palette, the value in the value column of the **Volume** parameter shows the computed volume of the room.

Cutting Openings in a Wall, Floor, Roof, and Ceiling

In Autodesk Revit, you can create an opening in the wall, floor, roof, structural floor, ceiling, and structural elements such as beams and braces using any of the tools available in it. To create an opening, invoke any of these tools from the **Opening** panel of the **Architecture** tab, as shown in Figure 7-23. From this panel, you can choose any of these five options, **By Face**, **Wall**, **Vertical**, **Shaft**, or **Dormer** to create an opening.

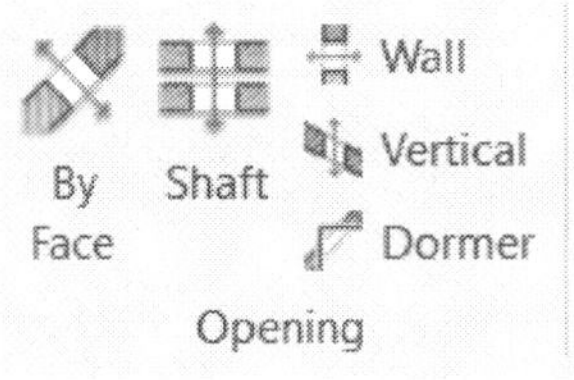

Figure 7-23 *Various opening tools in the* ***Opening*** *panel*

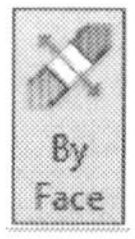

You can use the **By Face** tool to cut openings on the faces of roofs, floors, or ceilings in the building model. This tool is useful in a project when you need to cut an opening in a floor for stairway or skylight opening. You can also use this tool for creating any cut-out in the roof for a chimney. To do so, invoke this tool from the **Opening** panel; you will be prompted to select a planar face of a roof, floor, ceiling, beam, or column. Select the face on which you want to create an opening, as shown in Figure 7-24. Once you select the face, the **Modify | Create Opening Boundary** tab will be displayed. In this tab, you can use various tools to sketch the opening in any view. Now, open the required view and sketch the opening using the reference planes, as shown in Figure 7-25. After sketching the opening boundary on the selected face, choose the **Finish Edit Mode** button from the **Mode** panel to finish the sketch of the opening and then exit the **Modify | Create Opening Boundary** tab. Figure 7-26 shows the resulting opening perpendicular to the selected face.

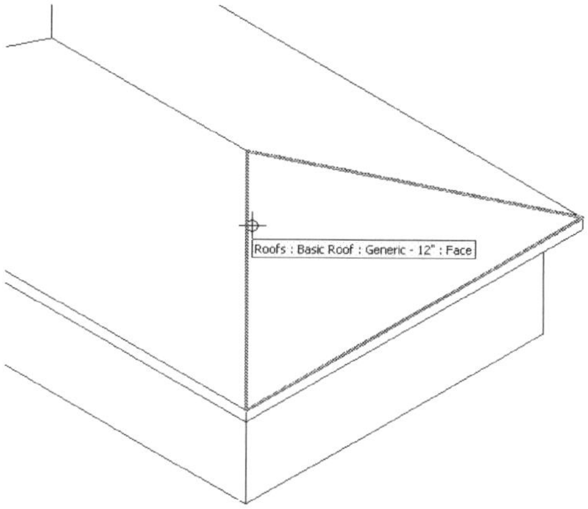

Figure 7-24 *Selecting the face to be cut of the roof*

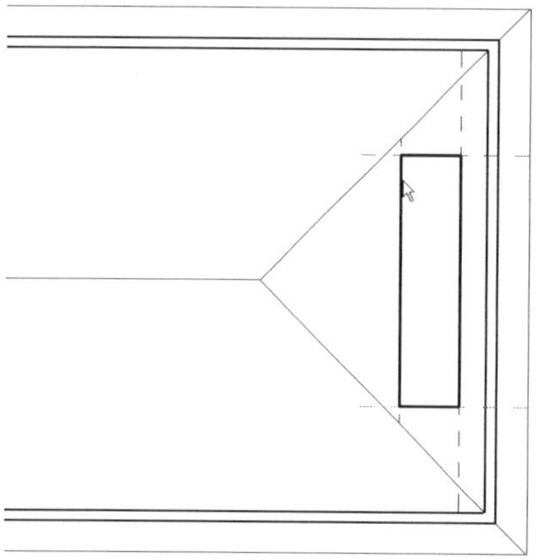

Figure 7-25 *Sketching the opening in the ceiling plan using the reference planes*

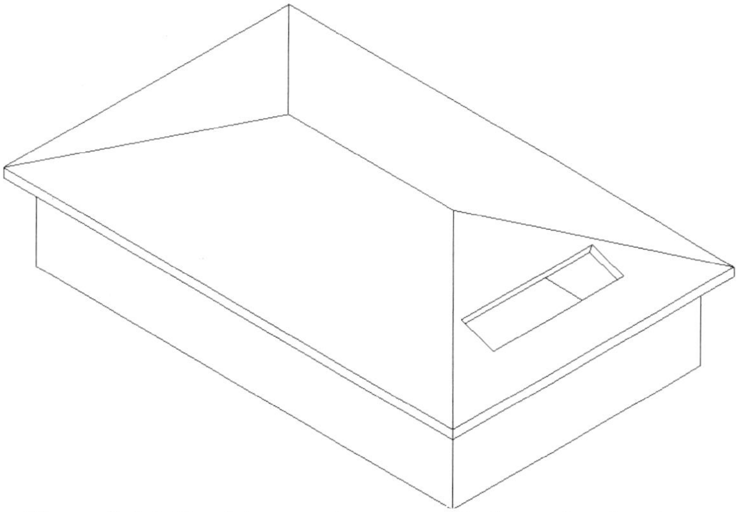

Figure 7-26 *Resulting opening perpendicular to the selected face*

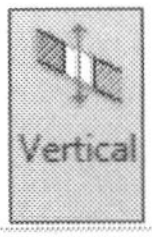

To cut a vertical opening, choose the **Vertical** tool from the **Opening** panel and select the required ceiling, roof, or floor; the **Modify | Create Opening Boundary** tab will be displayed. You can use various tools from this tab to sketch the boundary of the opening. Sketch the opening in the appropriate view using the sketching tools. After sketching the opening boundary, choose the **Finish Edit Mode** button from the **Mode** panel; a vertical opening will be created in the selected element. You can use the sketching tools to draw a sketch of appropriate size, as shown in Figure 7-27. The opening can also be viewed in the 3D view, as shown in Figure 7-28.

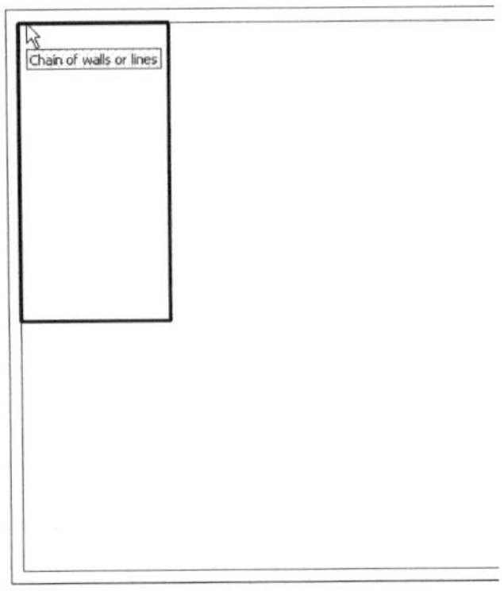

Figure 7-27 *Sketching the roof opening in the ceiling plan view*

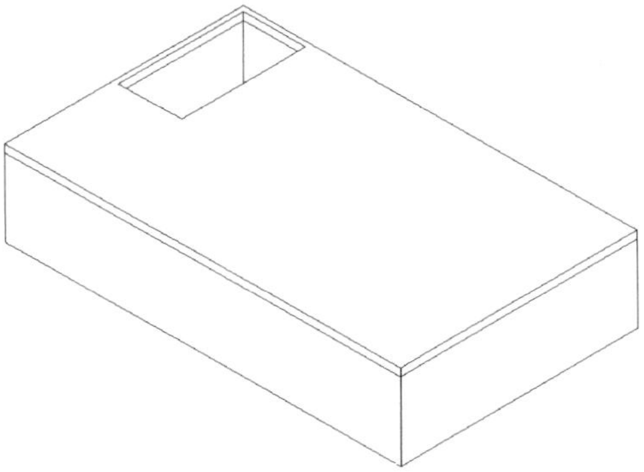

Figure 7-28 *Roof opening in the 3D view*

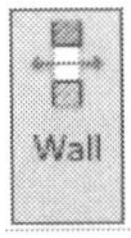

To create a rectangular opening in a wall, choose the **Wall** tool from the **Opening** panel in the **Architecture** tab; you will be prompted to select a wall. Select the wall in which you want to cut an opening and then sketch a rectangular opening of the required size by clicking and dragging the cursor.

To create a dormer opening in the dormer roof, choose the **Dormer** tool from the **Opening** panel; you will be prompted to select the roof in which you want to create a dormer opening. Select the roof; the screen will enter the sketch mode. Choose the **Pick Roof/ Wall Edges** tool from the **Pick** panel of the **Modify | Edit Sketch** tab and pick the boundary of the dormer to create the dormer opening. Next, choose the **Finish Edit Mode** button from the **Mode** panel; a dormer opening will be created.

Note

The boundary that you will pick for the dormer opening in a roof should be an edge of the selected roof, wall, or both, and should form a closed loop.

To cut an opening up to the entire height of a building, choose the **Shaft** tool from the **Opening** panel; the **Modify | Create Shaft Opening Sketch** tab will be displayed. Choose a suitable sketching tool from this tab to create an opening of the required shape. Next, choose the **Finish Edit Mode** button from the **Mode** panel; the opening will be created passing through the entire height of the building. Make sure that before sketching the opening, you select the required work plane and the view to sketch the opening.

You can also specify the levels that will be cut by the opening. It will help you to restrict the opening to a particular level. To specify the levels, select the opening; the **Modify | Shaft Openings** tab will be displayed. Select a level for the **Base Constraint** parameter in the **Properties** palette to start the opening. Next, select a level for the **Top Constraint** parameter to end the opening. The opening will be cut through the selected levels.

JOINING WALLS WITH OTHER ELEMENTS

In Autodesk Revit, walls are not attached to a floor and roof automatically. If needed, they can be attached using the **Attach Top/Base** tool. For example, using the **Attach Top/Base** tool, you can attach the structural layer of a wall to a floor without attaching its finish layer. Similarly, if you need to detach a wall, you can use the **Detach Top/Base** tool.

Using the Attach Top/Base and Detach Top/Base Tools

To attach walls to other elements, first select the walls to be attached; the **Modify | Walls** tab will be displayed. Next, choose the **Attach Top/Base** tool from the **Modify Wall** panel; the **Options Bar** will prompt you to choose whether you want to attach the element to the top or to the bottom of the wall. To attach the floor to the wall, select the **Base** radio button and to attach a roof or a ceiling, select the **Top** radio button. After selecting the required radio button, select the element(s) to be attached to the wall(s); they will be attached to the top/base of the selected wall(s) and the **Options Bar** will return to the previous options allowing you to attach or detach more elements to the already selected walls.

The wall attachment status can be checked from its instance properties. To do so, select the desired wall; the instance parameters of the selected wall will be displayed in the **Properties** palette. In this palette, the **Top is Attached** and **Base is Attached** instance parameters show the status of attachments for the selected walls, as shown in Figure 7-29. The selection of the check box corresponding to the **Top is Attached** parameter in the value column indicates that the selected wall is attached to an element at its top. Similarly, the selection of the check box displayed in **Value** column of the **Base is Attached** parameter indicates that the selected wall is attached to an element at its base. After you attach walls to other elements, you can detach them. To do so, invoke the **Detach Top/Base** tool from the **Modify** panel of the **Modify | Walls** tab. The usage of this tool is similar to the **Attach Top/Base** tool that has already been discussed.

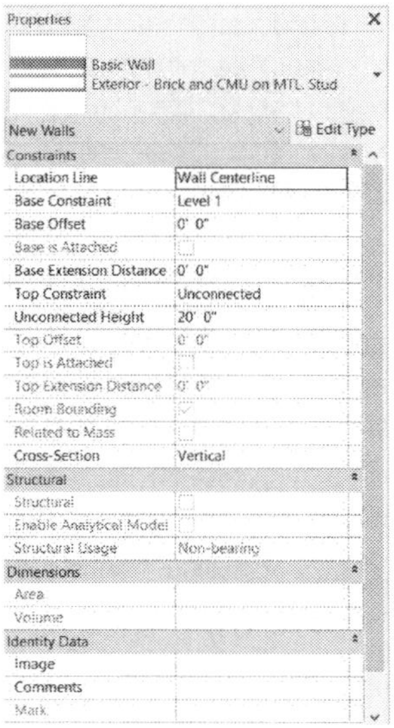

***Figure 7-29** Checking the status of attachments to walls in the **Properties** palette*

TUTORIALS

Tutorial 1 Apartment 1

In this tutorial, you will add a floor and a ceiling to the *c06_Apartment1_tut1.rvt* project created in Tutorial 1 of Chapter 6. Also, you will attach walls to the floor. To complete this tutorial, you need to use the following project parameters: **(Expected time: 30 min)**

1. Floor type -
   For Imperial **Floor: LW Concrete on Metal Deck**.
   For Metric **160mm Concrete With 50mm Metal Deck**
2. Ceiling type-
   For Imperial **Compound Ceiling: GWB on Mtl. Stud**,
   For Metric **Compound Ceiling: Plain**
   Level- 8'6" from the floor level.

The following steps are required to complete this tutorial:

a. Open the required file.
b. Hide the annotation tags such as the section line, grids, and the elevation tag.
c. Create the floor using the **Floor** too.
d. Create the ceiling using the **Ceiling** tool.
e. Attach walls to the floor.
f. Save the project using the **Save As** tool.
   For Imperial *c07_Apartment1_tut1.rvt*
   For Metric *M_c07_Apartment1_tut1.rvt*
g. Close the project.

Opening the Project File and Hiding Annotation Tags

To start with this tutorial, open the specified project file and then hide the annotation symbols and tags such as section line, grids, and elevation tags using the **Visibility/Graphics** tool.

1. Choose **Open > Project** from the **File** menu and then open the *c06_Apartment1_tut1.rvt* (*M_c06_Apartment1_tut1.rvt* for Metric) file created in Tutorial 1 of Chapter 6. You can also download this file from *https://www.CADSOFT.com*. The path of the file is as follows: *Textbooks > Civil/GIS > Revit Architecture> Exploring Autodesk Revit 2025 for Architecture*

2. Double click on **First Floor** plan from **Floor Plans** head in the project browser, if not displayed by default. Choose the **Visibility/Graphics** tool from the **Graphics** panel of the **View** tab; the **Visibility/Graphics Overrides for Floor Plan: First Floor** dialog box is displayed.

3. Choose the **Annotation Categories** tab and clear the check boxes under the **Visibility** parameter for **Elevations**, **Grids**, and **Sections**.

4. Now, choose the **Apply** button and then the **OK** button; the specified settings are applied to the plan view and you return to the drawing window.

Creating the Floor

In the following set of steps, you will add floor to the apartment plan using the **Floor** tool.

1. Choose the **Floor: Architectural** tool from **Architecture > Build > Floor** drop-down; the **Modify | Create Floor Boundary** tab is displayed along with the **Options Bar**. Notice that the drawing area fades when you invoke the **Floor** tool, which indicates that you are in the sketch mode.

2. In the **Options Bar**, select the **Extend into wall (to core)** check box, if it is not selected. Also notice that the **Pick Walls** tool is chosen by default in the **Draw** panel.

3. Move the cursor near the center of the north wall of the apartment plan, and press the TAB key when the wall is highlighted.

4. When the chain of walls is highlighted, click to sketch the floor boundary; the boundary is sketched and appears as a magenta rectangle, as shown in Figure 7-30.

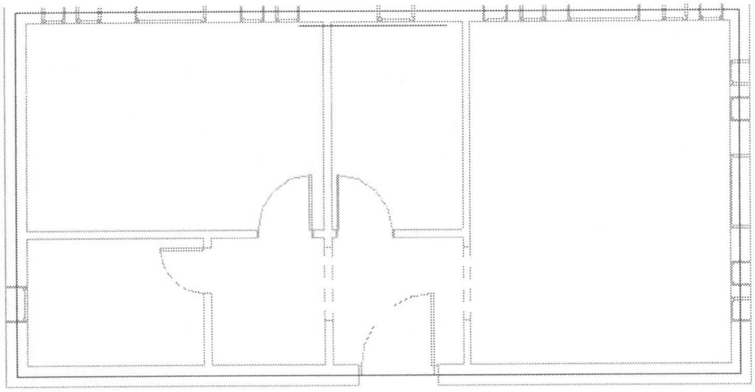

Figure 7-30 The floor boundary sketched using the ***Pick Walls*** *tool*

5. Choose the **Finish Edit Mode** button from the **Mode** panel to complete the sketching of the floor.

6. Next, you need to assign the type to the floor.

   Select the option from the **Type Selector** drop-down list in the **Properties** palette, as shown in Figure 7-31.

   For Imperial **Floor: LW Concrete on Metal Deck**
   For Metric **160mm Concrete With 50mm Metal Deck**

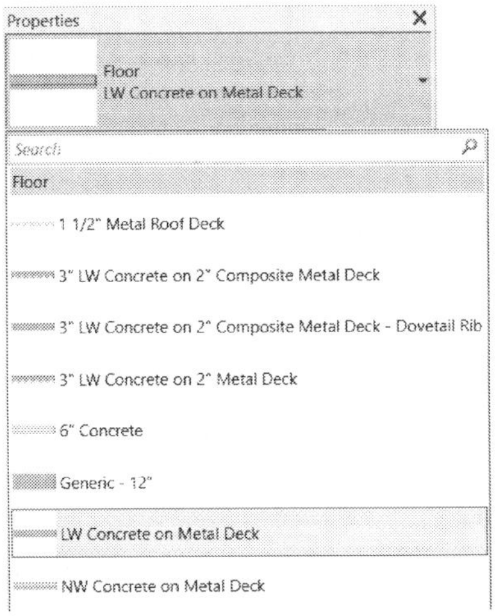

Figure 7-31 Selecting the ***LW Concrete on Metal Deck*** *floor type*

7. Choose the **Modify** button from the **Select** panel of the **Modify | Floor** tab to exit the **Floor: Architectural** tool. The floor is created for the *Apartment 1* project, as shown in Figure 7-32.

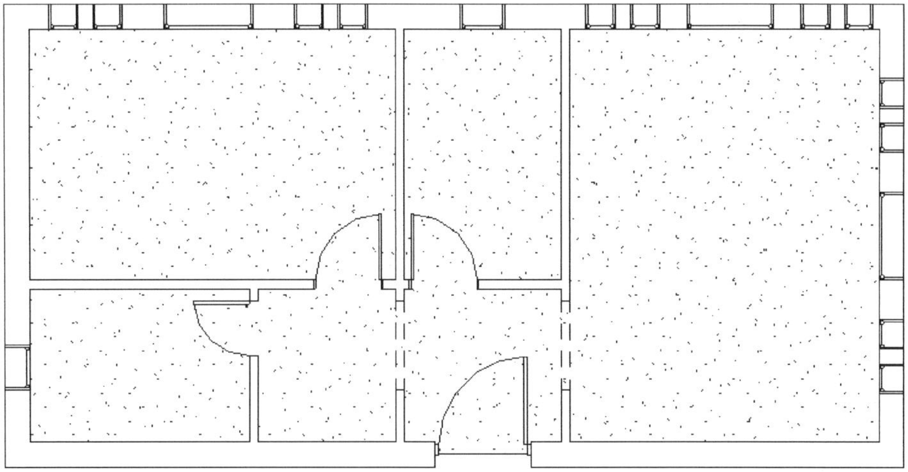

Figure 7-32 *The **LW Concrete on Metal Deck** type of the floor created*

Creating the Ceiling

After creating the floor for the apartment, you need to add a ceiling to it. For this you need to transform the current view to the ceiling plan view.

1. Double-click on **First Floor** from the **Ceiling Plans** head in the **Project Browser**.

   As you open this view, you will notice the appearance of the annotation tags. Repeat steps 2, 3, and 4 given in the **Opening the Project File and Hiding Annotation Tags** section in this tutorial to hide various grids, elevation, window, and section tags from the first floor ceiling plan view.

2. Invoke the **Ceiling** tool from the **Build** panel of the **Architecture** tab; the **Modify | Place Ceiling** tab is displayed.

3. Click on the **Type Selector** drop-down list in the **Properties** palette and then select the desired option from it.

   For Imperial **Compound Ceiling: GWB on Mtl. Stud**
   For Metric **Compound Ceiling: Plain**

   Next, you need to define the clear height of the ceiling.

4. In the **Properties** palette, click on the value field of the **Height Offset From Level** instance parameter and enter **8'6"(2590 mm)**. Then, choose the **Apply** button; the new height is assigned to the ceiling.

5. Move the cursor inside any room; the room boundary is highlighted. Now, click inside the highlighted boundary area; the ceiling is created. Notice that the created ceiling is not distinctly visible in the ceiling plan because the ceiling type selected has a plain GWB board finish.

6. Repeat step 5 to create individual ceilings for every room in the *Apartment 1* project. After creating all the ceilings, press ESC twice to view them in the 3D view of the project.

7. Click on the + symbol for the **3D Views** head in the **Project Browser** and double-click on **{3D}**; the 3D view of the apartment project with the created ceiling is displayed. You can move the cursor over the ceiling of any room to highlight it and display the ceiling type, as shown in Figure 7-33.

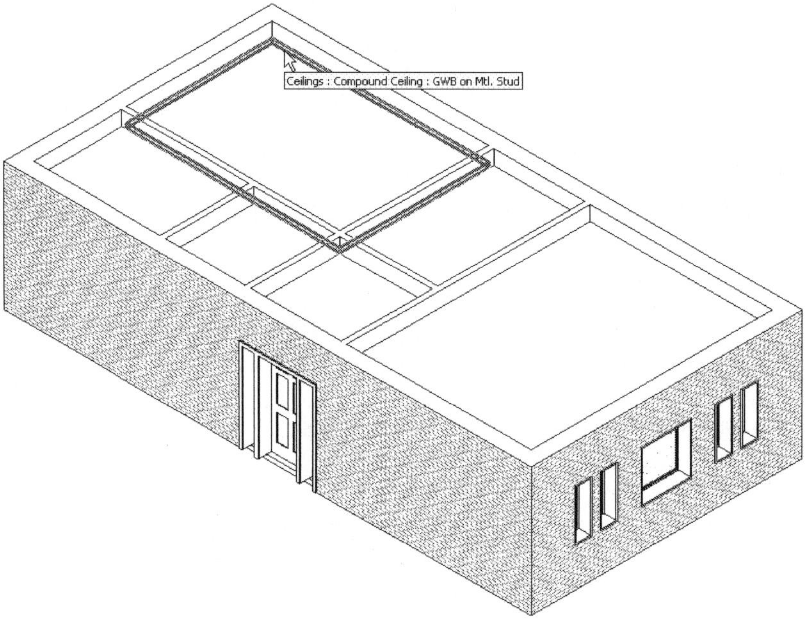

Figure 7-33 *Highlighting and displaying the ceiling type for bedroom*

Attaching Walls to the Floor

In this section, you will use the **Attach** tool to attach walls to the floor so that you get a clear intersection between the floor and the wall. You need to use the section view to attach the floor.

1. Double-click on **Section X** in the **Project Browser**.

2. Now, move the cursor near the external wall that is cut in the section, and when it is highlighted, press TAB to highlight all the exterior walls.

3. Click to select the chain of exterior walls in the section view and then choose the **Attach Top/Base** tool from the **Modify Wall** panel of the **Modify | Walls** tab; the **Options Bar** is displayed.

4. From the **Options Bar**, select the **Base** radio button.

5. Move the cursor near the floor and click when it is highlighted; the floor is attached to walls, as shown in Figure 7-34.

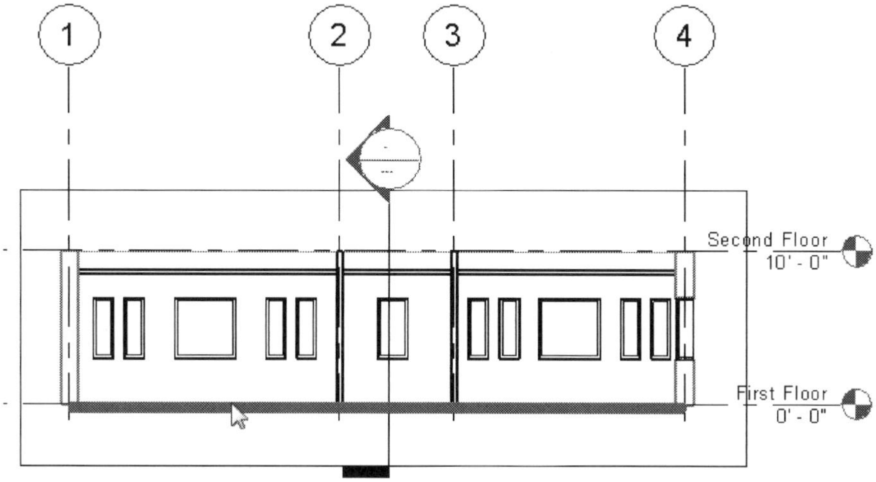

Figure 7-34 *Selecting the floor to attach it with the exterior walls*

6. Double-click on **First Floor** in the **Floor Plans** head in the **Project Browser**; the corresponding floor plan in the drawing window is displayed.

Note
In this section of the tutorial, as you attach walls with the floor, you may not find any difference in the wall and the floor intersections as compared to when they were not attached. To see the difference, attach the wall with the floor and then enter a positive value in the value column corresponding to the ***Height Offset From Level*** *instance parameter in the* ***Properties*** *Palette of the selected floor.*

Saving the Project

In this section, you will save the project file using the **Save As** tool.

1. Choose **Save As > Project** from the **File** menu; the **Save As** dialog box is displayed. Enter required name in the **File name** edit box
   For Imperial **c07_Apartment1_tut1**
   For Metric **M_c07_Apartment1_tut1**
   Choose **Save**; the project file is saved.

2. Now, choose **Close** from the **File** menu to close the file.

EXERCISE

Exercise 1 Apartment 2

Create a floor and a ceiling on the First Floor level of the *c06_Apartment2_ex1.rvt* (*M_c06_Apartment2_ex1.rvt* for Metric) project created in Exercise 1 of Chapter 6. Attach the walls to the floor. Use the following project parameters: **(Expected time: 30 min)**

1. Floor type -
   For Imperial **Floor: LW Concrete on Metal Deck**
   For Metric **160mm Concrete With 50mm Metal Deck**
   Extents to wall core.

2. Ceiling type-
   For Imperial **Compound Ceiling: GWB on Mtl Stud**
   For Metric **Compound Ceiling: Plain**
   Level- 8'6" (2591 mm) from the First Floor level.

3. Name of the file to be saved-
   For Imperial **c07_Apartment2_ex1**
   For Metric **M_c07_Apartment2_ex1**

Chapter 8

Using Basic Building Components-III

Learning Objectives

After completing this chapter, you will be able to:

- *Add components to a building model*
- *Create stairs using the Stairs tool*
- *Add railings to various locations in a project using the Railing tool*
- *Understand the procedure of adding ramps to a building model*
- *Add curtain walls with doors and mullions*
- *Copy elements from one level to another*

INTRODUCTION

This chapter describes the usage of building elements such as freestanding components, stairs, railings, ramps, and curtain walls. In addition, the chapter discusses how to copy elements from one level to another.

USING COMPONENTS IN A PROJECT (BIM OBJECTS)

Autodesk Revit provides various freestanding components or BIM objects that can be placed in a project. These include furniture items, plumbing fixtures, electrical fittings, trees, and many other family elements. Unlike other dependent elements, these freestanding components do not have any predefined associativity with other elements. However, the inserted components enables you to assign certain restrictions to their placement.

BIM CONCEPTS *Apart from what is present in the Autodesk library, you may need customized families while working in a project. For this you can download various components as families from various sites such as https://www.bimobject.com/, https://www.revitcity.com/, https://bimsmith.com/, and more.*

Adding Components

Ribbon: Architecture > Build > Component drop-down > Place a Component
Shortcut Key: CM

You can add components to a building model using the **Place a Component** tool. This tool can be invoked from the **Build** panel. On invoking this tool, the **Modify | Place Component** tab will be displayed. You can use the options in this tab to select the component type and specify the properties of the component to be inserted, load components from the predefined library, or create a new component in the drawing. Alternatively, type **CM** to invoke the **Place a Component** tool.

Revit provides different in-built component types. You can select them from the **Type Selector** drop-down list in the **Properties** palette, as shown in Figure 8-1. You can also load additional component types from the **English-Imperial** folder. To do so, choose the **Load Family** tool from the **Mode** panel of the **Modify | Place Component** tab; the **Load Family** dialog box will be displayed. In this dialog box, browse to the **English-Imperial** folder and select the required component from the specific sub-folder. After you select the required component and close the **Load Family** dialog box, the selected component will be added in the **Type Selector** drop-down list.

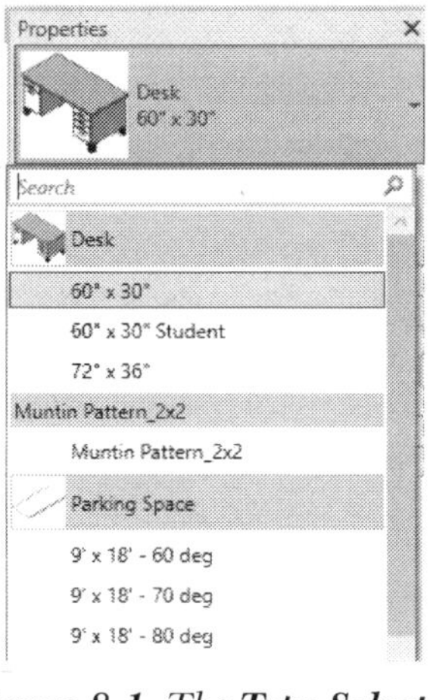

***Figure 8-1** The **Type Selector** drop-down list displaying different component types*

After selecting the component from the **Type Selector** drop-down list, select the **Rotate after placement** check box in the **Options Bar**, so that you can rotate the component after placing it. Now, move the cursor to the area where you want to add it. You will notice a component symbol attached to the cursor. Moreover, the temporary dimensions will display the distance of the component from the nearby elements. To place the component, click at the desired location; the temporary dimensions will be highlighted. Click on the appropriate dimension to enter a new value; the component will move to a new distance. As you have selected the **Rotate after placement** check box, the **Rotate** tool will be invoked when you add a component in the project. You can then rotate it graphically.

ADDING STAIRS

Ribbon:	Architecture > Circulation > Stair

In Autodesk Revit, you can create a variety of stair shapes such as straight, curved, and spiral stairs with ease. In Revit, stairs are parametric elements; therefore, you can create them by specifying the rules for risers, treads, and stringers. You can create stairs in your building design both in the plan and 3D views.

In Autodesk Revit, stairs can be added to individual components such as Run, Landing, and Supports. To do so, invoke the **Stair** tool from the **Circulation** panel; the **Modify | Create Stair** tab will be displayed. In the **Component** panel of this tab, you can choose the **Run**, **Landing**, or **Support** tool based on the requirement. Some of these tools are discussed next.

Run Tool

You can use the **Run** tool to create a run of stairs. To create the run of stair, invoke the **Stair** tool from the **Circulation** panel; the **Modify | Create Stair** tab will be displayed. In the **Components** panel of this tab, ensure that the **Run** tool is chosen. As a result, a list of tools to create the stair run is displayed in the list box displayed in the **Components** panel. The list box contains the **Straight**, **Full-Step Spiral**, **Center-Ends Spiral**, **L-Shape Winder**, **U-Shape Winder**, and **Create Sketch** tool. Some of these tools are discussed next.

Landing Tool

This tool is used to create a landing between two runs. To create landing, choose the **Stair** tool from the **Circulation** panel in the **Architecture** tab; the **Modify| Create Stair** tab will be displayed. From the **Components** panel of this tab, choose the **Landing** tool; the **Pick Two Runs** and **Create Sketch** tools are displayed in the list box next to it. You can use the **Pick Two Runs** tool to create a landing by selecting two runs of a stair. On invoking this tool, you will be prompted to select the first run. Click on the first run of the stair; the stair run will be selected and highlighted, as shown in Figure 8-2. Also, you will be prompted to select the second stair run. Click on the second stair run; a landing will be created between the first and second stair runs, as shown in Figure 8-3.

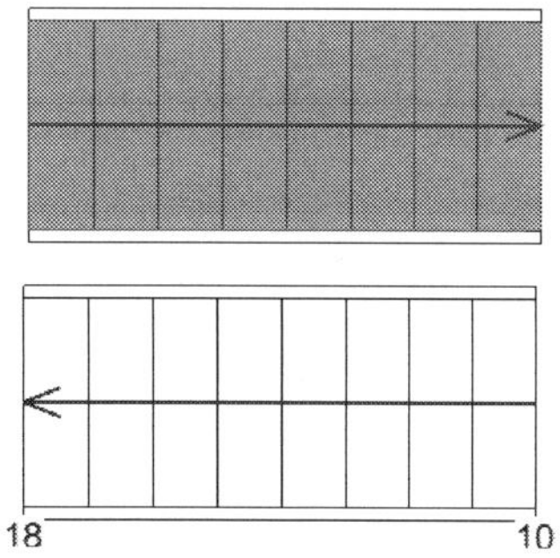

Figure 8-2 *The first stair run highlighted*

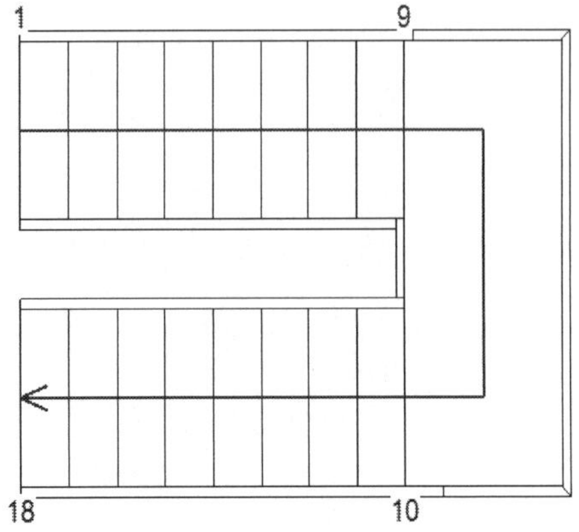

Figure 8-3 Landing created between the first and second stair runs

Support Tool

You can use this tool to add a stringer or a carriage by selecting an individual path on a run or landing. Invoke this tool from the **Components** panel of the **Modify | Create Stair** tab; you will be prompted to pick the edge of a stair run or landing to place the support. Choose the **Finish Edit Mode** button from the **Mode** panel; the support will be added to the stairs run, as shown in Figure 8-4.

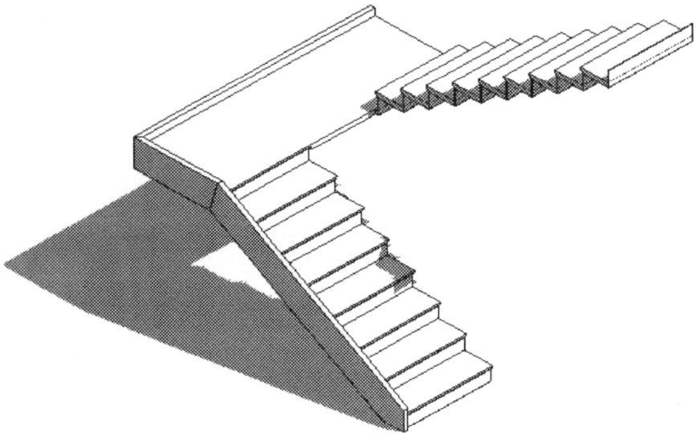

Figure 8-4 Support added to the stair runs

Modifying Stairs Properties

You can specify the properties of stairs before creating them and then you can modify its properties. To modify the stairs after you create them, select the stairs; the properties of the stairs will be displayed in the **Properties** palette, as shown in Figure 8-5. In the **Properties** palette, the **Base Level** parameter indicates the level from which the stairs will start, and the **Top Level** parameter indicates the level at which they will end. By default, the current level is set as the base level, and the next story level is set as the top level.

The parameters under the **Dimensions** section are used to specify different dimensional parameters for the stairs. You can determine the number of risers in the stairs by specifying a value in the **Value** column of the **Desired Number of Risers** parameter. In the **Value** column for the **Actual Tread Depth** parameter, you can specify a value for the depth of the treads in a stair. Other instance parameters can also be modified in the **Properties** palette.

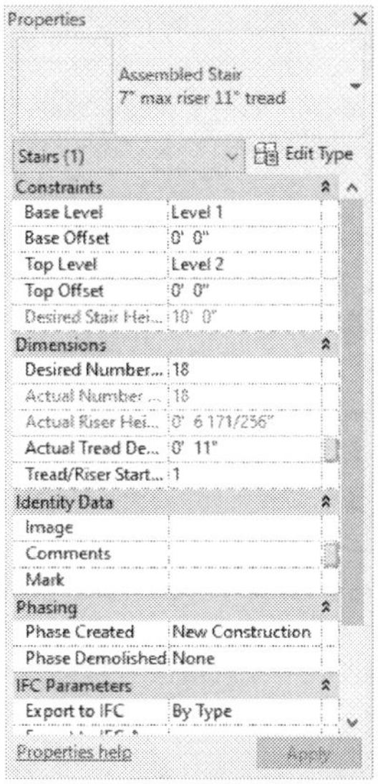

*Figure 8-5 The **Properties** palette displaying various instance properties of the selected stairs*

You can choose the **Edit Type** button from the **Properties** palette for viewing and modifying the type properties. On choosing this button, the **Type Properties** dialog box will be displayed, as shown in Figure 8-6. In the **Type Properties** dialog box, you can choose the **Edit** button for specifying the calculation rules. When you choose this button, the **Stair Calculator** dialog box will be displayed. This dialog box can be used to set new rules to calculate the slope of the stairs. In the **Type Properties** dialog box, you can enter the **Minimum Tread Depth**, **Minimum Run Width** and the **Maximum Riser Height** in their corresponding value columns. You can also specify other parameters in the **Type Properties** dialog box such as Tread Thickness, Riser Thickness, Nosing Length, and so on. You can select the **Monolithic Stairs** check box to create stairs made up of only a single type of material.

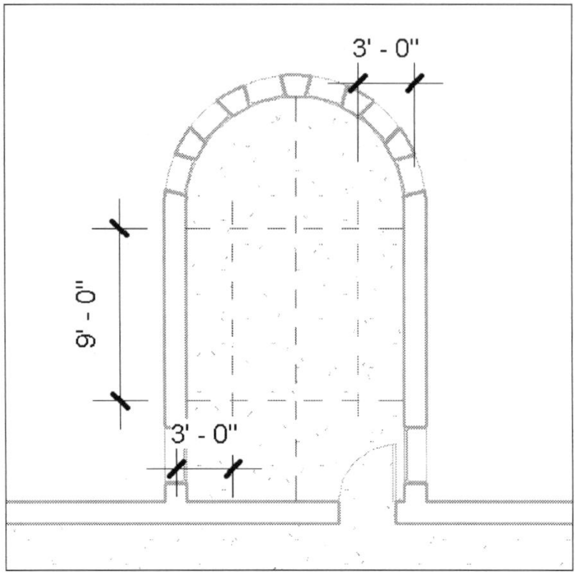

*Figure 8-6 The **Type Properties** dialog box for stairs*

ADDING RAILINGS

Railings are automatically generated with the stairs but they can also be created by specifying their profile. You can also create them for other locations in the building project such as terraces, passages, and so on. You can select and modify their properties to achieve the desired railing configuration. In Revit, you can add railing by sketching its path or by placing it in a stair or a ramp in the project. The methods of adding a railing by sketching a path and placing it on a host such as stair and ramp are discussed next.

Adding Railings by Sketching the Path

Ribbon: Architecture > Circulation > Railing > Sketch Path

You can add railing in a project by sketching its path. To do so, you can use the **Sketch Path** tool. Invoke this tool from **Architecture** > **Circulation** > **Railing** drop-down; the **Modify | Create Railing Path** tab will be displayed. This tab displays various sketching tools to sketch the railing path. You can sketch the railing path with or without a host as per the requirement of the project. The host to the railing can be a floor stairs, ramps and so on. To select a host, choose the **Pick New Host** tool from the **Tools** panel of the **Modify | Create Railing Path** tab; you will be prompted to select an element in the drawing on which the railing is to be created. Select the host if required. In the Properties palette, you can select the type of the railing. To assign a type to the railing, select an option from the Type Selector drop-down list. In the Properties palette, you can also assign various instance parameters such as **Base Level**, **Base Offset**, **Tread / Stringer Offset**, and so on to the railing. Now, to sketch the railing path, choose any of the drawing tools from the list box that will be displayed in the **Draw** panel of the

Modify | Create Railing Path tab to sketch. For example, you can use the **Line** tool to define a specific railing path, as shown in Figure 8-7. You can use the **Ref Plane** tool to assist you in sketching the railing boundary, if required. After completing the sketch of the railing path, choose the **Finish Edit Mode** button from the **Mode** panel; the railing will be created, as shown in Figure 8-8.

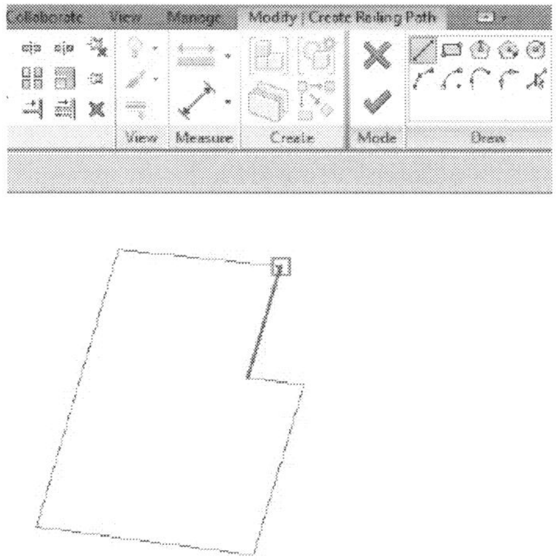

Figure 8-7 Using the ***Line*** *tool to create the boundary for a railing*

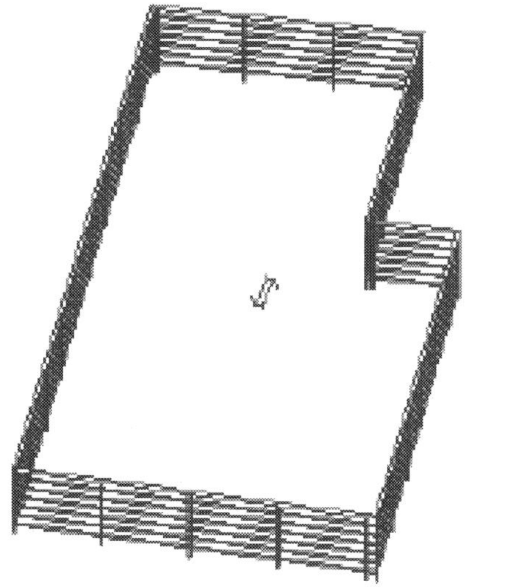

Figure 8-8 3D view of the railing created

Note

You can now sketch railings onto the top faces of floors, slabs, slab edges, walls, or roofs. Balusters and railings adjust to the slope of irregular surfaces.

Adding Railings by Placing On Stair/Ramp

Ribbon: Architecture > Circulation > Railing

The **Railing** tool allows you to add railings to the entire stair tower or an individual staircase. Such railings can be edited ranging from one instance to all the instances in the group. In Revit, you can attach railing to building staircase and the toposurface for model fencing, road barriers, or outdoor handrails. The example of railing added to building staircases and toposurface is shown in Figure 8-9.

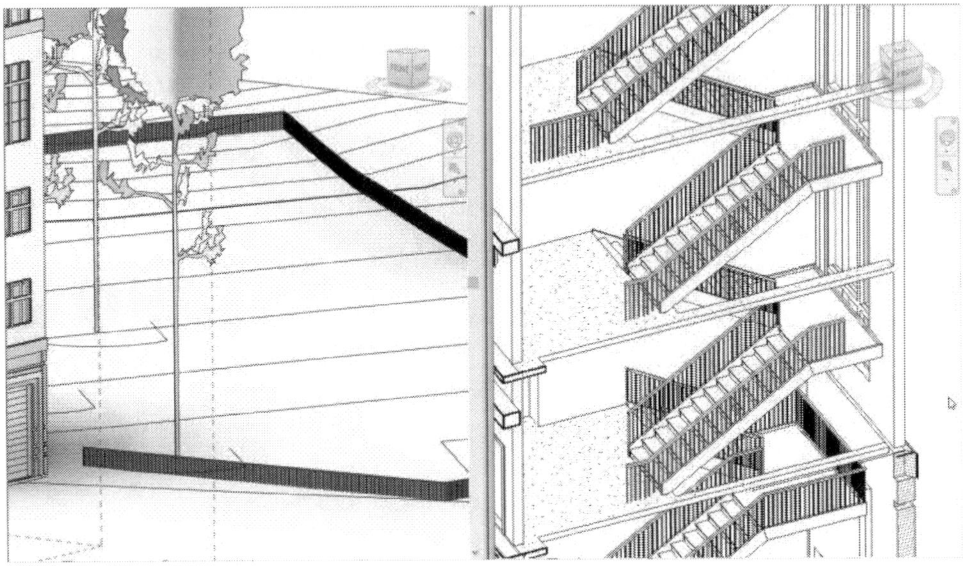

Figure 8-9 Railing added to building staircase and toposurface

Modifying Railing Properties

You can modify the properties and parameters of a railing before or after creating it. To specify the properties and parameters of a railing, you can use various instance properties displayed in the **Properties** palette. To specify the type properties of the proposed railing, choose the **Edit Type** button in the **Propertie**s palette; the **Type Properties** dialog box will be displayed. In this dialog box, the **Edit** buttons in the value column for the **Rail Structure** and **Baluster Placement** type parameters can be chosen to view and modify the parameters of the railing such as rail type, baluster type, baluster spacing, and so on. The **Baluster Offset** parameter indicates the offset distance of the baluster from the sketched rail boundary.

Apart from the preloaded baluster types, you can load additional ones by choosing the **Load Family** tool from the **Load from Library** panel in the **Insert** tab. On choosing the **Load Family** tool, the **Load Family** dialog box will be displayed. In this dialog box, open the **English-Imperial >Railings > Balusters** and select the desired file(s) for adding baluster type(s) in the project. After selecting the desired file(s), choose the **Open** button in the **Load Family** dialog box; the dialog box will close and the selected baluster type(s) will be loaded into the project.

Note

Railings are associated with their hosts and get modified with the changes in the host element. For example, if the stairs are deleted, the railing will also be deleted. You can, however, delete the railing independent of its hosts.

Modifying Railing Joints

You can modify the joint type of a joint in a railing. To do so, open the plan or the 3D view in which the railing you wish to edit is visible. Next, select the railing; the **Modify | Railings** tab is displayed. Choose the **Edit Path** tool from the **Mode** panel; the **Modify | Railings >**

Sketch Path tab will be displayed and the railing path will be displayed in the sketch mode. In the **Modify | Railings > Sketch Path** tab, choose the **Edit Joins** tool from the **Tools** panel and then move the cursor along the path of the railing. As you move the cursor along the path, you will notice a snap is displayed at the intersection of the railing path or at the point where the two paths join. Click when the snap is displayed, and then from the **Rail Join** drop-down list in the **Options Bar**, select an option to specify the type of join. Similarly, you can click on other joins in the railing path and change their type as per the project requirement. After you finish assigning the type to the joins, choose the **Finish Edit Mode** button from the **Mode** panel of the **Modify Railings > Edit Sketch** tab to finish editing of the railing joins.

ADDING RAMPS

Ribbon: Architecture > Circulation > Ramp

You can create ramps in Autodesk Revit in a manner similar to creating stairs. The **Ramp** tool is used to create ramps and can be invoked from the **Circulation** panel. On invoking this tool, the **Modify | Create Ramp Sketch** tab will be displayed. Also, the properties used for creating the ramp will be displayed in the **Properties** palette. You can use various tools and options in this tab to create the sketch of the ramp. In the **Properties** palette, you can assign properties and type to the ramp.

After sketching the ramp boundary for the proposed ramp, you need to view and assign properties to it. To do so, you can use various properties displayed in the **Properties** palette of the ramp. The instance properties of the ramp can be specified in this palette. In the **Properties** palette, the **Width** parameter indicates the width of the ramp, whereas the **Base Level** parameter indicates the level from where the ramp will start, and the **Top Level** parameter indicates the level at which the ramp will end. The slope and width of the ramp can be specified even before it is created. Next, to view and modify the type properties of the proposed ramp, choose the **Edit Type** button in the **Properties** palette; the **Type Properties** dialog box will be displayed. In this dialog box, the value of the **Maximum Incline Length** parameter indicates the maximum length of the ramp before a landing is required. Specify the maximum slope of the ramp in the **Ramp Max Slope (1/x)** parameter value. The **Shape** parameter is used to create a ramp with a solid base or a slab. Figure 8-10 shows an example of sketching a ramp by specifying the start point and the endpoint.

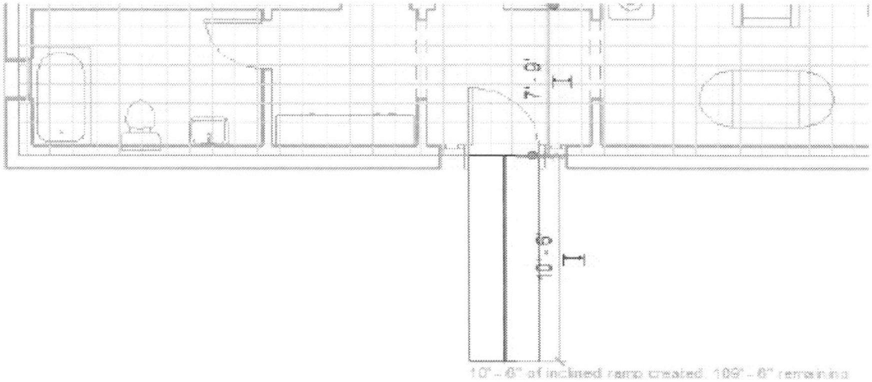

Figure 8-10 *Sketching the ramp run*

After specifying the instance and type properties of the ramp, you can specify the type of railing to be used for it. To do so, choose the **Railing** tool from the **Tools** panel of the **Modify | Create Ramp Sketch** tab; the **Railings** dialog box will be displayed. In this dialog box, select the railing type from the available drop-down list and choose the **OK** button; the **Railings** dialog box will close and the selected railing type will be assigned to the ramp.

Next, click in the drawing area to specify the start point of the ramp. On doing so, a rectangle shaped geometry will be displayed. This geometry will show the total size required for creating the ramp. The ramp size can be calculated based on the current ramp property settings. The required length of the ramp is also displayed on specifying the first point of the ramp run. Next, specify the endpoint of the ramp and choose the **Finish Edit Mode** button from the **Mode** panel of the **Modify | Create Ramp Sketch** tab. This creates the ramp with the given parameters, as shown in Figure 8-11. If the specified length of the ramp is less or more than the required length, Autodesk Revit will display a warning message box showing the error and its cause, as shown in Figure 8-12. You can ignore the error, if possible, or rectify the sketch or the properties to create the desired ramp using curtain systems in a project.

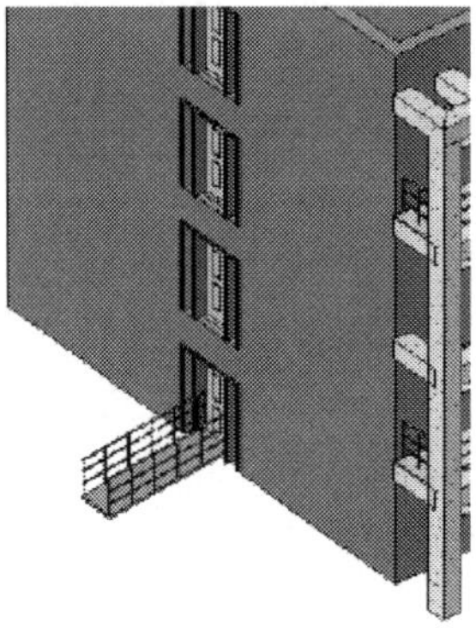

Figure 8-11 *3D view of resultant ramp*

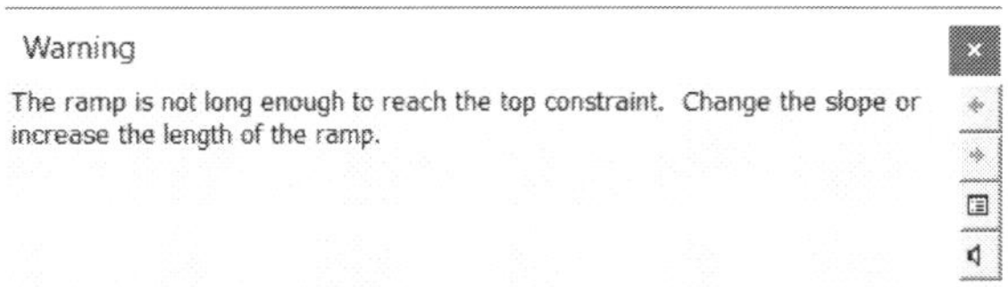

Figure 8-12 *The **Warning** message box showing the error type*

USING CURTAIN SYSTEMS IN A PROJECT

In Autodesk Revit, a typical curtain system comprises the panels, curtain grid, curtain wall, and mullions, as shown in Figure 8-13. Each of these elements can be modified individually. A curtain system can be added by using the **Wall: Architectural** tool or by using the specific curtain system tools. You can also generate the curtain systems of different shapes and configurations. A curtain system created using the **Wall: Architectural** tool is also referred to as a curtain wall. You can create curved, sloping, and store front curtain systems.

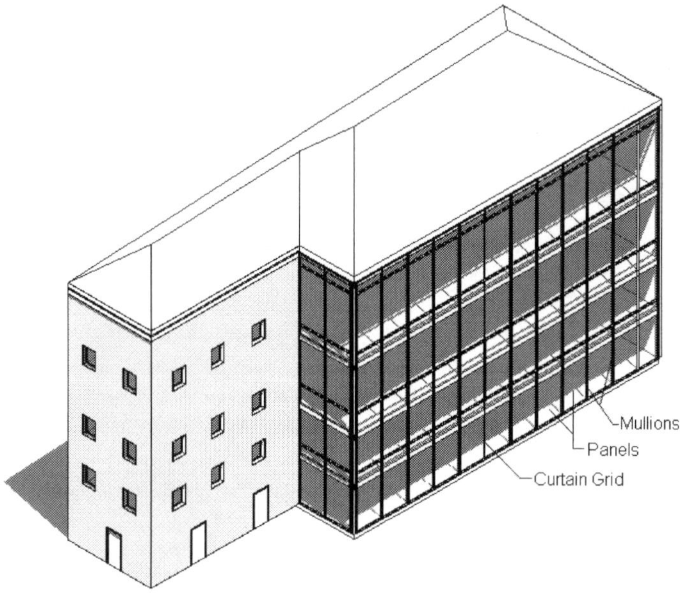

Figure 8-13 *Typical curtain system elements*

Creating a Curtain Wall Using the Wall: Architectural Tool

A curtain wall can be created by using the **Wall** tool. To create a curtain wall, invoke the **Wall** tool and select the required type from the **Type Selector** drop-down list in the **Properties** palette, as shown in Figure 8-14. Note that the **Type Selector** drop-down list displays three different types of curtain wall options such as **Curtain Wall 1**, **Exterior Glazing**, and **Storefront** under the **Curtain Wall** head. After selecting the desired curtain wall type from the **Type Selector** drop-down list, you can start sketching the proposed curtain wall by using the sketching tools available in the **Draw** panel.

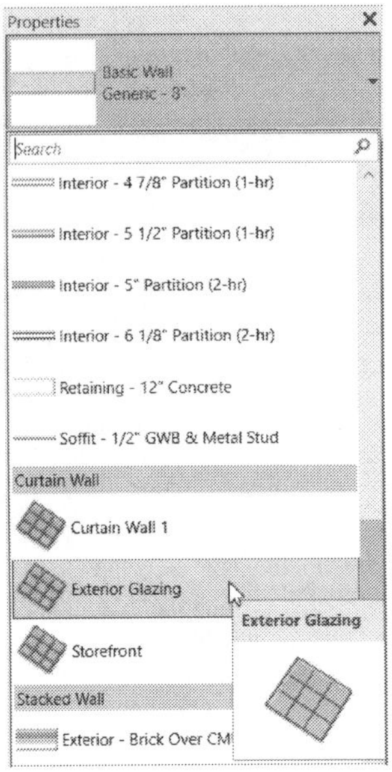

Figure 8-14 Selecting the curtain wall type from the ***Type Selector*** *drop-down list in the* ***Properties*** *palette*

Alternatively, select any wall and replace it with the predefined curtain wall using the **Type Selector** drop-down list. If you select the **Curtain Wall** option as a wall type from the **Type Selector** drop-down list, a curtain wall will be created with a single panel. Even a curved curtain wall is first created as a straight curtain wall. You can then use the **Curtain Grid** tool in the **Build** panel of the **Architecture** tab to add curtain grids at a specific distance to give the desired shape. The other two curtain wall types available in the **Type Selector** drop-down list have predefined curtain grids. They can, however, be modified to the desired spacing.

Tip
Autodesk Revit enables you to create elements with minimum faults by warning you about the errors in their creation. The parametric building elements associate with each other and help in creating designs that have the least discrepancies.

Creating a Curtain Wall by Picking Lines

You can create a curtain wall from existing lines in a drawing. The existing lines can be model lines or edges of elements such as roofs, curtain panels, and other walls. To create a curtain wall by using the existing lines, choose the **Pick Lines** tool from the **Draw** panel of the **Modify | Place Wall** tab; you will be prompted to select an edge or a line from the drawing. Select the desired edge or line from the drawing; the curtain wall will be created.

For example, you can select the edge of a floor slab or any line in a different level to create the curtain wall. Figure 8-15 shows a flat and a curved curtain system created by using the **Pick Lines** tool. Note that the curved curtain system appears flat when it is created and acquires the actual curved form when the curtain grids are added.

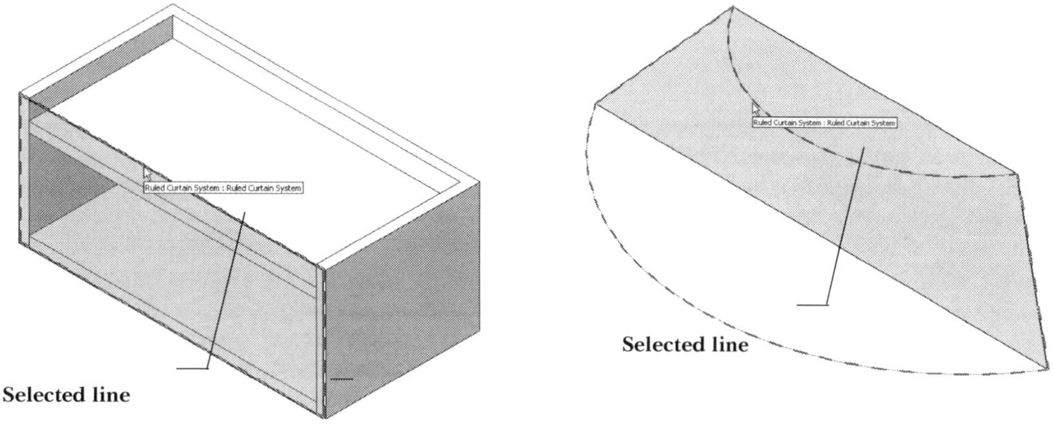

Figure 8-15 A flat and a curved curtain system created using the ***Pick Lines*** *tool*

Creating a Curtain System on a Face

Ribbon: Architecture > Build > Curtain System

You can create a curtain system on a face by using the **Curtain System** tool. This tool is effective for creating a curtain system on the solids and faces created by using the massing tools (the use of massing tools is described in Chapter 10). After creating a massing geometry, invoke this tool from the **Curtain Build** panel; the **Modify | Place Curtain System by Face** tab will be displayed and you will be prompted to select a face that will be used to create the curtain system. The selected face can be flat or with a single or double curved faces. After selecting the face, choose the **Create System** tool from the **Multiple Selection** panel in the **Modify | Place Curtain System by Face** tab; a curtain system will be created based on the face selected. The type of the curtain system will be based on the default value selected in the **Type Selector** drop-down list in the **Properties** palette.

Adding Curtain Grids

Ribbon: Architecture > Build > Curtain Grid

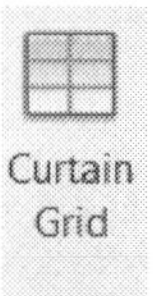

Curtain grids are added to a curtain system to define its geometry and form a grid for adding mullions. Invoke the **Curtain Grid** tool from the **Build** panel; a preview image of the grid will be displayed on the panels. Also, the **Modify | Place Curtain Grid** tab will be displayed. The **Placement** panel of this tab displays three tools: **All Segments**, **One Segment**, and **All Except Picked**. The **All Segments** tool is invoked by default and lets you place grid segments on all panels where the preview appears. The **One Segment** tool, if invoked, will create a grid segment only in the panel where the preview appears. Similarly, on invoking the **All Except Picked** tool, you can place grid segments on all panels except those you select to exclude.

After you select any tool from the **Placement** panel, move the cursor to a curtain wall or a curtain system where you wish to add the curtain grids. On doing so, a dashed line will appear, indicating the position to place the curtain grid. A horizontal dashed line appears when the cursor is near a vertical edge, and a vertical dashed line will appear when the cursor is close to a horizontal edge. Click at the desired locations to add the curtain grids. The object snap options displayed can also be used for this purpose. Figure 8-16 shows two examples of a curtain grid added to a curtain system by using the **All Segments** and **One Segment** tools.

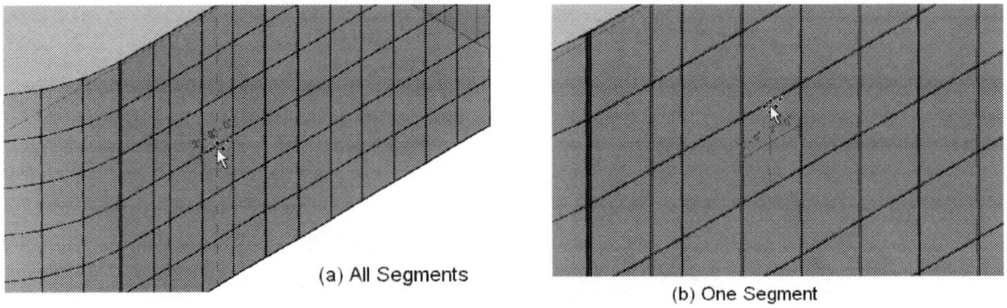

Figure 8-16 *Adding curtain grids to a curtain system by using the **All Segments** and **One Segment** tools*

Tip
*When you create curtain grids individually using the **Curtain Grids** tool, you can delete or drag each curtain grid to achieve the appropriate curtain grid pattern.*

Modifying Curtain System Panels

On creating a curtain grid, the curtain system is divided into curtain system panels. Each portion between the curtain grids is treated as a single curtain system panel. By default, each panel is treated as a glazed system panel. However, Autodesk Revit enables you to modify the properties of each panel. You can create a solid panel or an empty panel. To modify a system panel, select it using the crossing selection method and the **Filter** tool (if required); the **Properties** palette will display the type and properties of the system panel. In the **Properties** palette, the **System Panel: Glazed** option is selected by default in the **Type Selector** drop-down list. In this drop-down list, you can select the **Empty System Panel: Empty** or **System Panel: Solid** option to replace the selected panel(s) with an empty or solid panel, respectively. In this manner, you can create a combination of solid and glazed panels for a curtain system. Figure 8-17 shows the curtain panels of three different types added to a curtain system.

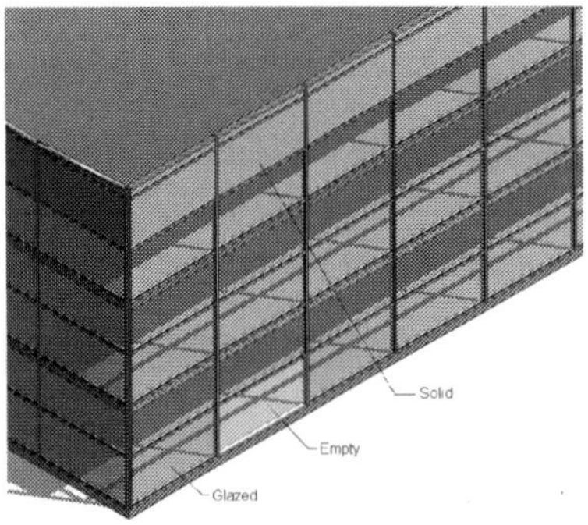

Figure 8-17 *Example of different curtain panels added to the curtain system*

Adding Doors and Awnings to a Curtain System

The doors and curtain wall awnings can be added to a curtain system by replacing the curtain system panel with them. To add doors and the awning to the curtain wall panel, first you need to load their family types. To do so, choose the **Insert** tab and then choose the **Load Family** tool from the **Load from Library** panel; the **Load Family** dialog box will be displayed. In this dialog box, browse the **English-Imperial** folder and load the curtain wall double glass and single glass doors, and awnings from the **Doors** and **Windows** sub-folders, respectively. However, before adding doors, ensure that the curtain panel to be replaced is of the desired door size. This may require you to remove certain curtain grid segments. To remove a grid segment from a grid system, select the grid and choose the **Add/Remove Segments** button in the **Curtain Grid** panel of the **Modify | Curtain Systems Grids** tab. Now, click on the segment to be removed and click anywhere outside the curtain system; the grid segment will be removed. Similarly, you can also add the removed curtain grid segment in a curtain grid system. To do so, select the grid from which a segment has been removed; the removed segment will be highlighted with a dashed line. Now, choose the **Add/Remove Segments** button from the **Curtain Grid** panel of the **Modify | Curtain Wall Grids** and click on the highlighted dashed line; the removed grid will be added. Once the panel of the desired size has been created in the curtain system, select it from the drawing; the **Modify | Curtain Panels** tab will be displayed. From the **Type Selector** drop-down list, select any loaded curtain wall doors and awnings. The selected curtain wall doors or awnings will be added to the selected panel.

Adding Mullions

Ribbon: Architecture > Build > Mullion

Once you have created the curtain grid, you can use the **Mullion** tool to add mullions to the segments of curtain grid segments. To use the **Mullion** tool, invoke it from the **Build** panel; the **Modify | Place Mullion** tab will be displayed. Also, the properties of the mullion will be displayed in the **Properties** palette. In this palette, you can select the mullion type from the **Type Selector** drop-down list. Autodesk Revit provides you with six main types of mullions: **Circular Mullion**, **L Corner Mullion**, **Quad Corner Mullion**, **Rectangular Mullion**, **Trapezoid Corner Mullion**, and **V Corner Mullion**. By default, the **Rectangular Mullion** type is selected in the **Type Selector** drop-down list. After selecting a type from the **Type Selector** drop-down list, you can change the type properties of the mullion you want to place. To do so, choose the **Edit Type** button from the **Properties** palette; the **Type Properties** dialog box will be displayed. In this dialog box, change the type parameters, where required, in their corresponding value field and then choose **OK**; the desired change will be applied and the **Type Properties** dialog box will close. Now, you can use the tools from the **Placement** panel in the **Modify | Place Mullion** tab to place the mullions using the curtain grids. The tools available in the **Placement** panel for placing mullions are **Grid Line**, **Grid Line Segment**, and **All Grid Lines**. You can use the **Grid Lines** tool to place a mullion across the entire grid line. Similarly, you can use the **Grid Line Segment** or **All Grid Lines** tool to place a mullion on the individual segment of the grid line or on all grid lines. After you invoke the desired tool, move the cursor near the curtain grid and click when it is highlighted; the mullion will be created on it. This step can be repeated to create mullions for the desired curtain grids, as shown in Figure 8-18. The length of the mullion created for the curtain grid will be equal to the length of the curtain grid. After the mullions are added on a curtain grid, you can control mullion joins. To do so, select the mullion; the **Modify | Curtain Wall Mullions** tab will be displayed. In the **Mullion** panel of this tab, you can choose the **Make Continuous** or **Break at Join** tool to control the mullion joins. You can choose the **Make Continuous** tool to extend the ends of mullions at a join so that they appear as one continuous mullion. Alternatively, you can choose the **Break at Join** tool to trim the ends of mullions at a join so that they appear as separate mullions.

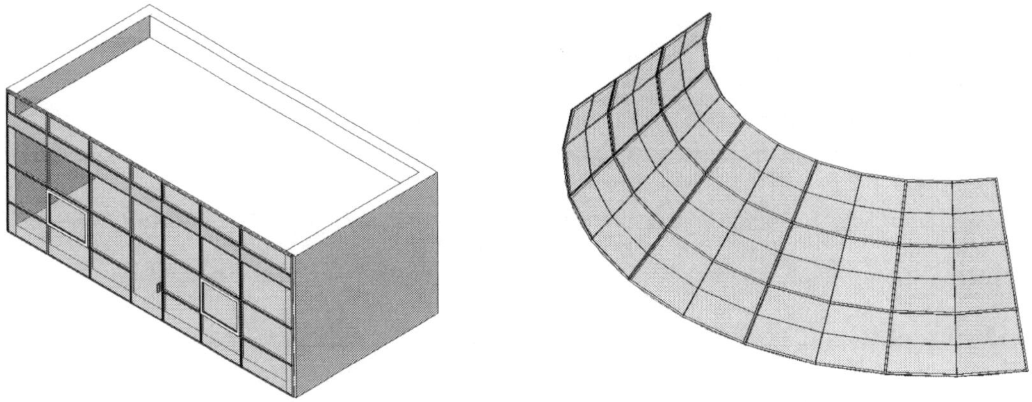

*Figure 8-18 Adding mullions to a curtain system using the **Mullion** tool*

COPYING ELEMENTS FROM ONE LEVEL TO ANOTHER

As described earlier, you can copy elements in the same work plane using the **Copy to Clipboard**, **Cut to Clipboard**, and **Paste** tools. For multistory building projects, you need to use the **Paste** tool to copy the elements such as doors, windows, furniture elements, and so on from one level to another level, exactly above or below their original location.

Select the elements to be copied to the other level. The **Filter** tool can be useful in selecting the elements of the same family. After you select the elements, invoke the **Copy to Clipboard** tool from the **Clipboard** panel of the **Modify | (Elements / Components)** tab; the selected elements will be copied to the clipboard.

Using the Pasting Tools

After copying the elements to the clipboard, choose the **Paste** drop-down from the **Clipboard** panel; a list of pasting tools will be displayed. Based on the required result, you need to choose the appropriate tool from the list. These tools are discussed next.

The **Paste from Clipboard** tool is used to paste elements from the clipboard into the current view.

The **Aligned to Current View** tool is used to paste elements into the current view which is different from the view in which the elements were copied. For example, you can copy elements from a plan view to a callout view.

The **Aligned to Same Place** tool is used to copy elements to the same place from they were copied from. This option is useful for developing different design options for the same location.

The **Aligned to Picked Level** tool is used to paste elements into levels by selecting their corresponding level line. You need to be in the elevation view for selecting the level line to paste the elements.

The **Aligned to Selected Levels** tool, the most frequently used tool, enables you to choose the level(s) for pasting the elements. You can choose the desired level(s) from the list of existing levels in the **Select Levels** dialog box.
The **Aligned to Selected Views** tool is useful for copying view specific elements such as dimensions from one level to another.

TUTORIALS

Tutorial 1 Apartment 1

In this tutorial, you will add furniture and sanitary elements to the first floor level of the *c07_Apartment1_tut1.rvt* project created in Tutorial 1 of Chapter 7. The approximate location of components is shown in Figure 8-19. The text has been given for reference and is not to be created. Extend the exterior walls to 3'0" above the top floor level and copy all interior walls, doors, windows, and other elements at the same location on all three upper floors. Also, create a flat roof for the top floor. Use the following parameters for different components to be added:

(Expected time: 30 min)

Alphabets in Figure 8-19 represent the furniture and plumbing components to be used, as given below.

1. Furniture Components

   For Imperial: Furniture components from the **Furniture** folder of the **English-Imperial**:

   A- **Bed-Shaker: Double 56" x 78"**
   B- **Entertainment Center-72" x 72" x 24"**
   C- **Dresser-72" x 60" x 18"**
   D- **Sofa-72"**
   E- **Table-Dining Oval-36" x 72"**
   F- **Table-Coffee-36" x 72"x 18"**
   G- **Chair-Corbu**
   H- **Table-End: 24" x 24"** (modify height to 15")

   For Metric: Furniture components from the **Furniture** folder of the **English**:

   A- **Bed-Shaker: Double 1422 x 1981 mm**
   B- **Entertainment Center- 1830 x 1830 x 610 mm**
   C- **Dresser- 1220 x 1525 x 0610 mm**
   D- **Sofa- 1830 mm**
   E- **Table-Dining Oval- 0915 x 1830 mm**
   F- **Table-Coffee- 0915 x 1830 x 0457 mm**
   G- **Chair-Corbu**
   H- **Table-End- 0610 x 0610 mm** (modify height to 381 mm)

2. Plumbing Components

   For Imperial: Plumbing components from **Specialty Equipment > Domestic** and **Plumbing** folders of **English-Imperial**

   J- **Kitchenette-Medium**: 8'6" (modified width)
   K- **Toilet-Domestic-3D**
   L- **Tub-Rectangular-3D**
   M- **Sink Vanity-Square**: **20" x 18"**

For Metric: Plumbing components from **Speciality Equipment > Domestic** and **Plumbing** folders of **English**

J- **Kitchenette-Medium**: 2591 mm (modified width)
K- **Toilet-Domestic-3D**
L- **Tub-Rectangular-3D**
M- **Sink Vanity-Square**: **500 x 440 mm**

3. Top floor roof type-
   For Imperial **Basic Roof: Generic 9"** - Flat
   For Metric **Basic Roof: Generic 400 mm** - Flat

The following steps are required to complete this tutorial:

a. Open the project file, invoke the **Place a Component** tool, and load the components from the additional libraries using the **Load Family** tool.
b. Add and position the components at their desired location in the building model using the **Rotate** and **Move** tools.
c. Extend the exterior walls to **3'0"(914 mm)** above the top most level.
d. Select the doors, windows, floor, ceiling, and components, and copy them to all upper floors using the **Aligned to Selected Levels** tool.
e. Create the roof of the topmost floor using the **Roof** tool.
f. Save the project using the **Save As** tool.
   For Imperial *c08_Apartment1_tut1.rvt*
   For Metric *M_c08_Apartment1_tut1.rvt*
g. Close the project.

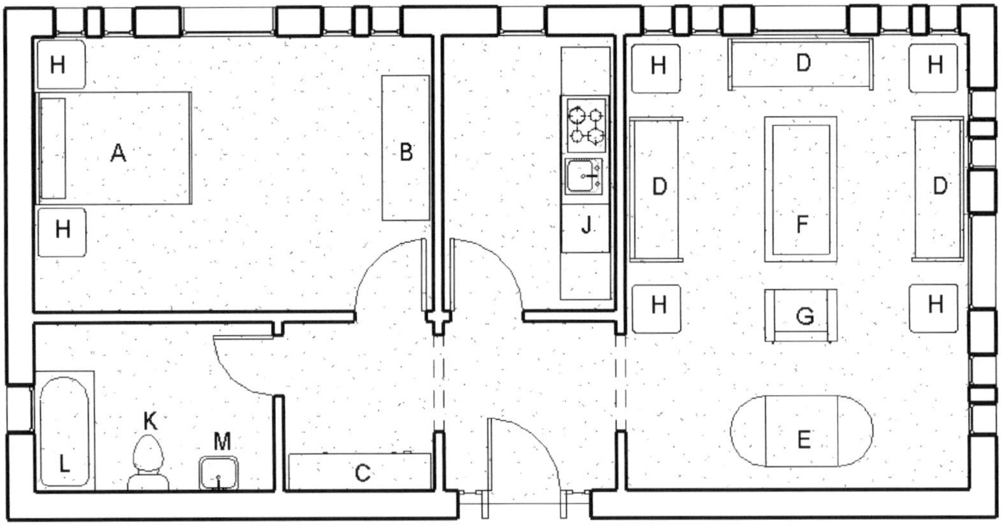

Figure 8-19 *Sketch plan for adding furniture and plumbing components to the Apartment 1 project*

Opening the Project File and Loading Components from Additional Libraries

In this section of the tutorial, you will first open the specified project file and add the components to the first floor plan view by using the **Component** tool. You will use the **Load Family** tool to load the components from the specified additional libraries.

1. Choose **Open > Project** from the **File** menu and open the *c07_Apartment1_tut1.rvt(M_c07_Apartment1_tut1)* file created in Tutorial 1 of Chapter 7 and make sure that the first floor plan is displayed. You can also download this file from *https://www.CADSOFT.com*. The path of the file is as follows: *Textbooks > Civil/GIS > Revit Architecture > Exploring Autodesk Revit 2025 for Architecture*.

Next, you need to place the components as specified in the project parameters.

2. Invoke the **Place a Component** tool from **Architecture > Build > Component** drop-down; the **Modify | Place Component** tab is displayed.

Next, you need to load additional components to the drawing.

3. Choose the **Load Family** tool from the **Mode** panel; the **Load Family** dialog box is displayed.

4. In the **Load Family** dialog box, select the desired folder to open it.
   For Imperial **English-Imperial > Furniture**
   For Metric **English >US > Furniture**

5. In this folder, select the following furniture types from their corresponding folders and load them one by one by repeating the procedure followed in steps 6 and 7.

   For Imperial: **Bed-Shaker** (Folder: Beds), **Chair-Corbu** (Folder: Seating), **Dresser** (Folder: Storage), **Entertainment Center** (Folder: Storage), **Sofa** (Folder: Seating), **Table-Dining Oval** (Folder: Tables), **Table-Coffee** (Folder: Tables), and **Table-End** (Folder: Tables).

   For Metric: **M_Bed-Shaker** (Folder: Beds), **M_Chair-Corbu** (Folder: Seating), **M_Dresser** (Folder: Storage), **M_Entertainment Center** (Folder: Storage), **M_Sofa** (Folder: Seating), **M_Table-Dining Oval** (Folder: Tables), **M_Table-Coffee** (Folder: Tables), and **M_Table-End** (Folder: Tables).

6. After selecting the required family, choose the **Open** button; the **Load Family** dialog box is closed and the selected type becomes available in the **Type Selector** drop-down list in the **Properties** palette. At this stage, it is recommended that you do not click in the drawing area or press ESC and repeat step 3 to load all the families mentioned in step 5.

7. After loading the desired families as mentioned in step 5, invoke the **Load Family** dialog box by choosing the **Load Family** tool from the **Mode** panel of the **Modify | Component** contextual tab.

In the **Load Family** dialog box, choose the **Fixtures** folder from **English-Imperial > Plumbing > Architectural** folder (For Metric, **English > US > Plumbing > Architectural** folder) and load the following fixture types from their corresponding folders:

For Imperial: **Sink Vanity- Square (**Folder: Sinks), **Toilet-Domestic-3D** (Folder: Water Closets), and **Tub-Rectangular-3D (**Folder: Bathtubs).

For Metric: **M_Sink Vanity- Square (**Folder: Sinks), **M_Toilet-Domestic-3D** (Folder: Water Closets), and **M_Tub-Rectangular-3D (**Folder: Bathtubs).

8. Repeat step 7 to invoke the **Load Family** dialog box. In this dialog box, access the **Domestic** folder from **English-Imperial > Speciality Equipment** (for Metric: **English > US > Speciality Equipment**) and load **Kitchenette-Medium** (for Metric **M_Kitchenette-Medium**) into the project file.

Adding and Positioning Components in the Building Model

You can now select the component type from the **Type Selector** drop-down list and add it to the building model. Different editing tools like **Rotate**, **Copy**, and **Move** can be used to position the components at their desired location and in the required direction. The steps below describe the procedure for a single component, **Bed-Shaker**. The following steps can then be used to add and position other furniture and plumbing components. As mentioned earlier, the exact location of the components is not important for this tutorial.

1. In the **Type Selector** drop-down list of the **Properties** palette, select the **Bed-Shaker: Double 56" x 78"** (**Bed-Shaker : Double 1422 x 1981 mm** for metric) option, as specified in the parameters for the project.
2. Select the **Rotate after placement** check box in the **Options Bar**.
3. Move the cursor near the west wall of the bedroom and click at the desired location, as shown in Figure 8-20; the component is temporarily placed and the **Rotate** tool is invoked, as shown in Figure 8-21.
4. Move the cursor vertically upward and click when the dashed rectangle (indicating the direction of the bed) is horizontally positioned, as shown in Figure 8-30; the bed is added. Now, press ESC twice. Depending on the point selected for the placement, you may need to use the **Move** tool to move the component to the desired position.

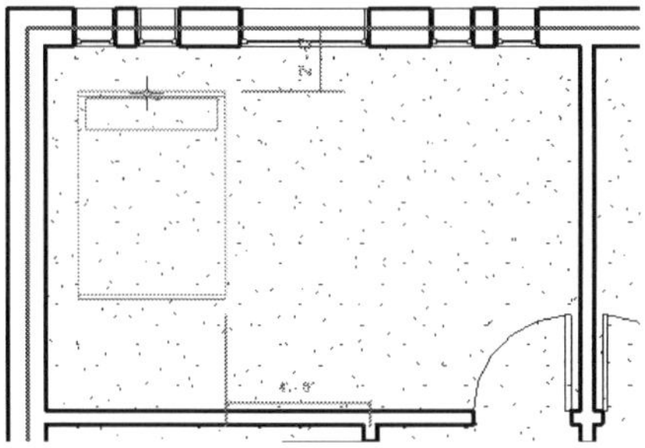

Figure 8-20 Placing a component temporarily

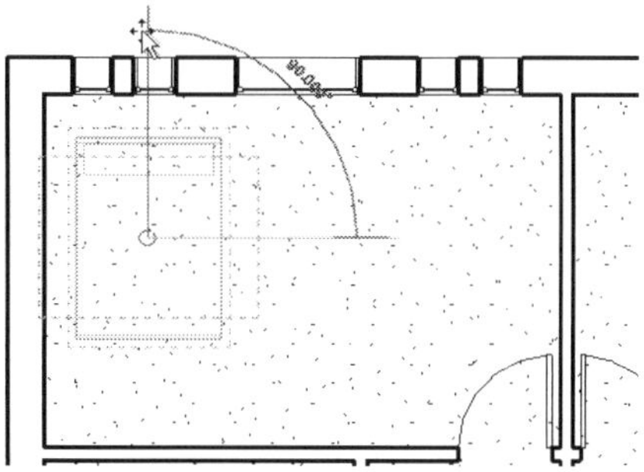

Figure 8-21 Rotating the component

5. Repeat steps 1 to 4 for adding and positioning the components in the building model at locations based on the given sketch plan, refer to Figure 8-19. The components are listed next according to their placements in sketch plan.
   For Imperial: **Entertainment Center: 72"x 72" x 24"**
   Dresser: 72" x 60" x 18"
   Sofa: 72"
   Table-Dining Oval: 36" x 72"
   Table-Coffee: 36" x 72" x 18"
   Chair-Corbu
   Table- End: 24" x 24"
   Kitchenette - Medium: Kitchenette - Medium
   Toilet-Domestic-3D
   Tub-Rectangular-3D
   Sink Vanity-Square: 20" x 18"

   For Metric: **Entertainment Center: 1830 x 1830 x 610 mm**
   Dresser: 1220 x 1525 x 0610 mm
   Sofa: 1830 mm
   Table-Dining Oval: 0915 x 1830 mm
   Table-Coffee: 0915 x 1830 x 0457 mm
   Chair-Corbu
   Table- End: 0610 x 0610 mm
   Kitchenette - Medium : Kitchenette - Medium
   Toilet-Domestic-3D
   Tub-Rectangular-3D
   Sink Vanity-Square : 0500 x 0440 mm
6. After adding all components, press ESC twice.

Tip

On the basis of dimensions, you can place the components from the neighboring elements. To do so, select the component to display its temporary dimensions. Select an appropriate temporary dimension and enter the new value. The component is moved to the exact location.

Modifying the Added Components

The width of the **Kitchenette - Medium** and the height of the **Table- End: 24" x 24"** components can be modified by using the **Type Properties** dialog box.

1. Select the **Kitchenette-Medium: Kitchenette - Medium** component placed in the building model; the **Modify | Specialty Equipment** tab is displayed.

2. Choose the **Edit Type** button from the **Properties** palette; the **Type Properties** dialog box is displayed.

Note

*Create a new type for the **Kitchenette-Medium: Kitchenette - Medium** component if it already does not exist.*

3. In the **Type Properties** dialog box, choose the **Duplicate** button; the **Name** dialog box is displayed.

4. Enter **8'6" (2591 mm)** in the **Name** edit box and choose the **OK** button to close it.

5. In the **Type Properties** dialog box, enter **8'6" (2591 mm)** as the value for the **Cabinet Width** type parameter.

6. Choose the **OK** button to close this dialog box. The **Kitchenette - Medium** component is modified to the new size. You may need to use the **Move** tool to reposition the component at the desired location.

7. Similarly, select all the **Table-End: 24" x 24"** (for Metric **Table- End: 0610 x 0610 mm**) component, access the **Type Properties** dialog box, and then modify the value of the **Height** parameter to **1'3" (381 mm)**. This modifies the height of all instances of the component.

8. Choose the **OK** button; the **Type Properties** dialog box is closed.

   The first floor plan after adding and positioning the components should appear similar to the sketch plan given for this tutorial.

Extending Walls to the Topmost Level

As described earlier, it is better to extend the exterior walls to the top story, instead of stacking them one above the other. To extend the exterior wall height, you need to set the **Top Constraint** parameter to the topmost level and then use the 3D view to view the changes.

1. Double-click on **3D** under the **3D Views** head in the **Project Browser**.

2. Now, move the cursor near the exterior wall and press the TAB key; the chain of exterior walls is highlighted.

3. Click to select the exterior walls; the **Modify | Walls** tab is displayed.

4. In the **Properties** palette for the **Top Constraint** instance parameter, select the **Up to level: Roof** option.

5. Replace the current value of the **Top Offset** parameter with **3'0" (915 mm)** as specified in the project parameters.

6. Choose the **Apply** button in the **Properties** palette. The exterior walls are extended to 3'0" (915mm) above the roof level, as shown in Figure 8-22. Press ESC to exit the tool.

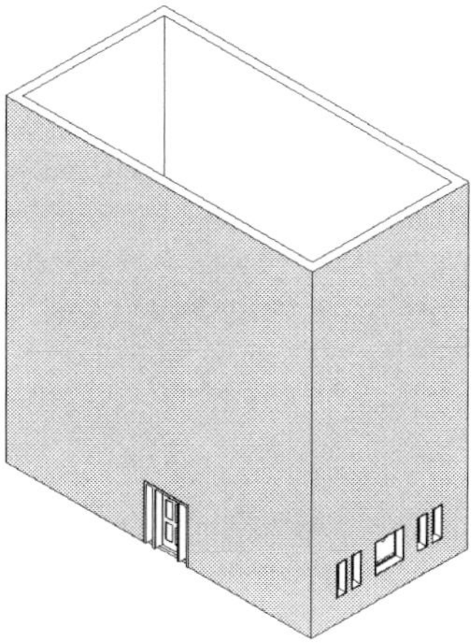

Figure 8-22 *Extending the exterior walls to the topmost level*

Copying Elements to Upper Levels

You can now select all elements except the exterior walls and use the **Aligned to Selected Levels** tool to copy them to the upper levels. The following steps describe the general procedure for copying elements to different levels by using the **Aligned to Selected Levels** tool.

1. In the 3D view, select all elements except the exterior walls and levels by creating a selection window from the left to right. All elements, including the interior walls, floor, ceiling, doors, windows, and components need to be selected.

Note

To exclude the exterior walls and levels, select all elements first and then press the SHIFT key and select the exterior walls and levels.

2. Choose the **Copy to Clipboard** tool from the **Clipboard** panel of the **Modify | Multi-Select** tab.

3. Now, choose the **Aligned to Selected Levels** tool from the **Modify | Multi-Select > Clipboard > Paste** drop-down; the **Select Levels** dialog box is displayed.

4. In this dialog box, select the **Second Floor**, **Third Floor**, and **Fourth Floor** levels by pressing and holding the CTRL button, as shown in Figure 8-23.

5. Choose the **OK** button to close the **Select Levels** dialog box. The selected elements and components are pasted on the three upper levels.

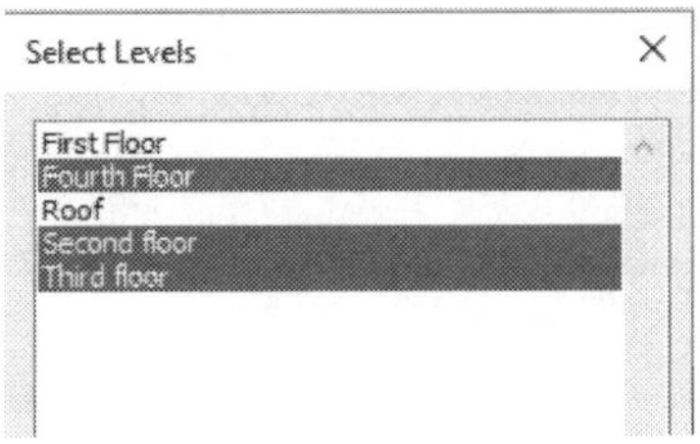

Figure 8-23 *Specifying levels to paste the selected elements*

Adding Roof to the Top Floor

The top floor is still without a roof. Therefore, you need to use the **Roof** tool to create a flat roof. It is recommended that you create the roof in its corresponding floor plan.

1. Double-click on **Roof** from the **Floor Plans** head in the **Project Browser** to display the roof plan.

2. Choose the **Roof by Footprint** tool from **Architecture > Build > Roof** drop-down; the **Modify | Create Roof Footprint** contextual tab is displayed.

3. Choose the **Pick Walls** tool from the **Draw** panel, if not chosen by default, and then clear the **Defines slope** check box in the **Options Bar**, if selected.

4. Move the cursor near the exterior wall of the plan, and when it is highlighted, press the TAB key to highlight the chain of exterior walls. Click to select them. The roof boundary is displayed. Ensure that it is displayed on the inner face. If it is not displayed on the inner face, click on the flip arrow to flip the selection.

5. Choose the **Finish Edit Mode** button from the **Mode** panel to create the roof boundary at the roof plan level.

6. Next, select the **Basic Roof: Generic - 9"** option (for Metric **Basic Roof: Generic - 400 mm**) from the **Type Selector** drop-down list in the **Properties** palette as specified in the project parameters.

7. Double-click on **3D** under the **3D Views** head in the **Project Browser** to display a 3D view of the multistory apartment. You can use the **ViewCube** tool from the drawing window to view the other sides of the building model, as shown in Figure 8-24.

Figure 8-24 Completed multistory apartment

8. Double-click on **First Floor** from the **Floor Plans** head in the **Project Browser** to return to the first floor plan.

Saving the Project

In this section, you will save the project file using the **Save As** tool.

1. Choose **Save As > Project** from the **File** menu and enter the desired name in the **File name** edit box of the **Save As** dialog box.

   For Imperial **c08_Apartment1_tut1**
   For Metric **M_c08_Apartment1_tut1**

   Next, choose **Save**; the project file is saved.

2. Choose **Close** from the **File** menu to close the project file.

EXERCISE

Exercise 1 Apartment 2

Add furniture and sanitary components to the *c07_Apartment2_ex1.rvt* project created in Exercise 1 of Chapter 7. Approximate location of the components is given in Figure 8-25. The text has been given only for reference and is not to be created. Extend the exterior walls to 3'0" above the roof level and copy all interior walls, interior doors, windows, and components at the same location on all the three upper floors. Also, create a flat roof at the roof level. Use the following parameters for various components to be added. The Northeast 3D view of the completed project should appear similar to the illustration shown in Figure 8-26.

(Expected time: 30 min)

Alphabets in Figure 8-25 represent the furniture and plumbing components to be used, as given below.

1. Furniture Components

   For Imperial: Furniture components from the **Furniture** folder of the **English-Imperial**:

   A- **Bed Shaker: Double 56" x 78"**
   B- **Entertainment Center: 72" x 72" x 24"**
   C- **Dresser- 72" x 60" x 18"**
   D- **Sofa: 72"**
   E- **Table-Ellipse: 72" x 36"**
   F- **Table-Coffee: 36" x 72" x 18"**
   G- **Table-End: 24" x 24"**

   For Metric: Furniture components from the **Furniture** folder of the **English**:

   A- **Bed Shaker: Double 1422 x 1981 mm**
   B- **Entertainment Center: 1830 x 1830 x 0610 mm**
   C- **Dresser- 1220 x 1525 x 0610 mm**
   D- **Sofa: 1830 mm**
   E- **Table-Ellipse: 1800 x 0900 mm**
   F- **Table-Coffee: 0915 x 1830 x 457 mm**
   G- **Table-End: 0610 x 0610 mm**

2. Plumbing Components

   For Imperial: Plumbing components from the **Plumbing Fixtures** folder of the **English-Imperial**:

   J- **Kitchenette-Medium**: **8'6"** (modified size)
   K- **Toilet-Domestic-3D**
   L- **Tub-Rectangular-3D**
   M- **Sink Vanity- Square: 24" x 19"**

   For Metric: Plumbing components from the **Plumbing Fixtures** folder of the **English**:

   J- **Kitchenette-Medium**: **2591 mm** (modified size)
   K- **Toilet-Domestic-3D**
   L- **Tub-Rectangular-3D**
   M- **Sink Vanity- Square 0610 x 0483 mm**

3. Roof type- **For Imperial** **Basic Roof: Generic 400mm** - Flat.

4. Name of the file to be saved-

   For Imperial **c08_Apartment2_ex1**

For Metric **M_c08_Apartment2_ex1**

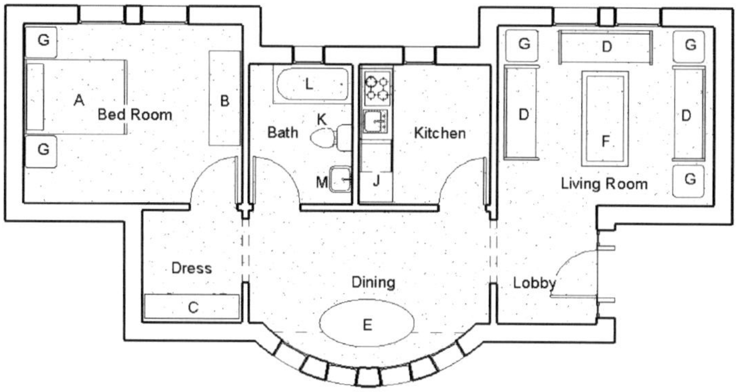

Figure 8-25 Sketch plan for adding components to the Apartment 2 project

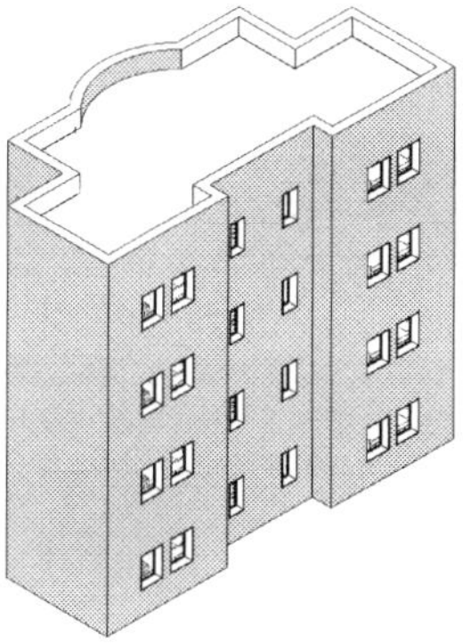

Figure 8-26 The Northeast 3D view of the Apartment 2 project

This page is intentionally left blank.

Chapter 9

Adding Site Features

Learning Objectives

After completing this chapter, you will be able to:
- *Create contoured sites using the Toposolid tool*
- *Modify the site parameters*
- *Add property lines to the site plan using the Property Line tool*
- *Add site components and plantations to the site plan*

INTRODUCTION

In the previous chapters, you learned the usage of different building elements such as walls, doors, windows, stairs, ramps, curtain wall, and so on. Now, in this chapter, you will learn to create contoured sites and modify the site parameters. Moreover, you will also learn to add property line, building pad, parking components, and so on to the site plan with the help of various tools available in Revit.

BIM
CONCEPTS

Creating a site or site feature is one of the important tasks in conceptual phase of the Building Information Modeling project. The site modeling helps in taking various types of decisions related to geometry of the building, drainage, environmental impact, and construction planning. Site modeling can be useful for 6D BIM process.

WORKING WITH SITE FEATURES

Autodesk Revit provides you with various tools to create site features. These tools can be accessed from the **Model Site** and **Modify Site** panels in the **Massing & Site** tab, refer to Figure 9-1.

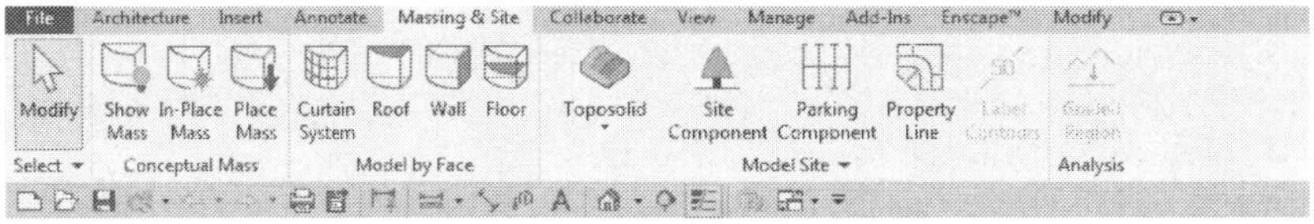

Figure 9-1 *Tools in the **Massing & Site** tab*

Note
*In the **Massing Site** tab, the tools in the **Conceptual Mass** and **Model by Face** panels are used for creating building massing and are not useful for creating site features.*

Tip
The visibility of topographic points can be controlled by using their different sub-categories: Boundary and Interior. Revit classifies points automatically based on their use.

Creating a Toposolid

Ribbon: Massing & Site > Model Site >Toposolid drop-down > Toposolid form Sketch

Autodesk has replaced the **Toposurface** category with a new category: the **Toposolid**. The toposolid exhibits similarities to shape-edited floors, effectively merging the characteristics of both floors and toposurfaces found in earlier Revit versions.

To create a toposolid, make sure the site plan view is displayed by default. Now, choose the **Create from Sketch** tool from th**e Massing&Site** > **Model Site** > **Toposolid** drop-down; the **Modify | Create Toposolid Boundary** contextual tab will be displayed.

In the **Draw** panel of the **Modify | Create Toposolid Boundary** contextual tab, the **Boundary Line** tool is chosen by default. As a result, the list of sketching tools will be displayed in the tool area next to it. You can choose a specific tool from the sketching area to sketch the boundary of the toposolid. For illustration, choose the **Line** tool from the sketching area and click in the model area to start sketching the boundary of the toposolid. Next, create a closed sketch by clicking more points, as shown in Figure 9-2.

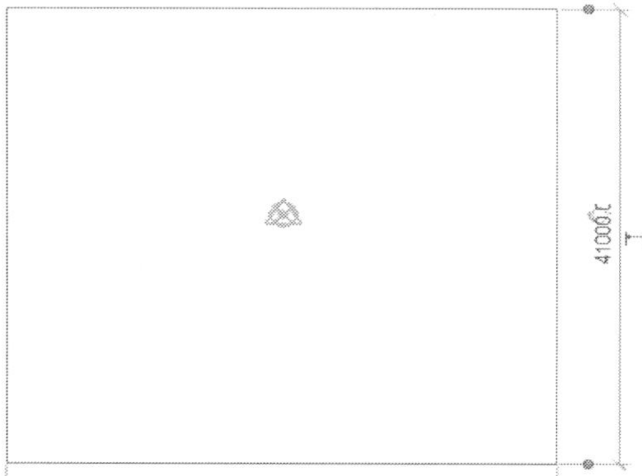

Figure 9-2 *Sketching the boundaries of the toposolid*

Next, in the **Properties** palette, select a specified type from the **Type Selector** drop-down list. For illustration, select the default type from the **Type Selector** drop-down list. In the **Properties** palette, make sure the current level is specified corresponding to the **Level** parameter. After specifying different parameters in the **Properties** palette, choose the **Finish Edit Mode** button; the **Modify | Toposolid** contextual tab will be displayed. You can use this tab to modify sub-elements of the toposolid, edit the boundary of the toposolid, and sub divide the toposolid. You will learn more about these options later in this chapter. Now, choose the **Modify** button from the **Select** panel; the chosen tool will exit and the toposolid will be created as shown in Figure 9-3.

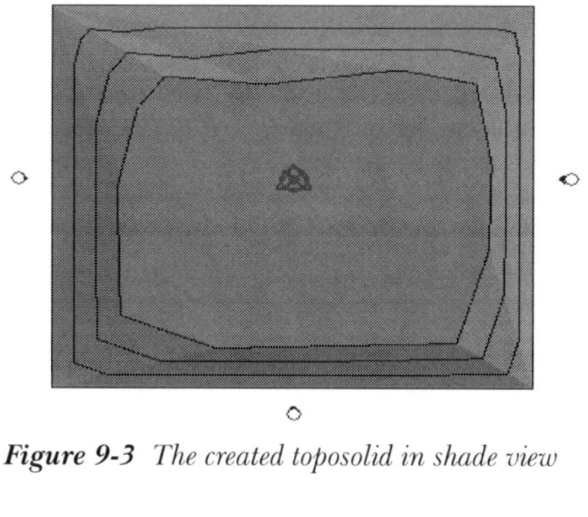

Figure 9-3 *The created toposolid in shade view*

Modifying Toposolid and Sub Elements

To modify the toposolid and add contours to it, choose the toposolid from the model area; the **Modify | Toposolid** contextual tab will be displayed. In the displayed tab choose the **Modify Sub Element** tool from the **Shape Editing** panel; the selected toposolid will turn into sub-editing mode with edges and grips displayed in its boundary, as shown in Figure 9-4.

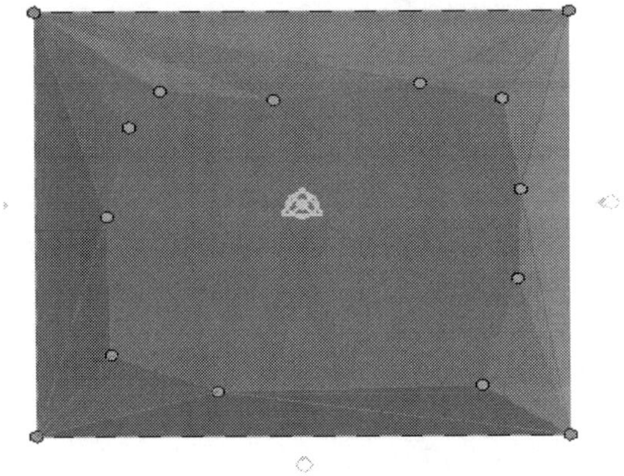

Figure 9-4 *The grips and edges displayed for sub-editing the toposolid*

In the **Shape Editing** panel of the **Modify | Toposolid** tab, you can choose the **Add Point**, **Add Split Line**, or **Pick Support** tool to reshape the selected toposolid and add more contour lines to it.

For the illustration, choose the **Add Point** tool; the **Add Point** panel will be displayed in the existing contextual tab. In the **Add Point** panel; the **Absolute** button will be chosen by default, as a result the points to be added in the toposolid will be referred from the absolute zero of the project.

Next, select a specified option from the **Elevation Base** drop-down list and specify a base elevation in the **Elevation** edit box. In the **Elevation Base** drop-down list, the **Current Level** option will be selected by default. As a result the elevation specified in the **Elevation** edit box will be referred from the current level. Specify an elevation value in the **Elevation** edit box and click at an appropriate area inside the toposolid to add a point of the specified elevation. Next, insert the desired points inside the toposolid to define the contour of the toposolid at the specified elevation. On doing so, the graded surface will be created in the toposolid to represent the site feature. Further, you can add more points in the toposolid to create the surface of the toposolid as per the site condition. After creating the toposolid, you can modify the instance and type properties of the toposolid, as discussed next.

Modifying the Instance Properties of the Toposolid

To modify the instance properties of the toposolid, select it from the model area; the properties will be displayed in the **Properties** palette, as shown in Figure 9-5. In the **Properties** palette, under the **Constraints** head, you can specify the associated level and the height offset from the associated level of the selected topology. The **Properties** palette for the selected topology also display option related to the slope, perimeter, area, and other measurements of the toposolid.

Modifying the Type Properties of the Toposolid

To modify the created toposolid, select it from the project. On doing so, the instance properties for the chosen toposolid will be displayed in the **Properties** palette. Inside the **Properties** palette, choose the **Edit Type** button; the **Type Properties** dialog box will be displayed, as shown in Figure 9-6.

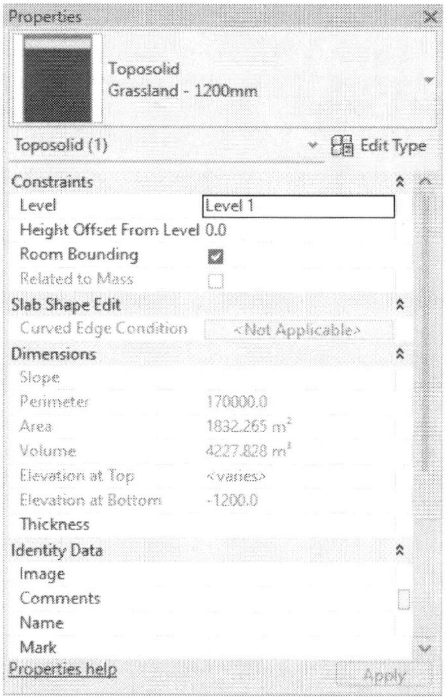

Figure 9-5 *Instance properties of the selected toposolid*

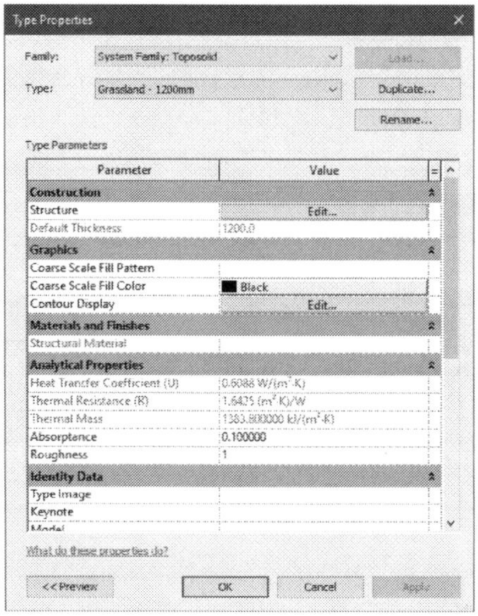

Figure 9-6 *The type properties of the selected toposolid*

In the **Type Properties** dialog box, navigate to the **Structure** parameter and choose the **Edit** button next to it; the **Edit Assembly** dialog box will be displayed, as shown in Figure 9-7.

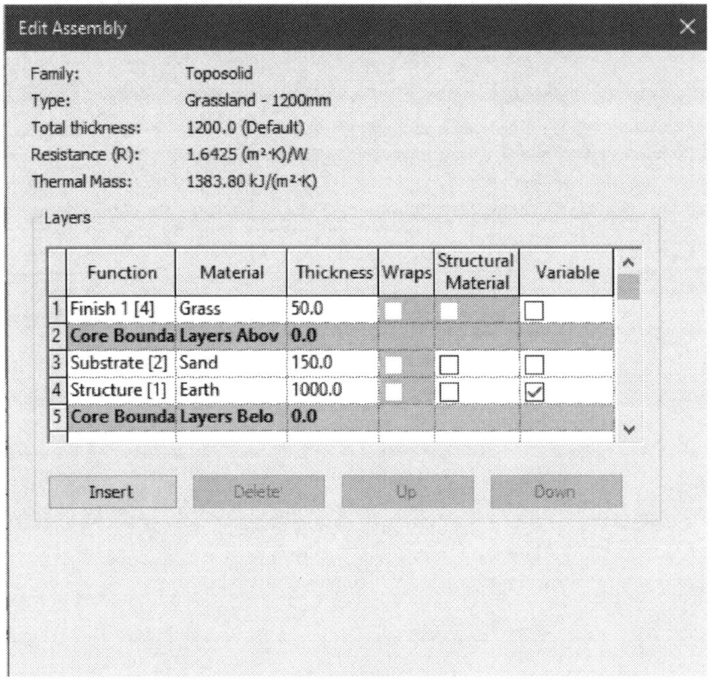

*Figure 9-7 The **Edit Assembly** dialog box*

In the **Edit Assembly** dialog box, you have the option to define different layers for the toposolid based on the cross-section of the groundwork. After setting up the layers, confirm your choices by choosing the **OK** button; the **Edit Assembly** dialog box will close, returning you to the **Type Properties** dialog box. In the **Type Properties** dialog box, navigate to the **Contour Display** parameter and click on its corresponding **Edit** button; the **Contour Display** dialog box will open, as shown in Figure 9-8.

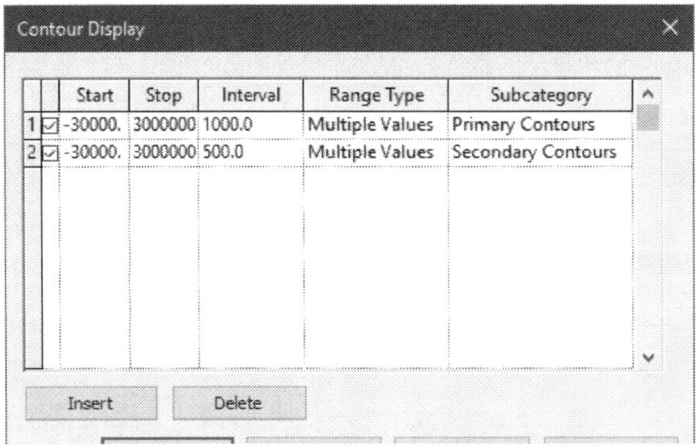

*Figure 9-8 The **Contour Display** dialog box*

In the **Contour Display** dialog box, you can define the start and end display values for both primary and secondary contours. Additionally, you can specify the intervals at which these contours will be displayed within the toposolid.

To include additional subcategories like folding lines, split lines, and common edges within the toposolid, you can choose the **Insert** button to specify the desired subcategory along with other relevant parameters. Once you have configured the contour display parameters and any additional subcategories, confirm your choices by choosing the **Apply** button, followed by the **OK** button. This will apply the settings to the toposolid and close the **Contour Display** dialog box.

In the **Type Properties** dialog box for the selected topology, you can input any other necessary parameters and then choose the **Apply** and **OK** buttons to apply the changes and close the dialog box.

Creating Sub-divisions

In a toposolid, you have the option to create additional subdivisions, such as roads, pavements, and water bodies, to define various landscape elements within it. To initiate the creation of a subdivision, begin by selecting the established toposolid from the project; the **Modify | Toposolid** contextual tab will be displayed. Within this tab, choose the **Sub-Divide** tool from the **Toposolid Shaping** panel. On doing so, the **Modify | Create Sub-Division Boundary** contextual tab will be displayed, accompanied by sketching tools displayed in the **Draw** panel.

You can use these sketching tools to create the desired sub-division boundary. After completing the sketch, choose the **Finish Edit Mod**e button from the **Mode** panel to complete the sketch. On doing so, the sketching mode closes and the options in the **Modify | Toposolid** contextual tab will be displayed. In that tab, you can choose the **Edit Sketch** tool from the **Mode** panel to make modifications to the sketch if necessary.

In the **Properties** palette, click in the edit box corresponding to the **Material** parameter and choose the **Edit** button that will appear; the **Material Browser** dialog box will be displayed. In this dialog box, you can make a suitable material selection for the designated sub-division. After selecting the appropriate material, confirm your selection by clicking **OK** to close the **Material Browser** dialog box.

In the **Properties** palette for the selected sib-division, you can specify the thickness of the sub-division by typing a desired value in the edit box corresponding to the **Sub-divide Height** parameter. Additionally, in the **Properties** palette, the **Inherit Contours** check box can be toggled to either enable or disable the inheritance of contours from the parent toposolid to the sub-division.

Simplifying Toposolids

You can reduce the number of points inside the toposolid to make it simpler and lighter. To do so, choose the **Simplify Toposolid** tool from the **Toposolid Shaping Panel** of the **Modify | Toposolid** contextual tab. On doing so, the **Simplify Toposolid** dialog box will be displayed. In this dialog box, enter a percentage value in the "**% points to keep** edit box to specify the simplification of the surface. Choose **OK** to simplify the surface according to your specifications and close the **Simplify Toposolid** dialog box.

Excavating Toposolids

You can excavate an existing toposolid by cutting it with an existing floor, roof, or another toposolid that intersects it. To excavate a toposolid, select it from the project; the **Modify | Toposolid** contextual tab will be displayed. Choose the **Excavate** tool from the **Excavate** drop-down in the **Toposolid Shaping** panel of the contextual tab, the selection will be grayed out, and you will be prompted to select element(s) to excavate the toposolid. Next, select an overlapping floor, roof, or toposolid from the project; the selected toposolid will be excavated. Choose the **Modify** button to exit the selection.

If you want to remove the excavation from a toposolid, select the toposolid and then choose the **Remove Excavation** tool from the **Excavate** drop-down in the **Toposolid Shaping** panel of the contextual tab, Revit will prompt you to select the element to stop excavating. Hover the cursor over the excavated area and click when the excavated solid is highlighted; the excavation will be removed from the toposolid.

Grading Toposolids

Ribbon: Massing & Site > Analysis> Graded Region

Grading is configuring the toposolid of the land by adding or removing earthen material to shape it for the project. Autodesk Revit provides you with the option of modifying the toposolid that indicates the shape change in the surface during construction. This can be done by using the **Graded Region** tool. This tool works only with the existing toposolid and is used to grade the topographical surface. To grade the plain topographical surface, choose the **Graded Region** tool from the **Analysis** panel of the **Massing & Site** tab; the **Edit Graded Region** dialog box will be displayed, as shown in Figure 9-9. This dialog box provides you with two options to edit the existing topography.

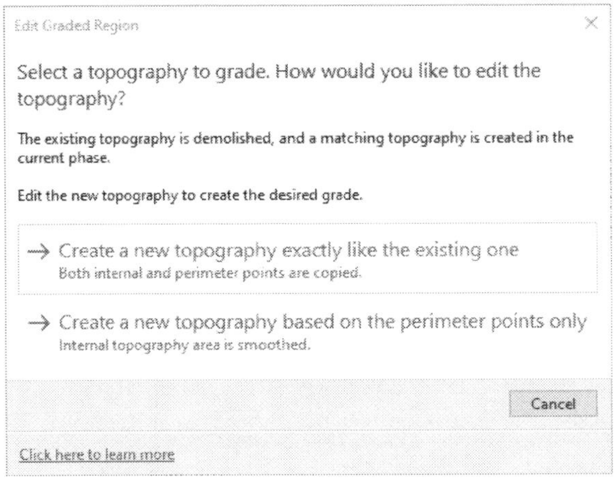

Figure 9-9 *The **Edit Graded Region** dialog box*

Select the **Create a new topography exactly like the existing one** option to create the new topography similar to the existing one but with the graded region. On selecting this option, you will be prompted to select the existing toposolid from the drawing area to create another toposolid as the existing one. Select the toposolid; the **Modify | Toposolid** contextual tab with the existing points of the surface will be displayed and the new toposolid will be generated using those points. Now, you can add more points in this toposolid by using the **Add Point** tool and also by adding the elevation value to it.

Select the **Create a new topography based on the perimeter points only** option to edit the existing toposolid. On selecting the option, you will be prompted to select the toposolid for which you want to edit the points. Select the toposolid; you will be prompted to add other points, if any, or edit the elevation of existing points.

After editing the toposolid, choose the **Modify** button; the new toposolid will be created and displayed with the graded region in the drawing area.

Tip
*The **Graded Region** tool is used to modify the toposurface. You can modify the toposurface phase assignment, define new elevation points on the contoured toposurface, and demolish the old toposurface by using this tool.*

Creating a Topography from Imported Data

Autodesk Revit provides you with the option of importing data and using it to create a toposolid. Based on the type of the data imported, Revit automatically creates a toposolid. The data imported can be a 3d CAD file or a point file in *.csv* file format.

Creating a Toposolid from CSV

To create a toposild from a CSV file, choose the **Create from Import** tool from **Massing & Site > Model Site > Toposolid** drop-down; the **Modify | Create Toposolid from Import** contextual tab will be displayed. In this tab, choose the **Create from CSV** tool from the **Toposolid** panel; the **Select File** dialog box will be displayed. In the **Files of type** drop-down list, select the required file format to import the data. You can select the **CSV Files** option to import the file in the CSV format or select the **Comma delimited Text** to import the file in the text form. Navigate to the required location using the **Look in** drop-down list and select the required file. Next, choose the **Open** button from the **Select File** dialog box; the **Format** dialog box will be displayed. In this dialog box, select an option from the **One unit in the file equals one** drop-down list to assign the drawing units to the imported points. After assigning the units, choose the **OK** button to close the **Format** dialog box. On doing so, the point data will be imported and the toposolid will automatically be created based on the coordinates of points in the imported point file.

Creating a Toposolid from CAD

Another method to create a toposurface is by using the 3D contour data. To do so, you need to import the 3D contour data by using the DWG, DXF or DGN format. You can import the required 3D contour data by choosing the **Import CAD** tool from the **Import** panel of the **Insert** tab.

After you have imported the required data, choose the **Create from Import** tool from **Massing & Site > Model Site > Toposolid** drop-down; the **Modify | Create Toposolid from Import** contextual tab will be displayed. In this tab, choose the **Create from CAD** tool from the **Toposolid** panel and select the imported 3D data from the drawing view; the **Add Points from Selected Layers** dialog box will be displayed. Select the layers to which you want to apply the elevation points and choose the **OK** button; a toposurface based on the elevations points will be placed along the contour lines.

Note

The data imported in the point file must contain the numeric x, y, and z coordinate values in the CSV or the TXT format. The other numeric information about the points should be given after the x, y, and z values in the file.

SETTING THE CONTOUR DISPLAY

You can change the settings of the contours of a toposolid. To do so, select the desired toposolid and then choose the **Edit Type** button from the **Properties** palette; the **Type Properties** dialog box will be displayed. In this dialog box, choose the **Edit** button corresponding to the **Contour Display** parameter; the **Contour Display** dialog box will be displayed, as shown in Figure 9-10.

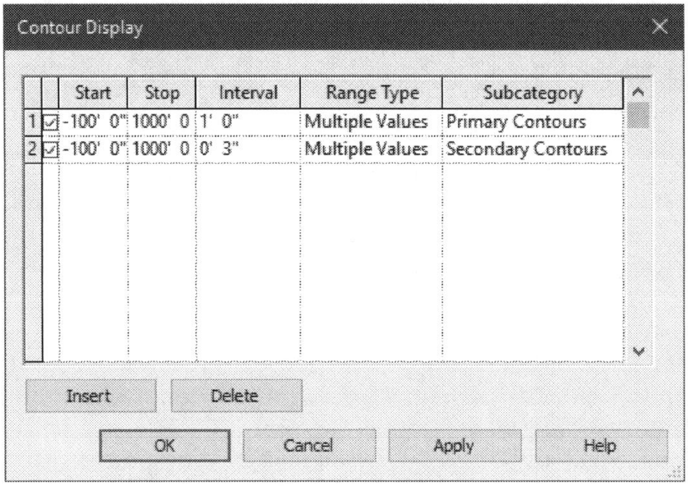

***Figure 9-10** The **Contour Display** dialog box*

In the dialog box, the **Start** column displays the elevation at which the contour lines become visible, and the **Stop** column displays the level at which they are no longer displayed. The **Interval** column value is used to set the interval for each contour. The **Range Type** column can be used to insert single or multiple additional contours. On selecting the **Single Value** option, the **Stop** and **Increment** edit boxes will be disabled. The **Subcategory** column is used to assign the subcategory under which the contours will be added. You can choose the **Insert** button to insert additional contours in the selected toposolid. Choose the **OK** button to close the **Contour Display** dialog box.

ADDING PROPERTY LINES

Ribbon: Massing & Site > Modify Site > Property Line

The **Property Line** tool is used to demarcate and define the extents of the property of a building project based on the survey data. Property lines can be sketched only in the plan views; therefore, these lines can be created only when a plan view, preferably a site plan view, is selected. Property lines can be added by two methods. The first method is by entering distances and bearings and the second method is by sketching. On invoking the **Property Line** tool from the **Massing & Site** tab, the **Create Property Line** window will be displayed, as shown in Figure 9-11. This window displays two options: **Create by entering distances and bearings** and **Create by sketching**. If you select the first option, you will be able to use a table having details of distances and bearings to create a property line. Alternatively, you can choose the second option to create property lines by sketching.

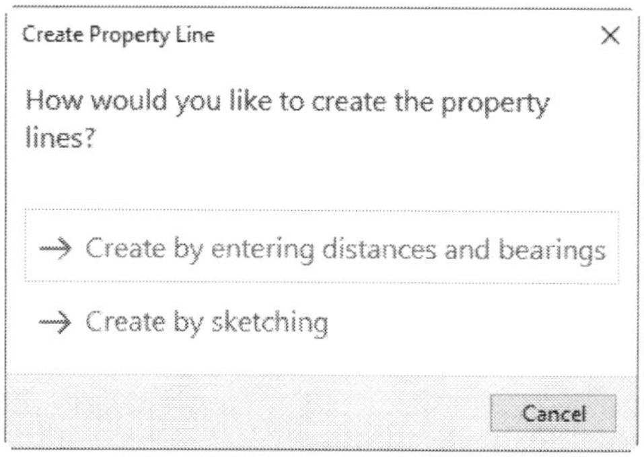

***Figure 9-11** The **Create Property Line** window showing the methods for creating the property line*

Sketching Property Lines

When you choose the **Create by sketching** option from the **Create Property Line** dialog box, the **Modify | Create Property Line Sketch** tab will be displayed. The **Draw** panel of this tab displays various sketching tools that can be used to draw the geometry of a property line. Choose the appropriate sketching tools and then the **Finish Edit Mode** button from the **Mode** panel to exit the sketching option and create the property line. The property line appears as a dashed line in the site plan view. Figure 9-12 shows an example of a property line created for a contoured site (highlighted for illustration).

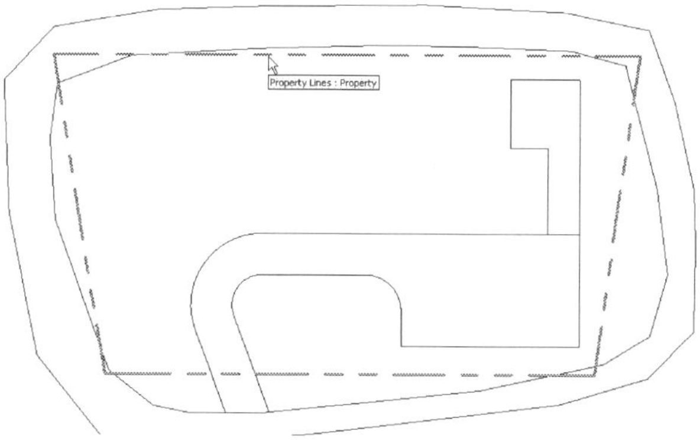

***Figure 9-12** Sketched property lines (highlighted)*

Creating Property Lines Using Distances and Bearings

You can also create property lines by entering survey data into the project. To do so, choose the **Create by entering distances and bearings** option from the **Create Property Line** window; the **Property Lines** dialog box will be displayed, as shown in Figure 9-13.

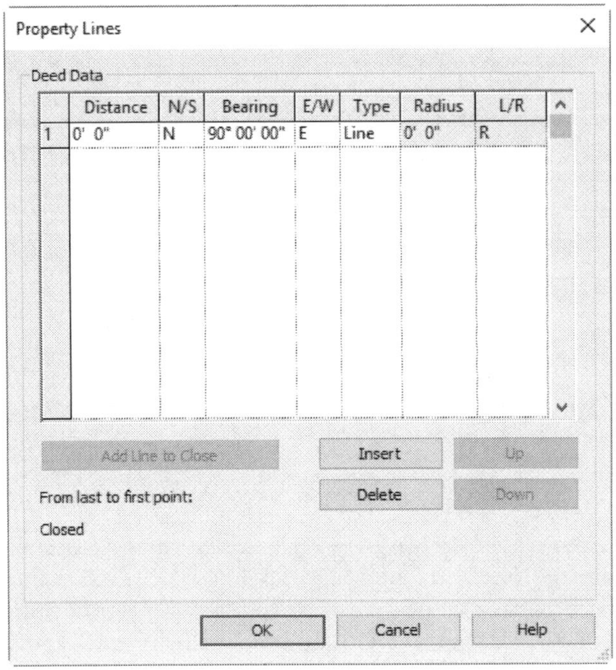

***Figure 9-13** The **Property Lines** dialog box showing the distance and bearing of various property line segments*

In the **Deed Data** area of this dialog box, you can view and enter the bearings and distances of the line segments that constitute the property line. The **Deed Data** area contains seven columns, namely **Distance**, **N/S**, **Bearing**, **E/W**, **Type**, **Radius**, and **L/R**. You can view and edit the bearing distance of the line segment in the **Distance** column.

The **N/S** and **E/W** columns specify the North/South and East/West directions, respectively. Click on the field of the **N/S** column; a drop-down list will be displayed. Select the **N** or the **S** option from the drop-down list to specify the direction reference. Similarly, you can select the **E** or **W** option from the drop-down list in the **E/W** column.

The **Bearing** column in the **Deed Data** area of the **Property Lines** dialog box displays the angle that the line segment of the property line subtends with the directions. To specify angle value of a line segment, click on the corresponding field of the **Bearing** column and enter a value in the edit box displayed. The value displayed in the **Type** column will specify whether a specified line segment will be an arc or a line. Click on the corresponding field of the **Type** column; a drop-down list containing two options, **Line** and **Arc** will be displayed.

To assign a type to the property line segment, select any of these options. Notice that on selecting the **Arc** option from the drop-down list, the fields corresponding to the **Radius** and **L/R** columns will be activated. Click on the field corresponding to the **Radius** column and enter a value in the edit box displayed. Similarly, click on the field of the **L/R** column; a drop-down list will be displayed. From this drop-down list, you can select any of the two options: **L** or **R**. You can select **L**, if you want the arc to appear on the left of the line segment and **R**, if you want the arc to appear on the right of the line segment.

You can choose the **Add Line to Close** button in the **Deed Data** area of the **Property Lines** dialog box to add a line segment such that the line connects the first and the previous line segment to close the geometry. This button will remain inactive in its default state, unless a positive value is entered for the first line segment in the **Distance** column.

You can choose the **Insert**, **Delete**, **Up**, or **Down** button in the **Deed Data** area to add, delete or reorder a line segment of the property line. After you specify the line segments of the property line, choose the **OK** button, the **Property Lines** dialog box will close. Now, click in the drawing area; a property line will be created.

In Autodesk Revit, you can convert a sketched property line to a table-based property line. To do so, select a property line from the drawing and choose the **Edit Table** tool from the **Property Lines** panel in the **Modify | Property Lines** tab. On doing so, Autodesk Revit displays a warning that the sketched property lines once converted into the table-based property lines cannot be converted back. It also warns that the constraints applied while sketching will be lost during the conversion process. You can choose the **Yes** button to proceed. The sketched property lines are converted into the table-based property lines.

The **Deed Data** table shows the distance and bearing of various line segments constituting the selected property line. These bearings are given with respect to the true north setting for the project. To set the true north for the project, ensure that you are in the **Site** view and then click in the value field corresponding to the **Orientation** parameter in the **Properties** palette and select the **True North** option from the drop-down list displayed. Next, choose the **Rotate True North** tool from **Manage > Project Location > Position** drop-down and rotate the site by clicking and specifying the angle in the drawing area.

Tip
*In a project, you can tag a property line. To do so, choose the **Load Family** tool and load the property line tag family file and then use the **Tag by Category** tool.*

ADDING SITE COMPONENTS

Ribbon:	Massing & Site > Model Site > Site Component

You can add different site components such as trees, bollards, building signs, planters, and so on to a site using the **Site Component** tool. The procedure for adding site components is similar to that of adding building components (for more information, refer to the Adding Components section in Chapter 8).

Invoke the **Site Component** tool from the **Model Site** panel in the **Massing & Site** tab; the **Modify | Site Component** tab will be displayed. In the **Properties** palette, select the component to be added from the **Type Selector** drop-down list and insert it into the drawing. You can add other components in your drawing apart from those available in the **Type Selector** drop-down list. To do so, choose the **Load Family** button in the **Mode** panel of the **Modify | Site Component** tab; the **Load Family** dialog box will be displayed. You can use this dialog box to load site components from the libraries available in the **English-Imperial > Site** folder location. This folder contains four subfolders: **Accessories**, **Logistics**, **Parking**, and **Utilities**. Each subfolder contains elements related to its title. For example, the **Utilities** subfolder contains elements such as **Fire Hydrant**, **Manhole**, and so on.

After choosing the desired components from the **Load Family** dialog box, choose the **OK** button; the components chosen will be added to the **Type Selector** drop-down list in the **Properties** palette. Now, you can choose the appropriate component type to add it to the site.

To add planting components to the site, load the plantation elements from the **Planting** subfolder in the **English-Imperial** folder using the **Load Family** dialog box. Next, select the tree type from the **Type Selector** drop-down list and add it to the site in the site plan view, as shown in Figure 9-14.

After you add trees to the site, they automatically get attached to the topographical level at the specified point of placement. The added components can be viewed in the section or 3D view of the site. The trees appear as simple lines unless the view is rendered. Figure 9-15 shows a rendered view of the contoured site with trees.

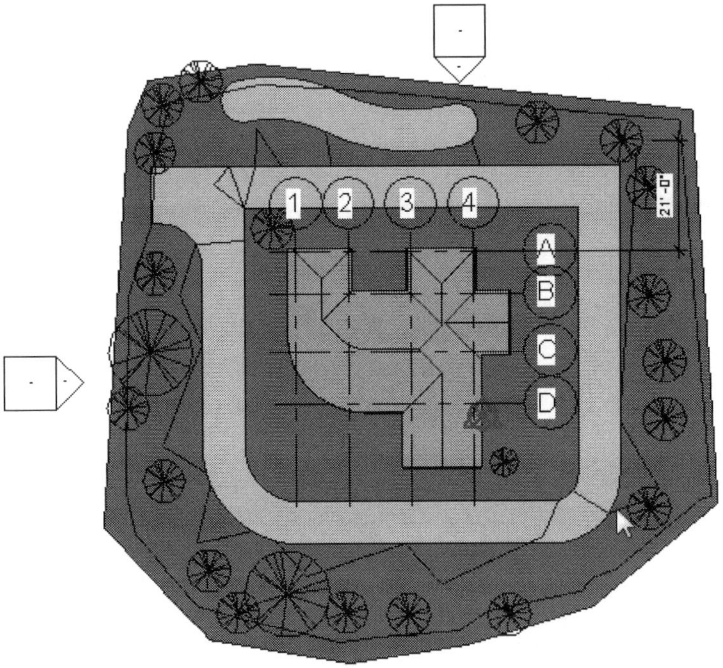

Figure 9-14 *Adding planting components to a contoured site*

Figure 9-15 *Rendered view of the contoured sites with trees*

ADDING PARKING COMPONENTS

Ribbon:	Massing & Site > Model Site > Parking Component

You can add parking components by using the **Parking Component** tool. To add parking components, invoke this tool from the **Model Site** panel; the **Modify | Parking Component** tab will be displayed and the type and properties for the proposed parking components will be displayed in the **Properties** palette. In this palette, select the appropriate type of parking space from the **Type Selector** drop-down list and add it to the site plan. Similar to a building pad, parking components are toposurface hosted components. You can add as many parking spaces as required. The **Rotate after placement** check box can be selected from the **Options Bar** to invoke the **Rotate** tool and rotate the parking component after you have placed it.

ADDING LABELS TO CONTOURS

Ribbon:	Massing & Site > Modify Site > Label Contours

The **Label Contours** tool is used to label the elevation levels of toposurfaces or contours. To invoke this tool and add labels to the contours or toposurfaces displayed in the drawing, choose the **Label Contours** tool from the **Modify Site** panel; the **Modify | Label Contours** tab will be displayed. In this tab, choose the **Edit Type** button from the **Properties** panel; the **Type Properties** dialog box will be displayed. In this dialog box, specify the font and size of the text. Next, choose the **OK** button to accept the specified values and close the dialog box.

To add labels, sketch a line that intersects the contour to be labeled. To create labels for multiple contours, sketch a line that intersects all of them. Autodesk Revit automatically adds labels to each contour based on their respective elevations. Figure 9-16 shows an example of labels added to alternate contours.

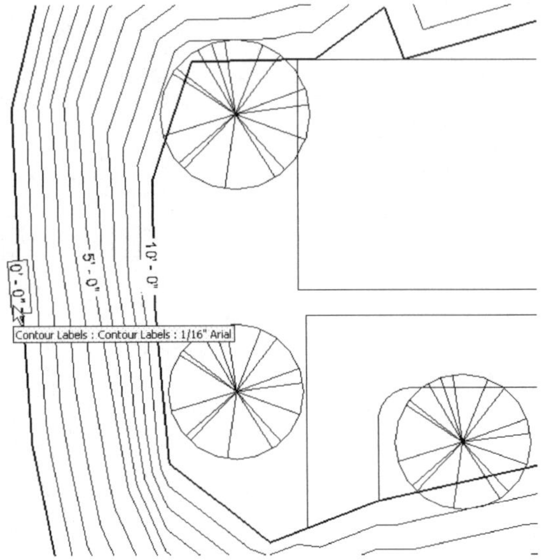

Figure 9-16 *Labels added to the contours*

TUTORIAL

The tutorial of this chapter is available in video format.

Tutorial 1 Site Plan

In this tutorial, you will create a contoured site plan for an apartment complex on the basis of Figure 9-17. To complete this tutorial, you need to download the *c09_SitePlan_tut_start.rvt* file from *www.Cadsoft.com*. Given below are the parameters for various components to be added.

(Expected time: 2hr 15 min)

1. Project file parameters:

   Project file
   For Imperial ***c09_SitePlan_tut_start.rvt***
   For Metric ***M_09_SitePlan_tut_start.rvt***

   File name to be assigned-
   For Imperial ***c09_SitePlan_tut1.rvt***
   For Metric ***M_c09_SitePlan_tut1.rvt***

2. Property size:
   For Imperial **600'0" X 500'0"** (Extents of property line)
   For Metric 180000 X 15000 mm

3. Width of the straight pedestrian walkway- **8'(2400 mm)**, circular- **12'(3600 mm)**.

4. Parking component:
   For Imperial **Parking Space: 9' X 18': 90 deg**
   For Metric **Parking Space: 4800 X 2400 mm- 90 deg**

5. Site component:
   For Imperial **Tree - Deciduous: Red Ash - 25'** placed on Level 2.
   For Metric **RPC Tree - Deciduous: Red Ash- 7.6 Metres**

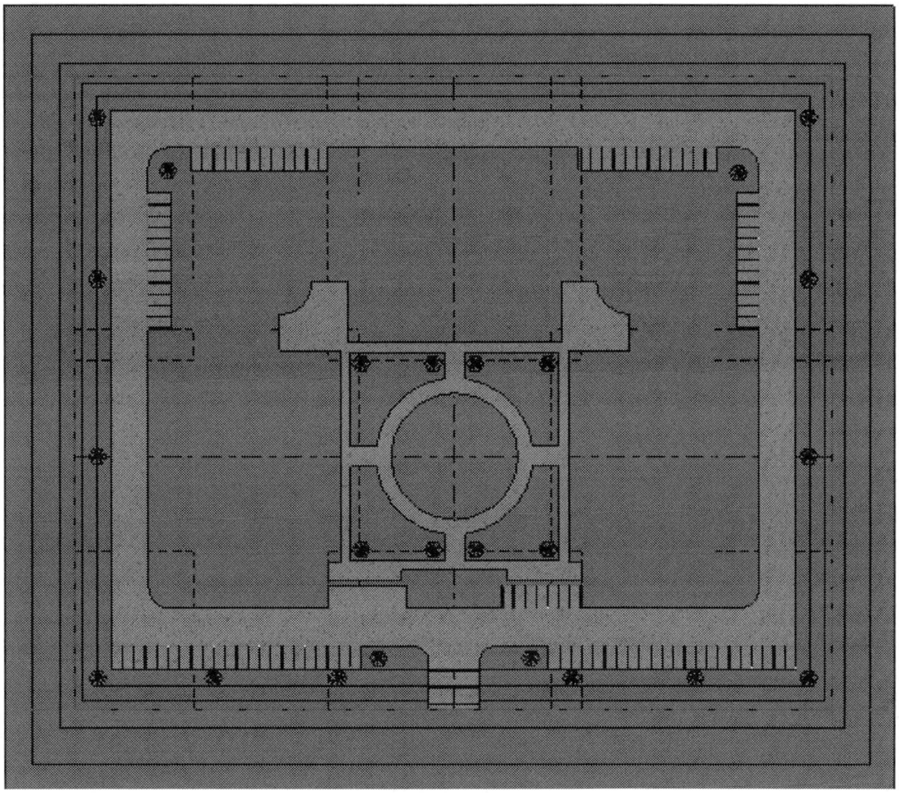

Figure 9-17 *Sketch plan for creating the site plan with contours, site components, and building pads*

The following steps are required to complete this tutorial:

a. Open the *c09_SitePlan_tut_start.rvt* file and open the **Site Plan** view.
b. Use the **Toposolid** > **Create from Sketch** tool to create the toposolid.
c. Use the **Sub-Divide** tool to create roads.
d. Add parking components using the **Parking Component** tool.
e. Add site components using the **Site Component** tool.
f. Change the visibility settings of the toposurface.
g. Save the project using the **Save As** tool.
   For Imperial *c09_SitePlan_tut1.rvt*
   For Metric *M_c09_SitePlan_tut1.rvt*
h. Close the project.

Opening a Project File and the Site Plan View

In this section, you will open an existing project file using the **Open** tool. As mentioned earlier, the toposolid should be created in the **Site Plan** view. Therefore, you will open the **Site Plan** view using the **Project Browser**.

1. Choose the **Open** tool from the **Model** area in the **Home Screen**; the **Open** dialog box is displayed.

2. In the **Open** dialog box, select the *c09_SitePlan_tut_start.rvt* file and then choose the **Open** button; the file opens with the **Site Plan** view, as shown in Figure 9-18. The view includes essential reference planes, model lines, and property lines. You can also download this file from *https://www.Cadsoft.com*. The path of the file is as follows: *Textbooks > Civil/BIM/GIS > Revit Architecture > Exploring Autodesk Revit 2025 for Architecture*.

Modifying the Site Settings

Before creating the toposolid, you need to modify the type properties of the toposolid type to achieve the desired contour layout. In this section, you will use the **Contour Display** dialog box to set parameters for creating additional contours. Then, you will create three elevation points to generate all site contours, as shown in the sketch plan.

1. In the **Project Browser**, navigate to **Families** > **Toposolid** > **Toposolid**, as shown in Figure 9-19. Next, select and right-click on the **Landscape-Surfac**e type; a flyout is displayed.

2. Choose the **Type Properties** option from the displayed flyout; the **Type Properties** dialog box is displayed.

3. In the dialog box, choose the **Edit** button corresponding to the **Contour Display** parameter; the **Contour Display** dialog box is displayed.

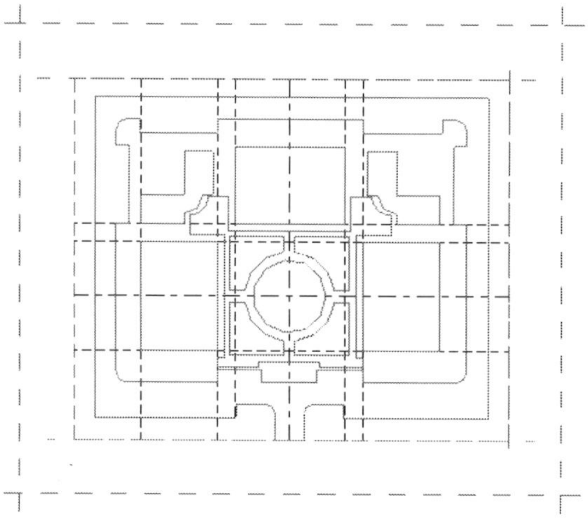

Figure 9-18 *The* ***Site Plan*** *view*

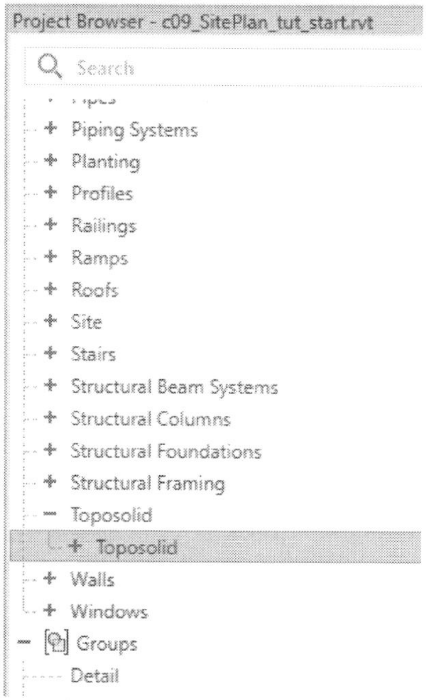

Figure 9-19 *The* ***Project Browser*** *showing* ***Toposolid*** *family type*

4. In the **Contour Display** dialog box, make sure that the interval for primary and secondary contours are set to **5'(1525 mm**) and **1'(304mm**) respectively, as shown in Figure 9-20.

5. Choose **Apply** and then **OK** to close the **Contour Display** dialog box.

6. Choose the **Edit** button corresponding to the **Structure** parameter; the **Edit Assembly** dialog box is displayed.

7. In the **Edit Assembly** dialog box, ensure that the **Material** parameter is set to **Site-Grass** and the **Thickness** parameter is set to **1'(305mm)**, respectively.

8. Choose the **OK** button to close the **Type Properties** dialog box.

Creating the Site Toposolid

In this section, you will use the **Topographies** tool and the **Point** tool and also pick points to generate the desired profile of contours. The contours should be created from the lowest to the highest level to be fully visible. You can first create a topographical surface at the base level and then create two more surfaces at **5'0"(1525 mm)** and **10'0"(3048 mm)** levels, respectively. On doing so, the Autodesk Revit automatically generates the intermediate contours at **1'0"** (**304.8 mm**) level.

1. Choose the **Create from Sketch** tool from the **Model Site > Toposolid** drop-down in the **Massing & Site** tab; the **Modify | Create Toposolid Boundary** contextual tab is displayed.

   Make sure to select the **Landscape-Surface** option as the type in the **Type Selector** drop-down list within the **Properties** palette.

2. In the **Properties** palette, select **Level 1** from the drop-down list next to the **Level** parameter, and make sure the **Height Offset From Level** parameter is set to **0 ' 0"**, by default.

3. In the **Draw** panel of the **Modify | Create Toposolid Boundary** contextual tab, choose the **Boundary Line** option and then choose the **Rectangle** tool from the sketch bar to activate the sketching mode.

4. Position the cursor on the upper-left corner of the site, close to the outer reference planes , as shown in Figure 9-20, and then click to assign the start point of the toposolid.

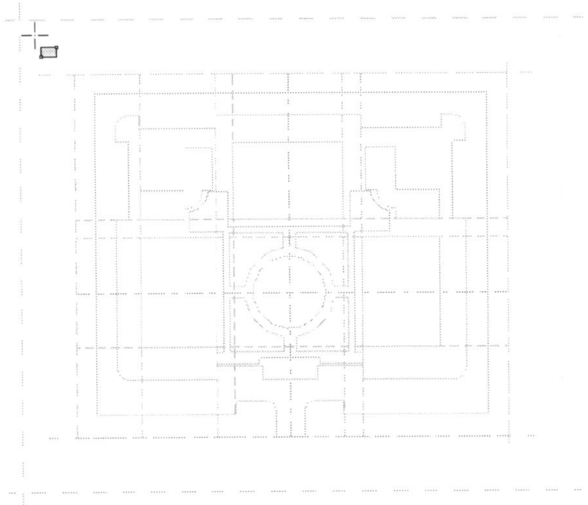

Figure 9-20 *The location to the start point of the toposolid*

5. Move the cursor towards the lower right corner near the intersection of the reference planes, as shown in Figure 9-21, and then click to locate the other corner point of the toposolid.

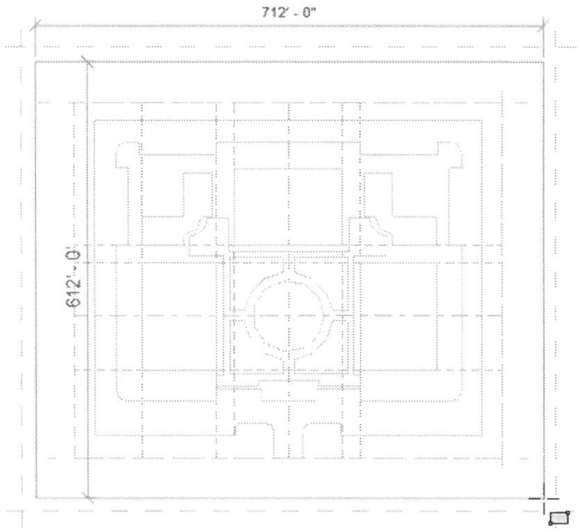

Figure 9-21 *The other corner point of the toposolid*

6. Choose the **Finish Edit Mode** button located in the **Mode** panel of the **Modify | Create Toposolid Boundary** contextual tab to create the toposolid.

Creating Additional Points in the Toposolid

1. Next, highlight the generated toposolid and choose the **Modify Sub Elements** tool within the **Shape Editing** panel located in the **Modify | Toposolid** contextual tab.

2. Choose the **Add Point** tool from the **Shape Editing** panel of the **Modify | Toposolid** contextual tab.

3. Select **Current Level** from the **Elevation Base** drop-down list, then enter **5'0"** (**1525 mm**) in the **Elevation** edit box and press the ENTER key.

4. Next, position the cursor at the top-left corner of the newly created toposolid, as shown in Figure 9-22, and initiate the second set of contour lines by clicking inside it.

5. Click on multiple points along the inner portion of the first toposolid to create the second topographical contour, as shown in Figure 9-23.

6. Choose the **Modify** button from the **Select** panel, Autodesk Revit automatically generates the intermediate contours at **5'0"** level, as shown in Figure 9-24.

7. Similarly, use the **Add Point** tool to generate the third topographical contours, located at the **10'0"** (**3048 mm**) elevation above the current level, as shown in Figure 9-25.

   Note that the solid topography shown in the figure can differ based on the quantity and arrangement of points.

8. After creating the toposolid, press ESC to exit from the creation of the toposolid.

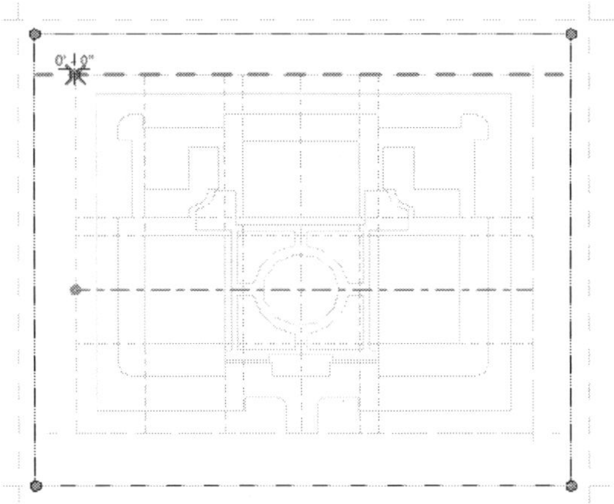

Figure 9-22 *The location of the first point for second set of toposolid*

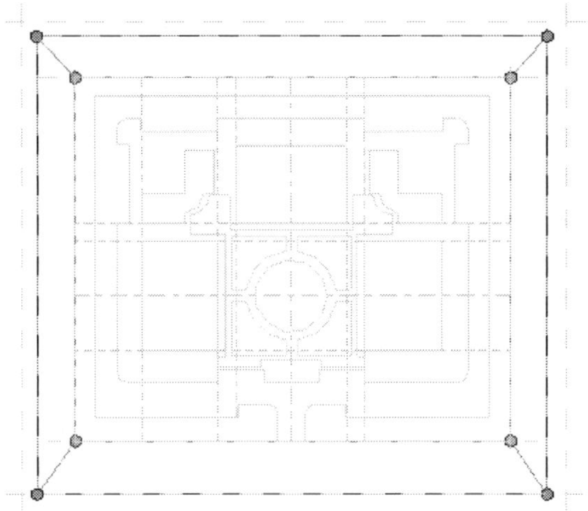

Figure 9-23 *The added points*

Creating Roads Using the Sub-Divide Tool

Now, create roads for the site plan using the **Sub-Divide** tool. You can create a solid on the toposolid in such a manner that a separate solid is created for the road. It can then be assigned a different material. You can use the sketching tools to sketch the road profile.

1. To create the road in the site, select the toposolid created and then choose the **Sub-Divide** tool from the **Toposolid Shaping** panel of the **Modify | Toposolid** contextual tab; the sketching tools are displayed in the **Draw** panel of the **Modify | Create Sub-Division Boundary** contextual tab.

2. Next, choose the **Pick Lines** tool from the sketch bar of the **Draw** panel and then move the cursor and place it over the line in the left side of the site, the line is highlighted, as shown in Figure 9-26.

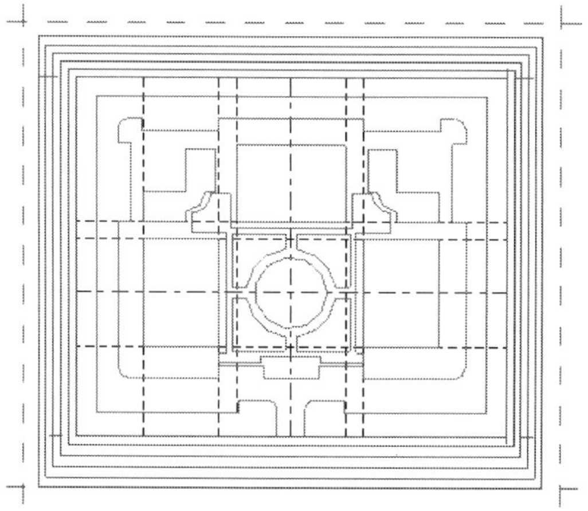

Figure 9-24 *Contours displayed at the* ***5'*** *elevation level*

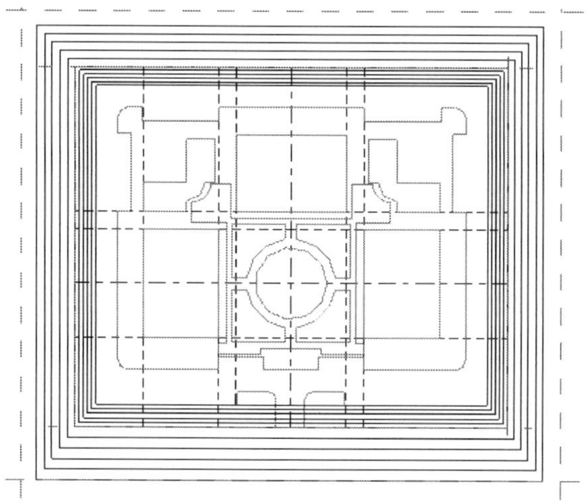

Figure 9-25 *The contours of the third toposolid at 10' elevation level*

3. Next, keeping the line highlighted, press the **Tab** key and then click when the closed region is highlighted.

4 Next, move the cursor inside the closed region to the line, as shown in Figure 9-27. When the line is highlighted press the **Tab** key, the inner closed region is highlighted. Now, click to complete the selection.

5. In the **Properties** palette, click on the edit box corresponding to the **Material** parameter; a button is displayed.

6. Choose the button; the **Material Browser** dialog box is displayed.

7. Click on the **Search** box located on the left in the **Material Browser** dialog box and then type **Concrete** in it; the concrete materials are displayed in the **Project Materials** area.

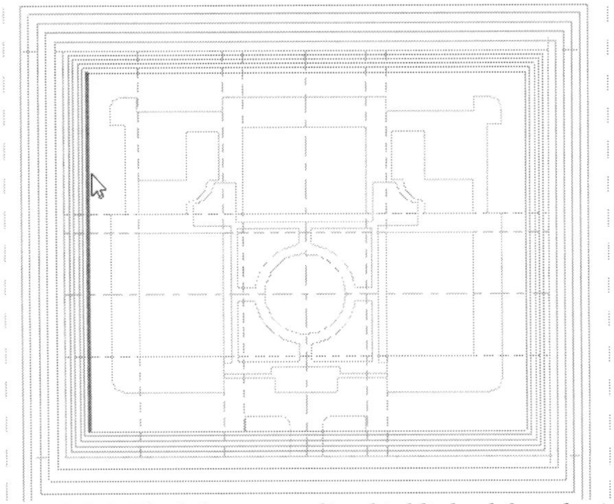

Figure 9-26 *The left property line highlighted for selection*

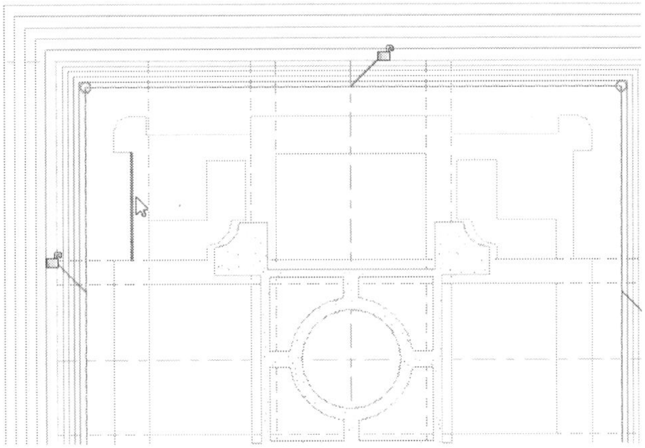

Figure 9-27 *Locating the inner line*

8. Select the **Concrete - Cast-in-Place Lightweight Concrete** material from the **Project Materials** area and choose **OK**; the **Material Browser** is closed and the material selected is assigned to the road.

9. In the **Properties** palette, specify the **Sub-divide Height** parameter to **0'-3"** and press ENTER.

10. Choose the **Finish Edit Mode** button from the **Mode** panel of the **Modify | Create Sub-Division Boundary** contextual tab; the sub-division is created. Press ESC to escape the **Toposolid** tool.

11. Repeat steps 1 to 6 to create the circular pavement inside the building, as shown in Figure 9-28

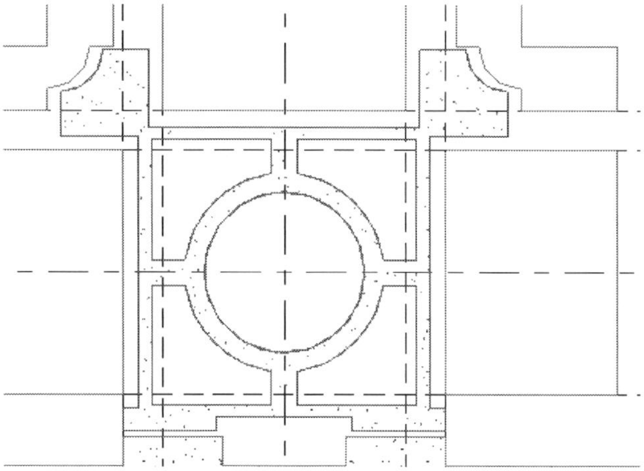

Figure 9-28 *The inner circular pavement created*

Adding Parking Components

In this section, you will add the parking components to the site plan using the **Parking Component** tool.

1. Choose the **Massing & Site** tab and then choose the **Parking Component** tool from the **Model Site** panel; the **Modify |Parking Component** tab is displayed.

2. In the **Properties** palette, select **Parking Space: 9' X 18'-90 deg (4800 X 2400 mm- 90 deg)** option from the **Type Selector** drop-down list, as specified in the component parameters for the project. Also, specify the **Offset from Host** parameter to **3"(75 mm)**.

3. Move the cursor near the entrance of the lower left corner of the site and click near the location shown in Figure 9-29. (You may need to use the **Zoom In Region** tool to enlarge the portion of the site). The selected parking component is added to the site plan. Depending upon the point selected for placement, you may need to use the **Move** or **Flip** tool to orient the component to the desired position.

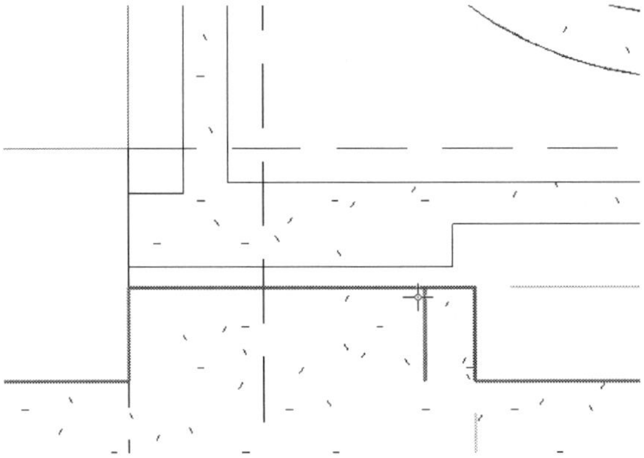

Figure 9-29 *Placing the first parking component*

4. Create multiple copies of the added parking component using the **Array** tool. Choose the **Linear** button from the **Options Bar** and create an array of 7 components placed consecutively, as shown in Figure 9-30.

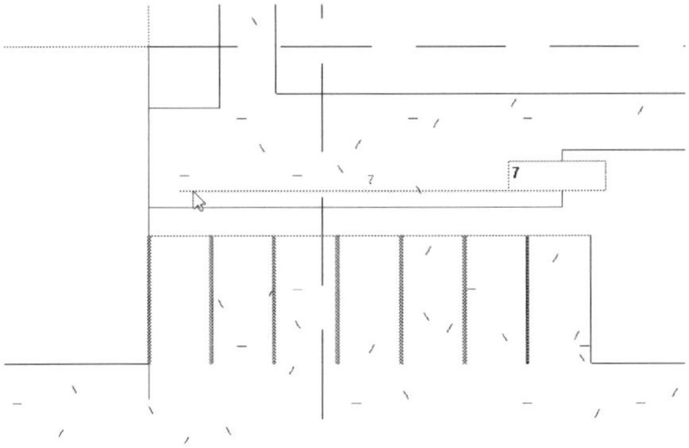

Figure 9-30 *Placing the 7 parking components*

5. Repeat steps 1 to 4 and add the same parking component to the other seven places in the project. The site plan appears similar to the illustration shown in Figure 9-31.

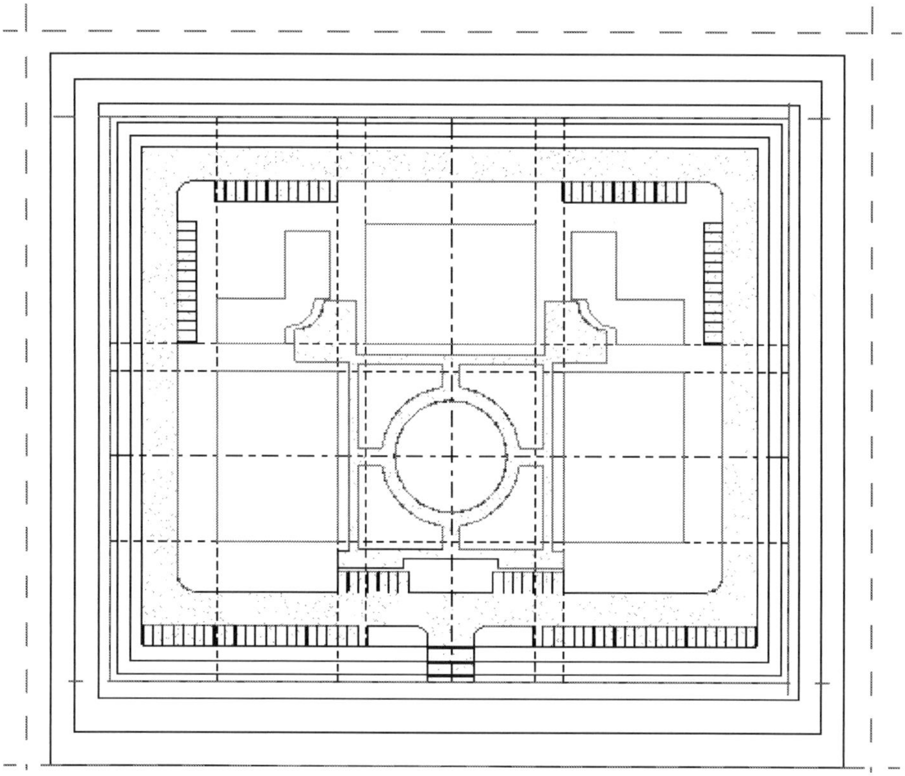

Figure 9-31 All parking components added

Adding Site Components

In this section, you will add site components from the in-built library of Autodesk Revit. The procedure is the same as that of adding parking components.

1. Choose the **Site Componen**t tool from the **Model Site** panel in the **Massing & Site** tab.

2. Next, you need to load the desired site component from the library. To do so, choose the **Load Family** tool from the **Mode** panel; the **Load Family** dialog box is displayed.

3. Now, browse to **English-Imperial > Planting** and select **RPC Tree - Deciduous** and then choose **Open**.

4. Next, in the **Properties** palette, select the required option from the **Type Selector** drop-down list, as specified in the site component parameters for the project.
   For Imperial **RPC Tree - Deciduous: Red Ash-25'**
   For Metric **RPC Tree - Deciduous: Red Ash- 7.6 Meters**

5. Move the cursor to the site plan and click to add the planting component. Add multiple trees by clicking points on the site plan. The trees can be placed at an approximate location, as given in the sketch plan.

6. Next, select all trees from the site plan; the instance properties of the selected trees are displayed in the **Properties** palette.

7. Now, select **Level 2** from the drop-down list corresponding to the **Level** parameter.

   The complete site plan with site components appears similar to the illustration given in Figure 9-32.

Creating Sections through the Site

You can use the **Section** tool to create section lines horizontally and vertically through the site.

1. Choose the **View** tab and then invoke the **Section** tool from the **Create** panel.

2. Create horizontal section line from the left to the right across the center of the design scheme.

3. Similarly, create a vertical section line from the bottom to the top across the center. The two section lines are created and their corresponding views are added under the **Sections (Building Section)** head in the **Project Browser**.

4. Using the **Project Browser**, rename the horizontal section **Section1** as **Longitudinal Section** and the vertical section **Section 2** as **Transverse Section**.

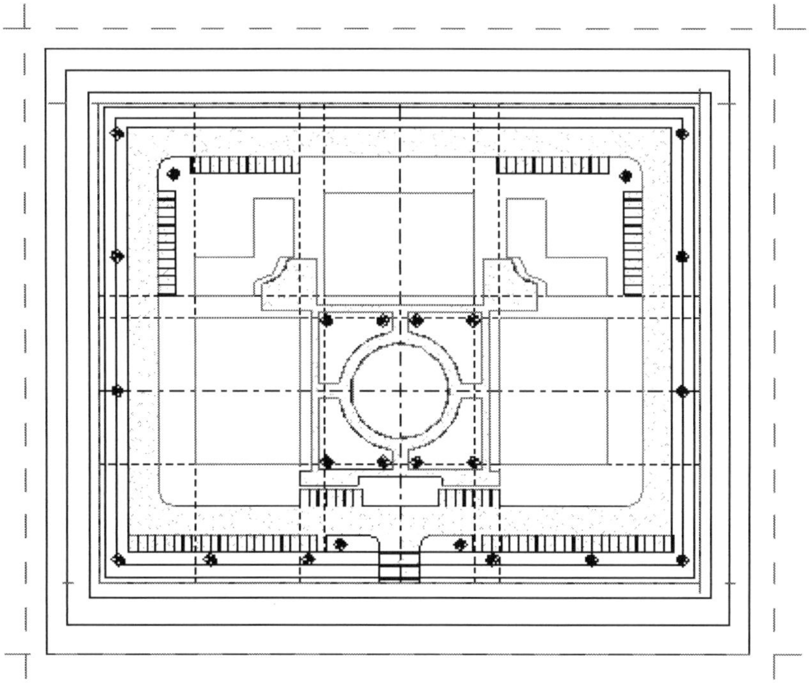

Figure 9-32 *Complete site plan with all components added*

5. Double-click on **Longitudinal Section** under the **Sections (Building Section)** head in the **Project Browser** to view the longitudinal section.

6. Double-click on **Transverse Section** under the **Sections (Building Section)** head in the **Project Browser** to view the transverse section.

   You may need to increase the view range box by dragging the section box from the top or the bottom. Also, drag the levels to the sides of the section view. Ensure that the scale of the view shown in the **View Control Bar** is 1"=20'0" (1: 100).

7. Choose **Visual Styles > Shaded** from the **View Control Bar** to display the shaded view of the Longitudinal and Transverse sections, as shown in Figure 9-33 and Figure 9-34.

8. Similarly, open the site plan and choose **Visual Styles > Consistent Colors** from the **View Control Bar** to display the shaded view of the site plan, as shown in Figure 9-35.

Setting the Visibility of the Sub-Category of Toposurface

In this section, you will change the topographical setting and visibility of primary and secondary contours in the toposurface.

1. Select the toposolid and then choose the **Edit Type** button from the **Properties** palette; the**Type Properties** dialog box is displayed.

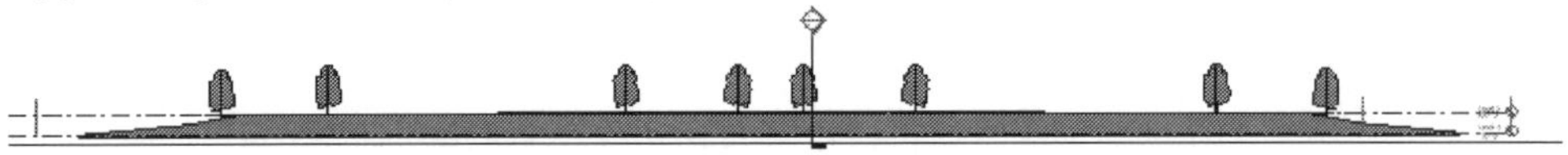

Figure 9-33 *The longitudinal section*

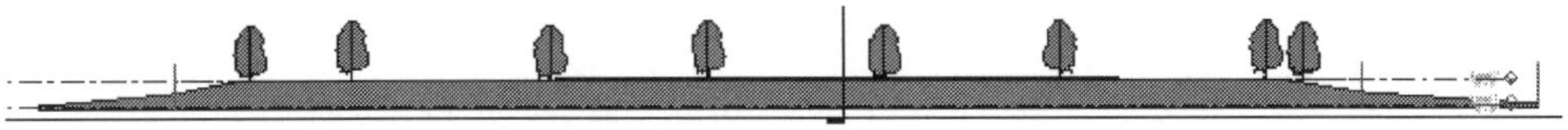

Figure 9-34 *The transverse section*

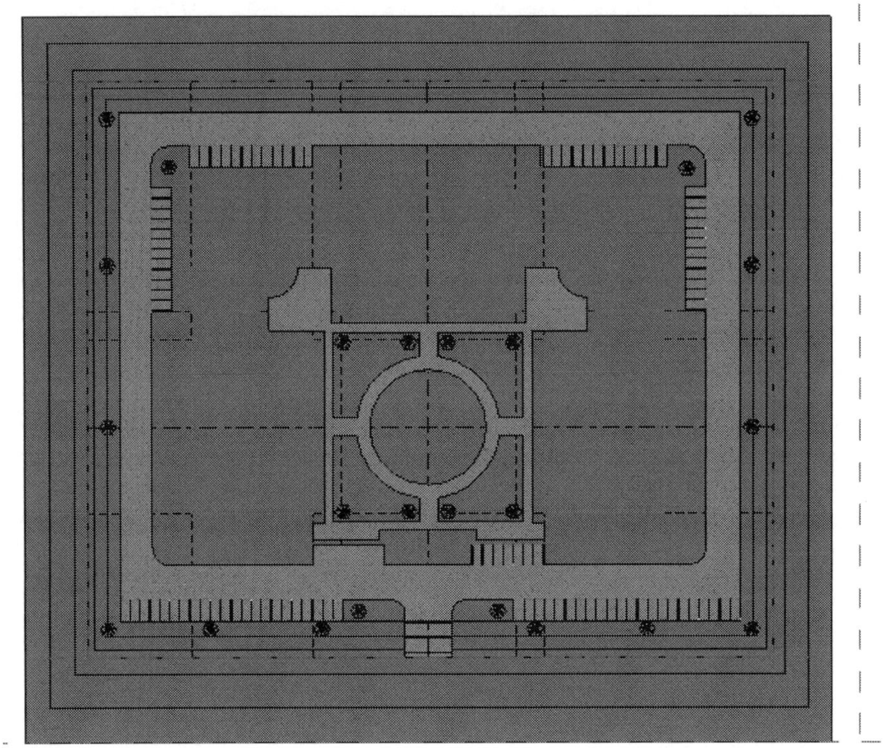

Figure 9-35 *Complete site plan with consistent colour visual style*

2. In the **Type Properties** dialog box, choose the **Edit** button corresponding to the **Contour Display** parameter; the **Contour Display** dialog box is displayed.

3. Click on the cell displayed in the **Interval** column of the **2** row and change the current value to **3"** (**76mm**).

4. Choose **Apply** and then **OK**; the **Site Settings** dialog box is closed.

5. Next, choose the **Visibility/Graphics** tool from the **Graphics** panel of the **View** tab; the **Visibility /Graphic Overrides for Floor Plan: Site** dialog box is displayed.

6. Ensure that the **Model Categories** tab is chosen in the **Visibility /Graphic Overrides for Floor Plan: Site** dialog box, expand the **Topography** category, and then select the **Primary Contours** sub-category; the selected sub-category is highlighted in blue.

7. Now, click on the button in the **Line** column under the **Projection/Surface** heading; the **Line Graphics** dialog box is displayed.

8. Select the **Color** button in the **Lines** area of the **Line Graphics** dialog box; the **Colors** dialog box is displayed.

9. Select the blue color displayed under the **Custom** colors area and choose the **OK** button; the **Color** dialog box is closed.

10. Next, choose the **OK** button from the **Line Graphics** dialog box; the dialog box is closed.

11. Now, choose **Apply** and **OK** button to close the **Visibility/Graphics Overrides for Floor Plan: Site** dialog box.

Note

After performing the steps as mentioned in the section, you can now see the secondary contours at a specified spacing and also see the change of the color of primary contours to blue.

Saving the Project

In this section, you will save the project file using the **Save As** tool.

1. Choose **Save As > Project** from the **File** menu to save the project. Specify the file name in the **File name** edit box and choose **Save**.

   For Imperial **c09_SitePlan_tut1**
   For Metric **M_c09_SitePlan_tut1**

   The project file is saved.

2. Choose **Close** from the **File** menu to close the project file.

Tip

*The modified material properties are not visible in the hidden lines view type. You can choose the **Visual Style** button in the **View Control Bar** and then choose the **Shaded** option from the menu to graphically view the modified materials.*

Self-Evaluation Test

Answer the following questions and then compare them to those given at the end of this chapter:

1. Which of the following tools helps to insert the entourages in the project?

   (a) **Parking Component** (b) **Topographies**
   (c) **Site Component** (d) **Split Surface**

2. Which of the following tools helps to define the extents of the project?

   (a) **Property Line** (b) **Graded Region**
   (c) **Subregion** (d) **Building Pad**

3. The __________ tool is used to create a road.

4. The depth of the cross-section of the toposolid can be specified in the ________ edit box of the **Edit Assembly** dialog box.

5. The __________ tool is used to add a point in the toposolid.

6. The __________ option can be chosen to invoke the **Rotate** tool immediately after placing a parking component.

7. After creating a toposolid, you can choose the __________ button from the **Modify | Toposolid** tab to edit the sketch of it.

8. You can pick points at different elevation levels while creating a toposolid. (T/F)

9. When you pick points to create a toposolid, they automatically join to form a loop. (T/F)

10. A toposolid can be split along a straight line only. (T/F)

Review Questions

Answer the following questions:

1. Which of the following parameters can be used to specify the add elevation points in the toposolid from the base level?

   (a) **Add Point** (b) **Add Split Line**
   (c) **Reset Shape** (d) **Pick Support**

2. Which of the following tools is used to create a road on a toposolid?

   (a) **Sub-division** (b) **Sub Region**
   (c) **Sub-Divide** (d) **None**

3. Which of the following contour setting parameters can be used to create intervals in contours?

   (a) **Interval** (b) **Start**
   (c) **Stop** (d) **None**

4. While creating a toposolid, the **Current Level** option for **Elevation Base** parameter is used to pick points relative to the base level. (T/F)

5. The properties of a toposolid cannot be modified after its creation. (T/F)

EXERCISES

Exercise 1 Site Plan

Create the peripheral pedestrian walkway for the *Site Plan* project created in Tutorial 1 of this chapter, as shown in Figure 9-36. The sketch plan shows the pathway in white. You can assume the width of the walkway at various locations. Assign the material **Site - Sand** to the created pathway surface. Save the file as *c09_SitePlan_ex1.rvt (M_c09_SitePlan_ex1.rvt).*

(Expected time: 30 min)

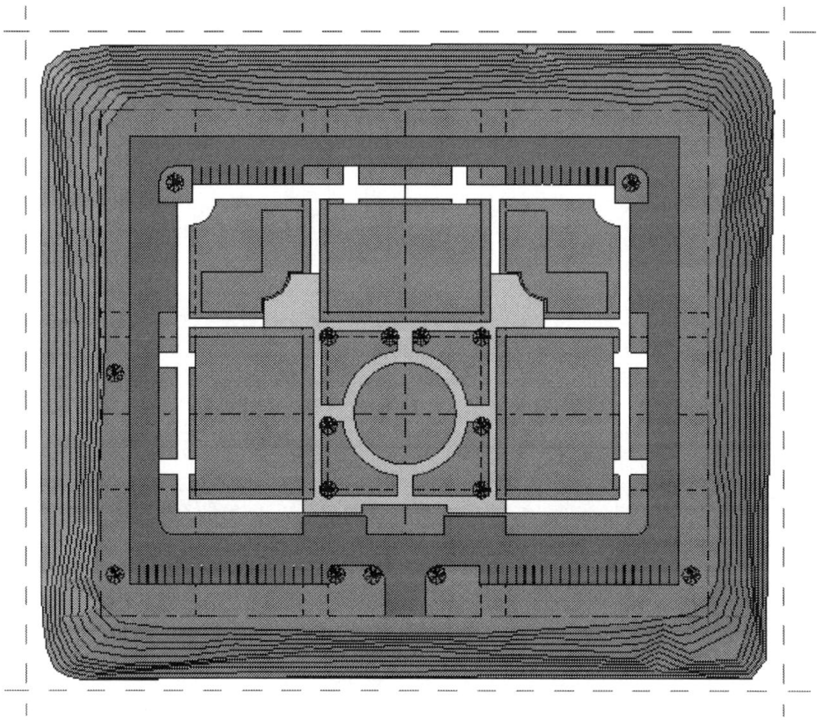

Figure 9-36 *Sketch plan for adding the peripheral pedestrian walkway to the Site Plan project*

Exercise 2 Museum Site Plan

Create the site plan for a new project, a museum, by adding a site toposurface, road subregion, pedestrian walkway subregion, parking components, and trees, based on the illustration given in Figure 9-37. The dimensions and texts have been given only for reference and are not to be added. The site plan is symmetrical along the diagonal axis and you can assume the missing dimensions. **(Expected time: 30 min)**

1. Project file parameters:
   For Imperial Template File- *default.rte*
   For Metric Template File- *DefaultMetric.rte*

   File name to be assigned:
   For Imperial *c09_MuseumSitePlan_ex2.rvt*
   For Metric *M_c09_MuseumSitePlan_ex2.rvt*

2. Property size:
   For Imperial **150'0" X 150'0"** (Extents of property line)
   For Metric **45720 X 45720 mm**

3. Site settings:
   Contours at **1'0" (304.8 mm)** intervals, top of site level- **5'0" (1525 mm)**
   Site material- **Site - Grass**, Road material- **Asphalt**, Walkway material- **Sand**

4. Parking component:
   For Imperial **Parking Space: 9' X 18'-90 deg**
   For Metric **Parking Space: 4800 X 2400- 90 deg**

5. Site component:
   For Imperial **RPC Tree - Deciduous: Red Ash-25'**
   For Metric **RPC Tree - Deciduous: Red Ash-7.6 Meter**

Chapter 10

Using Massing and Family Tools

Learning Objectives

After completing this chapter, you will be able to:
- *Understand the concept of massing in Revit*
- *Create massing geometries using the Massing tool*
- *Cut massing geometries using the Void Form tools*
- *Convert a massing geometry into building elements*
- *Add other building elements to a converted geometry*
- *Create conceptual massing in conceptual design environment*
- *Create a family and its types*

INTRODUCTION

In the earlier chapters, you learned to use different building elements and components to create a building model. These building elements are parametrically associated and enable you to generate a building model based on specific design requirements such as wall types, door width, window height, and so on. Each of the elements must be assigned specific properties to achieve the desired element parameters, thereby making a building model accurate. Needless to say, this is a fairly time-consuming procedure. Autodesk Revit provides you with an alternative and much easier method to create a building model, known as massing.

BIM
CONCEPTS

Creating a building mass is an essential step to perform coneptual design in the BIM process. A conceptual design is an early phase of the design process, in which the broad outlines of function and form of a building is articulated. It provides building managers with a reliable assessment of likely performance, possible looks, basic understanding of the scope of the development effort, including marketability, labor requirements, and expected costs.

UNDERSTANDING MASSING CONCEPTS

At the conceptualization stage of a project, you may want to study it in terms of its building volumes and shapes. You may also want to convey the basic idea of the structure of a building in a three-dimensional form without putting in a lot of detailing. This can be achieved by using various tools for creating massing geometries.

The tools for creating massing geometry not only enable you to conceive and create a variety of building shapes and volumes with relative ease but also convey the potential design in terms of building masses and geometric shapes. You can create and edit geometric shapes and amalgamate them to form a building structure. This process can be compared to the creation of building a model using foam blocks. You have the freedom of choice to add or cut geometric shapes and join different blocks or masses to form an assembly. Figure 10-1 shows a group of volumes that can be created to represent the building volume and mass.

Figure 10-1 *Building blocks created using the massing geometry*

CREATING THE MASSING GEOMETRY

In Autodesk Revit, you can create the massing geometry in any of these three environments: Family Editor, Conceptual Design, and Project.

To create the massing geometry in the Family Editor environment, choose **New > Family** from the **File** menu; the **New Family - Select Template File** dialog box will be displayed. In this dialog box, choose the *Generic Model.rft* file from the **English-Imperial** folder [for Metric the *Metric Generic Model.rft* file from the **English** folder] and then choose the **Open** button; a new file will open in the Family Editor environment. In the new file, create the massing geometry using various tools available in the ribbon.

The Conceptual Design environment is a type of Family Editor that provides the advanced modeling tools and techniques for creating massing families. To start creating a mass in this environment, choose **New > Conceptual Mass** from the **File** menu; the **New Conceptual Mass - Select Template File** dialog box will be displayed. In this dialog box, select the **Mass** template file from the **Conceptual Mass** folder and then choose **Open**; a new file will be displayed in the Conceptual Design environment. In the new file, you can create the massing geometry using the tools available in the **Create** tab of the ribbon, as shown in Figure 10-2.

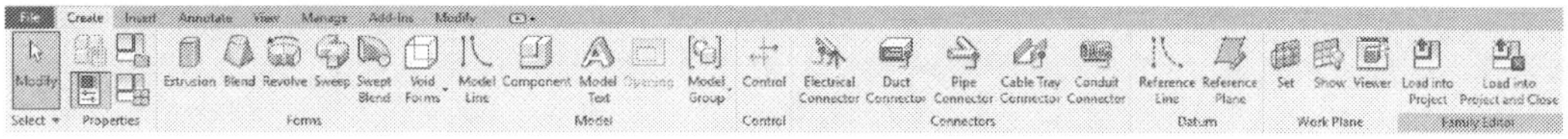

Figure 10-2 *Various tools in the **Create** tab of the ribbon*

Creating a Massing Geometry in the Family Editor

As discussed earlier, the Family Editor environment provides tools to create massing geometry. To create a new mass, choose **New > Family** from the **File** menu; the **New Family - Select Template File** dialog box will be displayed. In this dialog box, select the *Generic Model.rft* file from the **English-Imperial** folder [for Metric the *Metric Generic Model.rft* file from the **English** folder] and choose **Open**; a new file will open using the selected template file.

Editing a Massing Geometry in the Family Editor

In Autodesk Revit, you can easily edit a massing geometry. It can be edited using drag controls or by editing massing parameters.

Creating Cuts in a Massing Geometry by Using the Family Editor

You can cut a massing geometry by creating a void form in it. This void form is cut or subtracted from the massing geometry it intersects. You can create void forms by using the tools displayed in the **Void Forms** drop-down in the **Forms** panel of the **Create** tab. The **Void Forms** drop-down displays five tools: **Void Extrusion**, **Void Blend**, **Void Revolve**, **Void Swept Blend**, and **Void Sweep**. You can choose an appropriate tool to generate the shape and volume of the void form. The method of creating a void form using these tools is similar to that of creating a solid form.

When any of the tools from the **Void Forms** drop-down is invoked, a contextual tab is displayed.

For example, if you invoke the **Void Extrusion** tool, the **Modify | Create Void Extrusion** tab will be displayed. You can use the options in this tab to sketch the profile for the extruded void geometry. After sketching the profile of the extruded void geometry, choose the **Finish Edit Mode** button from the **Mode** panel of the **Modify | Create Void Extrusion** tab; the **Modify | Void Extrusion** tab will be displayed. From this tab, you can use various editing options to modify the extruded void geometry. Next, click in the drawing area or press ESC; the void form will automatically cut its shape and volume from the intersecting massing geometry.

Placing the Massing Geometry in a Project

In the Project environment of Autodesk Revit, you have the option to add a predefined massing geometry to the project using the **Place Mass** tool. To add a predefined geometry, invoke this tool from the **Conceptual Mass** panel of the **Massing & Site** tab; the **Massing - Show Mass Enabled** window will be displayed. Choose the **Close** button from the window; the **Revit** message box will be displayed. Choose the **Yes** button from this message box; the **Load Family** dialog box will be displayed. In this dialog box, browse to the **Mass** subfolder in the **English-Imperial** folder and then select a massing family. After selecting the massing family, choose the **Open** button; the **Load Family** dialog box will be closed. Also, the **Modify | Place Mass** tab will be displayed and the selected family will be loaded into the project. Now, click in the drawing area to place the loaded massing geometry.

Creating the In-Place Mass in a Project

In the Project environment, you have the option to create an in-place massing geometry by using the **In-Place Mass** tool. Invoke this tool from the **Conceptual Mass** panel of the **Massing & Site** tab; the **Massing - Show Mass Enabled** dialog box will be displayed if you are invoking the **In-Place Mass** tool for the first time in your project. Click on the **Close** option given at the bottom right corner of the dialog box; the **Name** dialog box will be displayed. In this dialog box, enter the name of the mass in the **Name** edit box and choose **OK**; the **Create** tab will be displayed. You can use the options in this tab to sketch the massing profile and convert it into a solid or void form. To sketch the profile for the mass, choose the **Model** tool from the **Draw** panel; the **Modify | Place Lines** tab will be displayed. You can use various sketching tools available in the **Draw** panel of this tab. While sketching, you can also use the **Reference Plane** tool from the **Draw** panel to draw references for the sketch. After sketching the profile, you can use any of the tools displayed in the **Create Form** drop-down in the **Form** panel. The **Create Form** drop-down displays two tools: **Solid Form** and **Void Form**. You can use the **Solid Form** tool to create a solid form and the **Void Form** tool to create a void form. To use any of the tools from the **Create Form** drop-down, you need to select the sketched profile from the drawing and then invoke the **Solid Form** or **Void Form** tool. After invoking any of these tools, the **Modify | Form** tab will be displayed. In this tab, you specify the settings to change the instance property of the mass created, divide the surfaces of the mass, and modify geometrical elements of the mass.

MASSING IN CONCEPTUAL DESIGN ENVIRONMENT

Conceptual design is the very first phase of a design process. The primary objective of a conceptual design is to create a representation of the idea generated for creating the building mass of a project. For creating a building project, conceptual design is very important for architects, engineers, and designers as it helps in finding out the final representation of the design intent. It thus enables them to create more specific sets of plans.

Interface of the Conceptual Design Environment

To open the interface of the Conceptual Design Environment, choose **New > Conceptual Mass** from the **File** menu. On doing so, the **New Conceptual Mass-Select Template File** dialog box will be displayed. In this dialog box, select the **Mass.rft** file from the **Conceptual Mass** folder in the **English-Imperial** folder (for Metric the **Metric Mass.rft** file from the **Conceptual Mass** folder in the **English** folder) and then choose the **Open** button; the interface of the Conceptual Design Environment along with the settings of the selected template file will be displayed, as shown in Figure 10-3. From the interface of the **Conceptual Massing Environment**, you can use various tools in the ribbon to create the conceptual mass.

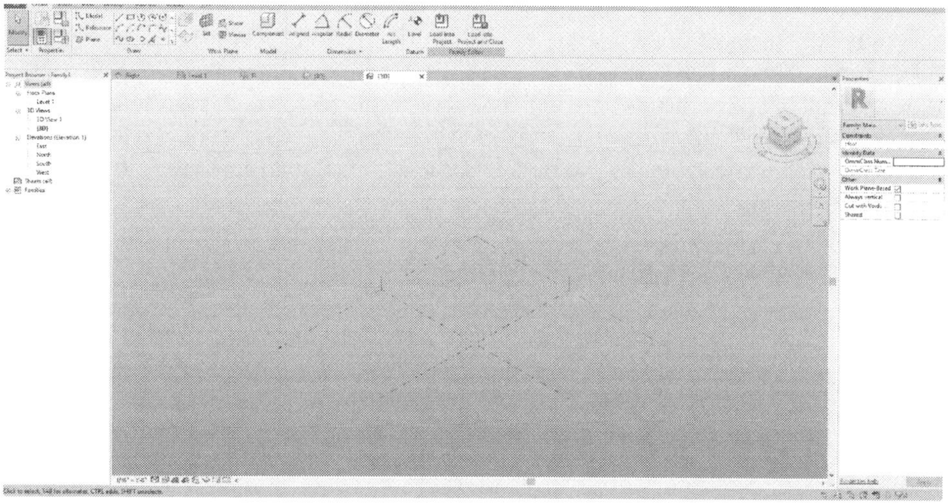

Figure 10-3 *The Conceptual Design Environment interface*

Creating Masses in Conceptual Design Environment

In the Conceptual Design Environment, you can create the conceptual mass by using various tools from the ribbon. The ribbon of the Conceptual Design Environment, consists of the following tabs: **Create**, **Insert**, **View**, **Manage**, and **Modify**. The **Create** tab is chosen by default. From the **Create** tab, you can access the tools required to create the Conceptual Mass. From the **Workplane** panel of the **Create** tab, you can choose any of the tools such as **Set**, **Show**, and **Viewer** that can be used for setting, showing, and viewing workplane to be used to create the conceptual mass. You can choose the **Level** tool in the **Datum** panel to add levels that will assist in the creation of its conceptual mass. You can add levels in 3D and elevation views in the Conceptual Design Environment. The **Draw** panel of the **Create** tab contains tools that can be used to sketch conceptual mass. In this panel, you can use the **Model**, **Reference**, and **Plane** tool to sketch the model lines, reference lines and reference planes to create the conceptual mass.

CREATING BUILDING ELEMENTS FROM THE MASSING GEOMETRY USING BUILDING MAKER TOOLS

After generating the shape and volume of massing forms, you can convert them into building elements such as walls, floors, roofs, and curtain systems. You can select the faces of the generated massing geometry and replace them with the desired building element. To convert faces into building elements, you can use various tools from the **Model by Face** panel of the **Massing & Site** tab in the ribbon.

Creating Walls by Selecting Faces

Ribbon: Massing & Site > Model by Face >Wall

You can convert the faces of the placed mass into walls. To do so, choose the **Wall** tool from the **Model by Face** panel of the **Massing & Site** tab; the **Modify | Place Wall** tab will be displayed. In this tab, select the wall type to be used for conversion from the **Type Selector** drop-down list. In the **Draw** panel of this tab, make sure that the **Pick Faces** option is chosen. In **Options Bar**, select the appropriate values of the **Level**, **Height**, and **Location Line** from their respective drop-down lists.

Now, to convert the faces into walls, move the cursor over the massing geometry. You will notice that the faces that can be converted are highlighted. Next, click on the desired face to make the conversion. You can use the TAB key to select all the faces of the massing geometry that can be converted into walls. Figure 10-4 shows the selected faces of the massing geometry and Figure 10-5 shows the walls created.

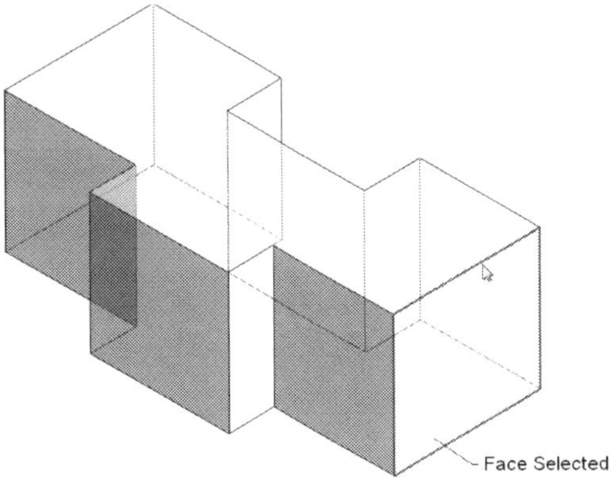

Figure 10-4 *Selected wall face*

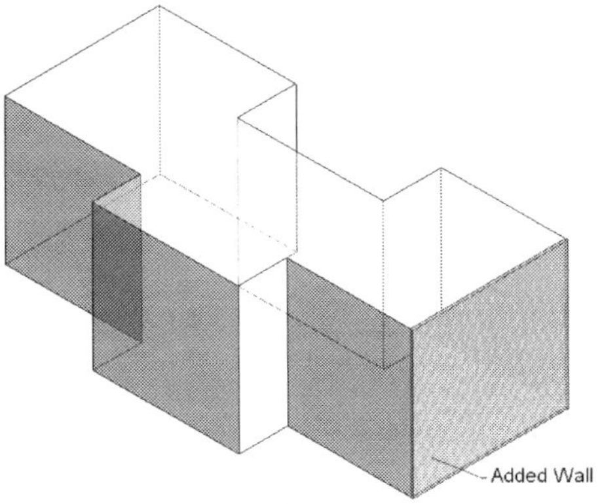

Figure 10-5 *Wall face converted into walls*

Creating Floors by Selecting Faces

Ribbon: Massing & Site > Model by Face > Floor

To create a floor in a massing geometry, first you need to define the floor area faces by selecting the massing geometry and then choosing the **Mass Floors** tool from the **Model** panel of the **Modify | Mass** tab. On doing so, the **Mass Floors** dialog box will be displayed. Select the levels at which the floor area faces are required and choose the **OK** button. Autodesk Revit will automatically generate the floor area faces based on the massing geometry and the defined levels.

Creating Roofs by Selecting Faces

Ribbon: Massing & Site > Model by Face > Roof

You can use the **Roof** tool in the **Massing & Site** tab to convert the massing faces into roofs. Faces can be horizontal planes of any shape or those created by using the **Extrusion**, **Revolve**, **Blend**, **Sweep**, or **Swept Blend** tool.

Invoke the **Roof** tool from the **Model by Face** panel; the **Modify | Place Roof by Face** tab will be displayed. From this tab, select the roof type to be used from the **Type Selector** drop-down list. Next, move the cursor over the massing geometry and click to select the highlighted face. Next, choose the **Create Roof** button in the **Multiple Selection** panel to convert it into a roof. Figures 10-6 and 10-7 show the top face of a multistory building converted into a roof.

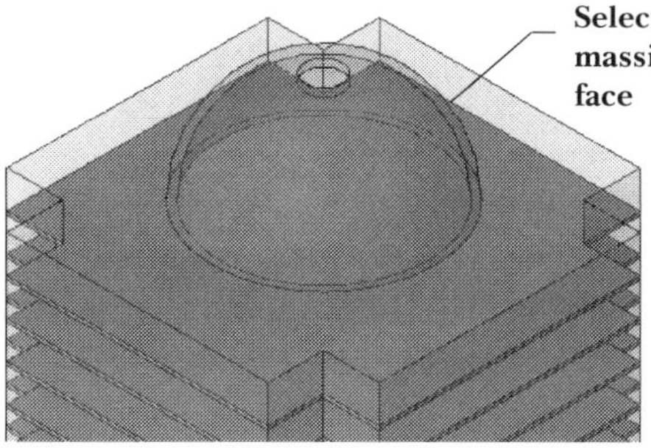

Figure 10-6 The top face selected for conversion into a roof

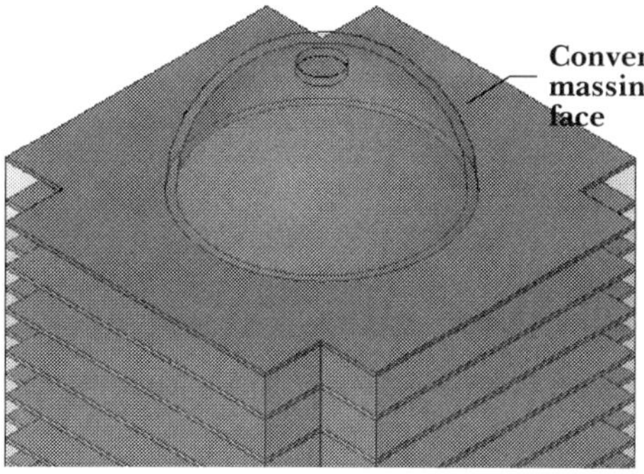

Figure 10-7 The top face of the massing geometry converted into a roof

Creating Curtain Systems by Selecting Faces

Autodesk Revit provides you with the **Curtain System** tool to convert the planar and non-planar faces into curtain systems. Invoke this tool from the **Model by Face** panel; the **Modify | Place Curtain System by Face** tab will be displayed. In this tab, select the curtain system type from the **Type Selector** drop-down list. In the **Properties** palette, you can modify the instance and type properties of the curtain system before creating it. Next, move the cursor over the face that needs to be converted into the curtain system and click to select it. Choose the **Create System** tool from the **Multiple Selection** panel to create the curtain system. Figure 10-8 shows the example of a curtain system of a dome. You can then add mullions to the curtain system using the **Mullions** tool, as shown in Figure 10-9. Figure 10-10 shows another example of using the **Curtain System** tool to convert non-planar massing faces into curtain systems.

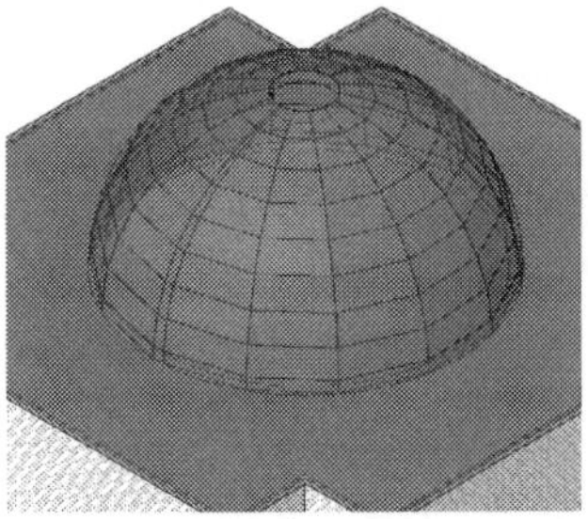

Figure 10-8 Creating a curtain system using the ***Curtain System*** *tool*

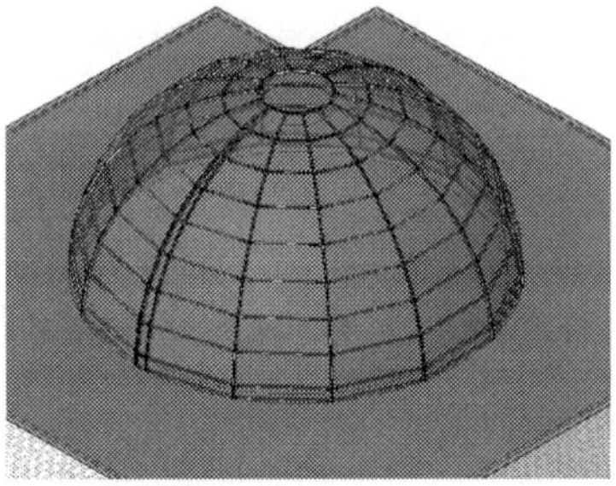

Figure 10-9 Mullions added to the curtain system using the ***Mullions*** *tool*

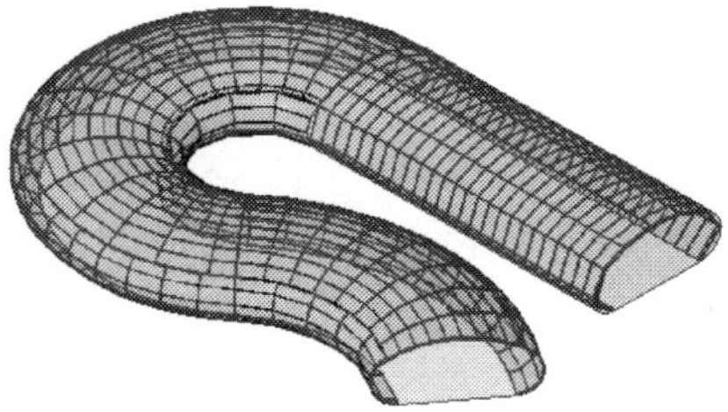

Figure 10-10 Converting non-planar massing faces into curtain systems with mullions

Controlling the Visibility of a Massing Geometry

To hide or display a massing geometry, choose the **Show Mass by View Settings** tool from **Massing & Site > Conceptual Mass > Show Mass by View Settings** drop-down. However, the use of this tool has a temporary effect on the visibility of the massing forms. When a project file is closed, the show mass settings are not saved in it. Therefore, next time when you will open the project file, the massing forms will not be visible.

To display the massing form in a project file, invoke the **Visibility/Graphics** tool from the **Graphics** panel of the **View** tab; the **Visibility/Graphics Overrides** dialog box for the current view will be displayed. In this dialog box, select the check box for the **Visibility** parameter of the **Mass** model category.

Adding other Building Elements

After converting the massing geometry into basic building elements, you can add other building elements such as doors, windows, roofs, and so on to the basic building elements. You can also use various editing tools to copy, edit, and delete the converted building elements.

Figure 10-11 shows the example of a building model in the plan view with modified wall type and added windows. You can use the editing tools such as **Mirror**, **Array**, and **Paste Aligned** to copy elements and create a building model of desired shape with specified parameters, as shown in Figure 10-12.

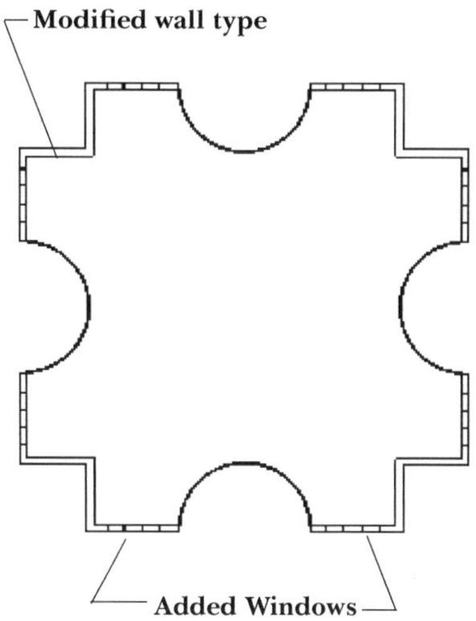

Figure 10-11 *Plan view showing the modified wall type and added windows*

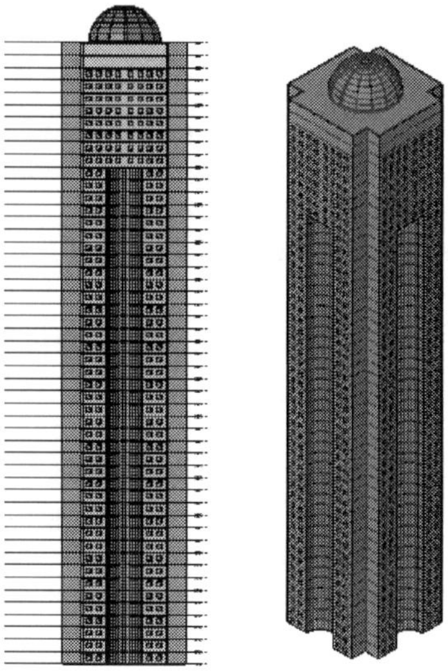

Figure 10-12 *A multistory building with building components*

Tip

With the ***Massing*** *tool, Autodesk Revit enables you to create a variety of shapes and volumes. Based on the desired geometry, you can use a combination of tools. The massing geometry can also be used to generate the rendered views of conceptual building volumes. This can prove to be an effective tool for communicating the design intent to the entire project team before starting the detailed project plans.*

CREATING FAMILIES

Revit family is defined as a collection of parametric components or elements having similar graphical representation, utility, and common parameters. By nature, all the elements used in Revit are family-based. A Revit family has a wide range of collections starting from a wall to a window and extended up to two-dimensional annotations or symbols required for your project. In Autodesk Revit, each family inherits the character of some specific type. Each type is a member of Revit family that consists of certain parameters known as **Type Parameters**. These parameters are constant for identical types. For example, a **Single-Flush** door family can have different types such as **Single-Flush: 36" x 84"** in Imperial **(Single-Flush : 0915 x 2134 mm** in Metric), **Single-Flush : 30" x 84" (Single-Flush : 0915 x 2032 mm** in Metric), and others. You can access these types from the **Type Selector** drop-down list.

Creating In-Place Families

This method is used to create custom families such as the furniture for reception and workstations in a building model. You can create a family while working on a project. The family created is saved within the project and gets updated according to the changes made in the project.

To create an In-Place family, choose the **Model In-Place** tool from **Architecture > Build > Component** drop-down; the **Family Category and Parameters** dialog box will be displayed. In the **Family Category** area, select an option from the **Filter list** drop-down list to filter the list of categories that will be displayed in the list box displayed below it. Double-click on the required category from the list box; the **Name** dialog box will be displayed. In this dialog box, enter a name for the family in the **Name** edit box and then choose the **OK** button; the sketch mode will be invoked and the options in the **Create** tab will be displayed. Sketch the reference planes and set the work plane to help you sketch the geometry for the family using the required sketching tools from the **Model** panel and the **Forms** panel. After creating the massing geometry, choose the **Finish Model** button from the **In-Place Editor** panel; a family will be created. After creating the geometry of a family, you can assign materials to the family and create family types with different dimensions.

Creating Families Using Standard Family Templates

In this method, you can create a new family by using certain standard in-built family templates as platform. These templates provide you with some parameters, reference planes, and dimensions in different views that are already saved in templates. You have to create the geometry for a family in the required view. To do so, you need to sketch reference planes, dimension reference planes, label dimensions, and then sketch the profile for the family using the required sketching tools. Once you have created the profile, apply materials to it and create family types by assigning different dimensions to the geometry. At the end, give a name to the template, save it, and load the family into the project to use it in your drawing.

The steps to create a family are explained next with an example of an arch-shaped panel window, as shown in Figure 10-13. The window is created using a standard template.

Figure 10-13 *An arch-shaped window with panels*

TUTORIALS

Tutorial 1 Office Building 2

In this tutorial, you will create massing geometry for a five-story office building based on the shape shown in Figure 10-14. It consists of a 50'0" X 50'0" (15240 X 15240 mm) central hall that is 60'0" (18288 mm) high. It has a 30'0" (9144 mm) diameter cylindrical atrium with a hemispherical dome at the top level. The central hall is flanked by the right and left wings that are 40'0" (12192 mm) high. The entrance area is 30'0" (9144 mm) high. Use the dimensions of the building given in the floor plan shown in Figure 10-15 and the dimensions of the elevation shown in Figure 10-16. The dimensions and the text have been given for reference and are not to be created. After creating the massing geometry, convert it into building elements with the parameters given next. **(Expected time: 1 hr 15 min)**

1. Project parameters:
   For Imperial Template File- *default.rte*,
   File name to be assigned- *c10_Office-2_tut1.rvt*
   Floor to floor height of the building - **10'0"**

   For Metric Template File- *DefaultMetric.rte*,
   File name to be assigned- *M_c10_Office-2_tut1.rvt*
   Floor to floor height of the building - **3048 mm**

2. Building Element types for the converted shell elements:
   For Imperial Floor: **Floor: LW Concrete on Metal Deck**
   Walls: Central Hall- **Basic Wall: Exterior - Brick on CMU**
   Left and Right Wing, and Entrance- **Curtain Wall: Exterior Glazing**
   (Spacing 5'0" horizontal, 10'0" vertical)
   Flat Roof: **Roof: Generic 9"**

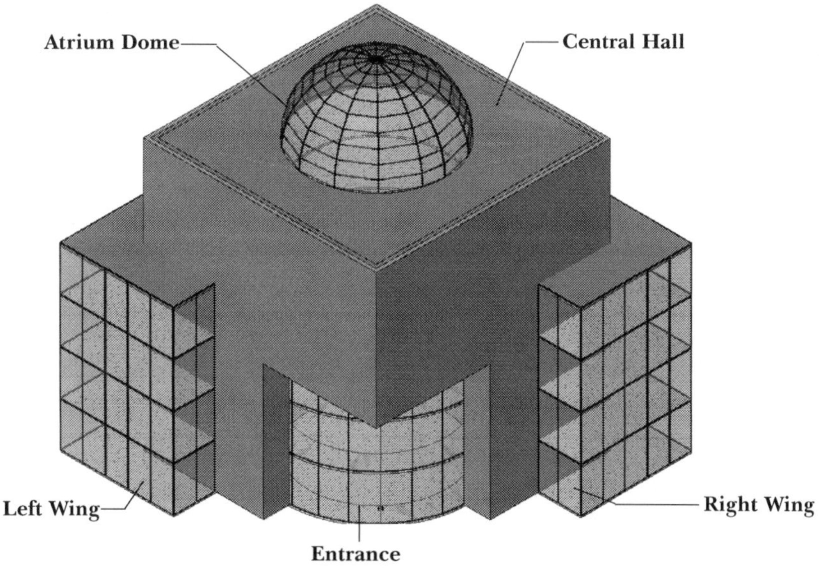

Figure 10-14 *3D sketch view of the office building*

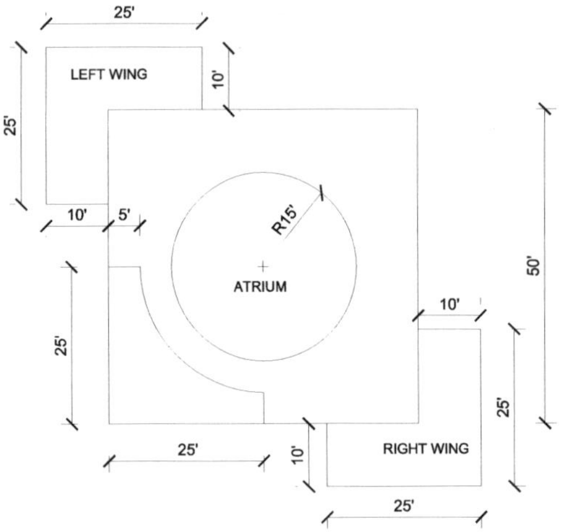

Figure 10-15 *Sketch plan for the office building*

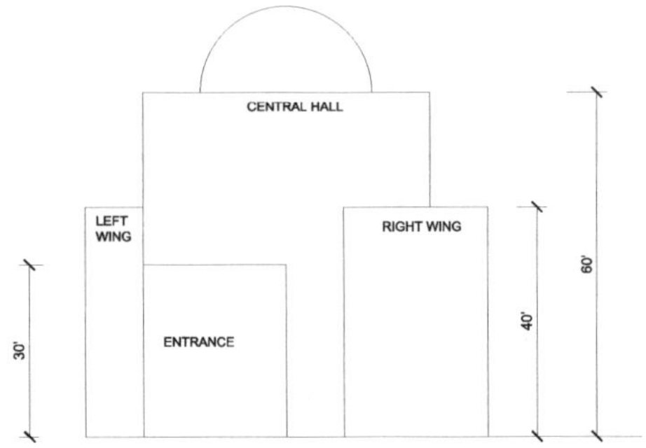

Figure 10-16 *Elevation for the office building*

Curtain system for dome: Spacing **3'6"** horizontal, **7'0"** vertical; Mullion type

For Metric Floor: **Floor: LW Concrete on Metal Deck**
Walls: Central Hall- **Basic Wall: Exterior - Brick on CMU**
Left and Right Wing, and Entrance- **Curtain Wall: Exterior Glazing**
(Spacing 1524 mm horizontal, 3048 mm vertical)
Flat Roof: **Roof: Generic 400 mm**
Curtain system for dome: Spacing **1067 mm** horizontal, **2134 mm** vertical; Mullion type

The following steps are required to complete this tutorial:

a. Open a new project file by using the default template file and adding levels.
b. Create a massing geometry using the **Form** tool.
c. Cut the geometry based on the sketch plan to create the atrium and the entrance using the **Void Form** tool.
d. Use the **Solid Form** tool to generate the atrium dome.
e. Convert the massing geometry into the specified building elements.
f. Save the project using the **Save As** tool.
For Imperial *c10_Office-2_tut1.rvt*
For Metric *M_c10_Office-2_tut1.rvt*
g. Close the project.

Opening the Project File and Adding Levels

You need to first open a new project file using the **New** tool. As the building is five-story high, you need to create four additional levels and rename them according to the floor levels.

1. Choose the **New** option under the **Projects** head from the interface screen; the **New Project** dialog box is displayed.
2. Choose the **Browse** button; the **Choose Template** dialog box is displayed.
3. In the **Choose Template** dialog box, select the *default* template file from the **English-Imperial**

folder (For Metric, select the *DefaultMetric* template file from the **English** folder) and choose **Open**; the **New Project** dialog box is displayed. Next, choose the **OK** button; the dialog box closes and a new project file opens in the drawing window.

4. In the **Project Browser**, double-click on **South** in the **Elevations (Building Elevation)** head to display the corresponding view.

5. Choose the **Level** tool from the **Datum** panel in the **Architecture** tab.

6. Add four levels above level 2 at an elevation of **10'0" (3048 mm)** each. (For more information on adding levels, refer to Adding Levels section in Chapter 6)

7. Rename the levels as follows:
   Level 1- First Floor
   Level 2- Second Floor
   Level 3- Third Floor
   Level 4- Fourth Floor
   Level 5- Fifth Floor
   Level 6- Roof
   Choose the **Yes** button in the **Revit** conformation box to rename the corresponding views.

Creating the Massing Geometry

Now, you can start creating the massing geometry using the tools for massing. The central hall and the two wings can be generated using the **Form** tool.

Note

The massing geometry can be created using different methods and tools. The steps given below describe a general procedure to create massing geometry. The steps and methods may vary depending on the design intent.

1. Double-click on the **First Floor** under the **Floor Plans** head in the **Project Browser** to display the corresponding plan.

2. Choose the **In-Place Mass** tool from the **Conceptual Mass** panel of the **Massing & Site** tab; the **Massing - Show Mass Enabled** message box is displayed.

3. Choose **Close** from this message box; the **Name** dialog box is displayed.

4. In the **Name** edit box of the **Name** dialog box, enter the name **Central Hall** and then choose the **OK** button.

5. Invoke the **Rectangle** sketching tool from the **Draw** panel of the **Create** tab.

6. Move the cursor in the area lying in the mid of the four elevation tags and sketch a square with **50'0"(15240 mm)** side. After sketching the square, choose the **Modify** button.
7. Now, select the created square and choose the **Solid Form** tool from **Modify|Lines > Form > Create Form** drop-down; the **Modify |Form** tab is displayed.

8. Next, choose the **Default 3D** tool from the **Create** panel of the **View** tab; the 3D view of the mass created is displayed.

9. Now, move the cursor over the top face of the box created, refer to Figure 10-17, and click when the edges of the faces are highlighted. On doing so, two temporary dimensions are highlighted.

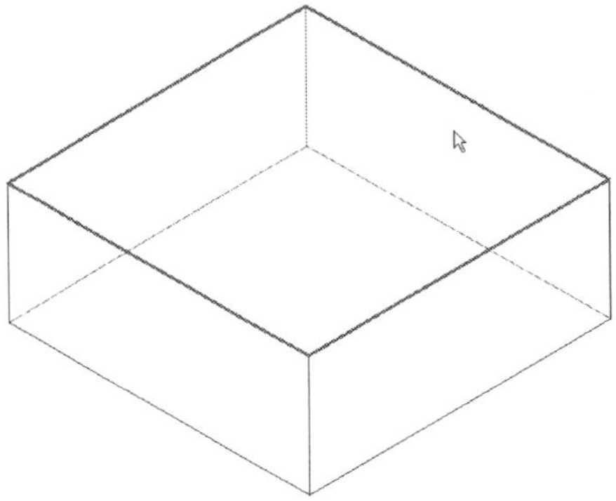

Figure 10-17 *Selecting the top face of the central hall*

10. Click on the value of the lower dimension (dimension that specifies the height of the box) and enter **60' (18288 mm)** in the edit box. Now, press ENTER; the box is resized to the height 60' (18288 mm).

11. Choose the **Modify** button and then right-click in the drawing area; a shortcut menu is displayed. Choose **Zoom To Fit** from the shortcut menu; the box created can now be viewed completely.

12. Now, choose the **Finish Mass** button from the **In-Place Editor** panel of the **Modify** tab; the Central Hall massing is created.

    Similarly, you need to create the left and right wings of the building using the **Form** tool. You will sketch the profile in the first floor plan view and extrude it to a height of **40'0"** (**12192 mm**), as specified in the sketch plan.

13. Create the mass for left and right wings of the building. Double-click on **First Floor** under the **Floor Plans** head in the **Project Browser** to display the corresponding plan.

14. Now, choose the **In-Place Mass** tool from the **Conceptual Mass** panel of the **Massing & Site** tab; the **Name** dialog box is displayed.

15. In the **Name** dialog box, enter the name **Wings** and choose the **OK** button.

16. Invoke the **Line** tool from the **Draw** panel of the **Create** tab and ensure that the **Chain** check box is selected in the **Options Bar**.

17. Move the cursor near the lower right corner of the square and sketch the profile based on the dimensions given in the sketch plan, refer to Figure 10-15. The profile for the right wing is sketched, as shown in Figure 10-18.

    Note that the sketch for the wings should be a closed sketch.

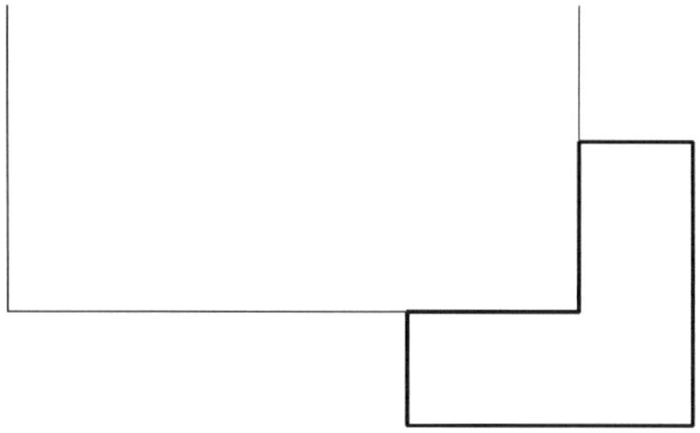

Figure 10-18 *Sketched profile for the right wing*

18. Choose the **Modify** button from the **Select** panel to clear the current selection.

19. Now, select the sketched profile for the right wing and choose the **Mirror - Draw Axis** tool from the **Modify** panel of the **Modify | Lines** tab.

20. Draw a line from the top right corner to the bottom left corner of the central square to define the mirror axis line. The profile is mirrored at the diagonally opposite corners of the square, as shown in Figure 10-19.

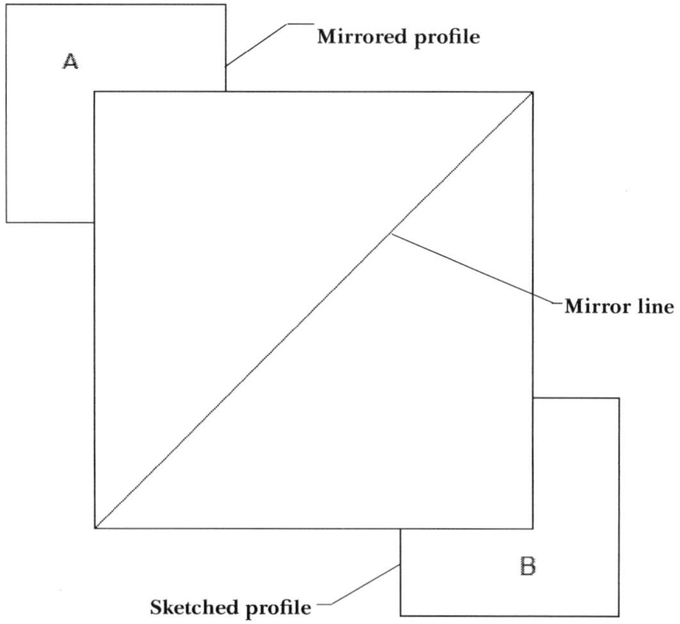

Figure 10-19 *Creating a mirror copy of the sketched profile*

21. Now, ensure that the mirrored profile is selected and the **Modify | Lines** tab is chosen.

22. Choose the **Solid Form** tool from the **Modify | Lines > Form > Create Form** drop-down.

23. Next, select the other profile at the lower right corner and repeat step 22.

24. Choose the **Modify** button from the **Select** panel and then choose the **Default 3D** tool from the **Create** panel of the **View** tab; the 3D view of the mass created is displayed.

25. Now move the cursor over the top face of the mass marked as A (refer to Figure 10-19) and then click on the face when it is highlighted. On doing so, two temporary dimensions are displayed.

26. Click on the value of the lower displayed dimension and enter **40' (12192mm)** in the edit box.

27. After entering the value in the edit box, press ENTER; the selected face of the mass will be extended to a height of 40' from the First Floor level and the mass corresponding to it is resized.

28. Next, select the top face of the other mass marked B and repeat steps 25 to 27.

29. Now, after you create the masses for the wings, choose the **Modify** button in the **Select** panel; the current selection is cleared.

30. Now, choose the **Finish Mass** button from the **In-Place Editor** panel of the **Modify** tab; the masses for the wings are created.

31. Now, move the cursor over the **ViewCube** tool and right-click; a shortcut menu is displayed. Choose **Orient to a Direction > Northeast Isometric** from the shortcut menu; the 3D view of the created mass geometry is displayed, as shown in Figure 10-20.

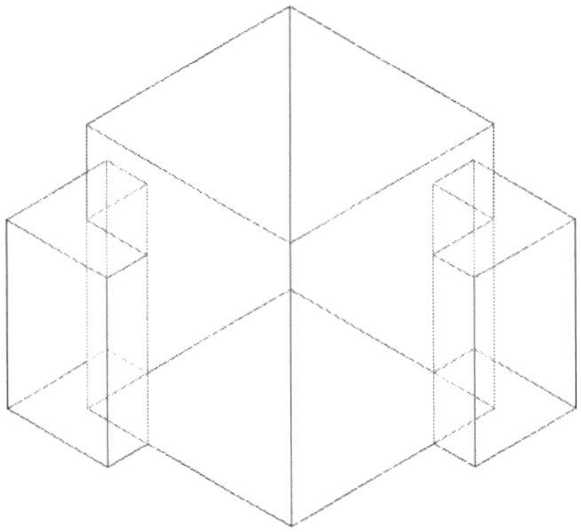

Figure 10-20 *The* ***Northeast Isometric*** *view of the central hall and the two wings*

Cutting the Massing Geometry

After creating the desired building blocks, you can now cut them to create the atrium and the entrance. The **Void Form** tool can be used to create the desired void geometry.

1. Double-click on **First Floor** under the **Floor Plans** head in the **Project Browser** to display the first floor plan view.

2. Choose the **Central Square** hall; the **Modify | Mass** tab is displayed.

3. Choose the **Edit In-Place** tool from the **Model** panel of the **Modify | Mass** tab; the **Modify** tab is displayed.

4. Next, choose **Visual Style > Wireframe** from **View Control Bar**.

5. Draw reference planes diagonally to locate the center of the square by using the **Plane** tool from the **Draw** panel of the **Modify** tab. After creating the reference planes, choose the **Modify** button from the **Select** panel to clear the current selection.

6. Now, invoke the **Circle** tool from the **Draw** panel of the **Modify** tab.

7. Move the cursor at the center of the square. Use the reference planes to locate the center and click to specify it.

8. Move the cursor to the right and enter the value **15'0" (4572mm)**. Next, press ENTER; the circle of **30'0"** diameter is created. Now, press ESC twice to exit and then select the circle created.

9. Choose the **Void Form** tool from the **Modify | Lines > Form > Create Form** drop-down; two circular images are displayed.

10. Choose the left image and then choose the **View** tab. Next, choose the **Default 3D** tool from the **Create** panel; the 3D view of the mass created is displayed.

11. Now, select the two segments of the circle profile of the void created by using the CTRL key, as shown in Figure 10-21. On selecting the segments, the corresponding pivot points are displayed. Hold and drag the blue segments till they cross the upper face of the central hall. Choose the **Modify** button.

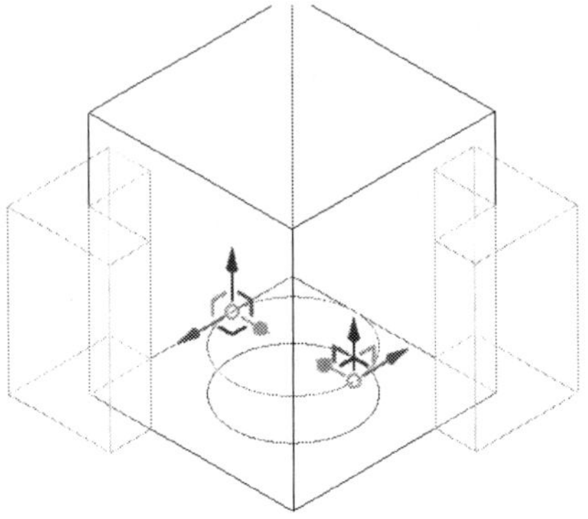

Figure 10-21 *The segments selected for the circular void*

12. Now, double-click on **First Floor** under the **Floor Plans** head in the **Project Browser** to display the corresponding plan.

13. Next, choose the **Modify | Form** tab and invoke the **Center-ends Arc** tool from the **Draw** panel.

14. Move the cursor to the center of the atrium circle and then click. Next, move the cursor horizontally toward the left and enter the value **20'0"** (**6096mm**) to specify the radius. Now, press ENTER.

15. Move the cursor counterclockwise and click to create a quarter arc, as shown in Figure 10-22.

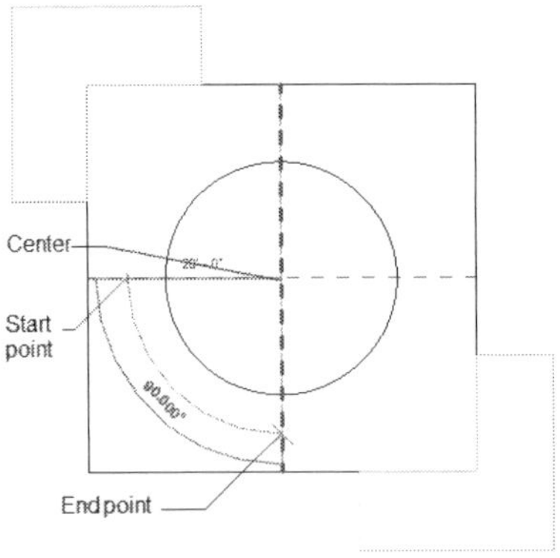

Figure 10-22 *Sketching an arc to create a cutting geometry*

16. Choose the **Line** sketching tool and ensure that the **Chain** check box is selected.

17. Sketch the profile of the cutting geometry, as shown in Figure 10-23.

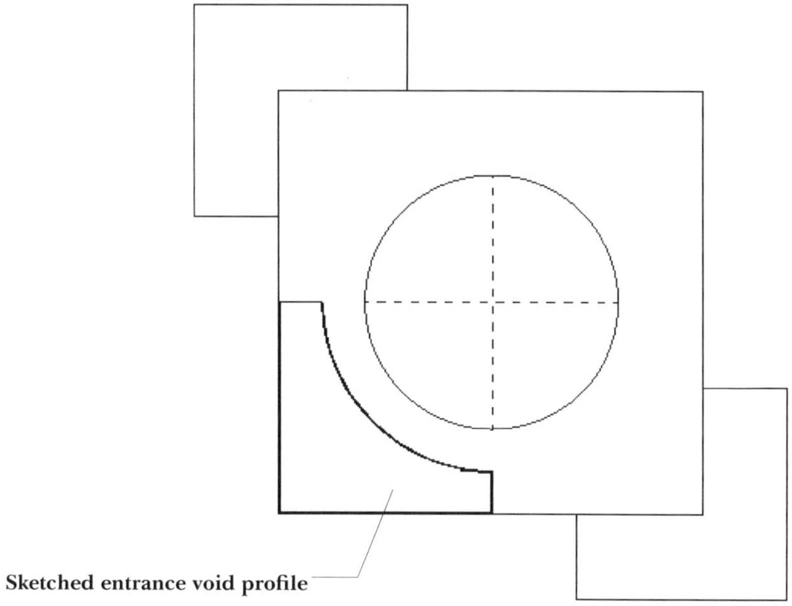

Figure 10-23 *Profile of the cutting geometry*

18. Choose the **Modify** button from the **Select** panel and then select the profile that was created recently; the **Modify|Lines** tab is displayed.

19. Choose the **Void Form** tool from the **Modify | Lines > Form > Create Form** drop-down; the void mass is created.

20. Next, choose the **View** tab and then choose the **Default 3D** tool from the **Create** panel; the 3D view of the created mass is displayed.

21. Now, move and place the cursor over the **ViewCube** tool and right-click; a shortcut menu is displayed.

22. Choose **Orient to a Direction > Front Left Isometric** from the shortcut menu.

23. Next, select the top face of the void geometry created, as shown in Figure 10-24.

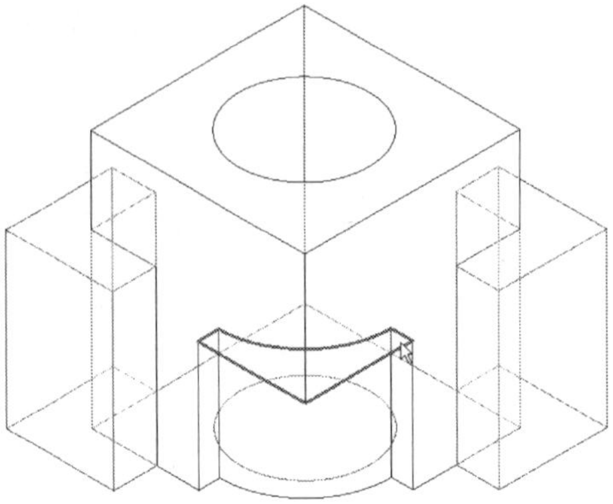

Figure 10-24 Selecting the face of the void geometry to edit the height

24. After selecting the top face, click on the dimension value of the temporary dimension that is displayed at the bottom. Enter **30'** in the displayed edit box and press ENTER.

25. Now, choose the **Modify** button from the **Select** panel and then choose the **Finish Mass** button from the **In-Place Editor** panel of the **Modify** tab.

26. Next, right-click on the drawing area and choose **Zoom To Fit** from the shortcut menu.

Creating the Atrium Dome

The atrium dome can be added using the **Form** tool. It is a hemispherical dome having a radius of **20'0"**(**6096 mm**) which is same as the radius of the atrium.

1. Choose the **In-Place Mass** tool from the **Conceptual Mass** panel of the **Massing & Site** tab; the **Name** dialog box is displayed.

2. In this dialog box, enter **Dome** in the **Name** edit box and choose the **OK** button.

3. Choose the **Set** tool from the **Work Plane** panel of the **Create** tab; you are prompted to pick a plane.

4. Select the right side entrance wall as the work plane, as shown in Figure 10-25. Note that to confirm that the selected face has been assigned the current work plane, choose the **Show** button from the **Work Plane** panel of the **Create** tab; the selected plane is highlighted.

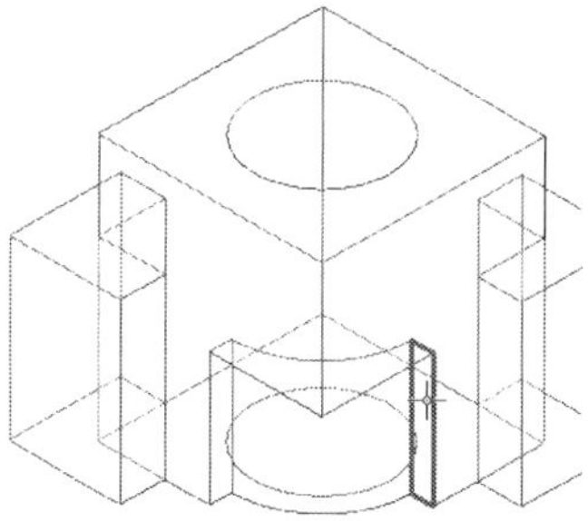

Figure 10-25 *Selecting the right side entrance wall as the work plane*

5. Now, choose the **Reference** tool from the **Draw** panel of the **Create** tab and invoke the **Line** tool.

6. Sketch the reference lines for the dome, as shown in Figure 10-26.

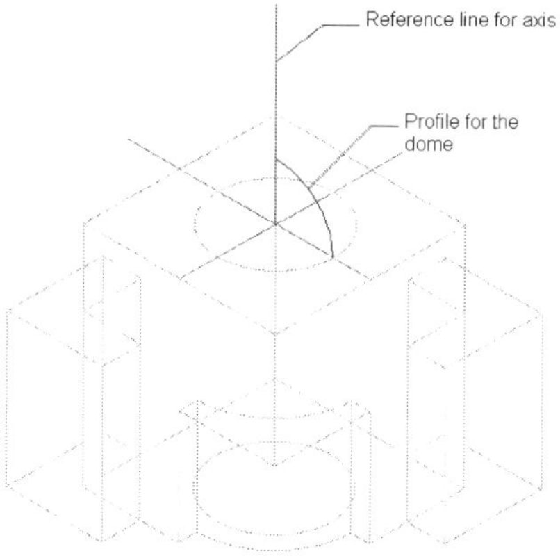

Figure 10-26 *The reference line and the profile for the dome*

7. Now, choose the **Model** tool from the **Draw** panel of the **Modify** tab and then invoke the **Center-ends Arc** sketching tool from the **Draw** panel.

8. Using the center of the atrium roof as the center of the curve, sketch a quarter arc, refer to Figure 10-26. Next, choose the **Modify** button to clear the current selection.

9. Select the reference line that defines the vertical axis and then select the arc using the CTRL key, refer to Figure 10-26.

10. Now, choose the **Solid Form** tool from the **Form** panel; the dome is created, as shown in Figure 10-27.

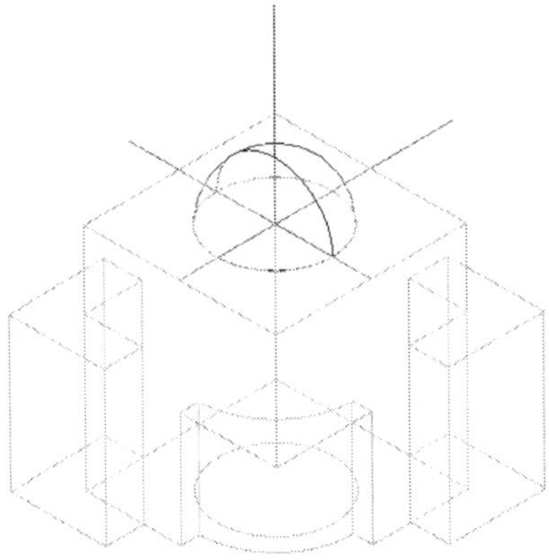

Figure 10-27 The dome created for the Atrium

11. Choose the **Finish Mass** button from the **In-Place Editor** panel of the **Modify** tab to finish the creation of the dome geometry. Ignore the warning message displayed, if any. Now, choose the **Modify** button from the **Select** panel.

Converting Massing Geometry into Building Elements

You can now convert the massing geometry into building elements using the **Floor**, **Wall**, and **Roof** tools. In order to create floors, you must first create different floor area faces.

1. Select the **Central Hall** and **Wings** massing geometries by holding the CTRL key; the **Modify | Mass** tab is displayed.

2. Choose the **Mass Floors** tool from the **Model** panel; the **Mass Floors** dialog box is displayed.

3. In the **Mass Floors** dialog box, select the check boxes for all the floors and choose the **OK** button; the floor area faces are created for the selected floors. Now, choose the **Modify** button.

4. Next, choose the **Floor** tool from the **Model by Face** panel of the **Massing & Site** tab; the **Modify | Place Floor by Face** tab is displayed.

5. In the **Properties** palette, select **Floor: LW Concrete on Metal Deck** (for Metric **160mm Concrete with 50mm Metal Deck**) from the **Type Selector** drop-down list.

6. Using the crossing window, select all the floor faces and then choose the **Create Floor** tool from the **Multiple Selection** panel; Autodesk Revit creates floors at the floor area faces.

Now, choose **Visual Style > Shaded** from the **View Control Bar**. The shaded view of the model is shown in Figure 10-28.

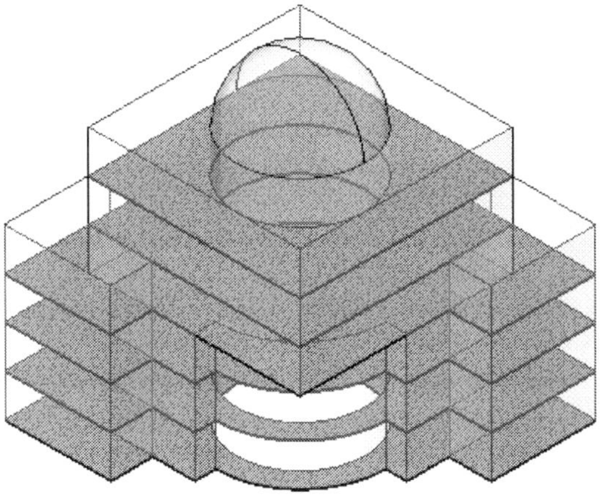

Figure 10-28 *Floor areas of the massing converted into floors*

7. Now, choose the **Wall** tool from the **Model by Face** panel of the **Massing & Site** tab; the **Modify | Place Wall** tab is displayed.

8. In the **Properties** palette, select the **Basic Wall: Exterior - Brick on CMU** wall type from the **Type Selector** drop-down list.

9. Now, move the cursor over the central hall massing geometry and when an exterior wall is highlighted, click to convert the wall face into the selected wall type.

10. Similarly, click on the other exterior walls of the central hall including the two side walls of the entrance. The resultant 3D view is shown in Figure 10-29.

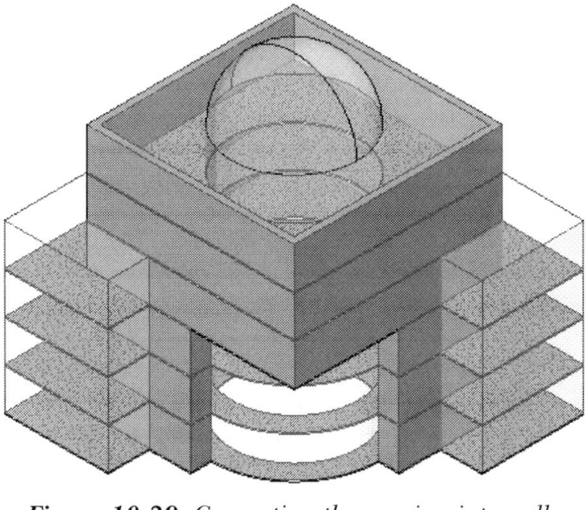

Figure 10-29 *Converting the massing into walls*

11. Next, in the **Properties** palette, select the **Curtain Wall: Exterior Glazing** wall type from the **Type Selector** drop-down list.

12. Choose the **Edit Type** button in the **Properties** palette; the **Type Properties** dialog box is displayed.

13. In the **Type Properties** dialog box, enter **5'0" (1524 mm)** in the **Value** field of the **Spacing** parameter in the **Vertical Grid** category and **10'0"(3048 mm)** for the **Spacing** parameter in the **Horizontal Grid** category.

14. Now, choose **Apply** and **OK**; the **Type Properties** dialog box is closed and the specified parameters are applied to the selected type.

15. Next, highlight and click on the exterior walls of the right and left wings to convert them into the selected exterior glazing walls.

16. Similarly, convert the curved entrance wall into the exterior glazing wall type. The 3D view now appears similar to the illustration shown in Figure 10-30. Now, choose the **Modify** button.

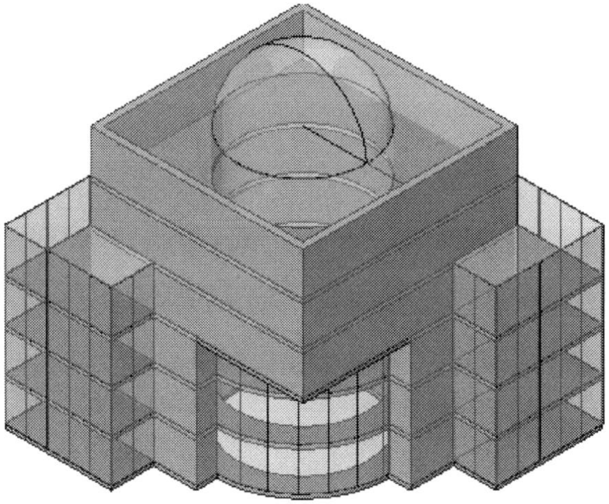

Figure 10-30 *Converting the entrance wall massing into walls*

17. Next, choose the **Roof** tool from the **Model by Face** panel of the **Massing & Site** tab; the **Modify | Place Roof by Face** tab is displayed.

18. In the **Type Selector** drop-down list, select the specified roof type.
    For Imperial **Basic Roof: Generic - 9"**
    For Metric **Basic Roof: Generic - 300 mm**

19. Highlight the roof face of the central hall and click.

20. Now, choose the **Create Roof** tool from the **Multiple Selection** panel of the **Modify | Place Roof by Face** tab; the roof is created on the selected face.

21. Similarly, create the roof of the same type for the two wings. After creating the roofs, the 3D view appears similar to the one shown in Figure 10-31. Next, you will create the curtain wall system for the dome.

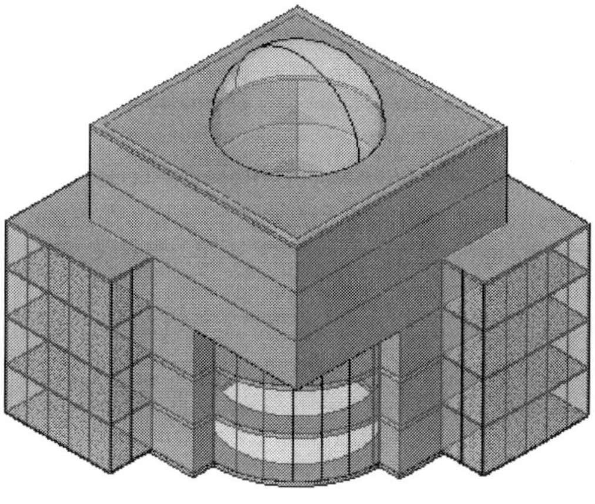

Figure 10-31 *Converting the massing into roof*

22. To create a curtain system for the dome, choose the **Curtain System** tool from the **Model by face** panel of the **Massing & Site** tab; the **Modify | Place Curtain System by Face** tab is displayed.

23. Choose the **Edit Type** tool from the **Properties** panel; the **Type Properties** dialog box is displayed.

24. Choose the **Duplicate** button; the **Name** dialog box is displayed. Enter **3'6" X 7'0" (1067 x 2134 mm)** in the edit box and then choose the **OK** button.

25. Enter the values **7'0"** (**2134 mm**) and **3'6" (1067 mm)** in the **Value** fields of the **Spacing** parameters under the **Grid 1 Pattern** and **Grid 2 Pattern** categories, respectively.

26. Choose the **OK** button; the **Type Properties** dialog box is closed.

27. Now, select both the curved faces of the dome and then choose the **Create System** tool from the **Multiple Selection** panel of the **Modify | Place Curtain System by Face** tab. The curved surfaces of the dome are converted into curtain systems. Now, choose the **Modify** button.

28. Next, you need to add a door to the entrance. To start with, choose the **Insert** tab and then choose the **Load Family** tool from the **Load from Library** panel; the **Load Family** dialog box is displayed.

29. In this dialog box, browse to the **English-Imperial > Doors** folder and then choose the **Door-Curtain-Wall-Double-Glass** (for Metric M_ **Door-Curtain-Wall-Double-Glass**) family file. After choosing the file, choose **Open**; the selected file is loaded in the drawing.

30. Choose the **Modify** button and then select the two curtain panels of the exterior glazing by the window selection process.

31. Invoke the **Filter** tool from the right end of the Status Bar; the **Filter** dialog box is displayed.

32. In the **Filter** dialog box, clear all the check boxes except the **Curtain Panel** check box and then choose the **OK** button; the **Filter** dialog box is closed and selection is done, as shown in Figure 10-32.

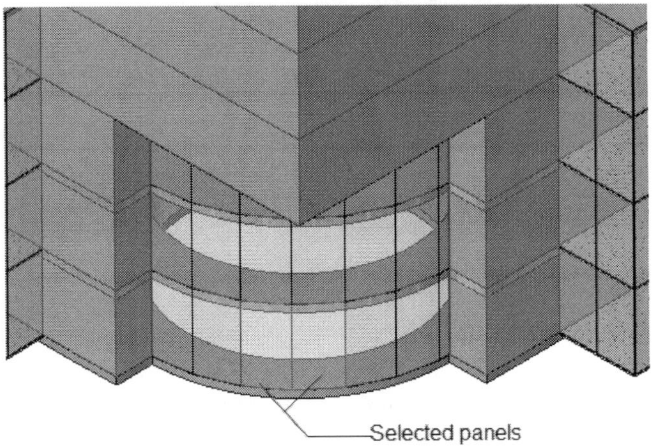

***Figure 10-32** Selecting panels from the glazing*

33. From the **Type Selector** drop-down list, select the **Curtain Wall Door-Curtain-Wall--Double-Glass** option. Next, choose the **Modify** button.

34. To add mullions to the exterior glazing, choose the **Mullion** tool from the **Build** panel of the **Architecture** tab; the **Modify |Place Mullion** tab is displayed.

35. In the **Type Selector** drop-down list, select the desired mullion type.
    For Imperial **Rectangular Mullion: 1" Square**
    For Metric **Rectangular Mullion: 30 mm Square**.

36. Next, choose the **All Grid Lines** tool from the **Placement** panel and move the cursor over the exterior glazings. Click on each face of the glazing when it is highlighted. Add all the curtain systems to the project, including the exterior wing walls, the curved entrance wall, and the dome. Use the **Grid Line Segment** tool from the **Placement** panel to add grids, if required.

The completed 3D view of the project appears similar to the one shown in Figure 10-14.

Saving the Project

In this section, you will save the project file using the **Save As** tool.

1. Choose **Save As > Project** from the **File** menu; the **Save As** dialog box is displayed.
2. Enter **c10_Office-2_tut1** in the **File name** edit box and choose **Save**; the file is saved.
3. Now, choose **Close** from the **File** menu to close the Autodesk Revit session.

EXERCISE

Exercise 1 Office

Create a massing geometry for an office building that consists of two identical 400'0" high towers with a connecting passage at 200'0" height. Each tower has a 50'0" X 50'0" base with an offset of 5'0" at 200'0" and 300'0" levels. There are two masts on each tower with a base radius of 2'6" and the top radius of 0'6" The connecting passage is 150'0" long and has a width and height of 25'0" with a vault roof. The plan view of the building is shown in Figure 10-33. The 3D, elevation, and shaded views of the building are shown in Figures 10-34, 10-35, and 10-36, respectively. The dimensions and text have been given for reference and are not to be created. Assume the missing dimensions proportionate to the building design. After creating the massing, convert it into the building elements with the parameters given next.

(Expected time: 45 min)

1. Project file parameters:
   Template File- **Architectural Template**,
   File Name to be assigned- *c10_OfficeTowers_ex1.rvt*

2. Building element types for the converted shell elements:
   Floor: **Floor: LW Concrete on Metal Deck**
   Walls: Towers- **Curtain Wall: Exterior Glazing**
   (horizontal and vertical spacing- 10'0")
   Connecting Passage- **Curtain Wall: Curtain Wall 1**
   Roof: **Basic Roof: Generic 9"**

After completion, the building model will appear similar to the illustration given in Figure 10-36.

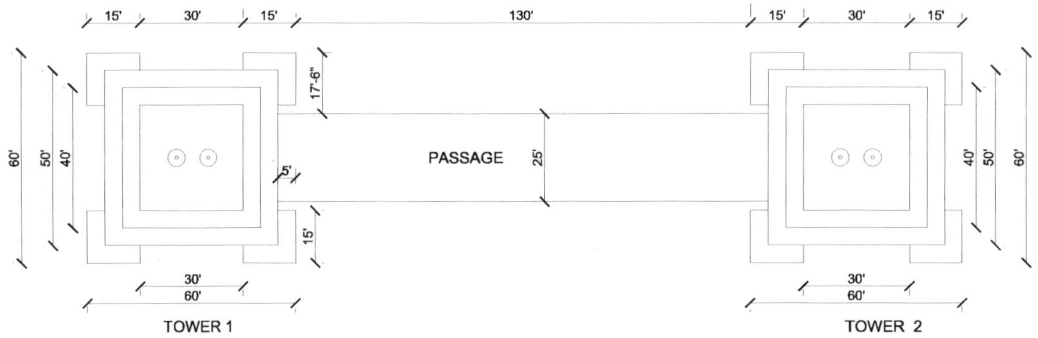

Figure 10-33 *Sketch plan for a multistory office building*

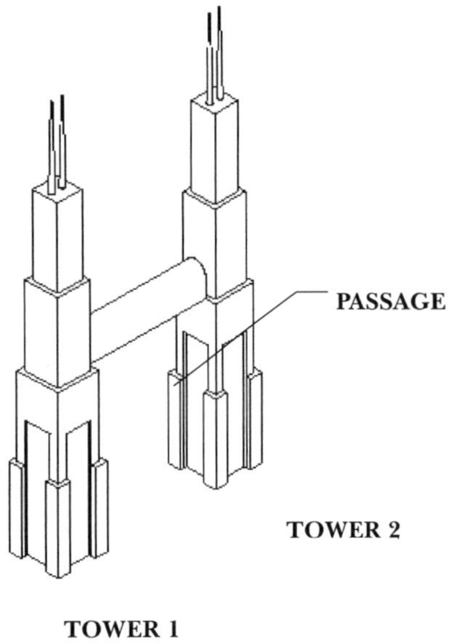

Figure 10-34 *3D view of the building*

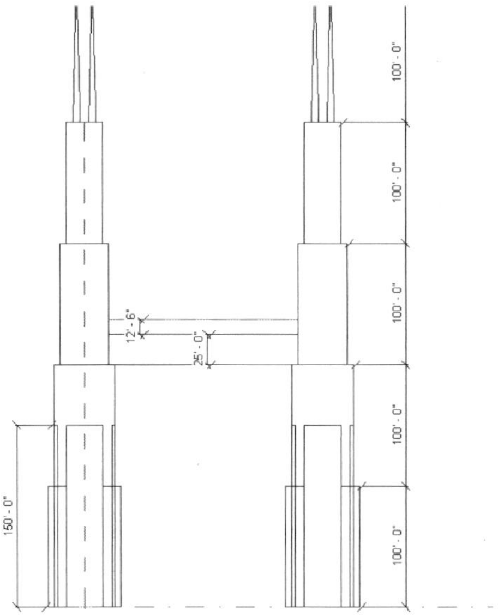

Figure 10-35 *Elevation view of the building*

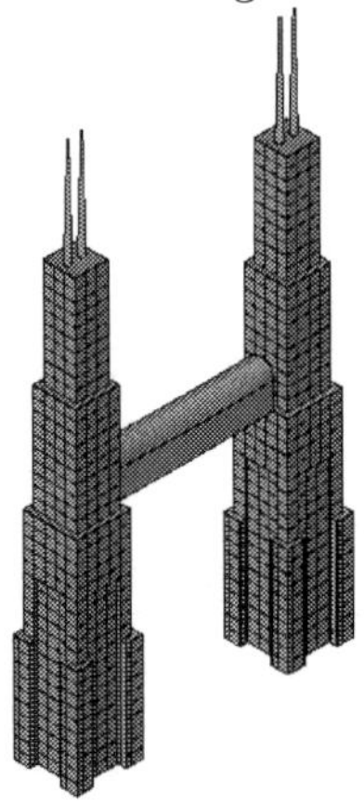

Figure 10-36 *Shaded 3D view of the building*

Chapter 11

Adding Annotations and Dimensions

Learning Objectives

After completing this chapter, you will be able to:

- *Add tags to building elements using the Tag tool*
- *Add room tags to the interior spaces using the Room Tag tool*
- *Add symbols to project view using the Symbol tool*
- *Create dimensions using various dimensioning tools*
- *Modify dimensions based on the given parameters*
- *Add spot elevation to a building model using the Spot Elevation tool*

INTRODUCTION

In the previous chapters, you learned to create and add different building elements to a project. This help you to convey the design intent of the project, graphically. However, you can execute a building or an interior project by adding documentation to a drawing. In this chapter, you will learn to add tags to elements and spaces, create dimensions, use annotations, and add dimensions to the elements. In later chapters of this book, you will learn about the tools that are used to document an Autodesk Revit project and add content to it. You will also learn to compose and plot sheets in the later chapters.

BIM
CONCEPTS

In a BIM project, adding annotations such as tags, dimensions, and symbols is an essential task in the detailed design and construction documentation phase. The amount of detailing and tag depends on the goals of the BIM project.

ADDING TAGS

You can easily tag various elements in a building model. A tag is a useful annotation that assists in identifying the tagged building element. When you design complex building models using Autodesk Revit, tags play an important role in arranging various elements in schedules. You can then add necessary description for each tagged element in a tabulated form. Figure 11-1 shows the usage of various tags in an interior layout.

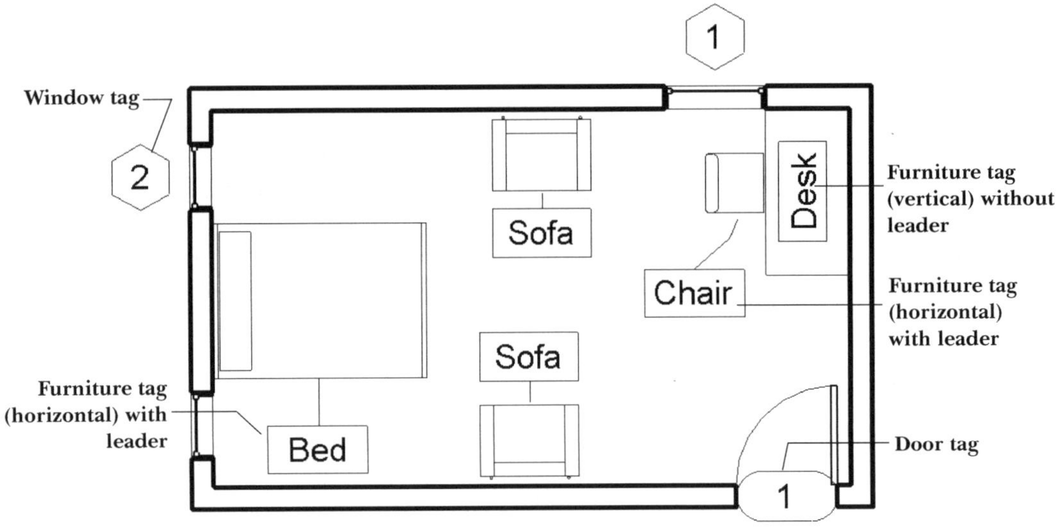

Figure 11-1 *Various tags in an interior layout*

Tagging Elements by Category

Ribbon: Annotate > Tag > Tag by Category

To attach a tag to an element based on the element category, choose the **Tag by Category** tool from the **Tag** panel of the **Annotate** tab, as shown in Figure 11-2. On doing so, the **Modify |Tag** contextual tab will be displayed and the parameters related to placing and orienting will be displayed in the **Options Bar**. In the **Options Bar**, you can select the **Horizontal** or **Vertical** option from the drop-down list displayed on the left, based on the desired direction of the text in the tag, refer to Figure 11-1. The **Leader** check box is selected by default in the **Options Bar**. As a result, a leader will be displayed with the tag. You can clear this check box, if you do not require a leader to be displayed with the tag. As the **Leader** check box is selected, the options in the drop-down list displayed on the right of it will be enabled. From this drop-down list, you can select two options: **Attached End** and **Free End**. On selecting the **Attached End** option from the drop-down list, you will not be able to move the end of the leader with the tag away from the element category it is attached to. As you select the **Attached End** option from the drop-down list, the edit box on its right gets enabled. In this edit box, you can specify the length of the leader or the distance of the tag from the attached element. Alternatively, if you select the **Free End** option from the drop-down list, the end of the leader can be moved away from the element it is attached to.

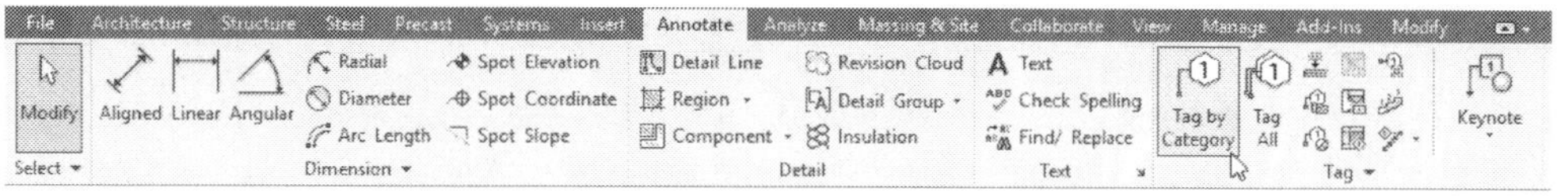

*Figure 11-2 Choosing the **Tag by Category** tool from the **Annotate** tab*

Tagging All Elements in a View

Ribbon: Annotate > Tag > Tag All

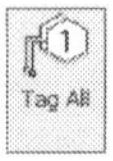

The **Tag All** tool is used to tag all elements that are visible in the current view. On choosing this tool, the **Tag All Not Tagged** dialog box will be displayed, as shown in Figure 11-3. In this dialog box, there are two radio buttons: **All objects in current view** and **Only selected objects in current view**. By default, the **All objects in current view** radio button will be selected. The **Only selected objects in current view** radio button will be active and ready for selection only when objects to be tagged are selected prior to invoking the **Tag All Not Tagged** dialog box. The selection of the **All objects in current view** radio button ensures that all objects present in the current view will be tagged, provided their categories are selected from the **Category** column of the **Tag All Not Tagged** dialog box. The table below the radio buttons displays the list of category tags of all elements present in the current view and their corresponding loaded tags.

Note

*If the category of element you wish to tag does not appear in this list, you need to load the corresponding tag family and then use the **Tag All** tool.*

You can select only one category at a time. Select the category of elements to be tagged and the tag type to be used from the **Tag All Not Tagged** dialog box, as shown in Figure 11-3. You can select the **Leader** check box to add a leader to the elements. You can also modify the length and orientation of the leader in the dialog box by entering a value in the **Leader Length** edit box and selecting an option from the **Tag Orientation** drop-down list, respectively. Next, choose the **Apply** button to apply the settings to the current view. For example, you can tag all furniture elements in a layout plan view by using the **Tag All** tool, as shown in Figure 11-4.

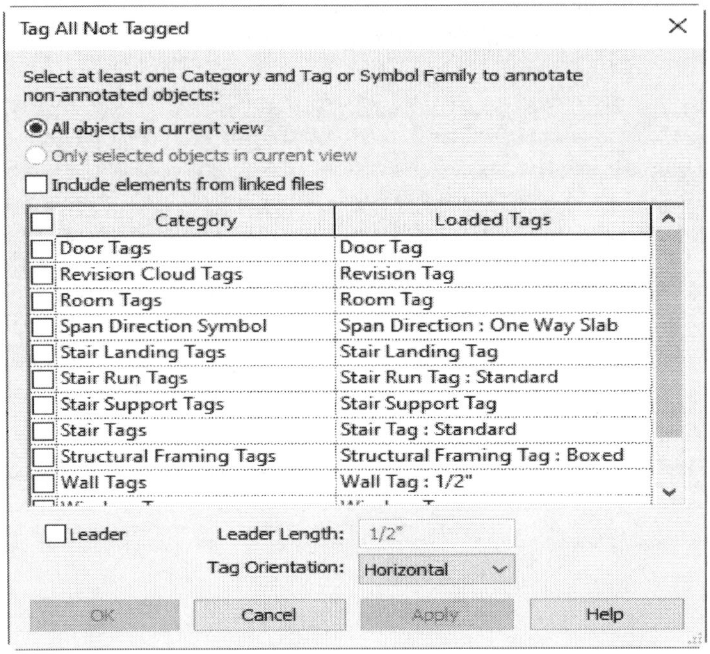

*Figure 11-3 Selecting the category of elements to be tagged from the **Tag All Not Tagged** dialog box*

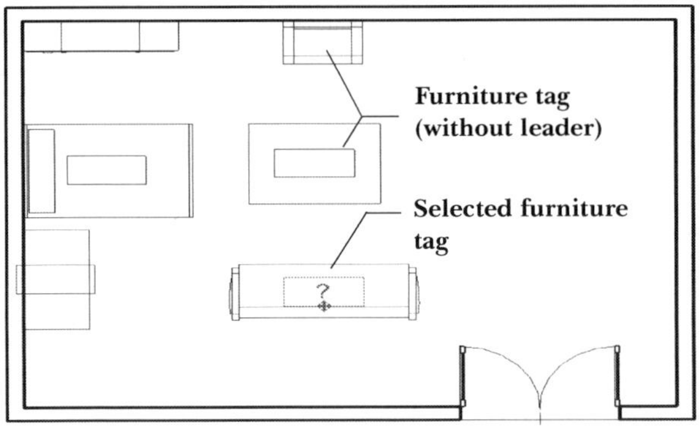

*Figure 11-4 The furniture elements tagged using the **Tag All** tool*

In Revit, you can use the **Tag All** tool to tag rooms or areas. If you tag the elements whose visibility has been turned off, the **Tag Visibility Enabled** message box will be displayed. This message box displays a message informing that the visibility of all elements, whose visibility is turned off, will be turned on while adding tags. Choose **Yes** to continue; the tags will be added to all elements belonging to the selected category. You can then edit them individually. This process needs to be repeated for tagging elements category-wise.

Tip

In Revit 2025, you can tag the mullions of the curtain wall.

Tagging Treads or Risers

In a project view, you can display the numbering of treads or risers for a run in a component based stair. The numbering for the run can be displayed in a plan, elevation, or in a section view. Like any other annotation, the numbering in the treads or risers is view-specific. To add the numbers of tread or risers in a plan view, choose the **Tread Number** tool from the **Tag** panel of the **Annotate** tab; you will be prompted to pick a reference line of a stair to place the Stair or Tread number. Move the cursor and place it over the stair at a location where you desire to place the numbers. You can place the numbers in any of the following locations of the stair run: Left, Right, Center, Left Quarter, and Right Quarter. Click on the desired reference line on the stair run: the riser or tread will be numbered along the stair path, as shown in Figure 11-5.

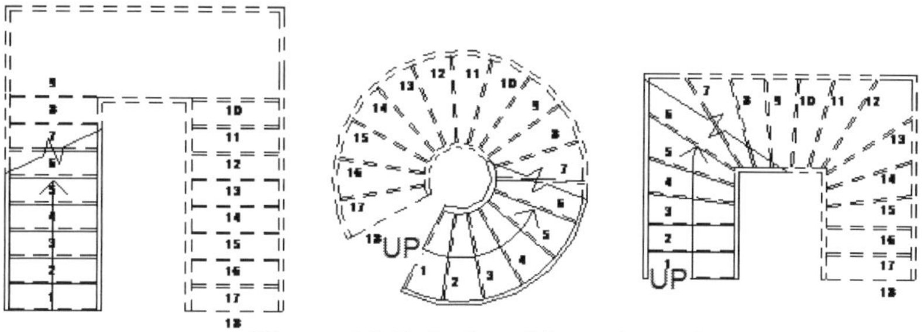

Figure 11-5 *Stairs with tread numbers*

Note

You cannot annotate the treads or risers in the sketch mode.

Tagging with Multiple Leaders

In Revit 2025, you can add multiple leader to an existing tag or multi-category tag and attach it to the other objects. To do so, select the existing multi-category tag from the project; the **Modify|Multi-Category Tags** contextual tab will be displayed, In this tab, choose the **Add/ Remove Host** button from the **Host** panel; you will be prompted to select one or more objects to which you want to attach the existing tag. Select the objects in the project that you want to attach to the selected tag; the leaders will be attached to the selected tags along with its value. In the displayed tag, you can display the count of the objects that will be attached to the tag. To do so, select the added tag and then choose the **Edit Family** tool from the **Mode** panel; the Family Editor will be displayed. Select the displayed label in the drawing area and then choose the **Edit Label** tool from the **Label** panel of the of the **Modify|Tag Label** contextual tab; the **Edit Label** dialog box will be displayed. Select the **Multi Leader Tag** option from the **Select available fields from** drop-down list available in the **Category Parameters** area; the **Host Count** field will be displayed in the text box area. Choose the **Add parameter(s)** button; the **Host Count** field will be added to the **Label Parameters** area. Choose **Apply** and then **OK**; the **Edit Label** dialog box will be closed. Now, choose the **Load into Project** button from the **Family Editor** panel; the **Load into Project** dialog box will be displayed. Select the desired project and then choose **OK**; the Project environment will be displayed and the selected tag will be modified to display the count of the host.

ROOM TAGS

Room tags are useful annotation tools for your drawing and they help you set information about the type of occupancy and define the enclosed area and volume. Autodesk Revit enables you to define and add a nomenclature to various interior and exterior spaces in a building project. This is important not only for identifying each space, but also for creating a room-wise project schedule. In Autodesk Revit, a room is also treated as an element of the project similar to the other elements such as a wall, door, or window.

In a building model, rooms and their tags are interrelated to each other. You can add a room with a tag. To do so, choose the **Room** tool from the **Room & Area** panel of the **Architecture** tab; the **Modify | Place Room** tab will be displayed. In the **Tag** panel of this tab, the **Tag on Placement** option is selected by default. As a result, the room added will have a tag. If you want to add a room without a tag, invoke the **Room** tool, deselect the **Tag on Placement** option from the **Tag** panel, and click at the desired place. You can add the room tag later by using the **Room Tag** tool from the **Tag** panel of the **Annotate** tab.

Autodesk Revit provides you with the option to keep the tag either in its original position or inside the room that has been moved. When you move a room that has a tag attached to it, an alert message will be displayed in the **Autodesk Revit 2025** dialog box, informing that the room tag is outside of its room. To enable the room tag such that it adjusts itself within the room at the new position, choose the **Move to Room** button from the displayed dialog box; the room tag will automatically move inside the room. Next, choose the **OK** button; a leader will be added to the room tag and the room tag will be placed outside the room that has been moved.

Room Separation

Ribbon: Architecture > Room & Area > Room Separator

Rooms are the defined spaces in a building and are used for specific utilities in a building project. They are divided into different parts to define the purpose of their utilization. They can be separated by a wall or without a wall depending upon the need of the project and the nature of the space. Revit defines a room by assessing the interior edges of the walls bounding the space. While working on large projects, you may need to define rooms by specifying the room boundary by using the **Room Separator** tool. This tool helps you add room-bounding lines to separate and adjust the room boundaries and create separate boundary enclosures without the intervention of a wall.

Invoke the **Room Separator** tool from the **Room & Area** panel; the **Modify | Place Room Separation** tab will be displayed. In this tab, the **Draw** panel displays various sketching tools that can be used to sketch the lines defining the separation boundary of the room. You can also use the **Pick Lines** tool from the **Draw** panel to pick the elements that define the room separation lines. To control the visibility of the separation lines, choose the **Visibility/Graphics** tool from the **Graphics** panel of the **View** tab; the **Visibility/ Graphics Overrides for <contextual view>** dialog box will be displayed. From this dialog box, choose the **Model Categories** tab and expand the **Lines** group by clicking on the **(+)** sign in the **Visibility** column of the **Model Categories** tab. Now, if you select the check box beside the **Room Separation** option, the separation lines will be visible in the plan view and can be printed. If you clear the check box, the separation lines will be hidden in the plan view but the room boundaries will be retained. There may be instances when you do not want to use a particular wall as a room-defining wall. In this case, set the property of that wall as non-room bounding by clearing the check box for the **Room Bounding** instance parameter. On doing so, the selected wall will not be considered while defining the room.

For example, Figure 11-6 shows a 3D view of a room that has an intermediate wall with an arched opening. Assume that you want to ignore it while defining the entire room. To do so, clear the check box corresponding to the **Room Bounding** instance parameter in the **Properties** palette. Autodesk Revit will ignore the intermediate wall and the room boundaries will be extended to the exteriors walls, refer to Figure 11-7. In this manner, you can add identification to each space and later create a schedule of areas. You will learn more about creating schedules in Chapter 12 of this textbook.

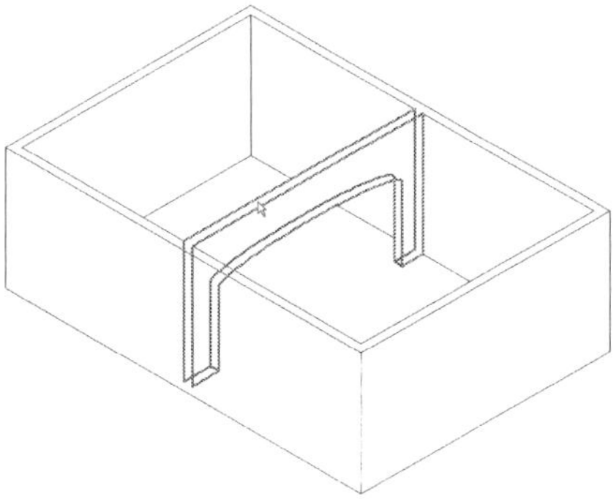

Figure 11-6 *An intermediate room with a wall having an arched opening*

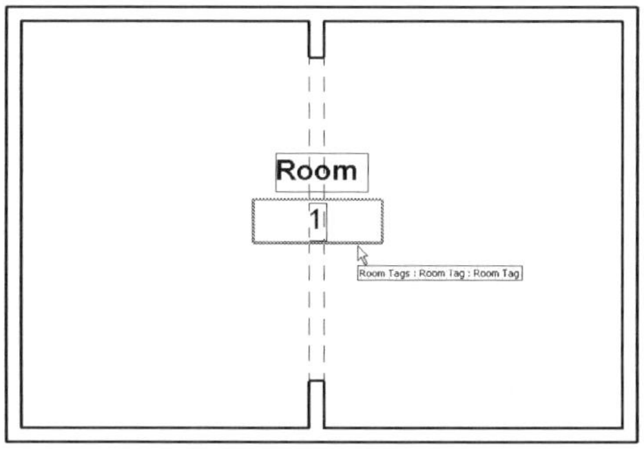

Figure 11-7 *Defining a room by changing the* ***Room Bounding*** *parameter of the intermediate wall*

Tagging Rooms

Ribbon: Architecture > Room & Area > Tag Room drop-down > Tag Room
Shortcut Key: RT

To add room tags, choose the **Tag Room** tool from the **Room & Area** panel; the **Modify | Place Room Tag** tab will be displayed. From the **Type Selector** drop-down list of the **Properties** palette, you can select any of these three options, **Room Tag**, **Room Tag With Area**, and **Room Tag With Volume**. To display the room tag along with its area, select the **Room Tag With Area** option in the **Type Selector** drop-down list. Similarly, you can display the volume of the room along with its tag by selecting the **Room Tag With Volume** option from the **Type Selector** drop-down list. By default, the **Room Tag** option is selected from the **Type Selector** drop-down list. As a result, the room tag is displayed without its area and volume. Now, after you select a specific type for the Room tag to be inserted, you can specify its instance and type properties from the **Properties** palette.

KEYNOTES

Keynote is a type of tag family. It is available for all model elements, detail components, and materials. Although you can add annotations to detail views using the **Text** tool, when it comes to labeling the components in detail, it is preferable to use keynotes. In keynoting, pre-defined list of notes is used. The notes are organized in a keyed list defined in a separate text file. The keynoting data provided in Revit is based on a separate text file containing the list of different keynotes. In Autodesk Revit, keynoting can be done by following any of the two systems, the 1995 CSI (Construction Specifications Institute) Master Format system, and the revised version of the 1995 CSI Master Format system, which was introduced in the year 2004 .

Loading Keynote File

Before you can use keynoting in a project, you must first load a keynote file. To do so, choose the **Keynoting Settings** tool from the **Annotate > Tag > Keynote** drop-down; the **Keynoting Settings** dialog box will be displayed, as shown in Figure 11-8. The **File locations** edit box in the **Keynote Table** area of this dialog box displays the path of the keynote file. To change the file path or to select a different keynote file, choose the **Browse** button next to the **File Location** edit box; the **Browse for Keynote File** dialog box will be displayed. Choose the desired keynote file from this dialog box and choose the **Open** button; the **Reload Successful** message box will be displayed showing that the keynote table reloaded successfully. Choose the **OK** button; the desired file will now replace the previous file and its path will be displayed in the **File Location** edit box.

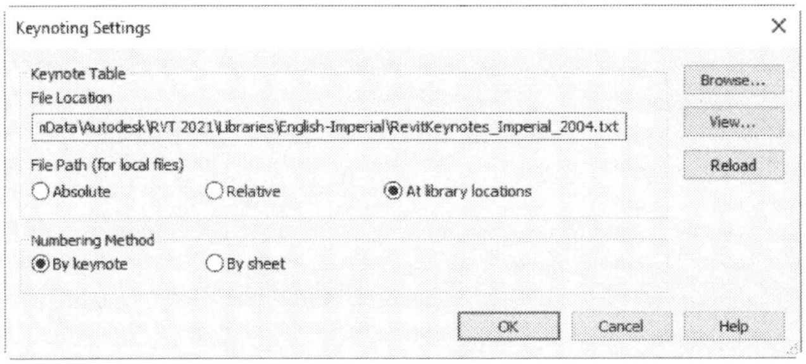

Figure 11-8 *The **Keynoting Settings** dialog box*

Placing Keynotes

In Autodesk Revit, there are three types of keynotes that you can use to detail your drawing. The three types of keynotes are: Element, Material, and User.

The Element keynote annotates building elements. The tag reads the keynote assigned to the element in the model such as the keynote assigned to a wall or door, not to the individual layers or components of the wall or door. To add the Element keynotes in the detail view of your drawing, choose the **Element Keynote** tool from **Annotate > Tag > Keynote** drop-down; the **Modify | Place Element Keynote** tab will be displayed. Now, in the **Properties** palette, click on the **Type Selector** drop-down list and select the desired type from the list displayed. From the **Type Selector** drop-down list, you can select any of these four options of Keynote Tags: for Imperial **Keynote Tag Keynote Number**, **Keynote Tag Keynote Number-Boxed-Large**, **Keynote Tag Keynote Number-Boxed-Small**, or **Keynote Tag Keynote Text** (for Metric **M_Keynote Tag Keynote Number**, **M_Keynote Tag Keynote Number-Boxed-Large**, **M_Keynote Tag Keynote Number-Boxed-Small**, or **M_Keynote Tag Keynote Text)**. If you select the **Keynote Number** option from the drop-down list, the added keynote will only display the keynote number in your drawing.

Adding Keynote Legends

Ribbon: View > Create > Legends drop-down > Keynote Legend

A keynote legend defines the keynote along with its property and functional description in tabulated format.

To create a new keynote legend, choose the **Keynote Legend** tool from the **Create** panel; the **New Keynote Legend** dialog box will be displayed. Specify a legend name in the **Name** edit box of the dialog box and choose the **OK** button; the **Keynote Legend Properties** dialog box will be displayed. To assign properties and parameters to the keynote legend, you can choose various tabs such as **Fields**, **Filter**, **Sorting/Grouping**, **Formatting**, and **Appearance** from the **Keynote Legend Properties** dialog box. Next, choose the **OK** button to exit the **Keynote Legend Properties** dialog box; a keynote legend will be added in the drawing. A keynote legend has two columns: **Key Value** and **Keynote Text**. **Key Value** in the legend lists all keynotes of the elements in the project that are tagged with a keynote and the **Keynote Text** column lists the description of the associated **Key Value** in the **Keynote Legend**.

A keynote legend gives information only about the elements that are tagged with a keynote. To view the keynote legend created by you, expand the **Legends** group in the **Project Browser**; the **Keynote Legend** will be displayed under the **Legends** node.

ADDING SYMBOLS

Ribbon: Annotate > Symbol > Symbol

The **Symbol** tool is used to add 2D annotation drawing symbols to a project view to make it more informative. For example, you can use the **Symbol** tool to insert a north symbol, graphical scale, centerline symbol, and so on into a specific project view. Like other annotations, symbols are also view-specific in character.

When you choose the **Symbol** tool from the **Symbol** panel, the **Modify | Place Symbol** tab will be displayed. Now, from the **Type Selector** drop-down list which shows the loaded symbols in the project, select the required symbol. The instance and type properties depend on the selected symbol. You can modify the instance and type properties of a symbol from the **Properties** palette.

The **Number of Leaders** spinner available in the **Options Bar** enables you to add multiple leaders to a single symbol. If the **Rotate after placement** check box is selected, the **Rotate** tool will be invoked as soon as you place the symbol. As a result, you can reorient the symbol after placing it. The selected symbol can be added to the project view by clicking at the desired location. Figure 11-9 shows an example of the centerline symbol added to a plan view. Multiple instances of the symbol can then be placed by clicking on the locations of other instances.

***Figure 11-9** Adding the centerline symbol to the plan view*

ADDING DIMENSIONS

Dimensions play a crucial role in the presentation of a project. Although the building model conveys the graphical image of the design, yet to materialize it at the site, the building view must provide information and statistics regarding each element. Since the design drawing is used for the actual construction of a project, it is essential to describe the building parts in terms of actual measurement parameters such as length, width, height, angle, radius, diameter, and so on. All these information can be added into the project using dimensions. The information conveyed through dimensions is, in most cases, as important as the project view itself. It ensures that the project drawings are read and interpreted in appropriate way. Adding dimensions also ensures that there are no discrepancies between various elements used in the generation of the building drawings.

Types of Dimensions

In Autodesk Revit, you can use two types of dimensions: temporary and permanent. By default, the temporary and permanent dimensions use the units settings specified in the initial start-up of the project. The temporary dimensions are not view-specific, whereas the permanent dimensions are view-specific. It means that if you change the view, the permanent dimensions will not be visible. These two types of dimensions are temporary dimensions and permament dimensions.

Dimensioning Terminology

Before using the dimensioning tools, it is important to understand the dimensioning terms that are used in Autodesk Revit. Figure 11-10 shows common dimensioning terminologies that are dimension line, dimension text, tick mark, and witness line.

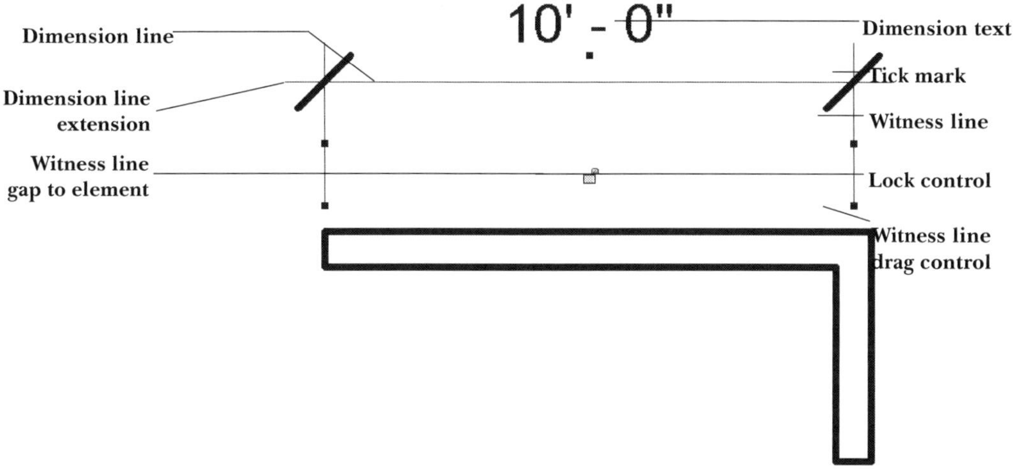

Figure 11-10 Various dimensioning terms

Adding Permanent Dimensions

Permanent dimensions are added specifically for a particular measurement. In Autodesk Revit, you can access various dimensioning tools from the **Dimension** panel in the **Annotate** tab, as shown in Figure 11-11. You can choose the appropriate dimension type and the dimension tool to add dimensions to the element. The annoted tab contains aligned tool, linear tool, angular tool, radial tool, diameter tool, and arc length tool.

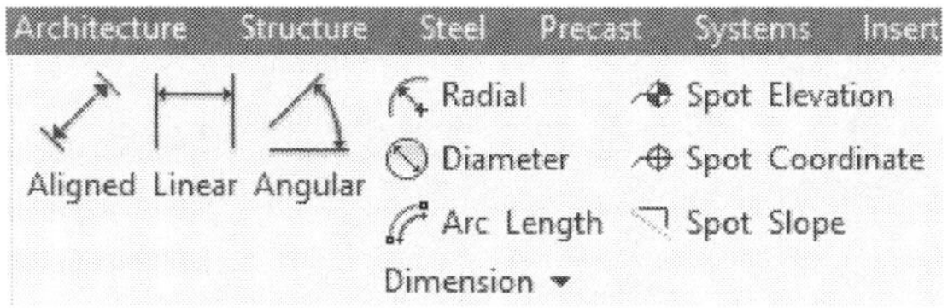

***Figure 11-11** Dimensioning tools in the **Dimension** panel of the **Annotate** tab*

Adding Alternate Dimension Units

In a project, besides the main dimensions, you can also display alternate dimensions in your model. These alternate dimensions can be both permanent and spot dimension types. The alternate dimension units display both imperial and metric units simultaneously in drawings. To add an alternate dimension to the main dimension in a project, select any of the dimension tool from the **Dimension** panel of the **Annotate** tab; the **Modify | Place Dimension** contextual tab and various related instance properties in the **Properties** palette will be displayed. Now, in the **Properties** palette, choose the **Edit Type** button; the **Type Properties** dialog box will be displayed. In this dialog box, choose the **Duplicate** button; the **Name** dialog box will be displayed. Enter the name **Alternate Units** in the **Name** edit box and choose the **OK** button to return to the **Type Properties** dialog box. Now, in the **Type Properties** dialog box, click on the **Value** field corresponding to the **Alternate Units** parameter; a drop-down list with the **None**, **Right**, and **Below** options will be displayed. You can select an option as per your requirement. On selecting the **Right** option, the alternate units will be displayed in-line to the primary units, and on selecting the **Below** option, the alternate units will be displayed below the primary units. Now, click on the **Value** field corresponding to the **Alternate Units Format** parameter; the **Format** dialog box will be displayed. In this dialog box, select the **Meters** option from the **Units** drop-down list, select the '**m**' option from the **Units Symbol** drop-down list and then choose the **OK** button; the **Format** dialog box will be closed. You can also assign square brackets '[]' to the **Alternate Prefix** and **Alternate Suffix** value fields. Now, choose the **OK** button; the **Type Properties** dialog box will be closed and the drawing window will be displayed.

Baseline and Ordinate Dimensions

You can dimension a series of elements and components simultaneously. This can be done by using the baseline and ordinate dimensioning techniques.

A baseline dimension is a stacked dimension that originates from a common baseline. It creates a referenced dimension from the first witness line which acts as a base for adding dimension, as shown in Figure 11-12. An ordinate dimension is generated to show the level of a particular point measured from a fixed point, called datum, as shown in Figure 11-13. The ordinate dimension of a point shows the measurement of its perpendicular distance from the datum or the origin point. Note that you can use the baseline and ordinate dimensioning techniques only for linear and aligned dimensions.

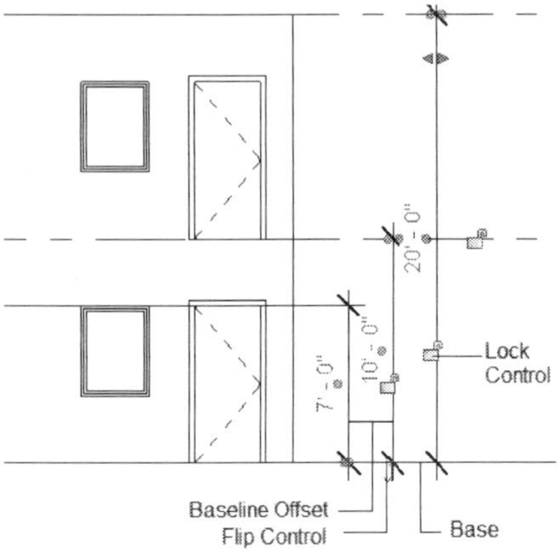

***Figure 11-12** The selected baseline dimension and its components*

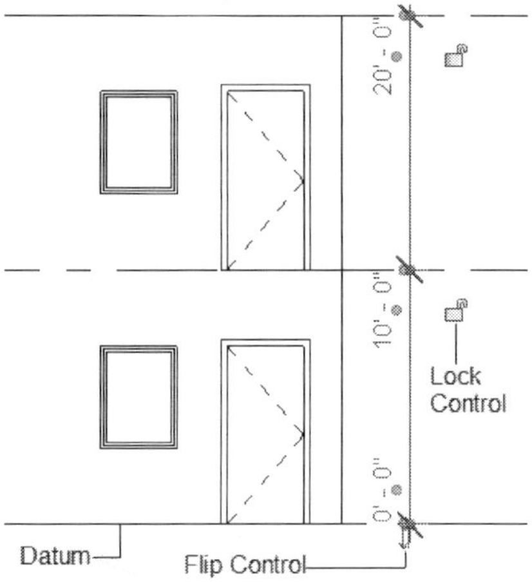

***Figure 11-13** The selected ordinate dimensions and their components*

Editing Dimensions

You can edit various properties of a dimension before or after creating it. To modify any dimension in the drawing, select it; the **Modify | Dimensions** tab will be displayed. From this tab, you can use various options to modify and edit the dimension selected. The **Type Selector** drop-down list displays various in-built dimension types available in Autodesk Revit.

Controlling the Display of Tick Marks and Dimension Arrows

Revit provides you the flexibility to control the display of tick marks and dimension arrows while dimensioning small segments. When the dimensioning segments are too small to accommodate the dimension arrow, Revit flips the dimension arrows on the exterior side and displays diagonal tick marks, as shown in Figures 11-14 through 11-16.

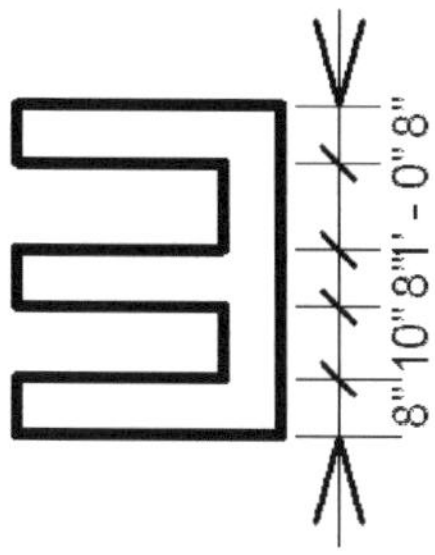

Figure 11-14 *Dimension arrows replaced by the tick marks*

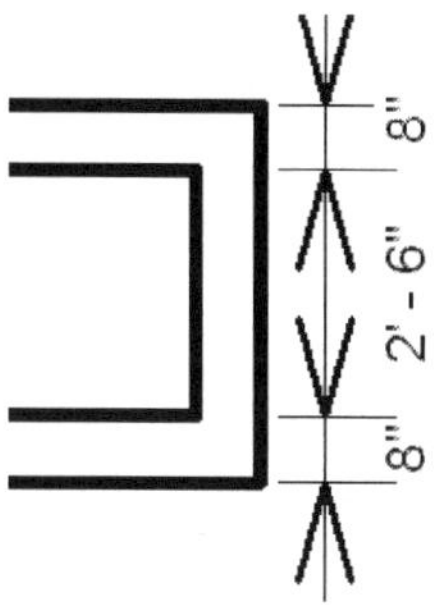

Figure 11-15 *Dimension arrows on the interior side of the witness lines*

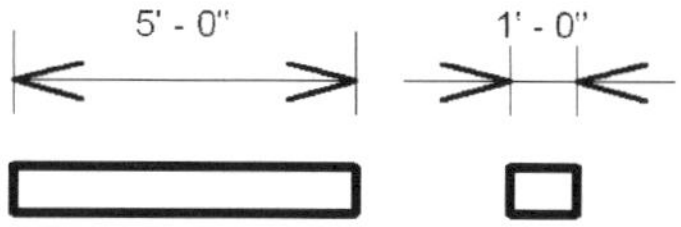

Figure 11-16 *The flipping of arrows on the exterior side in case of small segments*

Creating Linear Wall Dimensions Automatically

Autodesk Revit provides the option to create dimensions automatically for linear walls. To do so, invoke the **Aligned** tool from the **Dimension** panel of the **Modify | Place Dimensions** tab and then select the **Entire Walls** option from the **Pick** drop-down list in **Options Bar**. This option enables you to pick a linear wall and create the dimensions associated with it. Next, choose the **Options** button from the **Options Bar**; the **Auto Dimension Options** dialog box will be displayed. In this dialog box, you can select the required references to create the dimensions. You can add dimensions to openings or to the walls intersecting with the selected wall. You can also specify whether to use the center of openings or their widths for giving reference. After selecting the appropriate option, select the linear wall to add the dimensions. When the **Openings** option is selected from the **Auto Dimension Options** dialog box, Autodesk Revit will create the dimensions that refer to the center or widths of openings. When you select the **Intersecting Walls** option, the automatic dimensions refer to all perpendicular walls intersecting the selected wall.

ADDING SPOT DIMENSIONS

Besides adding spot elevations, Revit also allows you to add coordinates. The spot dimensions are the dimensions that display the elevation level and the coordinates of a point with respect to the base level. They can be placed in the plan, elevation, or 3D views. You can add spot dimensions on non-horizontal and non-planar surfaces. You can use the spot elevation to specify the elevation level of points on a ramp, road, topographical surface, stairs, and other features. The properties of the spot elevation are similar to the properties of dimensions.

Placing a Spot Dimension

To place a spot dimension, choose the **Spot Elevation**, **Spot Coordinate** or **Spot Slope** tool from the **Dimension** panel of the **Annotate** tab; the **Modify | Place Dimensions** tab will be displayed. From this tab, select the type of spot dimension symbol to be used from the **Type Selector** drop-down list. Move the cursor near the edge of the element and click to select the point of reference when the edge of the element is highlighted. If you have selected the spot dimension without a leader, it will be placed at the selected point. If the type of spot dimension selected is with a leader, an arrow will start from the specified point. Move the cursor and click to locate the elbow of the leader. Now, move the cursor to the desired location and click again to place the spot dimension; the spot dimension will be created. Figure 11-17 shows the spot elevations added to the plan view and Figure 11-18 shows the spot elevations and spot coordinates added to a 3D view.

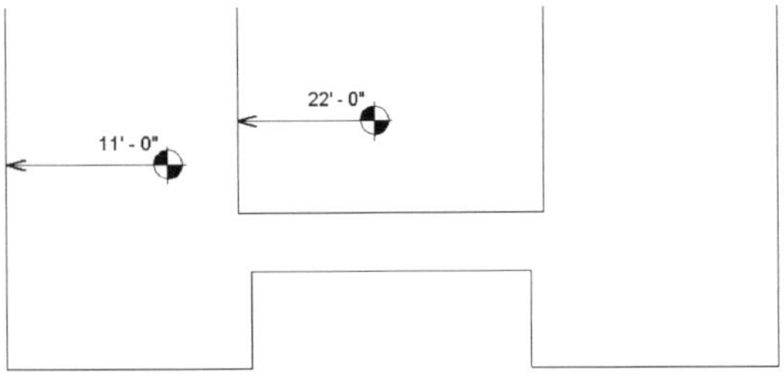

Figure 11-17 *Placing the spot elevation in the plan view*

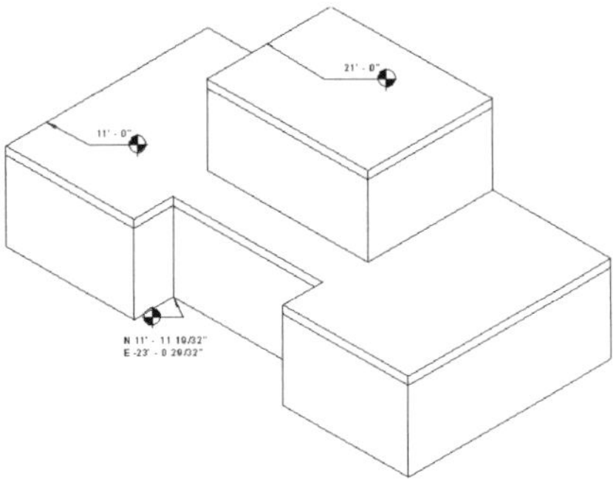

Figure 11-18 *Placing the spot elevation in the 3D view*

Modifying Spot Dimension Properties

Similar to other dimensions, you can also modify various properties of spot dimensions before or after creating them. To modify the instance parameters of a spot dimension, select it; the instance properties of the selected spot dimension will be displayed in the **Properties** palette. In this palette, you can select the check box in the value field corresponding to the **Leader** parameter to display the leader with the spot dimension. In the **Properties** palette, you can specify other instance parameters to specify the settings of the text and the leader. You can modify the type properties of the spot dimension. To do so, select a spot dimension and choose the **Type Properties** tool from the **Properties** panel; the **Type Properties** dialog box will be displayed, as shown in Figure 11-19. An important property of the spot dimension is the elevation origin. By default, Autodesk Revit assumes the project origin as the base level. It then calculates the distance or the elevation of the spot level from this origin. You can use the default settings or specify a different elevation origin to calculate the spot elevations. This can be done by setting the value of the **Elevation Origin** type parameter to **Relative**. You can also select a placed spot level and modify its relative base level.

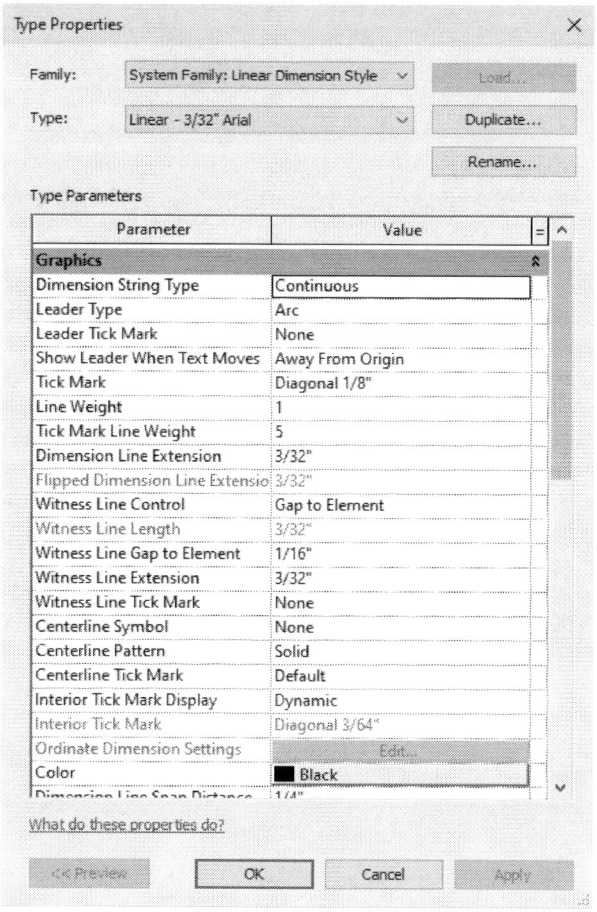

*Figure 11-19 The **Type Properties** dialog box for spot dimension*

Converting Temporary Dimensions to Permanent Dimensions

As described in the earlier chapters, temporary dimensions appear while you create various elements such as walls, doors, windows, and so on. They can be used to specify the size and location of the elements in the building model, refer to Chapter 3, Creating Walls. You can move or resize an element by clicking on the temporary dimension and entering a new value. When you create an element, a conversion control symbol will also be displayed along with the temporary dimension, as shown in Figure 11-20. When you click on this symbol, the temporary dimension will be converted into a permanent dimension, as shown in Figure 11-21.

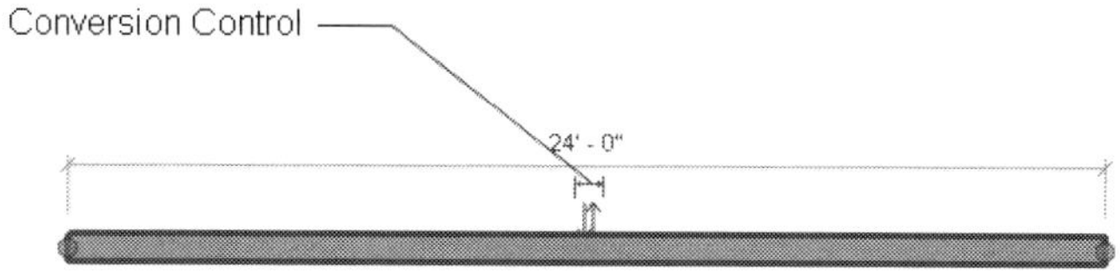

Figure 11-20 A temporary dimension with the conversion control

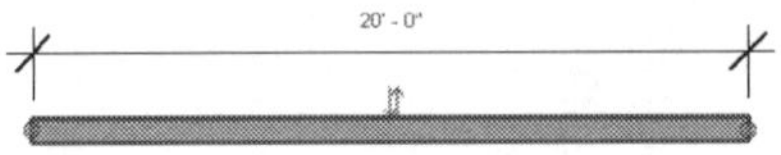

Figure 11-21 *Permanent dimension*

Tip

As described earlier, the temporary dimensions can be used to move or copy the elements to the desired locations. For example, when you modify the temporary dimension of a selected wall segment, it is modified as per the value of the length. Thus, in Autodesk Revit, building elements and their dimensions are parametrically associated with each other.

TUTORIAL

Tutorial 1 Apartment 1

In this tutorial, you will first add rooms to the enclosed spaces of the *Apartment 1* project created in Tutorial 1 of Chapter 8, and then name the rooms by using the **Room Tag** tool. You will also tag the furniture and then dimension the ground floor plan view based on the sketch plan shown in Figure 11-22. The dimensions shown on the top and right sides of the plan are referenced to the wall centerlines, whereas the dimensions on the bottom and left side are referenced to the wall faces. The exact location of the tags and dimensions is not important in this tutorial. Use the parameters given below for creating the dimensions.

(Expected time: 45 min)

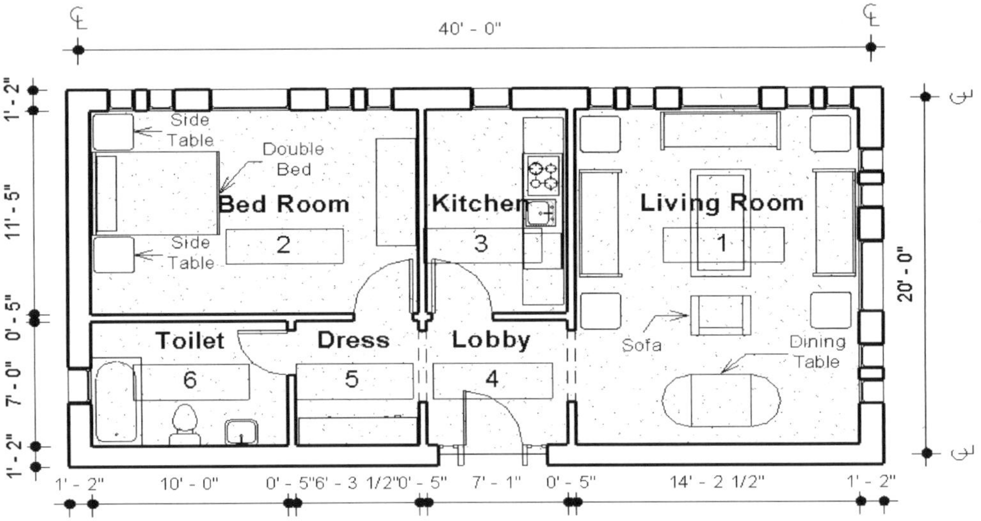

Figure 11-22 *Sketch plan for adding room tags, furniture tags, and dimensions to the Apartment 1 project*

1. Room tag- **Room Tag: Boxed**
2. Furniture tag- **Furniture Tag: Standard** with leader arrowhead- **Arrow 30 Degree**

3. Dimension Parameters:
   Dimension type to be based on the following options:
   For Imperial **Linear w Center - 3/32" Arial** and **Linear - 3/32" Arial**
   For Metric **Diagonal Center - 2.5 mm Arial** and **Linear - 3/32" Arial**

   New dimension type names- **Linear Dimension - CL** and **Linear Dimension - Faces**
   Tick Mark- For Imperial **Dot Filled 1/16"**
   For Metric **Filled Dot 3mm**
   Witness Line Gap to Element- For Imperial **1/4"**
   For Metric **6.35mm**

The following steps are required to complete this tutorial:

a. Open the specified project file and the first floor plan view.
b. Use the **Room** and **Room Tag** tools and add rooms and room tags, refer to Figures 11-23 through 11-27.
c. Invoke the **Tag** tool and then load and add the furniture tags, refer to Figures 11-28 through 11-31.
d. Invoke the **Dimension** tool and set the type parameters.
e. Add dimensions to the project plan view using the wall centerlines as the reference, refer to Figures 11-32 through 11-34.
f. Add dimensions using automatic dimensioning with wall faces as the reference, refer to Figure 11-35.
g. Save the project using the **Save As** tool.
   For Imperial *c11_Apartment1_tut1.rvt*
   For Metric *M_c11_Apartment1_tut1.rvt*
h. Close the project.

Opening the Project File

To begin this tutorial, open the file that you have created in Tutorial 1 of Chapter 8 for the *Apartment 1* project.

1. Choose **Open > Project** from the **File** menu and open *c08_Apartment1_tut1.rvt for imperial (M_c08_Apartment1_tut1.rvt for metric)*. You can also download this file from *https://www.Cadsoft.com*. The path of the file is as follows: *Textbooks > Civil/GIS > Revit Architecture > Exploring Autodesk Revit 2025 for Architecture*.

2. After you open the file, double-click on **First Floor** in the **Floor Plans** head from the **Project Browser**; the first floor plan is displayed.

Adding Room and Room Tags

In this section of the tutorial, you will first add rooms (without a tag) to the enclosed space. To add room tags to the interior spaces, you need to invoke the **Room Tag** tool. The room tag is preloaded in the default template file. As this template will be used in this tutorial, you do not need to load it. Next, you need to move the cursor inside the enclosed space and click to add the room tag. You can then click on the tag and rename it. Autodesk Revit sequentially numbers the room tags. Therefore, you need to add them based on the sequence given in the sketch plan.

Note

In a building project, you can add rooms with tags attached to them and then rename the tags according to the space functionality. In this section of the tutorial, the rooms are added without tags and then the tags are added to make you understand the use of the ***Room*** *tool and the* ***Room Tag*** *tool separately.*

1. Invoke the **Room** tool from the **Room & Area** panel in the **Architecture** tab; the **Modify | Place Room** tab is displayed. Now, as you move the cursor in the drawing area, you will notice a blue rectangle along with the default text **Room** attached to it. Next, you need to ensure that while adding rooms, the tags should not appear.

   Note that the **Tag on Placement** button is not chosen in the **Tag** panel.

2. Move the cursor inside the living room (for location, refer to Figure 11-40), and click in the enclosed space. Notice that when you move inside the enclosed space, it is shaded with blue color and the shading becomes transparent. Also notice that the room is marked by two diagonal lines and a rectangle showing the extent of the enclosed space.

3. Without exiting the **Room** tool, repeat step 2 to add room to the kitchen, toilet, bed room, dress, and lobby, refer to Figure 11-22. Note that after adding the last room, you need to press ESC to exit.

4. After adding the rooms to the spaces, you need to tag them. Choose the **Room Tag** tool from the **Tag** panel of the **Annotate** tab; the **Modify | Place Room Tag** tab is displayed. Notice that when you invoke the **Room Tag** tool, all rooms in the drawing are displayed in a blue transparent shade with the diagonal lines showing their extents.

5. Move the cursor inside the living room; a Room tag attached to the cursor is displayed, as shown in Figure 11-23. Click to add the tag, as shown in Figure 11-24. Now, press ESC to exit from the **Room Tag** tool

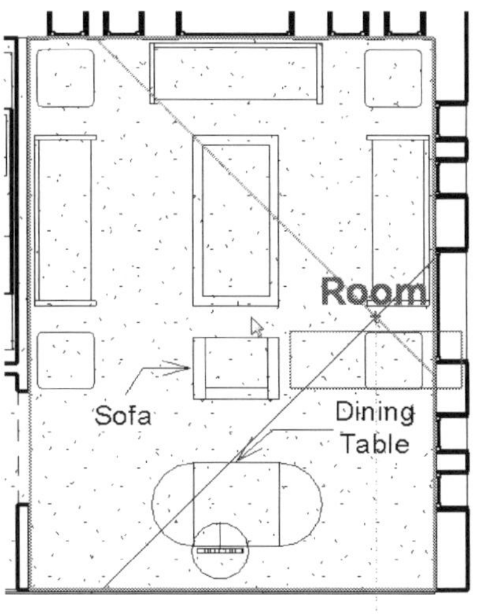

Figure 11-23 Preview of the room tag and the highlighted room boundary

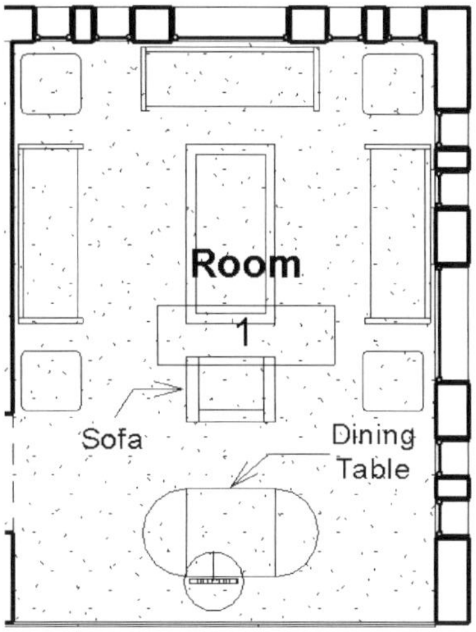

Figure 11-24 The room tag added

6. Select the room tag marked **1**(in the living room) and click on the text **Room**; an edit box is displayed.

7. Type **Living Room** in the edit box, as shown in Figure 11-25. Now, press ENTER and then ESC. The room tag is renamed, as shown in Figure 11-26.

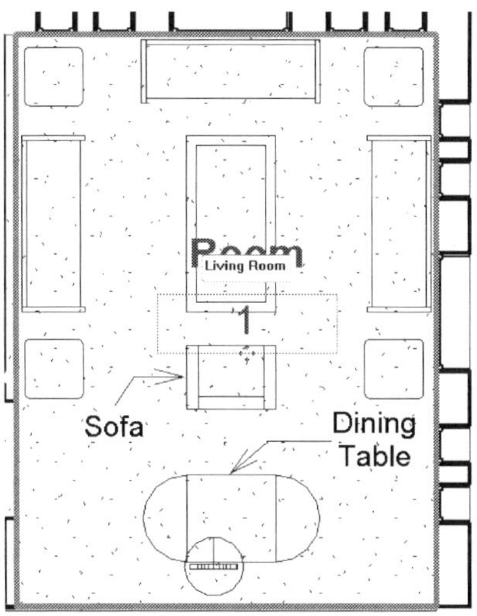

Figure 11-25 Renaming the room tag

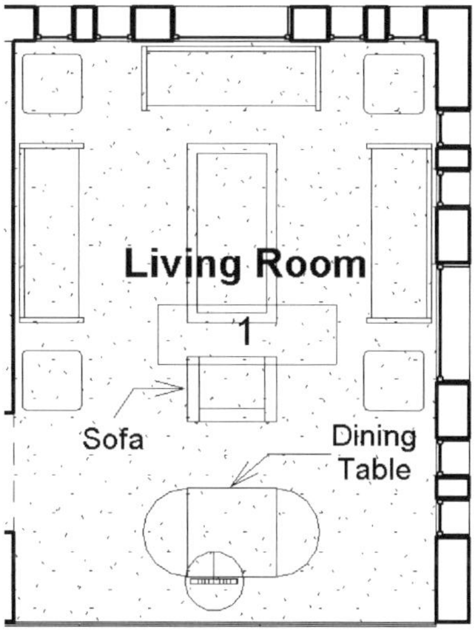

Figure 11-26 The renamed room tag

8. Repeat steps 5 to 7 to add and rename the room tags of the other rooms in the first floor plan in the sequence of the room tag numbers shown in the sketch plan. Also, refer to the

sketch, as shown in Figure 11-40, to name the respective rooms. Press the ESC Key to exit the tool.

Note

Depending on the view scale, the text height and other parameters of the room tag are automatically calculated by Autodesk Revit.

9. The completed room tags on the first floor level are shown in Figure 11-27.

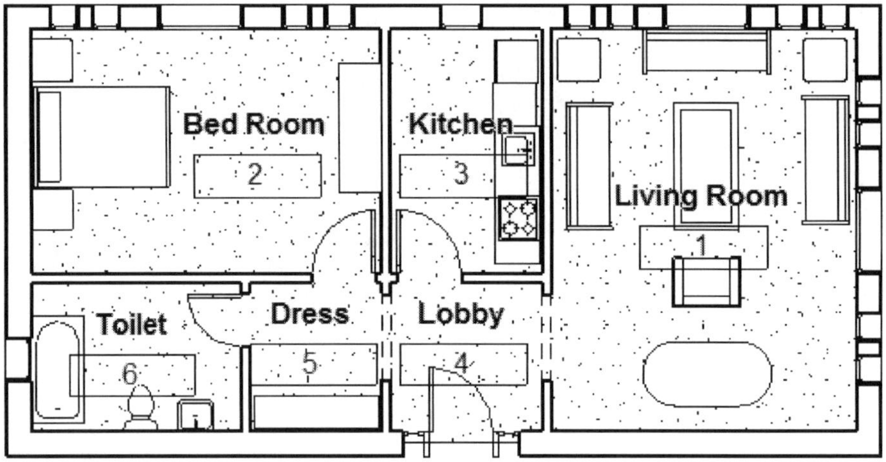

Figure 11-27 *The room tags added to all rooms in the first floor plan view*

Adding Furniture Tags

In this section of the tutorial, you will use the **Tag** tool to add tags to the furniture items. The furniture tags were not loaded into this project file initially. Therefore, when you select a furniture item, Autodesk Revit prompts you to load the furniture tag from its library. You can load the furniture tag and then add tags to the items. You can also set the properties of the leader as specified in the project parameters. The added tags can then be suitably renamed.

1. Choose the **Annotate** tab and then choose the **Tag by Category** tool from the **Tag** panel.

2. Move the cursor over the bed in the bedroom and click when it is highlighted.

   On doing so, the **No Tag Loaded** dialog box is displayed informing that no tag is loaded for the element type selected and prompts you to load the tag.

3. In the message box, choose the **Yes** button; the **Load Family** dialog box is displayed.

4. Browse to the **Annotations > Architectural** folder and select **Furniture Tag** from the list of family files.
5. Choose the **Open** button to load the furniture tag.

6. Move the cursor over the bed and click when the furniture tag is displayed at an appropriate location; the furniture tag is added without a name and a "?" symbol, as shown in Figure 11-28.

7. Press ESC twice and select the furniture tag added in the previous step. On doing so, the **Modify | Furniture Tags** tab is displayed.

8. From the **Type Selector** drop-down list, select **Furniture Tag: Standard** to replace the boxed tag with a standard furniture tag without a box, as given in the project parameters.

9. Next, click on the tag name marked **?** to display the edit box and enter the name **Double Bed** in it, as specified in Figure 11-29. Next, press ENTER; Autodesk Revit displays a message box informing that you are changing a type parameter and it could affect many elements.

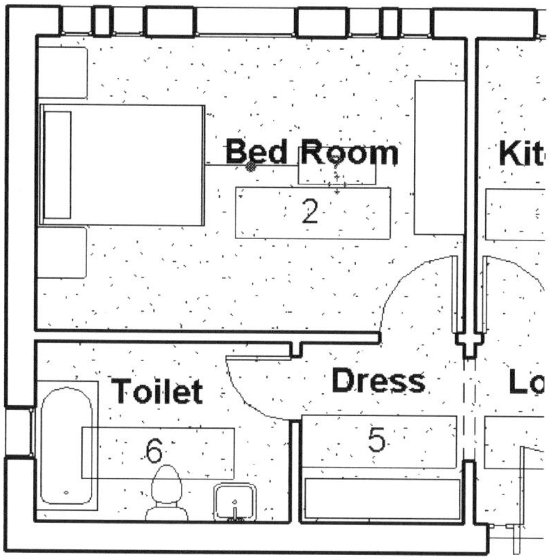

Figure 11-28 Using the ***Tag*** *tool to add a furniture tag*

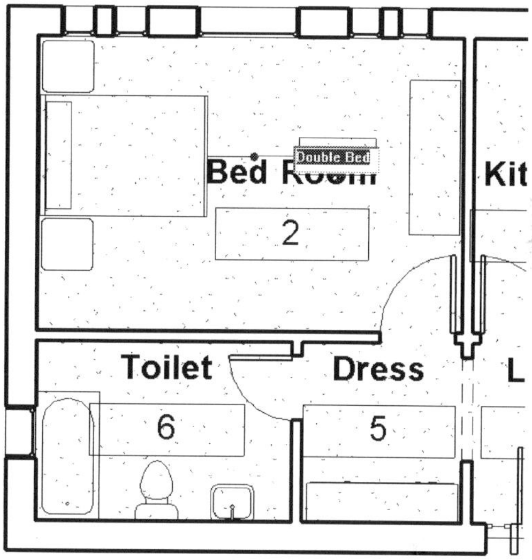

Figure 11-29 Renaming the furniture tag

10. Choose **Yes** in the message box; the furniture tag is renamed.

    Next, you need to add an arrowhead to the furniture tag leader as per the parameters given.

11. Choose the **Edit Type** button from the **Properties** palette; the **Type Properties** dialog box is displayed.

12. Click in the **Value** column for the **Leader Arrowhead** parameter and select **Arrow 30 Degree** from the drop-down list.

13. Choose **Apply** and then **OK**; the **Type Properties** dialog box closes and the arrowhead is added to the furniture tag leader.

14. Drag the furniture tag upward to the new location using the move control, as shown in Figure 11-30.

15. Drag the leader elbow control represented by the blue dot upward in such a way that the leader appears similar to that shown in Figure 11-31.

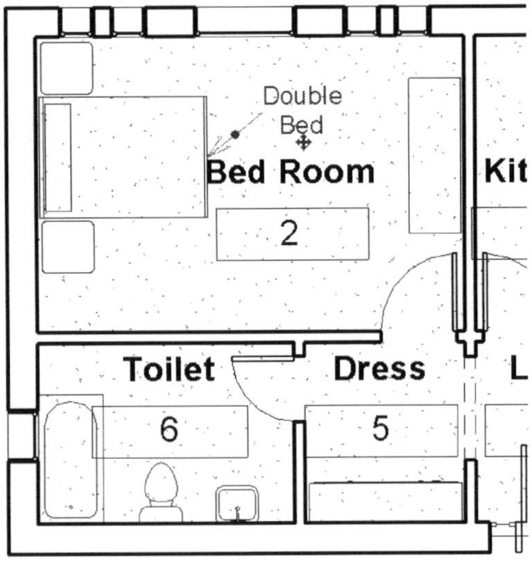

Figure 11-30 *Dragging the furniture tag to a new location*

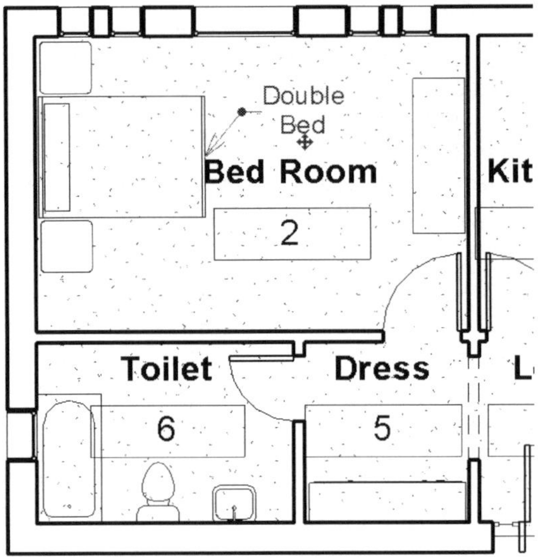

Figure 11-31 *Adjusting the tag leader at the new tag location*

16. Similarly, add other furniture tags to various furniture items in the **First Floor** plan view based on the given sketch plan, refer to Figure 11-22.

Invoking the Dimension Tool and Setting the Dimension Parameters

In this section of the tutorial, you will add dimensions to the project view. Note that the sketch plan shows two types of linear dimensions, centerline and wall faces. You will select appropriate dimension type from the **Type Selector** drop-down list. Further, you will use the **Type Properties** dialog box to create a new dimension style and set the parameters for the new dimension type as specified in the project parameters.

1. Start dimensioning the project view.

   Invoke the **Aligned** tool from the **Dimension** panel of the **Annotate** tab; the **Modify | Place Dimensions** tab is displayed.

2. Select the required dimension style from the **Type Selector** drop-down list.
   For Imperial **Linear Dimension Style: Linear w Centre - 3/32" Arial**
   For Metric **Linear Dimension Style: Diagonal_Centre - 2.5mm Arial**

3. Choose the **Edit Type** button from the **Properties** palette; the **Type Properties** dialog box is displayed.

4. Choose the **Duplicate** button to display the **Name** dialog box.

5. Enter **Linear Dimension - CL** in the **Name** edit box and choose the **OK** button.

6. Click in the **Value** column corresponding to the **Tick Mark** type parameter and select **Dot Filled 1/16" (Filled Dot 3mm)** from the drop-down list as specified in the project parameters.

7. Click in the **Value** column corresponding to the **Witness Line Gap to Element** type parameter and enter the value **1/4"(6 mm)**.

8. Choose the **Apply** button and then the **OK** button to close the **Type Properties** dialog box.

9. Repeat steps 2 to 5 to create another dimension style using the **Linear Dimension Style: Linear - 3/32" Arial** (**Linear Dimension Style: Diagonal - 2.5mm Arial** for Metric) and rename it to **Linear Dimension - Faces**. Modify the **Tick Mark** and **Witness Line Gap to Element** parameters to the values used for the **Linear Dimension - CL** dimension style.

Adding Dimensions to the Project View

Now, add dimensions to the project plan. As all the dimensions to be added are linear dimensions, use the **Linear** option to create them. In the following steps, you will add dimensions to the project plan using different methods. First, you will create centerline dimensions on the right and topsides of the plan view using the **Wall centerline** and **Individual References** options. Next, you will use the **Wall faces** and **Entire Walls** options to create interior dimensions of various rooms placed on the left and bottom of the plan view.

1. Select **Linear Dimension - CL** from the **Type Selector** drop-down list.

2. Ensure that the **Wall centerlines** option is selected from the drop-down list displayed in the **Options Bar**, and **Individual References** is selected in the **Pick** drop-down list.

3. Move the cursor over the south wall near the right corner of the plan view until the object snap for the wall centerline is highlighted, as shown in Figure 11-32. Click to start the dimension.

4. Move the cursor vertically upward until the object snap for the centerline of the north wall is displayed. Now, click to specify the endpoint of the dimension distance; the dimension line is displayed and the dimension moves with the cursor, as shown in Figure 11-33.

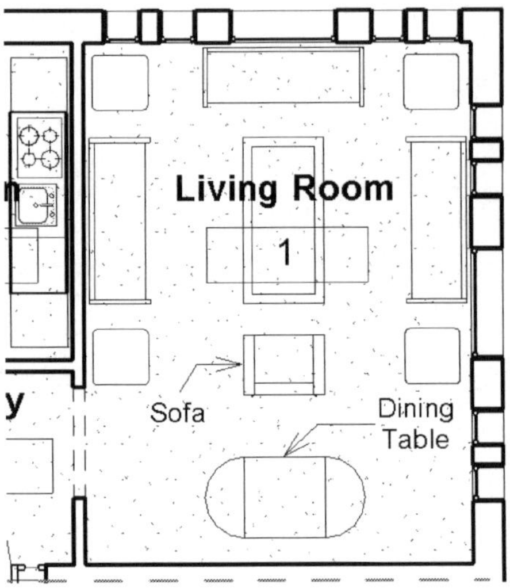

Figure 11-32 *Selecting the wall centerline reference to start a dimension*

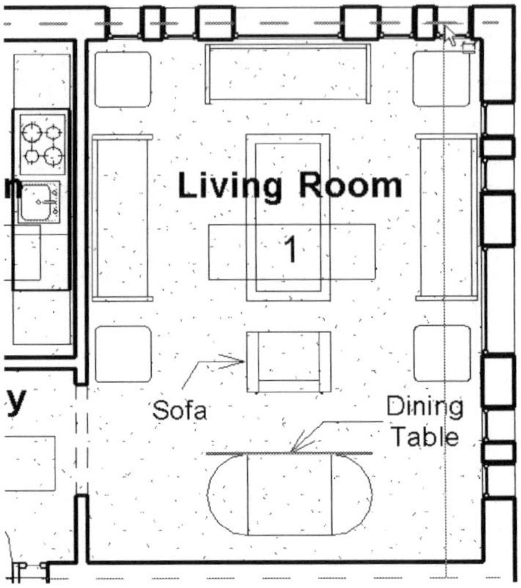

Figure 11-33 *Selecting the second wall centerline reference*

5. Move the cursor toward the right until it crosses the exterior wall and click to specify the location of the dimension. The dimension is created and its controls are highlighted, as shown in Figure 11-34.

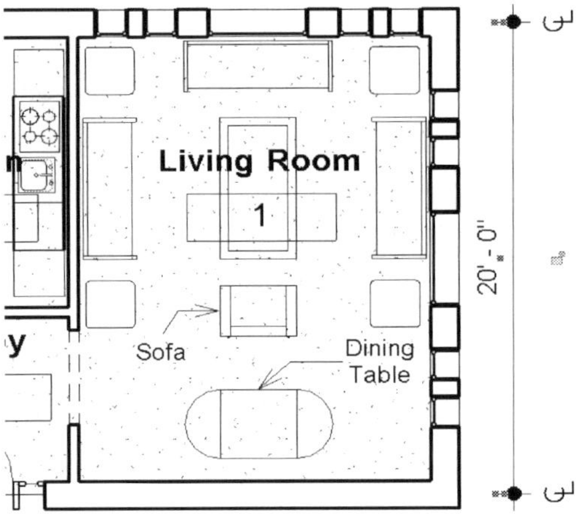

Figure 11-34 *The dimension with its controls*

6. To add dimensions to the north wall.

   Choose the **Entire Walls** option from the **Pick** drop-down list in the **Options Bar**.

7. Move the cursor over the north wall and click when it is highlighted.Autodesk Revit automatically selects the centerlines of the end walls and displays the dimension.

8. Move the cursor such that the dimension appears above the north wall, as shown in Figure 11-35, and click to place the dimension.

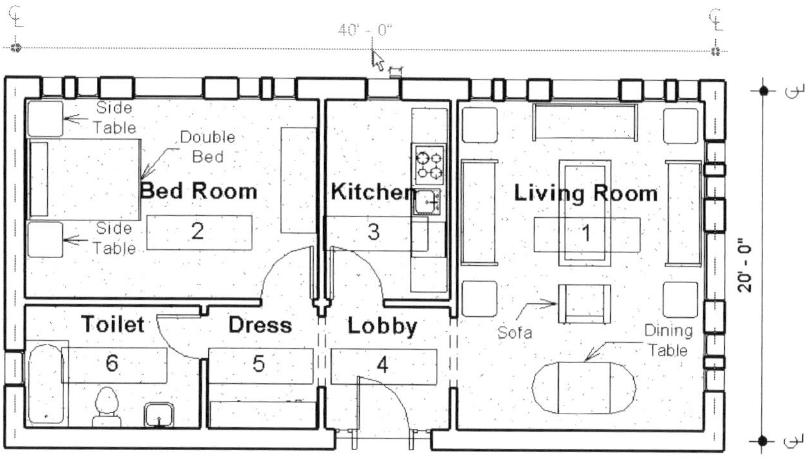

Figure 11-35 *Adding dimensions to the North wall*

9. Select the **Wall faces** option from the drop-down list on the left of the **Options Bar**.

10. Choose the **Options** button in the **Options Bar** to display the **Auto Dimension Options** dialog box.

11. Select the **Intersecting Walls** check box and choose the **OK** button.

12. Move the cursor over the west wall and click when it is highlighted; the dimensions are displayed.

13. Move the cursor such that the dimensions appear on the left side of the west wall and click to add them, as shown in Figure 11-36.

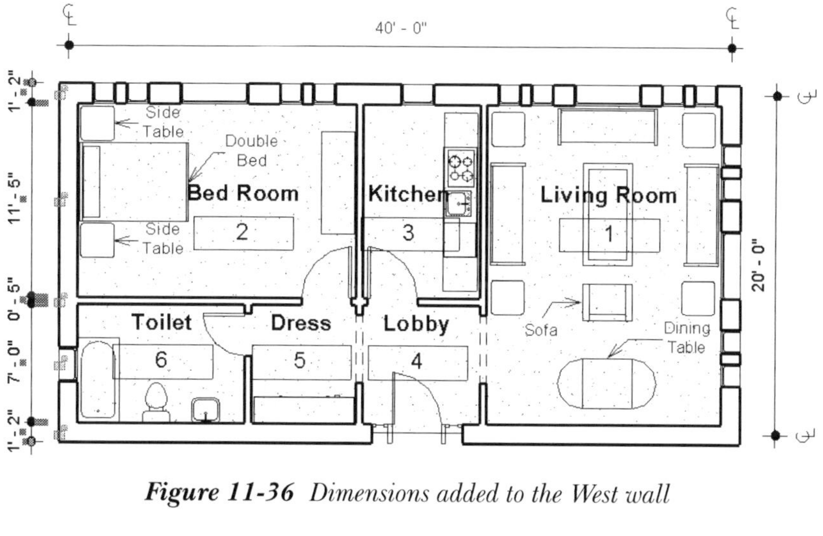

Figure 11-36 *Dimensions added to the West wall*

14. Using the above settings, add dimensions to the south wall.

15. Choose the **Modify** button from the **Select** panel of the **Modify | Place dimensions** contextual tab to exit the tool.

Saving the Project

In this section, you will save the project file using the **Save As** tool.

1. Choose **Save As > Project** from the **File** menu; the **Save As** dialog box is displayed. Enter **c11_Apartment1_tut1 (M_c11_Apartment1_tut1)** in the **File name** edit box and then choose **Save**.

2. Choose **Close** from the **File** menu to close the project file.

The completed *Apartment 1* plan view appears similar to the sketch plan given for this tutorial.

EXERCISES

Exercise 1 Club

Add rooms (without tags), room tags, and dimensions (linear, radial, and angular) to the first floor plan view of the *Club* project created in Tutorial 2 of Chapter 8, based on the sketch plan shown in Figure 11-37. Use the **Room Separator** tool to demarcate and tag the entrance lounge and passages. Note that dimensions for the east side hall are referenced to the wall centerlines, while those for the west hall are referenced to the wall faces. You can provide suitable names to the modified dimension styles. Ensure that the scale for the first floor plan view is set to 1/16"=1'0" (1: 200). The exact location of the room tags and dimensions is not important for this exercise. Use the following parameters for creating room tags and dimensions.

(Expected time: 30 min)

1. Room Tags:
   Room Tag with Area

2. Dimension Parameters:

For Imperial	Tick Mark- **Arrow Filled 30 Degree** Witness Line Gap to Element- **1/8"** Text Size- **1/8"** Text Offset- **1/16"**
For Metric	Tick Mark- **Arrow Filled 30 Degree** Witness Line Gap to Element- **3mm** Text Size- **3mm** Text Offset- **6mm**

3. Name of the file to be saved-

For Imperial	**c11_Club_ex1**
For Metric	M_**c11_Club_ex1**

Note

Dimensions and area values may deviate slightly from the one shown in the sketch plan.

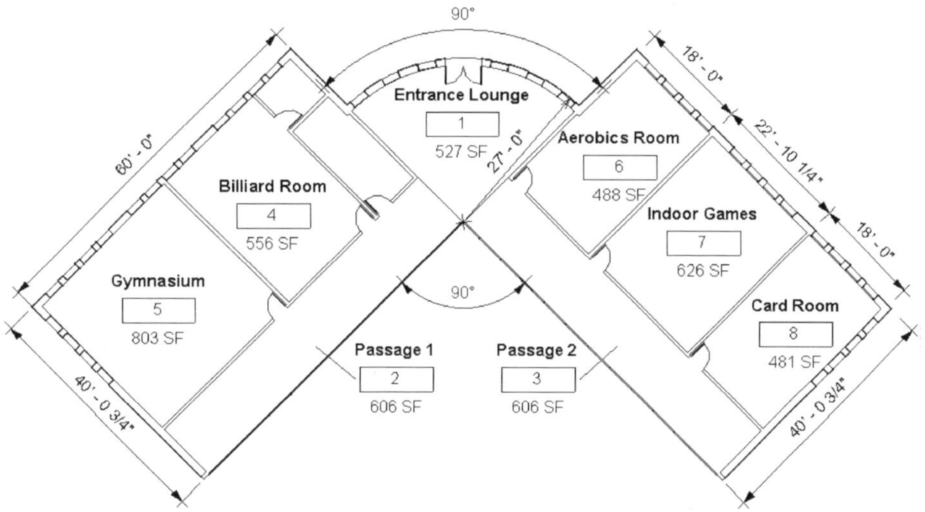

***Figure 11-37** Sketch plan for adding room tags and dimensions to the Club project*

Exercise 2 Elevator and Stair Lobby

Add text and dimensions to the first floor plan view of the *Elevator and Stairs Lobby* project created in Tutorial 3 of Chapter 8. Figure 11-38 shows partial first floor sketch plan of the project. The dimensions are to be added symmetrically on both sides of the vertical axis. Ensure that the view plan scale is set to 1/8"=1'0" (1: 400). All dimensions have wall centerlines as the reference. The exact location of the tags and dimensions is not important for this exercise. Use the following parameters for creating the dimensions.

(Expected time: 30 min)

1. Dimension Parameters:

   For Imperial — Dimension type to be based on: **Linear w Center 3/32" Arial**
   New dimension type name- **Linear 3/32" Arial-Stairs**
   Tick Mark- **Diagonal 1/8"**
   Witness Line Gap to Element- **1/8"**

   For Metric — Dimension type to be based on: **Diagonal Center 2.5 mm Arial**
   New dimension type name- **Linear 2.5 mm Arial-Stairs**
   Tick Mark- **Diagonal 3mm**
   Witness Line Gap to Element- **3mm**

2. Name of the file to be saved-

   For Imperial **c11_ElevatorandStairLobby_ex2**
   For Metric **M_c11_ElevatorandStairLobby_ex2**

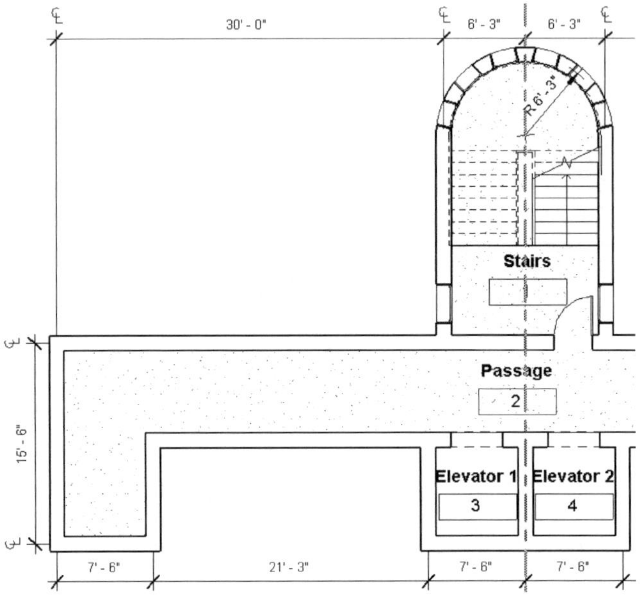

Figure 11-38 *Partial sketch plan for adding text and dimensions to the Elevator and Stair Lobby project*

Chapter 12

Creating Project Details and Schedules

Learning Objectives

After completing this chapter, you will be able to:

- *Create project details*
- *Create a callout view*
- *Add crop region to the model*
- *Create a drafting detail*
- *Add text notes to the project detail*
- *Add model text to a building model*
- *Add revision cloud and revision tag to the project detail*
- *Create project schedules*

PROJECT DETAILING IN Autodesk Revit

For a building or an interior project, details are as important as the building design. Apart from enabling architects and building industry professionals to convey their ideas about materializing a project on site, details describe the materials to be used for construction and their usage. The interconnection of various materials used in a project can be explained graphically using the enlarged views of specific locations. The extent of detailing provided for a project also describes the precision desired while using materials. It can also be used to describe the procedure to be followed for achieving the design intent. For example, you may want to describe the procedure for fabricating a furniture item or explain the interconnection of various materials on the floor or roof joint. The information provided through details helps the fabricator, builder, or contractor understand the architect's concept of achieving the desired design intent. The executing agency can then construct the project based on these details.

CREATING DETAILS IN A PROJECT

This method uses an enlarged portion of a building model view to create the project detailing. You can create a callout view by using plan, section, elevation, or detail view as its parent view. A new view is added to the **Project Browser** and is displayed under the same head as the parent view. The callout view is dependent on its parent view and is deleted if the parent view is deleted.

Callout View

A callout view is used to display an enlarged view of a part of a building model. This view is important for detailing. Creating a callout view is a common practice amongst engineers as it helps them to view the project more precisely and with a higher detail. In an architectural project, callout is used to show details of the basic building elements in a model. You can create callout in plan view or in elevation view. Note that, the callout tag added to these views will be linked to the callout view.

A callout tag consists of the following parts: callout bubble, callout head, and leader line, as shown in Figure 12-1. The callout bubble is the line drawn around that part of the model that you want to enlarge. The callout head is the symbol that represents the callout displaying detail number and sheet number of a model. The line connecting the callout bubble to the callout head is called the leader line.

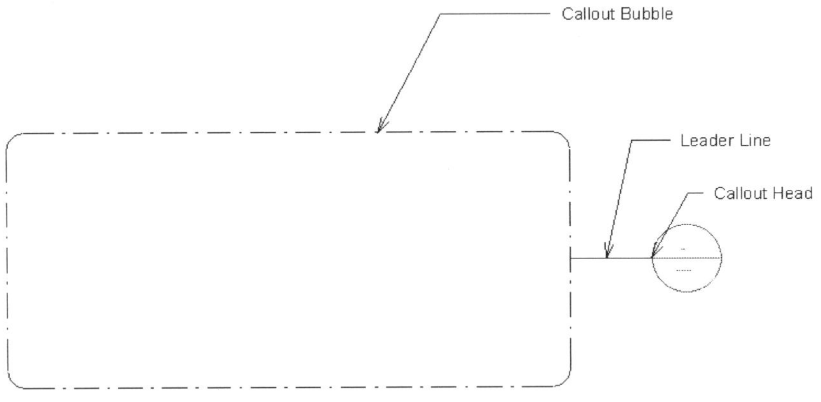

***Figure 12-1** A callout tag*

Displaying the Callout View

After you create a callout in the existing view, a new callout view is added to the **Project Browser** under the parent category. The parent category is the category in which the callout view is created. Therefore, if you create a callout view in the section view, the callout view will be added under the **Sections** heading in the **Project Browser**. Double-click on its name in the **Project Browser** under the **Sections** heading to display the corresponding callout view in the drawing window. Figure 12-2 shows an example of a callout view.

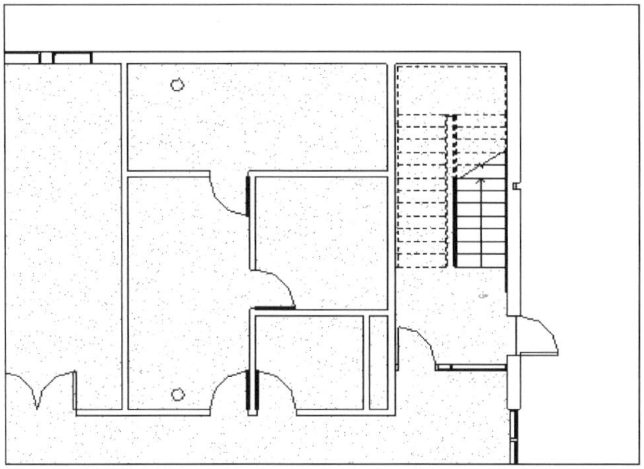

Figure 12-2 *A callout view displayed*

Alternatively, highlight the callout bubble and right-click to display the shortcut menu. Choose **Go To View** from the shortcut menu to display the callout view.

Modifying Callout View Properties

You can modify the appearance of a callout bubble in the parent view by using the bubble controls. These controls are displayed when the callout view is selected. The drag controls can be used to modify the extent of a callout view. The rotation control is used to rotate a callout bubble along with its leader and tag. The leader elbow control can be dragged to some desired location.

The instance properties of the callout view are different for different types of callouts. To modify the instance properties of a callout view, select it from the drawing; the instance properties of the selected callout view will be displayed in the **Properties** palette. In the **Properties** palette shown in Figure 12-3, the **View Name** instance parameter under the **Identity Data** head is used to specify a name for a callout view. The title for the sheet can be entered in the **Title on Sheet** parameter. The **Display Model** parameter under the **Graphics** head is used to set the display type for the view of a building model. By default, this parameter is set to **Normal**. You can set the value of the **Display Model** parameter to **Halftone** from its corresponding drop-down list to display the model element in the current view as a faded image. You can select the **Do not display** option from the drop-down list corresponding to the **Display Model** parameter to hide the model element in the current view. The **Range: Base Level** parameter under the **Underlay** head can be used to set the display of model elements that are displayed in other levels and are not visible in the current view. By default, the **Range: Base Level** parameter is set to **None**.

You can select a level under this head to make the elements of the selected level visible in the current view. The elements that are displayed as underlay appear dimmed. The **View Template** parameter under the **Identity Data** head can be used to assign a view template to each view in a project. By default, the **None** option is selected in this parameter. To assign an existing view template or to create a new template and assign it to the current view, choose the **<None>** button displayed corresponding to the **View Template** parameter; the **Assign View Template** dialog box will be displayed, as shown in Figure 12-4. In the **View templates** area of this dialog box, you can select the desired options from the **Discipline filter** and **View type filter** drop-down lists to limit the view templates that will be displayed in the **Names** list box. In the **Names** list box, you can select the desired template to be assigned to the view. On selecting a template from the **Names** list box, its corresponding parameters will be displayed in the **View properties** area. In the **View templates** area, you can select an existing view template and choose the **Rename** button; the **Rename** dialog box will be displayed. In the **New** edit box, enter the desired name and choose the **OK** button; the selected template is renamed and will be displayed in the **Names** list box. In the **View templates** area, you can choose the **Delete** and **Duplicate** buttons to delete an existing view template and create a copy of an existing view template, respectively. To apply a view template to the current view, select the view template from the **Names** list box and choose the **Apply** button. To close the **Assign View Template** dialog box, choose the **OK** button.

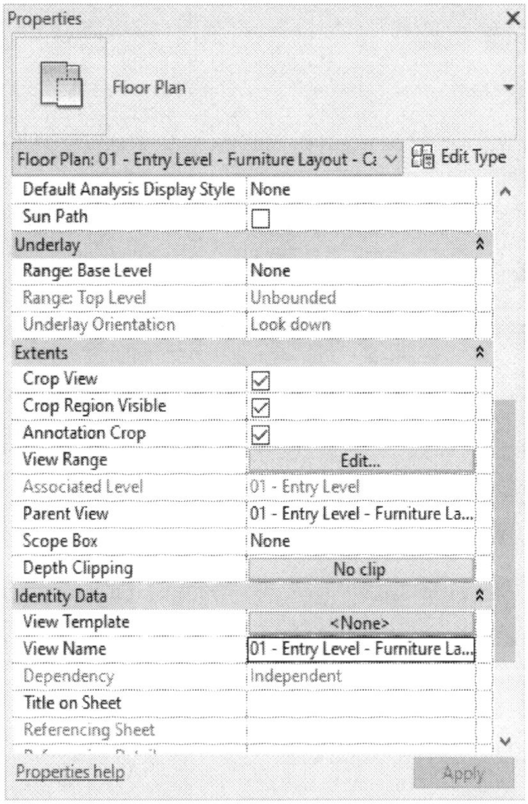

***Figure 12-3** The partial view of the **Properties** palette displaying the instance properties of a callout view*

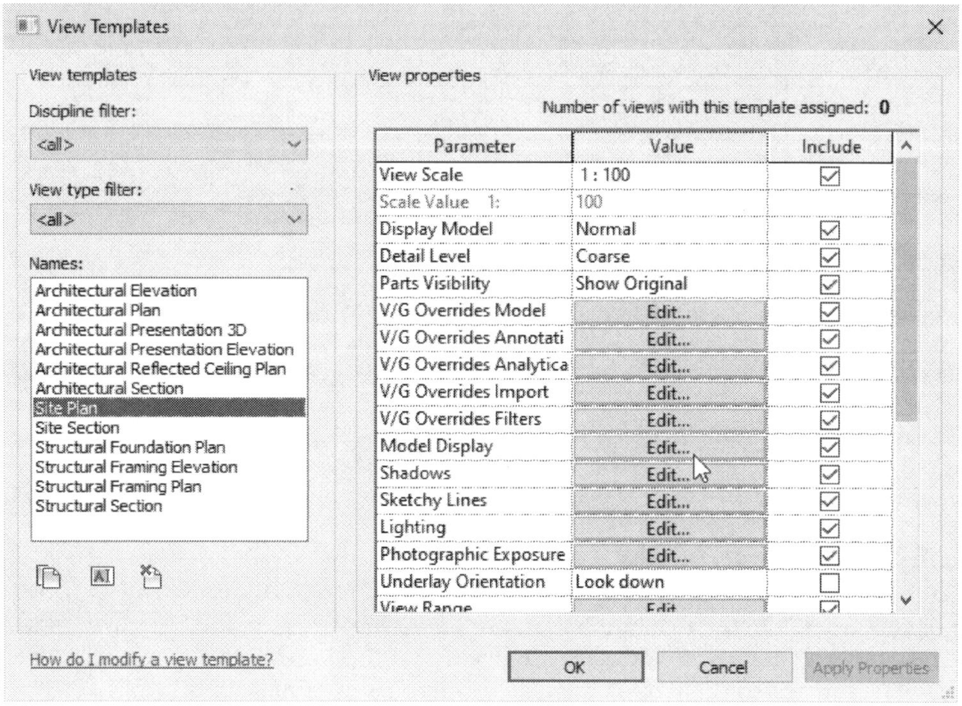

*Figure 12-4 The **Assign View Template** dialog box*

Adding Details to the Callout View

Autodesk Revit provides various tools to add details to a callout view. In the callout view, you can sketch lines using the **Detail Line** tool and add the detail components provided in the Autodesk Revit's library.

CROP REGIONS

Crop regions help to crop views to exclude the unwanted content thereby reducing the view size. Revit has two types of crop regions that allow you to crop views for the model and annotation categories. You can control the visibility of the crop region from **View Control Bar**. The two types of crop regions are discussed next.

Model Crop Region

You can use the model crop region to crop all model elements such as doors, windows, walls, and so on.

Annotation Crop Region

The annotation crop region crops all the annotation elements such as tags, dimensions, keynotes, and so on. All the elements that are touched partially by the annotation crop region boundary are cropped fully to avoid the partial display of annotations. Figure 12-5 shows both the annotation and model crop regions with drag controls.

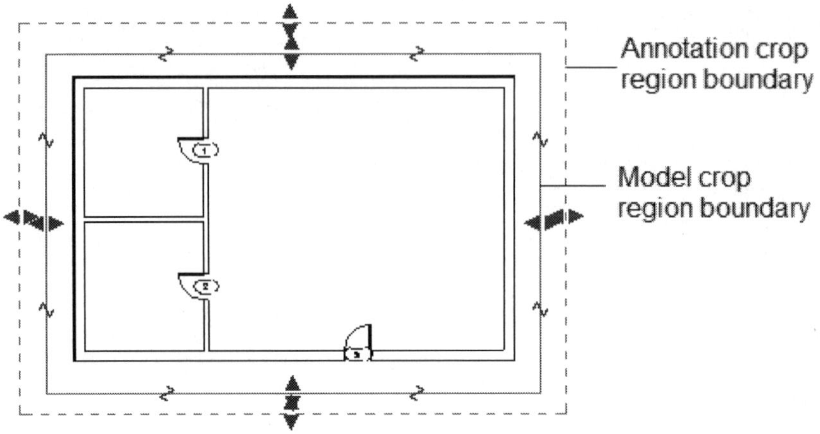

Figure 12-5 The model crop and annotation crop regions with drag controls

CREATING DRAFTED DETAILS

Drafted details are created for the details that are not referenced to the existing project views. These details are not linked to a building model, therefore, they do not get updated with it. To create a drafted detail, first create a drafting view and then use the drafting tools provided in Autodesk Revit to sketch the detail. You can also import in-built details from Autodesk Revit's detail library and use them. The created drafted detail can then be used as a reference detail.

Creating a Drafting View

To create a drafting view, invoke the **Drafting View** tool from the **Create** panel of the **View** tab; the **New Drafting View** dialog box will be displayed, as shown in Figure 12-6. This dialog box is used to set the name and scale for the drafting view.

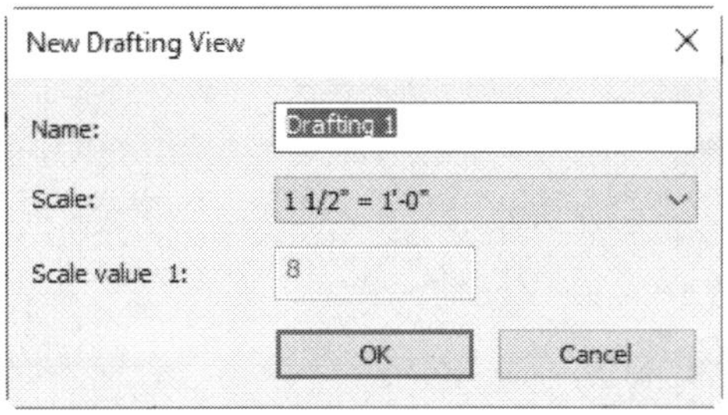

Figure 12-6 The ***New Drafting View*** *dialog box*

Drafting a Detail

The method of drafting a detail in the drafting view is similar to the one used in the callout view. In a callout view, lines are traced over the enlarged project view whereas in a drafting view, the project view is not available and a new detail is drafted as if you were drafting on a blank sheet.

Line Style Settings

Ribbon: Manage > Settings > Additional Settings drop-down > Line Styles

Autodesk Revit provides a variety of line styles that can be used not only in detail views but also in a building model. To view these line styles, choose the **Line Styles** tool from the **Settings** panel; the **Line Styles** dialog box will be displayed. This dialog box displays a list of predefined line styles, as shown in Figure 12-7.

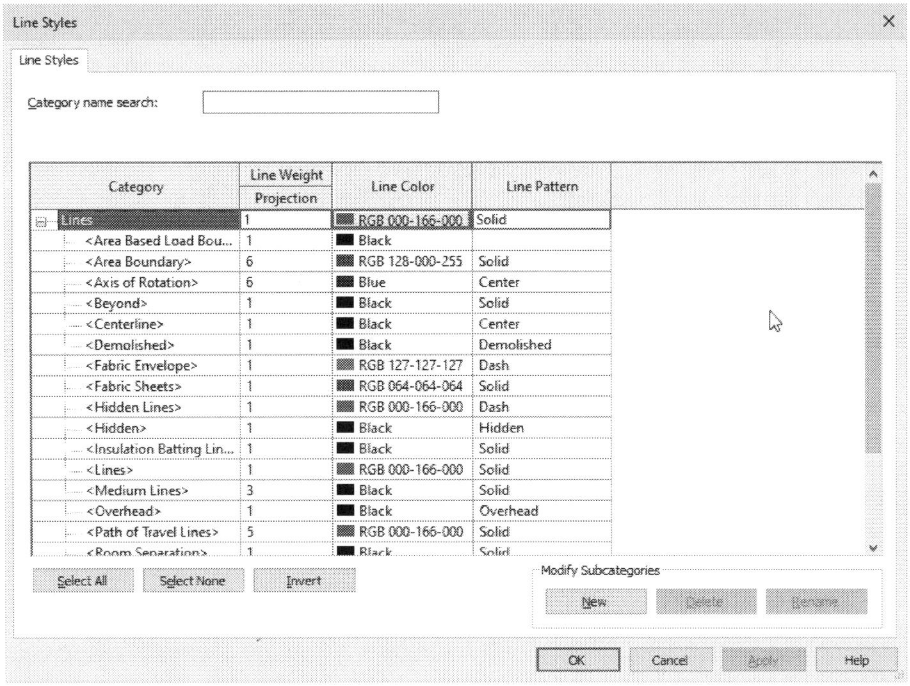

__Figure 12-7__ The __Line Styles__ dialog box displaying the list of in-built lines styles

Using Line Weights

Ribbon: Manage > Settings > Additional Settings > Line Weights

The **Line Weights** tool is used to control the display of line widths in various scales of a project view. To do so, invoke this tool from the **Settings** panel of the **Manage** tab; the **Line Weights** dialog box will be displayed. This dialog box has three tabs: **Model Line Weights**, **Perspective Line Weights**, and **Annotation Line Weights**, as shown in Figure 12-8.

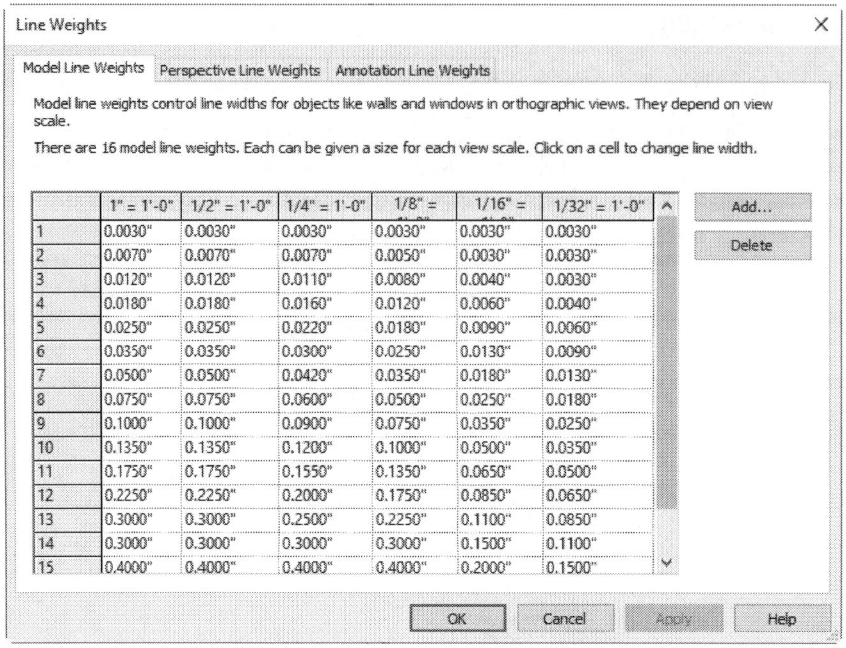

	1" = 1'-0"	1/2" = 1'-0"	1/4" = 1'-0"	1/8" =	1/16" =	1/32" = 1'-0"
1	0.0030"	0.0030"	0.0030"	0.0030"	0.0030"	0.0030"
2	0.0070"	0.0070"	0.0070"	0.0050"	0.0030"	0.0030"
3	0.0120"	0.0120"	0.0110"	0.0080"	0.0040"	0.0030"
4	0.0180"	0.0180"	0.0160"	0.0120"	0.0060"	0.0040"
5	0.0250"	0.0250"	0.0220"	0.0180"	0.0090"	0.0060"
6	0.0350"	0.0350"	0.0300"	0.0250"	0.0130"	0.0090"
7	0.0500"	0.0500"	0.0420"	0.0350"	0.0180"	0.0130"
8	0.0750"	0.0750"	0.0600"	0.0500"	0.0250"	0.0180"
9	0.1000"	0.1000"	0.0900"	0.0750"	0.0350"	0.0250"
10	0.1350"	0.1350"	0.1200"	0.1000"	0.0500"	0.0350"
11	0.1750"	0.1750"	0.1550"	0.1350"	0.0650"	0.0500"
12	0.2250"	0.2250"	0.2000"	0.1750"	0.0850"	0.0650"
13	0.3000"	0.3000"	0.2500"	0.2250"	0.1100"	0.0850"
14	0.3000"	0.3000"	0.3000"	0.3000"	0.1500"	0.1100"
15	0.4000"	0.4000"	0.4000"	0.4000"	0.2000"	0.1500"

***Figure 12-8** The **Line Weights** dialog box*

ADDING TEXT NOTES

Text notes form an important part of a project detail. They not only help in adding the specification for various building elements but also help in conveying the specific design intent. Autodesk Revit provides a variety of options to add text notes to a building model detail view by using the **Text** tool.

Creating Text Notes

Ribbon:	Annotate > Text > Text
Shortcut Key:	TX

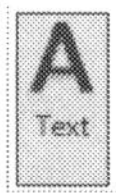

Invoke the **Text** tool from the **Text** panel; the **Modify | Place Text** tab will be displayed, as shown in Figure 12-9. The options in the **Alignment** panel are used to align text. You can align text to the left, center, or right of the text box. Autodesk Revit enables you to create text notes with or without a leader. A leader is a line that connects the text note to the referred portion of a project view. To attach a leader to a text note, you can use any of these four tools, **No Leader**, **One Segment**, **Two Segments**, and **Curved** available in the **Leader** panel. The **No Leader** tool is chosen by default in this panel and is used to create a text without a leader line. The **One Segment**, **Two Segments**, or **Curved** tool can be used to create a textnote with a single segment, two segments, or a curved leader, respectively, as shown in Figure 12-10. To use the **One Segment** tool, click at the location of the start point of the leader or the arrow. As you move the cursor, the leader extends with it. Next, click at the required location to place the text note and then type its content. If you choose the **Two Segments** tool, you need to specify one more point for the elbow of the leader. If you choose the **Curved** tool, you need to specify two points: one for the start point and the other for the endpoint of the arc leader.

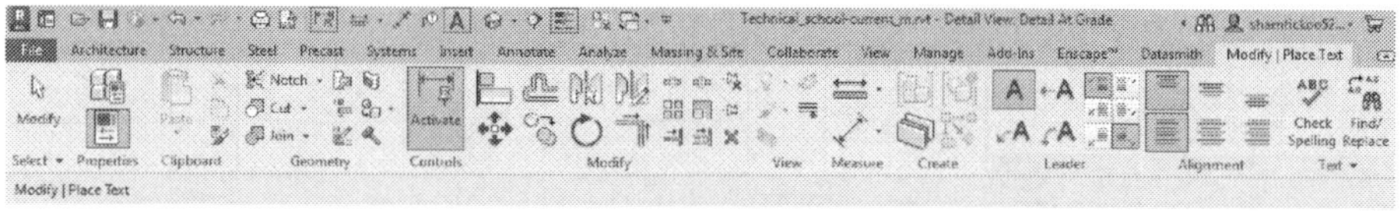

Figure 12-9 *Text options available in the* ***Modify / Place Text*** *tab*

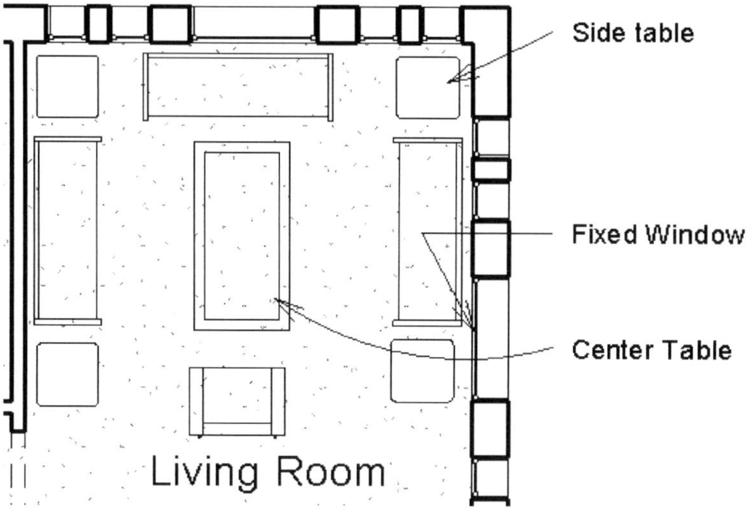

Figure 12-10 *Leaders created using the* ***One Segment***, ***Two Segments***, *and* ***Curved*** *options*

Editing Text Notes

You can either set the text note properties before creating the text note or can edit its properties afterwards. When you select a text note after it is created, its controls are displayed, as shown in Figure 12-11. You can use them to edit the parameters of a text note. In the selected text note, one blue dot on each side represents the stretch controls that can be used to modify its size. The rotation control placed at the top right corner of the text note is used to rotate it. The location of the leader arrow head remains in its original position and only the text note is rotated. The leader tail automatically adjusts to the rotated text note. You can also drag the location of the leader elbow and the leader head by using the drag control dots. The drag control is used to change the location of the text note.

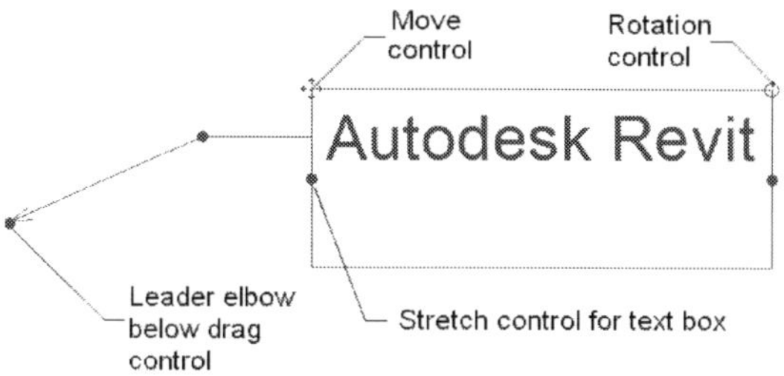

Figure 12-11 *Editing the text note using its controls*

As you select the created text note, the **Modify | Text Notes** tab will be chosen. In this tab, you can use various editing tools such as **Copy**, **Mirror**, **Rotate**, **Array**, and so on to edit and arrange the text note. Figure 12-12 shows various options in the **Modify | Text Notes** tab for editing and modifying the selected text.

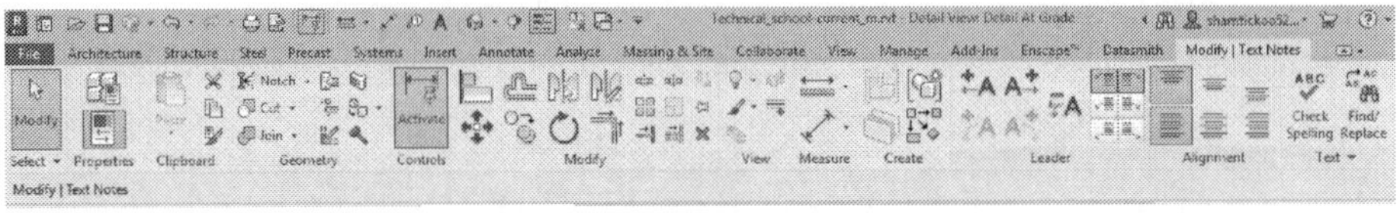

***Figure 12-12** Text note editing options in the **Modify / Text Notes** tab*

The alignment of the text note can be modified by choosing the appropriate alignment option from the **Alignment** panel. You can use the appropriate add leader options such as **Add Right Side Straight Leader**, **Add Left Side Straight Leader**, **Add Left Side Arc Leader**, and **Add Right Side Arc Leader** from the **Leader** panel to add a leader to a text note created without a leader. You can also use any of these options to add multiple leaders. The **Remove Last Leader** button in the **Leader** panel is used to remove the most recently added leader to a text note. To modify a text note, select the text inside the text box displayed with it, click inside the text box, and then enter the new note. You can also modify the instance properties of the text note such as font style and text size in the **Properties** palette. In this palette, the value assigned to the **Horizontal Align** parameter specifies the horizontal text alignment of the selected text. To change this value, click on the value column and select the desired alignment option from the drop-down list displayed. The **Arc Leaders** instance parameter specifies whether or not an arc leader will be attached to a text. The value column of this parameter displays a check box. By default, this check box is cleared. You can select this check box to attach an arc leader to the text note.

You can also edit the type properties of a text using the **Properties** palette. To do so, choose the **Edit Type** button from the **Properties** palette; the **Type Properties** dialog box will be displayed, as shown in Figure 12-13.

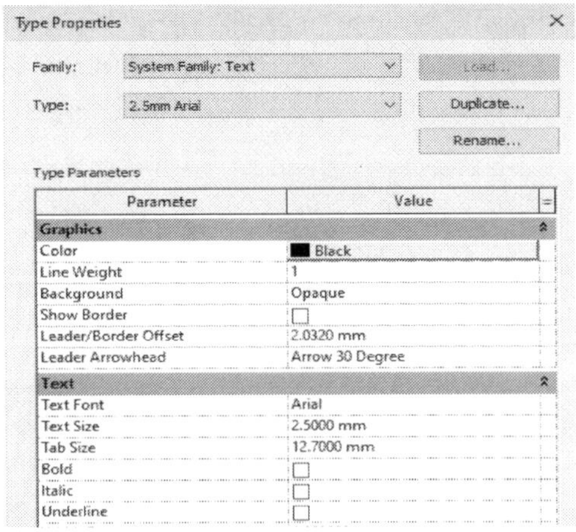

***Figure 12-13** Partial view of the **Type Properties** dialog box for text notes*

Creating a Model Text

Ribbon: Architecture > Model > Model Text

The **Model Text** tool enables you to create a text that acts as a 3D building element. The added texts are visible in all related views. They are rendered as a part of a building model and not as annotation symbols. The model text can be added to a selected level or a plane.

To add the model text to a building model, invoke the **Model Text** tool from the **Model** panel. On doing so, the **Edit Text** dialog box will be displayed and you will be prompted to enter the model text notes. After entering the text notes, choose the **OK** button to close the dialog box. The preview image of the text note entered will be displayed in the drawing window attached to the cursor. Click to place the model text at the desired location. Now, you can select the model text and modify its properties. Figure 12-14 shows an example of a model text added to the facade of a building.

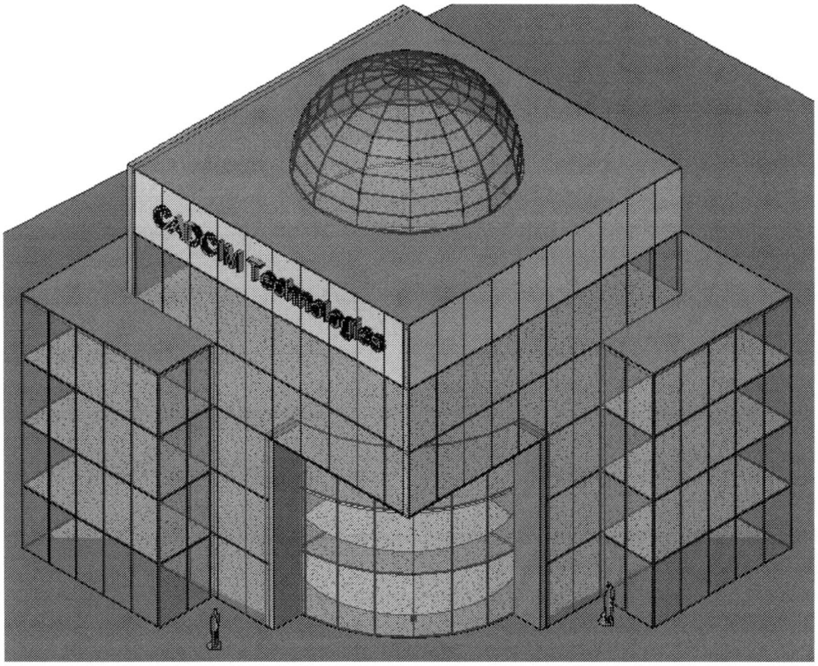

Figure 12-14 *Model text added to a building model*

Tip

You should set the work plane before invoking the ***Model Text*** *tool because model text can be placed only in the current work plane.*

REVISION CLOUDS

While creating project plans, you may need to revise a portion of a project view and issue revised drawings to the concerned authorities. The revision cloud is added to refer and to indicate the revised portion. Except 3D views, you can add the revision cloud to any project view such as floor plans, elevations, sections, and so on. Autodesk Revit enables you to create the revision clouds of the desired shape and size. You can also add a revision tag to the revision cloud for referring to the revisions made to a specified area.

Creating the Revision Cloud

Ribbon: Annotate > Detail > Revision Cloud

To create a revision cloud, invoke the **Revision Cloud** tool from the **Detail** panel; the **Modify | Create Revision Cloud Sketch** tab will be displayed. You can use various sketching tools in the tab to sketch, edit a sketch, and assign properties to the revision cloud. To sketch the revision cloud, choose any of the drawing tools from the **Draw** panel, move the cursor near the revised area, and click at the desired location to start creating the revision cloud. On moving the cursor, you will notice that a curved bubble-shaped line is displayed. Click to specify the next point of the cloud outline. In this way, you can click at multiple points to sketch the revision cloud of desired size and shape enclosing the portion of the view that needs to be referred for revision. You can choose other standard tools to create revision cloud such as line, rectangle, and so on. You can use SPACEBAR to flip the direction of the cloud. Next, choose the **Finish Edit Mode** button from the **Mode** panel of the **Modify | Create Revision Cloud Sketch** tab to create the revision cloud, as shown in Figure 12-15.

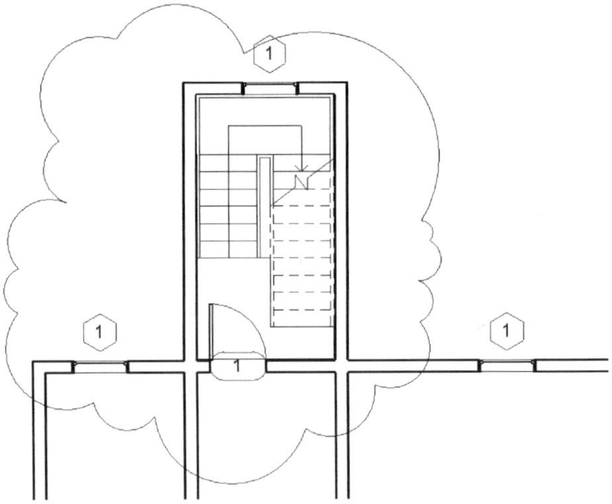

Figure 12-15 *The revision cloud created around the portion to be referred for revision*

Adding a Revision Tag

While working on projects, you may need to use multiple revision clouds in a project view. To identify and mark each revision cloud, you can add revision tags to it.

A revision tag can be added using the **Tag by Category** tool. To add a revision tag, invoke the **Tag by Category** tool from the **Tag** panel of the **Annotate** tab and then move the cursor near the revision cloud. Click in the drawing area when the cloud is highlighted; the revision tag will be placed. In case it is not loaded in the project file, Autodesk Revit will prompt you to load it. You can load the *Revision Tag.rfa* family file from the **English-Imperial > Annotations** folder. Now, you can move the cursor over the revision cloud and click to place the revision tag. A leader will automatically be added to the revision tag because the **Leader** check box in the **Options Bar** is selected by default. You can then use the drag controls to direct the leader toward the revision cloud, as shown in Figure 12-16. However, you can remove the leader from the revision tag by selecting it and then clearing the **Leader** check box in **Options Bar**. Once you have added the revision tags, Autodesk Revit will enable you to create revision schedules to be placed on the drawing sheet. You will learn more about creating schedules in the next section.

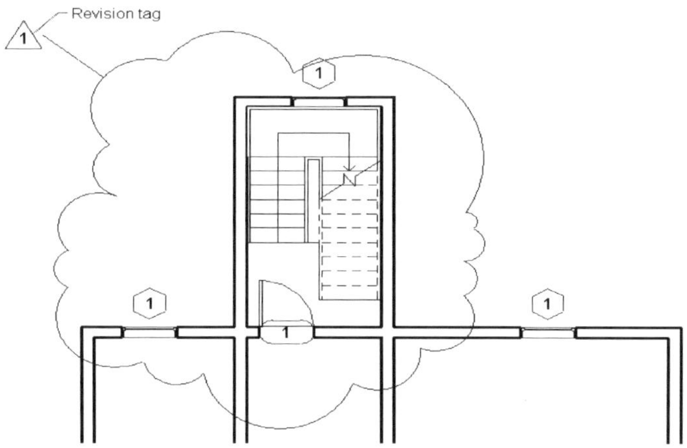

Figure 12-16 *A revision tag added to the revision cloud*

USING SCHEDULES IN A PROJECT

Schedule is another format for providing project information. In a building project, there may be different items that can be used a number of times at different locations. Schedules are primarily used to provide information regarding these items in a tabulated form. For each item, the schedule tables provide information regarding its size, material, cost, finish, level, and so on. These schedules can then be used by associated agencies for various purposes. They can assist the fabricating agency to manufacture all such items. The quantity surveyors and estimators can calculate the quantity and cost of the grouped items.

Generating a Schedule

Ribbon: View > Create > Schedules drop-down > Schedule/Quantities

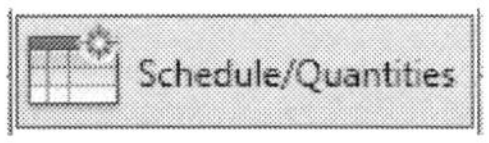

In Autodesk Revit, you can create a schedule by using the **Schedule/ Quantities** tool. You can invoke this tool from the **Create** panel. On doing so, the **New Schedule** dialog box will be displayed, as shown in Figure 12-17.

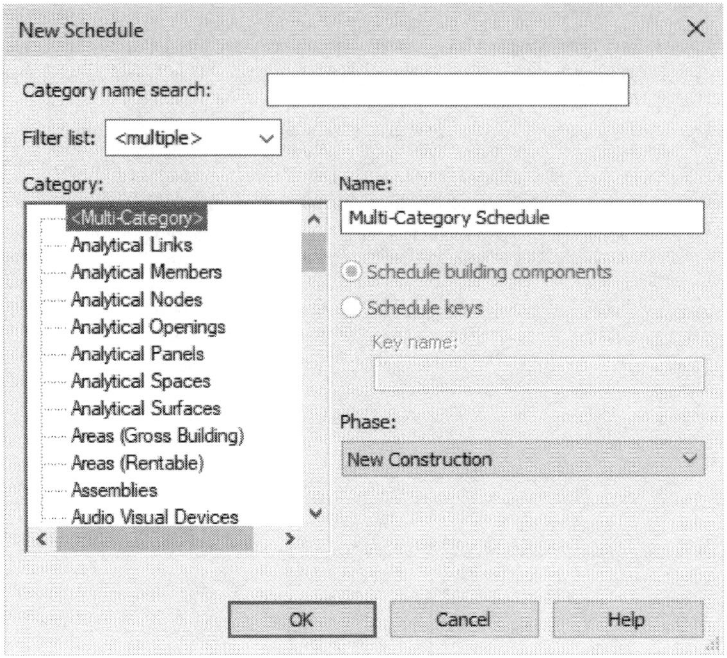

Figure 12-17 *The* ***New Schedule*** *dialog box*

In the **New Schedule** dialog box, various categories of building elements are listed in the **Category** list. By default, all the check boxes are selected in the **Filter List** drop-down. You can select one or more disciplines from the **Filter List** drop-down list. Then, you can select the required category from the **Category** list to create a schedule. For example, to generate a door schedule, you can select **Doors** from the **Category** list. The **Wall Sweeps** category in the **Category** list is used to create a schedule for wall sweeps. The category selected for creating a schedule is displayed in the **Name** edit box. The **Schedule building components** radio button is used to create a schedule of building elements. The **Filter List** drop-down list is used to display all the discipline categories. In this drop-down list, you can select the check box of the discipline you want to display. When you choose the **OK** button, the **Schedule Properties** dialog box will be displayed, as shown in Figure 12-18.

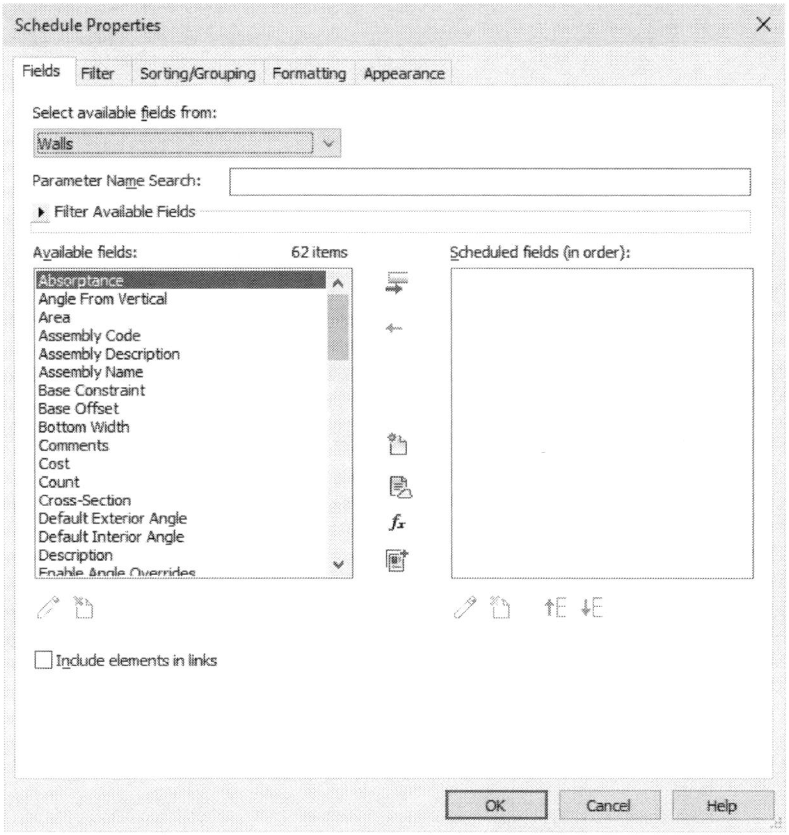

*Figure 12-18 The **Schedule Properties** dialog box*

Exporting Schedule to Excel Sheet

In Autodesk Revit, you can export a schedule to MS Office Excel to generate a spreadsheet. To do so, from the **File** menu, choose **Export > Reports > Schedule**; the **Export Schedule** dialog box will be displayed, as shown in Figure 12-19. Enter the name of file in the **File Name** edit box and select the **Delimited text (*.txt)** from the **File of type** drop-down list. Next, choose the **Save** button; the modified **Export Schedule** dialog box will be displayed, refer to Figure 12-20.

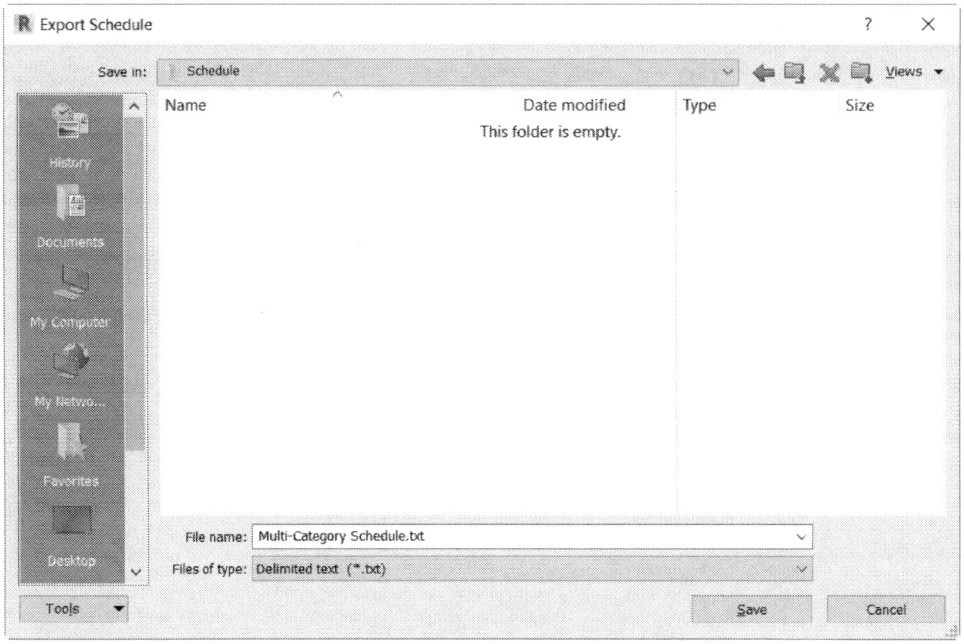

*Figure 12-19 The **Export Schedule** dialog box*

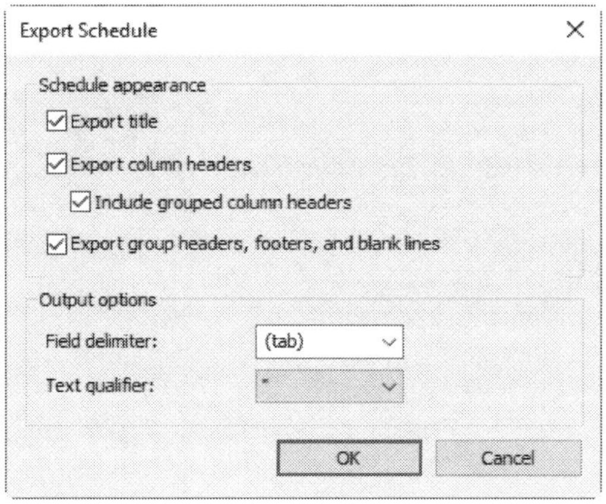

*Figure 12-20 The **Export Schedule** dialog box with schedule appearance and output parameters*

Creating a Legend View

Ribbon:	View > Create > Legends drop-down > Legend

Legends are created to list the symbols and annotations used in project views. You can create them for various elements used in a project such as wall types, tag type, symbols, and so on. A legend is created as a view.

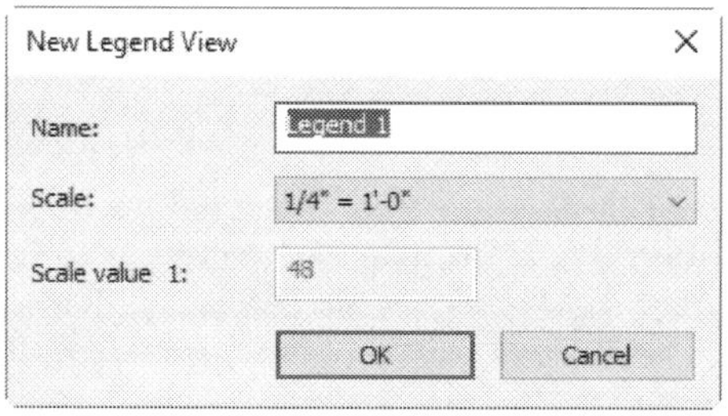

Figure 12-21 *The **New Legend View** dialog box*

To create a legend, choose the **Legend** tool from the **Create** panel; the **New Legend View** dialog box will be displayed, as shown in Figure 12-21. Enter a name for the legend in the **Name** edit box. You can set the scale for the legend by selecting the scale from the **Scale** drop-down list. To set a scale other than the options available in the drop-down list, select the **Custom** option and specify the scale in the **Scale value 1:** edit box. Once you choose the **OK** button in the **New Legend View** dialog box, a new view will open in the drawing window and will be added under the **Legends** head in the **Project Browser**. You can then use various tools to create the desired legend type. Next, you will add legend components to the opened legend view. To do so, choose the **Legend Component** tool from **Annotate > Detail > Component** drop-down; the **Options Bar** will display the associated tools and options. The **Family** drop-down list displays family elements. The required view for the legend can be selected from the **View** drop-down list. The **Host Length** edit box can be used to specify the length of the legend symbol for the selected family type. You can select the desired family element and create a legend. You can use the **Text** tool in the **Text** panel of the **Annotate** tab to add text notes to the legend view. The **Detail Lines** tool can be used to create a tabulated legend with lines.

Similarly, the **Symbol** tool can be used to create a legend of the symbols used in a project. For example, you can create a legend for all the tags used in the project documentation using the **Symbol** tool, as shown in Figure 12-22.

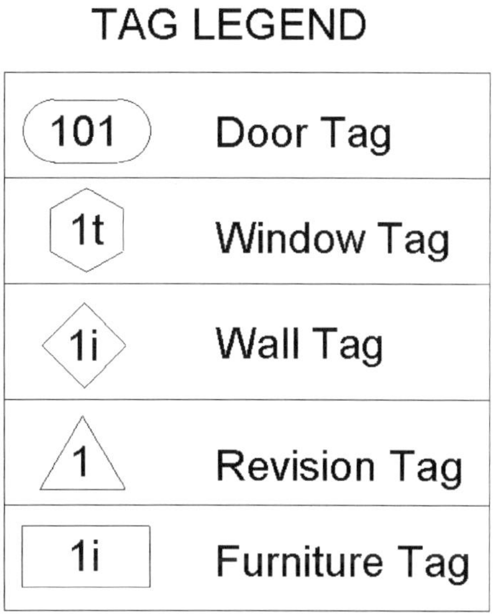

Figure 12-22 *A tag legend*

> **Tip**
> *Legends can now be copied from a sheet to another. You can select the "Paste Aligned To Same Place" feature to make sure legends are in the exact same spot on all sheets.*

TUTORIALS

Tutorial 1 Apartment 1 - Callout View

In this tutorial, you will create a callout view of the kitchen door jamb in the *Apartment 1* project represented by a highlighted rectangle with a cursor, as shown in Figure 12-23. You will use the *c11_Apartment1_tut1.rvt* file created in Tutorial 1 of Chapter 11. You need to add the detail lines, detail components, text notes, and break lines to complete the door jamb detail, as shown in Figure 12-24. Use the parameters given in the figure and those given below to create it.

(Expected time: 1hr)

1. Callout view parameters:
   Callout view name to be assigned- **Door Jamb Detail**
   View Scale- **1 1/2" = 1'-0"** View Scale- **1:10 (For Metric)**
   Detail Level- **Fine**

2. Detail component to be loaded from the **English-Imperial > Detail Items (Detail Components) > Div 06-Wood and Plastic > 061100-Wood Framing** and **English-Imperial > Detail Items (Detail Components) >Div 08-Openings > 081400-Wood Doors** folder locations. For **English > Detail Items > Div 06-Wood and Plastic > 06100-Rough Carpentry > 06110-Wood Framing** and **English > Detail Items >Div 08-Doors and Windows > 08200-Wood and Plastic Doors > 08210- Wood Doors.**

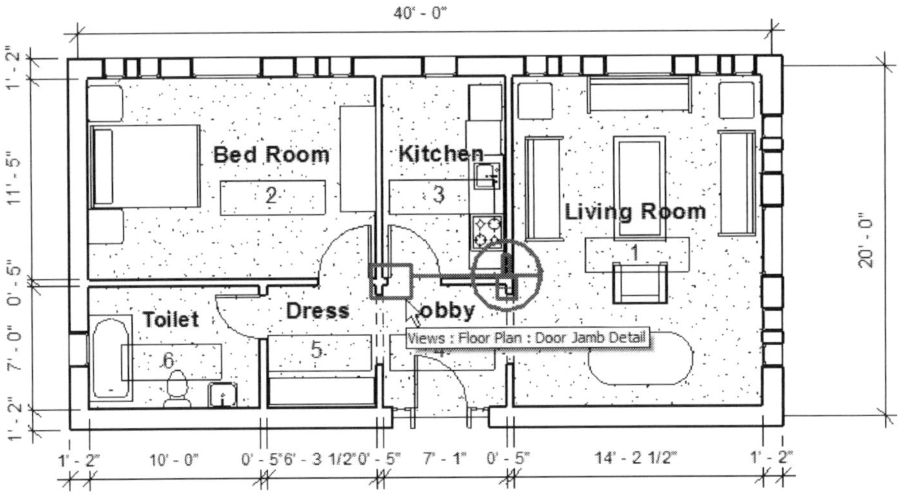

***Figure 12-23** Sketch plan for adding a callout view to the Apartment 1 project plan*

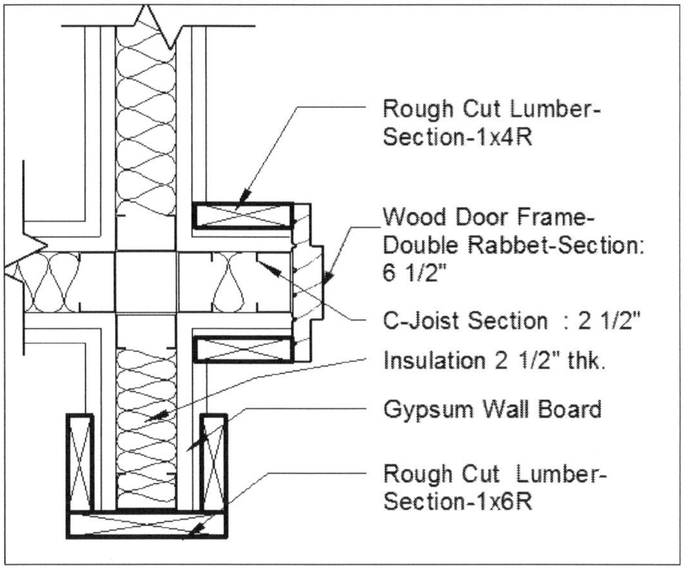

Figure 12-24 *The callout view of the Apartment 1 project*

3. Insulation instance parameters:
   Insulation Width: **2 1/2" (64 mm)**
   Insulation Bulge to Width Ratio: **3**

4. Text note parameters:
   Text Size: **3/32" (2.4 mm)**
   Leader Arrowhead: **Arrow Filled 15 - Degree**

5. Load the break line from **English-Imperial > Detail Items (Detail Components)** folder (For Metric **English > Detail Items (Detail Components)** folder) with the following modified parameters:

   Values for the **Right** and **Left** instance parameter: **0' 4" (102 mm)**
   Values for the **Jag Depth**, **Jag Width**, and **Masking depth** parameters: **0' 2" (51 mm)**

   The following steps are required to complete this tutorial:

a. Open the specified project file and then create the callout view using the **Callout** tool.

b. Display the view and set its parameters.

c. Use the callout view as an underlay and add detail components using the **Detail Component** tool.

d. Add detail lines to the callout view using the **Detail Lines** tool.

e. Use the **Insulation** tool to add insulation to the detail.

f. Invoke the **Text** tool and add the text notes after setting the parameters.

g. Add break lines to complete the detail callout view.

h. Save the project using the **Save As** tool.
For Imperial *c12_Apartment1_tut1.rvt*
For Metric *M_c12_Apartment1_tut1.rvt*

i. Close the project.

Opening the Project File and Creating the Callout View

1. Choose **Open > Project** from the **File** menu and open the *c11_Apartment_tut1.rvt* file created in Tutorial 1 of Chapter 11. The first floor plan is displayed. You can also download this file from *https://www.Cadsoft.com*. The path of the file is as follows: *Textbooks > Civil/GIS > Revit Architecture > Exploring Autodesk Revit 2025 for Architecture*

2. Choose the **Zoom in Region** tool from the **Navigation Bar** and zoom near the area inside the rectangle, refer to Figure 12-36.

3. Next, choose the **Rectangle** tool from the **View > Create > Callout** drop-down; the **Modify | Callout** tab is displayed.

4. Click at two points to define the callout view rectangle, as shown in Figure 12-25.

   The callout view created is displayed as **First Floor - Callout 1** under the **Floor Plans** head in the **Project Browser**.

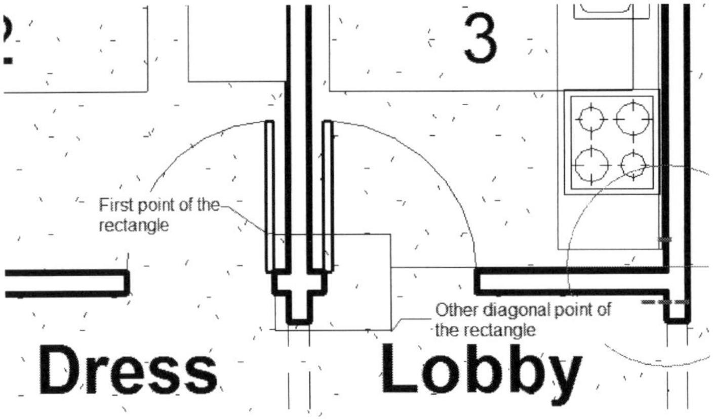

Figure 12-25 *Defining the rectangle to create a callout view*

Displaying the Callout View and Setting View Parameters

1. Double-click on the callout view name **First Floor- Callout 1** under the **Floor Plans** head in the **Project Browser**.

   Now, select the door element in the callout view and right-click; a shortcut menu is displayed. Choose **Hide in view > Category** from the shortcut menu; the doors in this view are now hidden, as shown in Figure 12-26.

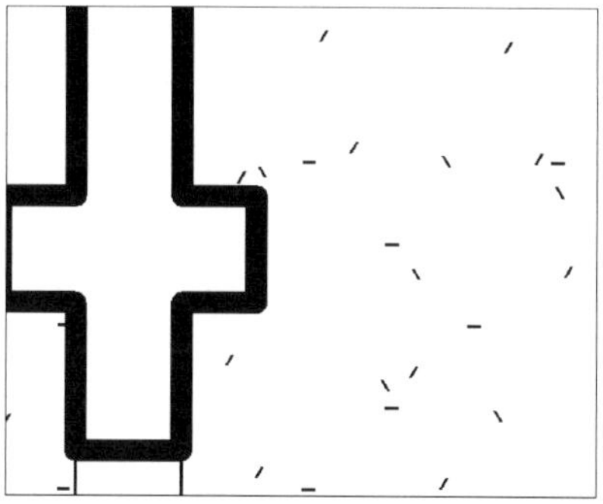

Figure 12-26 *Callout view displayed in the viewing area*

2. In the **Properties** palette, click in the **Value** field for the **View Name** instance parameter and enter **Door Jamb Detail**.

3. For the **View Scale** parameter, click in its value field and select the desired option from the corresponding drop-down list.
   For Imperial **1 1/2" = 1'- 0"**
   For Metric **1: 10**

4. Similarly, click in the value fields for the **Display Model** and **Detail Level** parameters and then select **Halftone** and **Fine** from the corresponding drop-down lists.

Adding Detail Components

In this section of the tutorial, you will add detail components by using the callout view as an underlay. Further, you will load the desired components, place them at the required location, and then modify their sizes.

1. Choose the **Annotate** tab and then choose the **Detail Component** tool from the **Component** drop-down of the **Detail** panel; the **Modify | Place Detail Component** tab is displayed.

   Next, you need to load the components according to the project parameters.

2. Choose the **Load Family** tool from the **Mode** panel; the **Load Family** dialog box is displayed.

3. Navigate to **English-Imperial > Detail Items > Div 08-Openings > 081400-Wood Doors** and select the **Wood Door Frame-Double Rabbet-Section** file (**English > Detail Items >Div 08- Doors and Windows > 08200-Wood and Plastic Doors > 08210- Wood Doors** and select the **M_Wood Door Frame-Double Rabbet-Section** file). Now, choose **Open**; the selected file is loaded and will be available in the **Type Selector** drop-down list.

4. Choose the **Type Properties** tool from the **Properties** panel; the **Type Properties** dialog box is displayed.

5. Choose **Duplicate**; the **Name** dialog box is displayed. Enter **6 1/2" (165 mm)** in the **Name** edit box and choose **OK**. Now, in the **Value** fields for the **Width** and **Stop** parameters, enter **0' 6 1/2" (165 mm)** and **0' 3" (76 mm)**, respectively.

6. Now, choose **OK**; the **Type Properties** dialog box will close.

7. Choose the **Load Family** tool from the **Mode** panel; the **Load Family** dialog box is displayed. In this dialog box, navigate to **English-Imperial > Detail Items > Div 06-Wood and Plastic > 061100-Wood Framing** and select the **Rough Cut Lumber-Section** file (for Metric **English > Detail Items > Div 06-Wood and Plastic > 06100- Rough Carpentry > 06110- Wood Framing** and select the **M_Rough Cut Lumber-Section** file). Next, choose **Open**; the **Specify Types** dialog box is displayed.

8. Choose **1x4R** and **1x6R** (for metric **25x100mm** and **25x165mm)** in the **Type** column using the CTRL key and then choose **OK**; the **Specify Types** dialog box will close and the selected types will now be available in the **Type Selector** drop-down list in the **Properties** palette.

Note

*You can create the others sizes of the sections by choosing the **Duplicate** button from the **Type Properties** dialog box.*

9. From the **Type Selector** drop-down list, select **Wood Door Frame-Double Rabbet-Section: 6 1/2" (165 mm)**.

10. Select the **Rotate after placement** check box available in the **Options Bar**.

11. Move the cursor to the viewing area and click near the door jamb location to place the door frame. Rotate it 90 degrees in counterclockwise direction to achieve the correct orientation and then click again. Next, choose the **Modify** button from the **Select** panel. Then, select the placed door frame and use the **Move** tool and the **Midpoint** snap option to move the door frame to the location shown in Figure 12-27. Choose the **Modify** button from the **Select** panel.

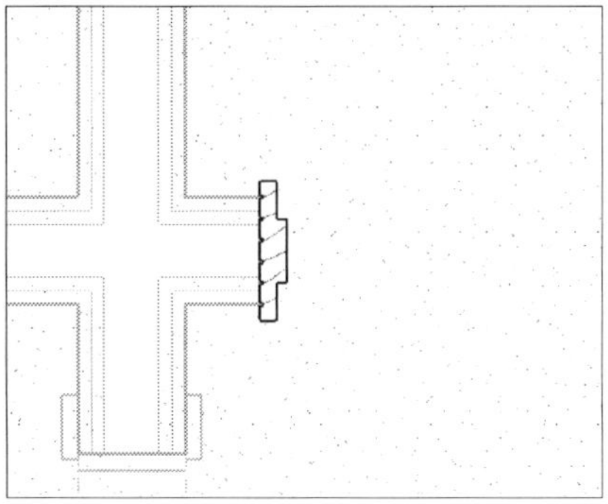

Figure 12-27 *Adding the door frame as a detail component*

12. Similarly, using the **Detail Component** tool and the editing tools, place the **Rough Cut Lumber-Section: 1x4R** component at the locations shown in Figure 12-28.

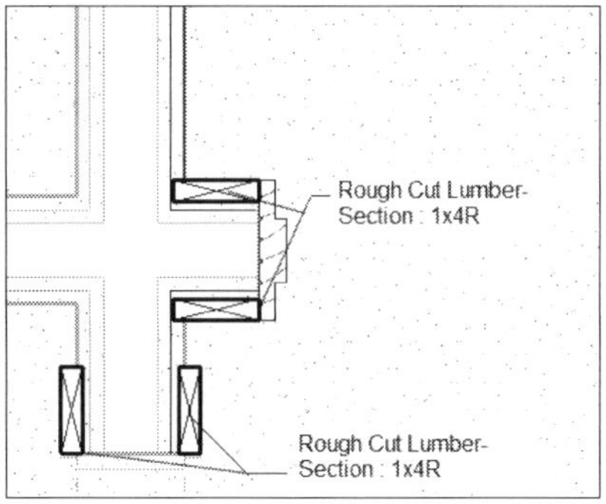

Figure 12-28 *Adding the wood board as a detail component*

You need to place another wood board of **6 1/2"** (165 mm) length. You can select and modify the properties of a similar component to create a new one.

13. Next, invoke the **Detail Component** tool and select the **Rough Cut Lumber-Section: 1x6R** option from the **Type Selector** drop-down list.

14. Choose the **Edit Type** button from the **Properties** palette; the **Type Properties** dialog box is displayed.

15. Choose the **Duplicate** button; the **Name** dialog box is displayed. Enter **1x 6.5R** in the **Name** edit box.

16. In the **Type Properties** dialog box, enter the value **0' 6 1/2" (165 mm)** in the **Value** column for the **Depth** dimension parameter.

17. Choose **OK**; the **Type Properties** dialog box is closed.

18. Ensure that the detail component **Rough Cut Lumber-Section : 1x6.5R** is selected in the **Type Selector** drop-down list and place it at the location, as shown in Figure 12-29.

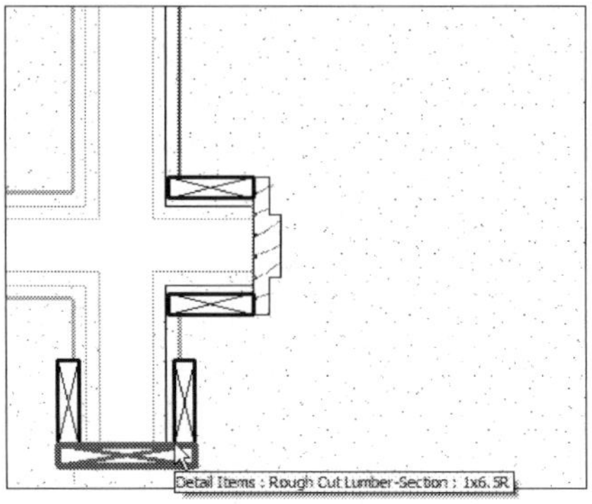

Figure 12-29 *The modified wood board added as a detail component*

Adding Detail Lines to the Project Detail

1. Choose the **Detail Line** tool from the **Detail** panel of the **Annotate** tab; the **Modify | Place Detail Lines** tab is displayed.

2. Ensure that the **Thin Lines** option is selected in the **Line Style** drop-down list in the **Line Style** panel. Now, use the snap options to sketch thin lines at 5/8" distance, depicting the double layer gypsum wall board in the callout view. You can also use the **Mirror**, **Offset**, and **Copy** tools to create detail lines. After adding the thin lines, the detail view will appear similar to the graphic shown in Figure 12-30.

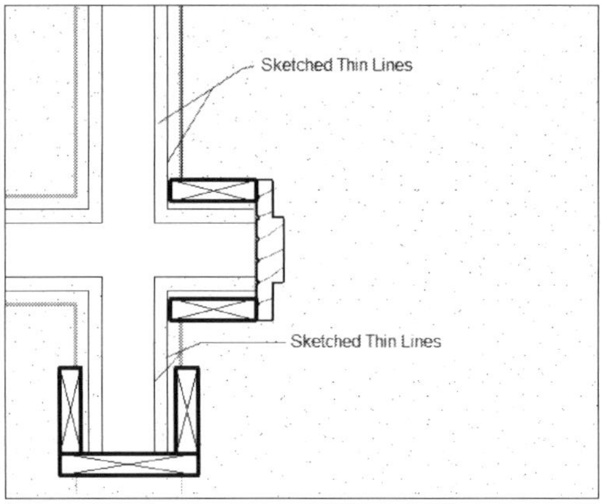

Figure 12-30 *The detail lines added to the callout view*

3. After you create the details lines, you will add metal studs to support the gypsum boards. To do so, choose the **Detail Component** tool from **Annotate > Detail > Component** drop-down; the **Modify | Place Detail Component** tab is displayed.

4. Choose the **Load Family** tool from the **Mode** panel; the **Load Family** dialog box is displayed. Now, browse to the **English-Imperial > Detail Items > Div 05-Metals > 054200-Cold-Formed Metal Joist Framing** folder and select the **C Joist-Section** file (for Metric **English > Detail Items > Div 05-Metals > 05400-Cold-Formed Metal Framing > 05420-Cold-Formed Metal Joists** folder and select the **M_C Joist-Section** file). Now, choose **Open**; the selected file will now be loaded and can be selected from the **Type Selector** drop-down list.

5. Ensure that the **C Joist-Section: 5 1/2" (140mm)**is selected in the **Type Selector** drop-down list and then choose the **Edit Type** button from the **Properties** palette; the **Type Properties** dialog box is displayed.

6. Choose **Duplicate**; the **Name** dialog box is displayed. Enter **2 1/2" (64 mm)** in the **Name** edit box and then choose **OK** to close the **Name** dialog box.

7. Now, in the **Type Properties** dialog box, click in the **Value** fields of the **Width** and **Depth** parameters and enter **2 1/2" (64 mm)** and **1 1/2" (38 mm)**, respectively.

8. Choose **OK**; the **Type Properties** dialog box closes and the new type of joist section is selected in the **Type Selector** drop-down list.

9. Move the cursor in the drawing area and then use the editing tools to add the selected type of joist to the detail view, as shown in Figure 12-31.

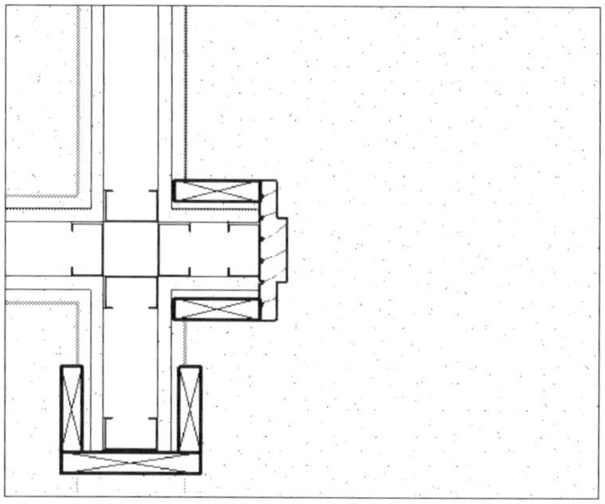

Figure 12-31 *The medium detail lines added to the detail view*

Adding an Insulation to the Detail

After you have added detail components and detail lines, you will add insulation between the metal joist framing members by using the **Insulation** tool.

1. Choose the **Insulation** tool from the **Detail** panel of the **Annotate** tab; the **Modify | Place Insulation** tab is displayed.

2. Ensure that the **Line** tool is chosen in the **Draw** panel. Next, enter **0' 2 1/2" (64 mm)** in the **Width** edit box and choose **to near side** from the drop-down list in the **Option bar**.

3. Move the cursor to the drawing window and click to specify the location of the start point and end point of the insulation, as shown in Figure 12-32. Now, press ESC twice; the insulation is created.

4. Next, select the insulation that you have recently created; the **Modify | Insulation Batting Lines** tab is displayed.

5. In the **Properties** palette, click in the value column of the **Insulation Bulge to Width Ratio (1/x)** parameter and enter **3**. Now, choose **Apply**; the type property of the selected component gets modified.

6. Similarly, to add other insulations, repeat steps 1 to 5 of this section. Refer to Figure 12-33 for location.

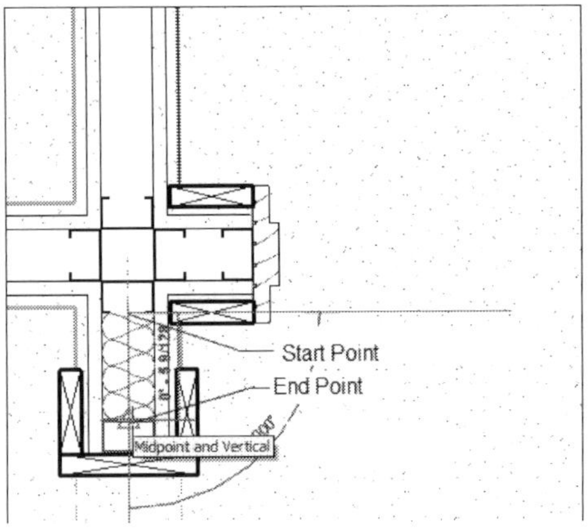

Figure 12-32 *Sketching the insulation*

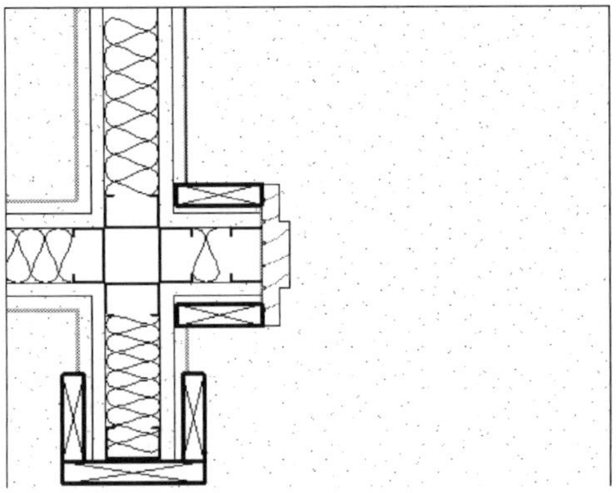

Figure 12-33 *The insulations added to the detail view*

Adding Text Notes to the Detail

You need to add text notes to the detail. Use an existing text note style as a template and create a new text type by modifying its parameters, as given in the project parameters.

1. Choose the **Text** tool from the **Text** panel of the **Annotate** tab; the **Modify | Place Text** tab is displayed.

2. From the **Type Selector** drop-down list, select **Text: 3/32" Arial (2.5mm Arial)**.

3. In the **Properties** palette, choose the **Edit Type** button; the **Type Properties** dialog box is displayed.

4. Choose **Duplicate** from the dialog box; the **Name** edit box is displayed.

5. Enter the name **Text Notes** in the **Name** edit box and choose **OK**.

6. In the **Type Properties** dialog box, ensure that the value for the **Text Size** type parameter is **3/32" (2.5 mm)**.

7. For the **Leader Arrowhead** parameter, select **Arrow Filled 15 Degree** from the drop-down list.

8. Choose **Apply** and **OK**; the settings are applied to the text notes and the **Type Properties** dialog box closes.

9. Choose the **Two Segments** tool from the **Leader** panel of the **Modify | Place Text** tab.

   If you want to create a text note with two segment leaders, you need to specify three points. The first point specifies the location of the leader head, the second point specifies the leader elbow, and the third point specifies the start point of the text.

10. Click on the door panel, refer to Figure 12-34, to specify the first point and move the cursor upward in the right direction and then click to specify the elbow point. Next, move the cursor horizontally toward right and click to specify the endpoint of the leader (start point of the text), as shown in Figure 12-34.

11. Enter the text **Wood Door Frame - Double Rabbet-Section**: **6 1/2** "(for Metric **Wood Door Frame - Double Rabbet-Section**: **165 mm**) in the edit box, as shown in Figure 12-35. Click outside the text box to complete the text note.

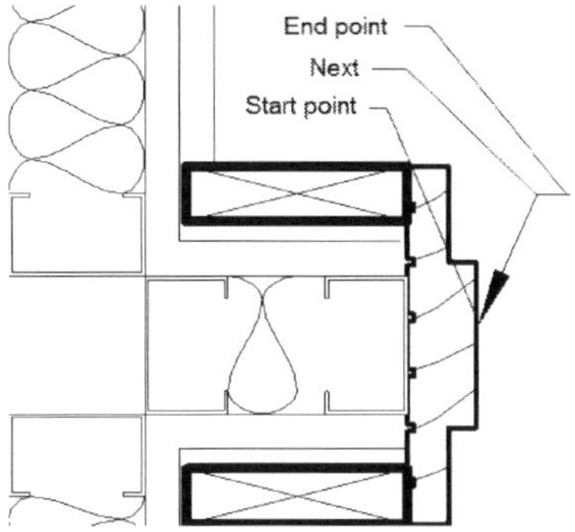

Figure 12-34 *Specifying three points for creating the text note with two segment leaders*

Figure 12-35 *Writing the text for the text note*

12. Similarly, add other text notes at appropriate locations, as shown in Figure 12-36, and press ESC when you have completed entering the text.

Adding Break Lines

In this section of the tutorial, you will add break line symbols to the continuing walls at the left and right by using the **Detail Component** tool.

1. Choose the **Detail Component** tool from **Annotate >Detail > Component** drop-down; the **Modify | Place Detail Component** tab is displayed.

2. Choose the **Load Family** tool from the **Mode** panel; the **Load Family** dialog box is displayed. Now, browse to **English-Imperial/English > Detail Items (Detail Components) > Div 01 - General** folder and select the **Break Line** file. Choose **Open**; the **Load Family** dialog box closes, and the **Break Line** option gets selected in the **Type Selector** drop-down list.

3. Move the cursor in the drawing view and add the first break line horizontally at the top of the callout view, refer to Figure 12-36.

4. Choose the **Modify** button from the **Select** panel and then select the break line you have recently added; the **Modify | Detail Items** tab is displayed.

5. In the **Properties** palette, click in the value fields of the **right** and **left** instance parameters and enter **0' 4" (102 mm)**.

6. Similarly, click in the value fields of the **Jag Depth**, **Jag Width**, and **Masking depth** parameters and enter **0' 2" (51 mm)**. Now, choose **Apply** so that the specified instance parameters are assigned to the selected break line. Next, you need to add a vertically aligned break line on the left of the callout view, refer to Figure 12-36.

7. Copy the first break line, place it at the specified location, and then rotate it. Use the **Copy** and **Rotate** tools to add this line.

8. In the **Project Browser**, click on the **Door Jamb Detail** sub-head under the **Floor Plans** head; the instance properties of the callout view are displayed in the **Properties** palette.

9. In the **Properties** palette, click in the value field of the **Display Model** instance parameter and select the **Do not display** option from the drop-down list to restrict the display of the building model. Next, choose the **Apply** button.

   The completed tutorial resembles the detail sketch plan, as shown in Figure 12-36.

Saving the Project

In this section, you will save the project file using the **Save As** tool.

1. Now, choose **Save As > Project** from the **File** menu; the **Save As** dialog box is displayed. Enter **c12_Apartment1_tut1** (for Metric **M_c12_Apartment1_tut1**) and choose **Save**; the project file is now saved.

2. Now, close the project file by choosing **Close** from the **File** menu.

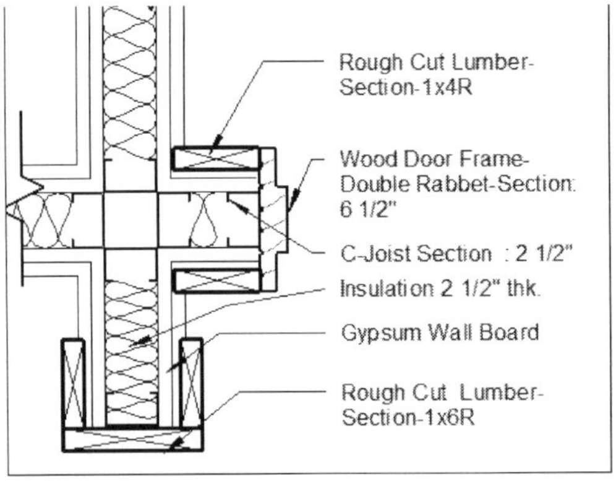

Figure 12-36 *Text added to the callout detail*

EXERCISES

Exercise 1 Club - Drafted Detail

In the *Club* project created in Exercise 1 of Chapter 11, create a drafted detail based on the sketch given in Figure 12-37. Load the required detail items (detail components). Use wide and thin lines to sketch the detail lines. Assume the parameters for adding text notes, dimensions, and break lines. The exact placement of the elements is not important for this exercise. Save the project file as *c12_Club_ex1.rvt (for Metric M_c12_Club_ex1.rvt).*

(Expected time: 30 min)

To add the Window jamb detail component in the Project, you need to download the *Window Jamb - Metal Fixed Flush.rfa* file from the website*: www.Cadsoft.com*

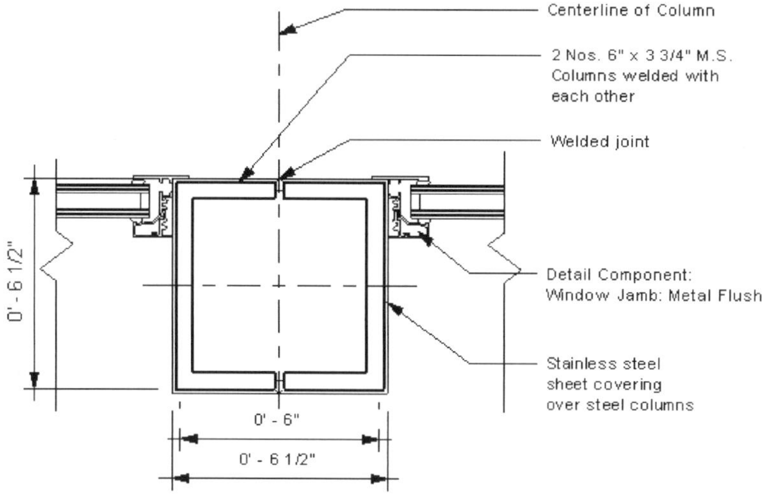

Figure 12-37 *Sketch plan for creating a drafted detail for the Club project*

This Page Is Intentionaly Left Blank

Chapter 13

Creating and Plotting Sheets

Learning Objectives

After completing this chapter, you will be able to:

- *Add drawing sheets to a project*
- *Add views to a drawing sheet*
- *Modify view properties of a drawing sheet*
- *Pan the viewport*
- *Add schedules to a drawing sheet*
- *Modify a building model in a drawing sheet*
- *Create dependent views*
- *Add matchline to dependent views*
- *Add view reference*
- *Print drawing sheets and project views*
- *Select and modify the printer settings*

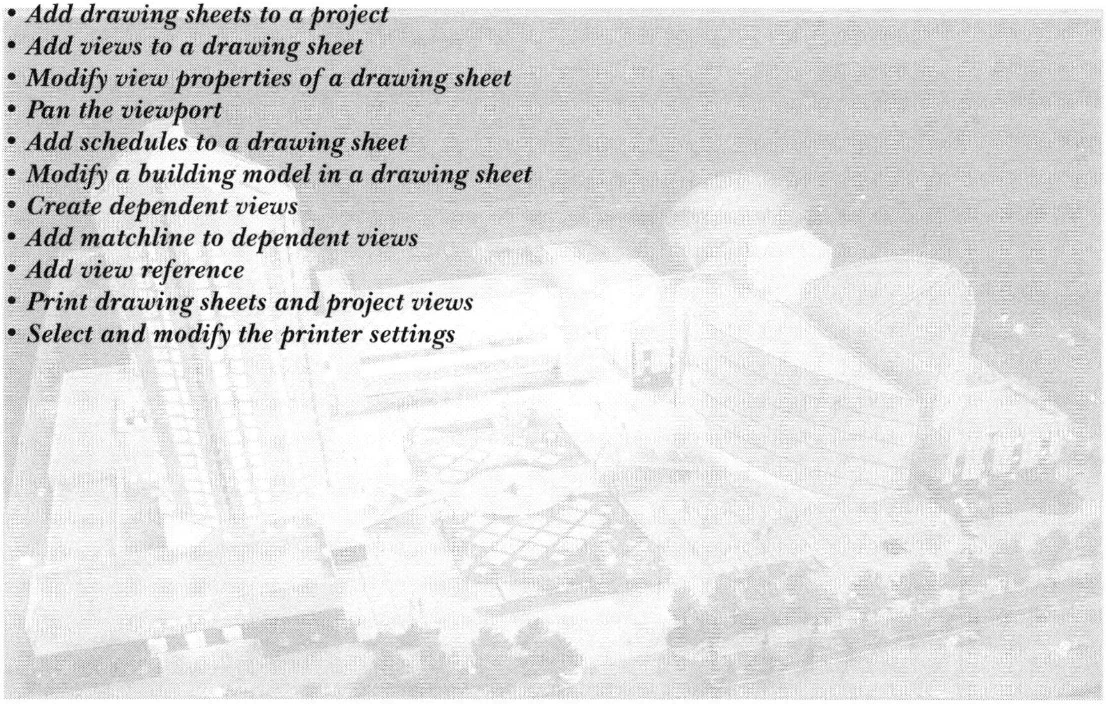

INTRODUCTION

In a BIM project, creating sheets is an essential task for creating a construction document set. To create a construction document set, you create sheets, and add drawings and schedules to the sheets. Further, you can print them or publish them in other formats such *.pdf*, *.dwf*, and more. You can also send the construction documents to Autodesk Docs or BIM 360, where clients or reviewers can mark up the designs for revisions.

CREATING DRAWING SHEETS

In Autodesk Revit, drawing sheets are treated like views in a project. They are defined by a border and a title block. Therefore, to create a drawing sheet, you first need to add a sheet view to the project.

Adding a Drawing Sheet to a Project

Ribbon: View > Sheet Composition > Sheet

To add a drawing sheet to a project, choose the **Sheet** tool from the **Sheet Composition** panel; the **New Sheet** dialog box will be displayed. This dialog box has two areas: **Select titleblocks** and **Select placeholder sheets**. Select the required title block for the drawing sheet from the list of title blocks. To load drawing sheets other than the ones in the list of title blocks in the **Select titleblocks** area, choose the **Load** button from the **New Sheet** dialog box; the **Load Family** dialog box will be displayed. Next, choose **English-Imperial > Titleblocks** (for Metric **English > US >Titleblocks**); the list of titleblocks family files will be displayed, as shown in Figure 13-1. Select the desired family file from the list to load it into your project and choose the **Open** button; the title block corresponding to the selected family file will be added to the list of title blocks in the **Select titleblocks** area. Select a titleblock from the list and choose the **OK** button; a titleblock of the selected family will be displayed in the model window.

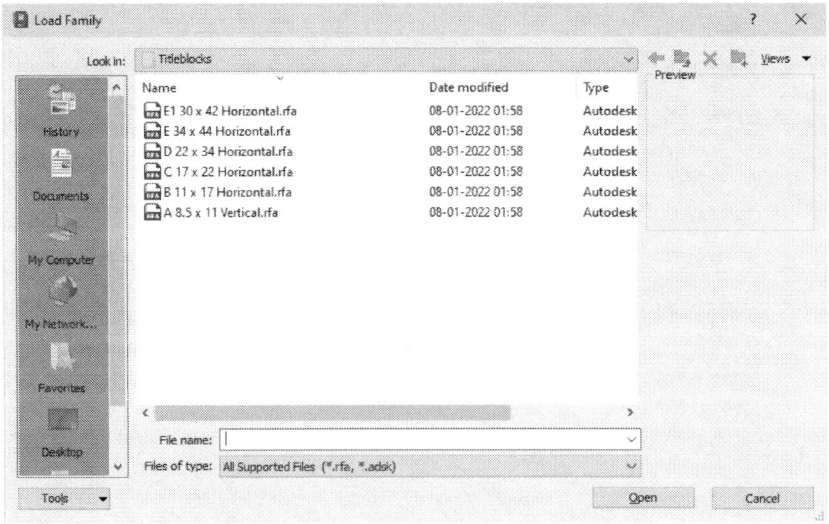

***Figure 13-1** The **Load Family** dialog box displaying the list of families for titleblocks*

Adding Views to a Sheet

Ribbon: View > Sheet Composition > Place View

Before you add views to a sheet, make sure that all the views are set and are prepared to be added to the sheet to create construction documentations. To add views to a sheet, choose the **Place View** tool from the **Sheet Composition** panel; the **Views** dialog box will be displayed with a list of views to be added to the drawing sheet, as shown in Figure 13-2. Select the required view and choose the **Add View to Sheet** button. On doing so, the selected view will appear as a viewport represented by a rectangle attached to the cursor. Move this cursor to the drawing area in the sheet and click in the preferred location to add the selected view. Next, in the **Properties** palette, change the **View Scale** parameter by selecting a suitable scale from the drop-down list in its **Value** field and choose **Apply**. The added view will adjust to the specified scale and appear in the sheet,

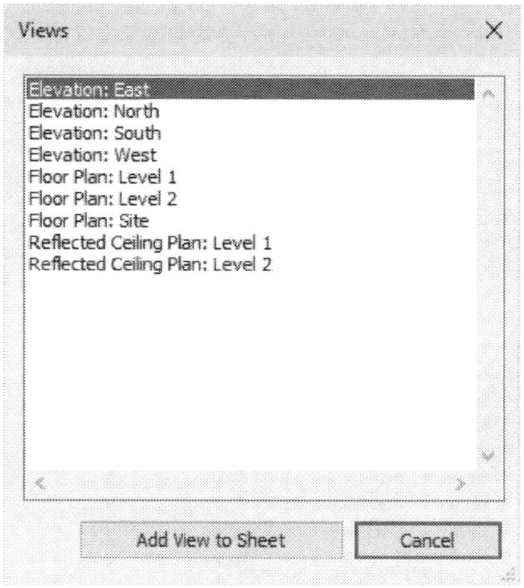

Figure 13-2 *The* ***Views*** *dialog box with the list of views to be added to the drawing sheet*

The viewport of the inserted view displays a label line at the bottom left corner. The label line shows the view name, view scale, and view number, as shown in Figure 13-3. In this manner, you can add a number of views to the same drawing sheet and create a complete set of drawing sheets for a project. You can modify various properties associated with the viewport in the **Properties** palette.

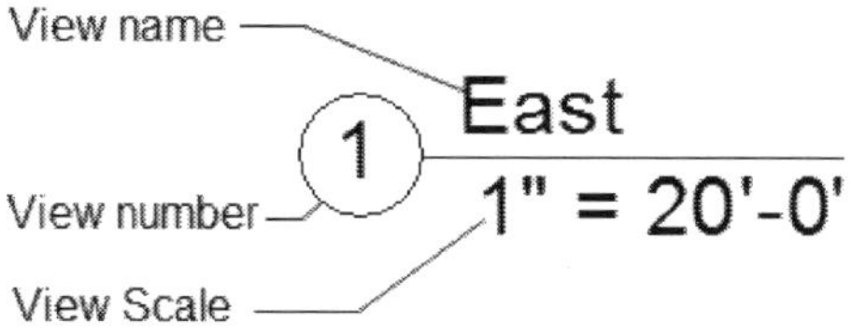

Figure 13-3 *Parameters in the label line of the viewport*

Modifying Viewport Properties

To modify the properties of a viewport associated with a view in a sheet, select the viewport corresponding to it; the **Properties** palette and the **Options Bar** will be displayed. The **Properties** palette for the viewport displays various instance parameters associated with the viewport, as shown in Figure 13-4. Based on the requirement, you can modify these values. In the **Properties** palette, click on the value column corresponding to the **View Scale** parameter and select an option from the drop-down list displayed to modify the scale of the view. In the palette, you can select the **Coarse**, **Medium**, or **Fine** option from the drop-down list corresponding to the value column of the **Detail Level** instance parameter to define the level of details to be displayed in a view. To control the visibility of different categories of elements in each viewport, choose the **Edit** button for the **Visibility/Graphics Overrides** parameter; the **Visibility/Graphics Overrides for Elevation (current view)** dialog box will be displayed for the corresponding view. In this dialog box, you can control the visibility of various model and annotation elements. In the **Properties** palette, you can rename the corresponding view of the selected viewport by entering a name in the value column corresponding to the **View Name** parameter. In the **Properties** palette, you can click on the value column corresponding to the **Rotation on Sheet** instance parameter to rotate the view clockwise or counterclockwise. Alternatively, you can rotate the view by selecting an option from the **Rotation on Sheet** drop-down list in the **Options Bar**. You can edit the type properties of the selected view in the sheet.

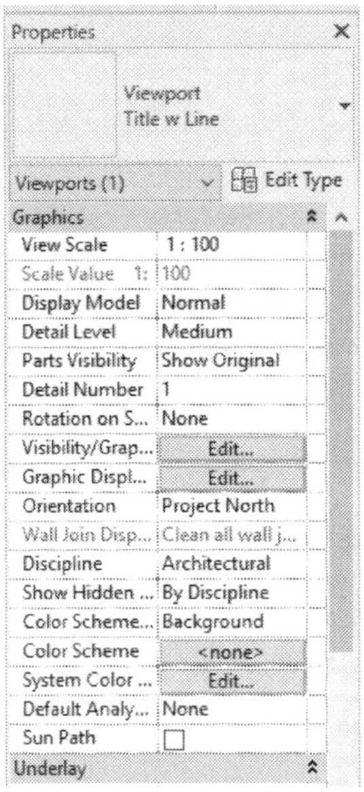

Figure 13-4 *Partial view of the* ***Properties*** *palette for the viewport*

Panning the Viewport Added to the Sheet

In Revit, you adjust the views added to the sheet by panning or moving them. To pan a view, select the viewport, right-click, and then choose **Activate View** from the shortcut menu displayed. On doing so, you will notice that only the selected view will be activated (You will learn more about it in the next section). Now, right-click again to display a shortcut menu. Choose the **Pan**

Active View option from the displayed shortcut menu; the cursor will convert into a symbol with four arrows. Press and hold the left mouse button and drag the mouse to move the entities in the viewport. Once you have panned and placed the entities at the appropriate location in the sheet, choose **Deactivate View** from the shortcut menu to revert to the sheet.

Replacing a View in the Viewport

In Revit, you can replace a view in a viewport. To do so, select the viewport that needs to replaced from the sheet; the **Modify | Viewports** contextual tab will be displayed. In the **Positioning & View** panel of the contextual tab, select the **View Origin** or **Viewport Center** option from the **Viewport Positioning** drop-down list to specify the base point of the viewport. Next, select the view from the **View** drop-down list to replace the existing view. On doing so, the selected view will be replaced by the new view in the sheet.

Adding Schedules to a Drawing Sheet

You can easily add the schedules created in a project to a drawing sheet. To do so, drag the schedules created (required schedules) from the **Project Browser** and drop them into the drawing sheet; the preview image of the schedule will be attached to the cursor. Move the cursor and click at the desired location in the drawing to place the schedule. The selected schedule will appear with blue triangles for each column and a break line on the right side, as shown in Figure 13-5.

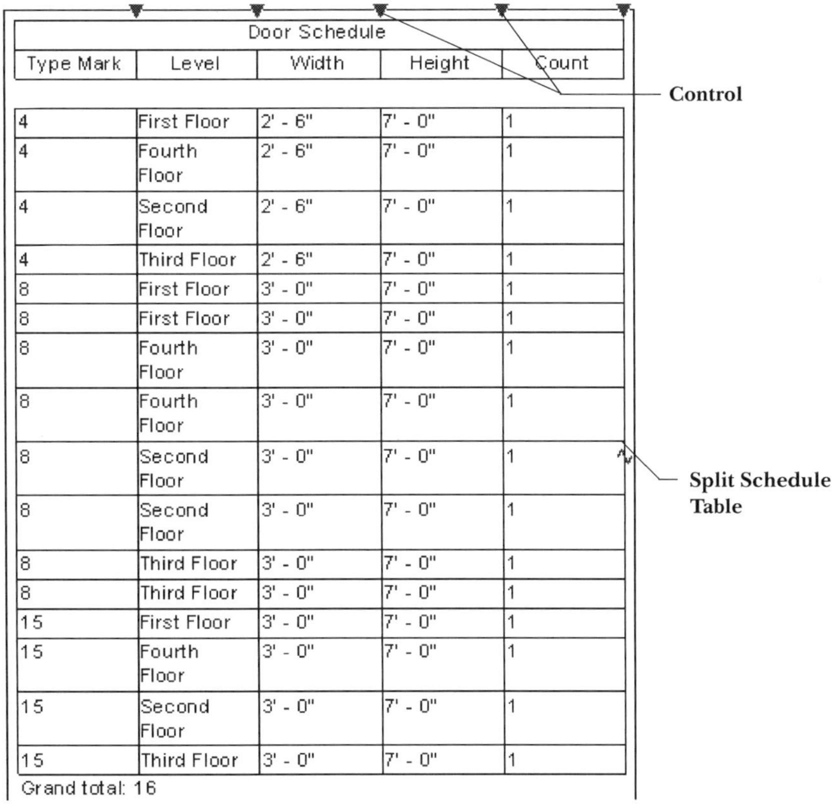

Door Schedule				
Type Mark	Level	Width	Height	Count
4	First Floor	2' - 6"	7' - 0"	1
4	Fourth Floor	2' - 6"	7' - 0"	1
4	Second Floor	2' - 6"	7' - 0"	1
4	Third Floor	2' - 6"	7' - 0"	1
8	First Floor	3' - 0"	7' - 0"	1
8	First Floor	3' - 0"	7' - 0"	1
8	Fourth Floor	3' - 0"	7' - 0"	1
8	Fourth Floor	3' - 0"	7' - 0"	1
8	Second Floor	3' - 0"	7' - 0"	1
8	Second Floor	3' - 0"	7' - 0"	1
8	Third Floor	3' - 0"	7' - 0"	1
8	Third Floor	3' - 0"	7' - 0"	1
15	First Floor	3' - 0"	7' - 0"	1
15	Fourth Floor	3' - 0"	7' - 0"	1
15	Second Floor	3' - 0"	7' - 0"	1
15	Third Floor	3' - 0"	7' - 0"	1
Grand total: 16				

Figure 13-5 *Schedule with controls added to a drawing sheet*

Modifying a Building Model in a Drawing Sheet

You can modify a building model directly from the views added to a drawing sheet. To do so, you need to activate a view and make the desired modifications in it. To activate a view from a drawing sheet, select the viewport; the **Modify | Viewports** tab will be displayed. In this tab, choose the **Activate View** button from the **Viewport** panel; the view in the selected viewport will be activated. In the activated view, you can work on the building model as in any other project view. The desired elements can be edited using the editing tools. New elements can also be added to the model. When the building model is edited, the parametric change engine of Autodesk Revit modifies other project views immediately. As a result, all the project views are updated automatically. For example, the modifications made to the location of a furniture item in an activated viewport of the drawing sheet are also made in the corresponding floor plan, sections, and other associated views.

Tip

*Another method to activate or deactivate a viewport is to move the cursor over the project view and right-click; a shortcut menu will be displayed. From the shortcut menu, you can choose the **Activate View** or **Deactivate View** option to activate or deactivate the specified viewport.*

CREATING GUIDE GRIDS

Guide Grid is a customized grid available in Revit sheets. It is used to place the drawing view in place in the sheet. To create a Guide Grid, choose the **Guide Grid** tool from the **Sheet Composition** panel of the **View** tab; the **Assign Guide Grid** dialog box will be displayed, as shown in Figure 13-6. In this dialog box, the **Create new** radio button is selected by default. You can enter a name for the Guide Grid in the **Name** edit box and choose the **OK** button; the Guide Grid will be displayed over the sheet in the drawing area, as shown in Figure 13-7. On creating a Guide Grid, it gets added as an instance property under the **Other** head in the **Properties** palette. Now, select the Guide Grid from the drawing area; the drag controls will be displayed. You can use the drag controls to resize the Guide Grids in the drawing area. The **Properties** palette displays the instance properties of the selected Guide Grid. In the **Properties** palette, you can specify guide spacing in the **Guide Spacing** edit box to increase or decrease the spacing of the grids. You can also rename the guide name under the identity data in the **Name** edit box in the **Properties** palette.

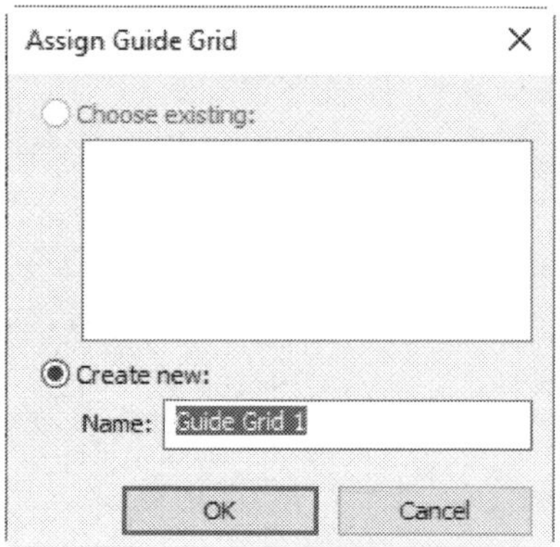

***Figure 13-6** The **Assign Guide Grid** dialog box*

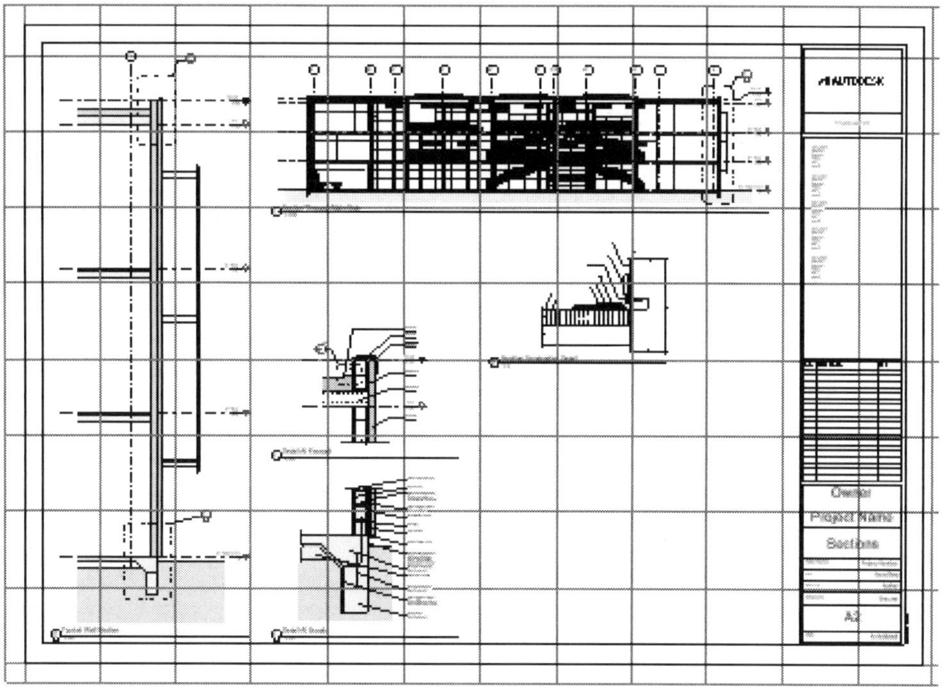

Figure 13-7 The Guide Grid displayed over the sheet

DUPLICATING DEPENDENT VIEWS

You can create multiple copies of the project views in Revit. These multiple views are called duplicate views. All the duplicate views are dependent on the original view from which they are created. Therefore, these dependent views are automatically updated if any changes are made in the original view. Dependent views are useful when you want to use or place the same view on more than one sheet. It is also useful in case of huge projects when the overall project view is too large to fit in a single sheet. In such cases, you can crop the parent view into small segments using the crop regions, create dependent views from them, and then place the cropped dependent views on the sheet.

Creating Dependent Views

You can create a dependent view from a plan, elevation, section, and callout view. To create a dependent view, open the view from which you want to create duplicate view and choose the **Duplicate as Dependent** tool from **View > Create > Duplicate View** drop-down. Alternatively, select the required view name in the **Project Browser** and right-click; a shortcut menu will be displayed. Choose **Duplicate View > Duplicate as Dependent** from the shortcut menu; the primary view will display a crop region boundary which will be added to the **Project Browser** as a dependent view under the primary view. Select the crop region boundary to display the drag controls. Now, using the drag controls, resize and crop the primary view to include only the required portion of the view in the dependent view. Next, select the name of the dependent view in the **Project Browser** and right-click; a shortcut menu will be displayed. Choose the **Rename** option from the shortcut menu, an edit box will displayed. Enter a name for the dependent view in the edit box and choose the **OK** button; the duplicate view will be renamed. Again, open the primary view and choose the **Show Crop Region** button from **View Control Bar** to display the crop region boundary. Similarly, create another dependent view. In this way, you can crop the primary view and create multiple dependent views. Figure 13-8 shows primary view and Figures 13-9 and 13-10 show two dependent views with crop boundaries.

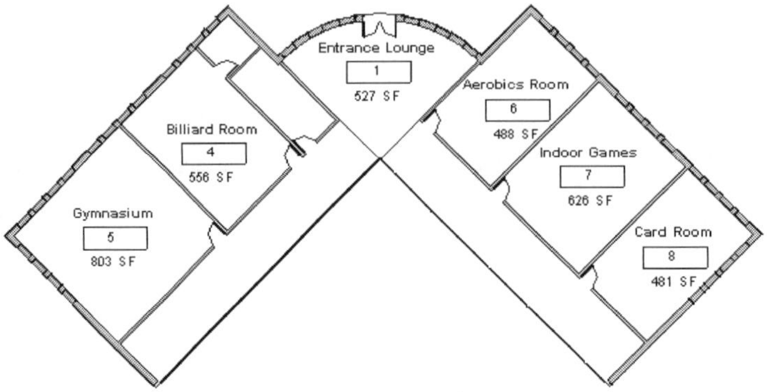

Figure 13-8 *A primary view*

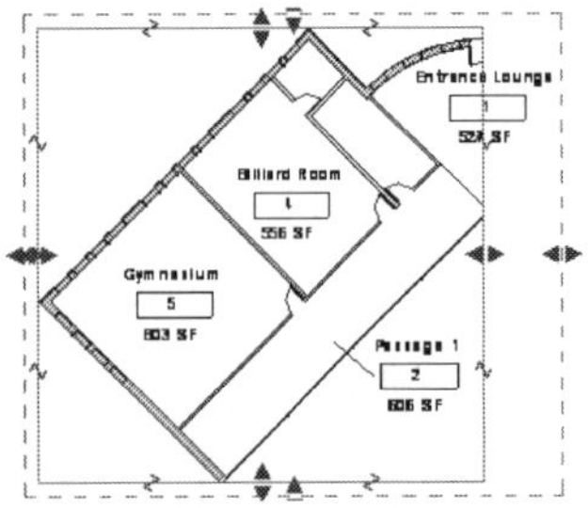

Figure 13-9 *The left dependent view showing the left portion of the drawing with crop regions*

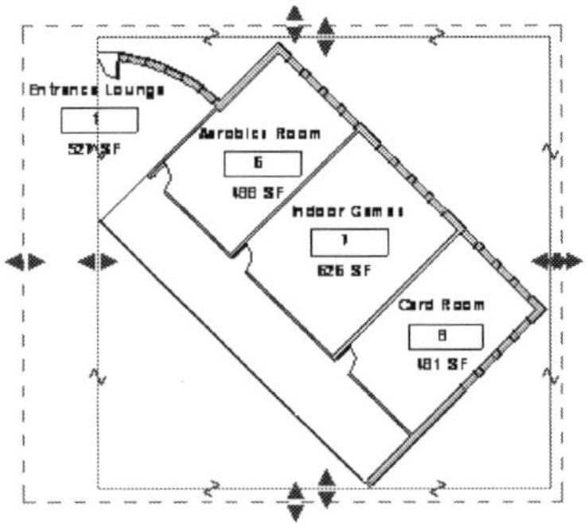

Figure 13-10 *The right dependent view showing the right portion of the drawing with crop regions*

Adding Matchline to Dependent Views

Matchlines are the graphical representation of a split in views. These are the sketch lines that indicate the area where a view is split. To add a matchline, open the primary view from which the dependent views are created. Next, choose the **Show Crop Region** button from the **View Control Bar**; the crop regions of both the primary and dependent views will be displayed. Choose the **Matchline** tool from the **Sheet Composition** panel in the **View** tab; the **Modify | Create Matchline Sketch** tab will be displayed and the screen will enter the sketch mode. Sketch a matchline using the sketching tools in the **Draw** panel of the **Modify | Create Matchline Sketch** tab to show the split. After sketching the matchline, choose the **Finish Edit Mode** button from the **Mode** panel of the **Modify | Create Matchline Sketch** tab; the matchline will be displayed, as shown in Figure 13-11. You can also edit the length and shape of the matchline. To do so, select the matchline; the **Modify | Matchline** tab will be displayed. In this tab, choose the **Edit Sketch** button from the **Mode** panel; the **Modify | Matchline > Edit Sketch** tab will be displayed. On invoking this tab, the screen will enter the sketch mode. Modify the sketch of the matchline as required and then choose the **Finish Edit Mode** button from the **Mode** panel in the **Modify | Matchline > Edit Sketch** tab.

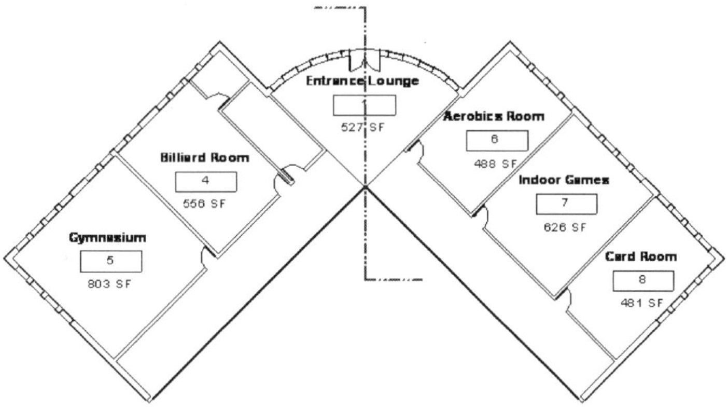

Figure 13-11 *Matchline showing the area where you want to split the drawing*

Adding View Reference

You can add a view reference in the primary view after you have placed the cropped dependent views on the sheet. A view reference is a symbol that is used to refer the dependent view available in the same or different sheet.

To add a view reference, select the primary view on the sheet and right-click; a shortcut menu will be displayed. Now, choose the **Activate View** option from the shortcut menu; the selected view will be highlighted. Next, invoke the **View Reference** tool from the **Tag** panel of the **Annotate** tab. Now, select the dependent view to be referred to from the **Target View** drop-down list available in the **View Reference** panel and then click in the selected drawing view to add the view reference. After adding the required view references, right-click, and then choose **Deactivate View** from the shortcut menu to deactivate the selected view. Now, if you double-click on the view reference tag, the respective referred view will open.

PRINTING IN Autodesk Revit

The created building model and project views can be stored in the computer's hard disk or on a temporary media. It also becomes imperative to print these views as a hard copy at various stages of the project. During the initial stages of the project development, you may need to print certain project views and discuss them with the others in your organization or with the clients. As the project progresses, other professional agencies such as structural, electrical, plumbing, HVAC, landscape, and so on become involved with the project. The printed drawings assist in a proper coordination between these agencies. The project may also require approval from statutory bodies and other departments. Printed drawings and project views explain the project and can be submitted as a complete set to obtain their approval.

Printing Drawing Sheets and Project Views

The **Print** tool is used to print project views and drawing sheets. You can invoke this tool from the **File** menu. On invoking the **Print** tool, the **Print** dialog box will be displayed, as shown in Figure 13-12.

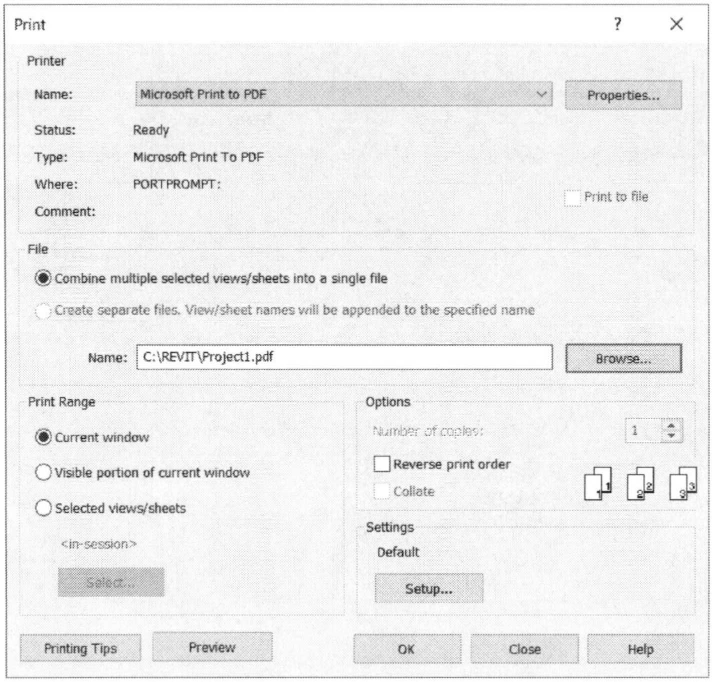

*Figure 13-12 The **Print** dialog box*

You have the option of using the default values in this dialog box. In case, the values conform to your requirements, you can choose the **OK** button and start printing directly. Otherwise, you can alter the print specifications using the options provided in the **Print** dialog box. The print options available in this dialog box are explained next.

Note

*The appearance and options available in the **Print** dialog box depend on the type of printer selected and the operating system used.*

Selecting and Modifying the Printer Settings

You can select a printing device from the **Name** drop-down list in the **Print** dialog box. This dialog box displays the information regarding the selected printer in the **Status**, **Type**, **Where**, and **Comment** fields below the printer name. The **Properties** button is used to set various printer properties. If you choose this button, a dialog box with a set of printer properties corresponding to the selected printer will be displayed, refer to Figure 13-13.

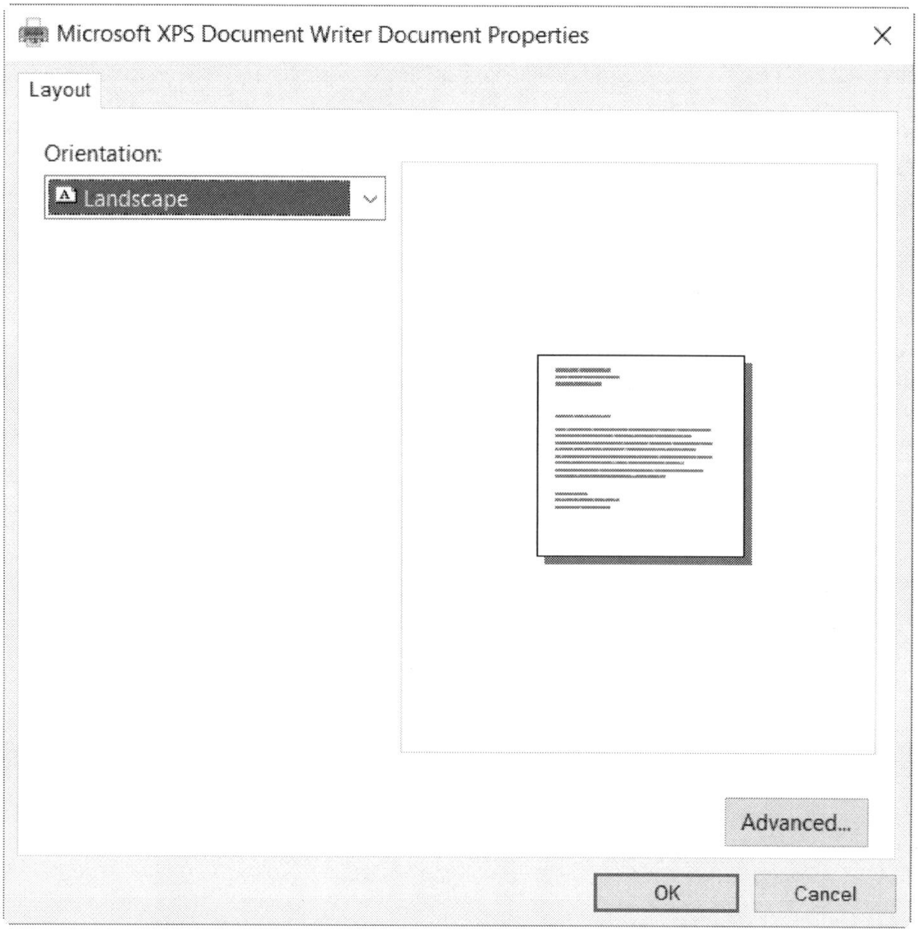

***Figure 13-13** The **Microsoft XPS Document Writer Document Properties** dialog box*

Using the Print Setup Dialog Box

File menu: Print > Print Setup

In Revit, you can specify various options for printing a drawing sheet. To do so, invoke the **Print Setup** dialog box by choosing **Print > Print Setup** from the **File** menu or choose the **Setup** button in the **Settings** area of the **Print** dialog box; the **Print Setup** dialog box will be displayed, as shown in Figure 13-14.

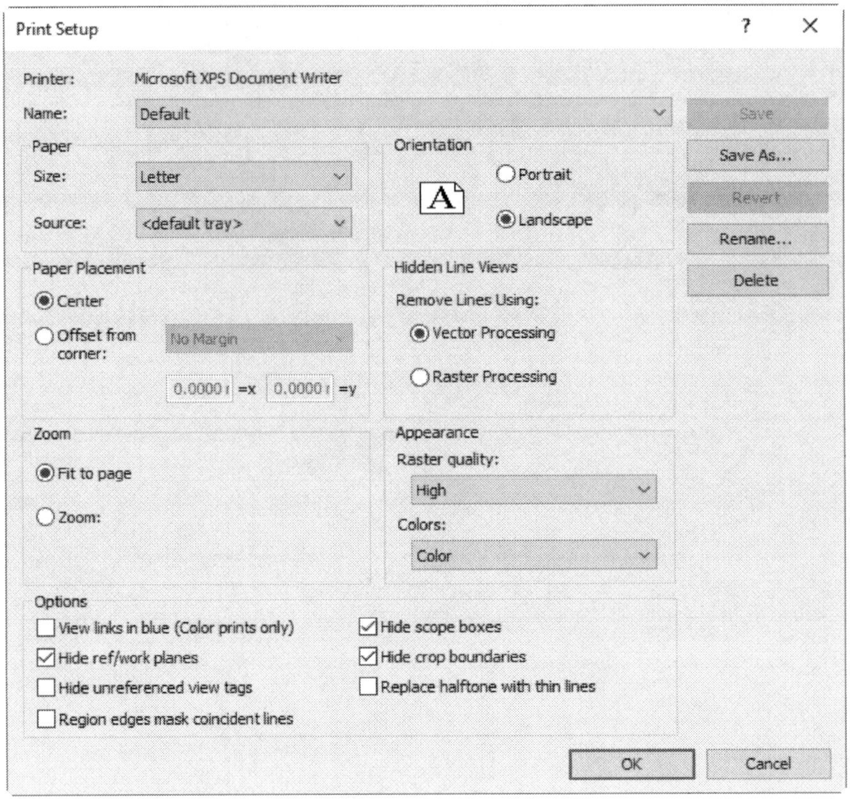

Figure 13-14 *The* ***Print Setup*** *dialog box*

Previewing the Print Setup

Before printing a project view or a drawing sheet, you can preview it using the **Print Preview** tool. This tool can be invoked by choosing the **Preview** button in the **Print** dialog box. When you invoke this tool, Autodesk Revit displays a draft version of the drawing sheet in the preview window, as shown in Figure 13-15. The displayed image acts as a preview of the hard copy that will be printed and is based on the set print parameters. If the displayed image does not conform to the desired result, you can use the **Close** button to close the preview and make the necessary modifications in the print setup till the desired result is achieved. While in the preview window, you can use the **Zoom In** and **Zoom Out** buttons to enlarge or reduce the view. On achieving the desired preview image, proceed to print the drawing using the **Print** tool.

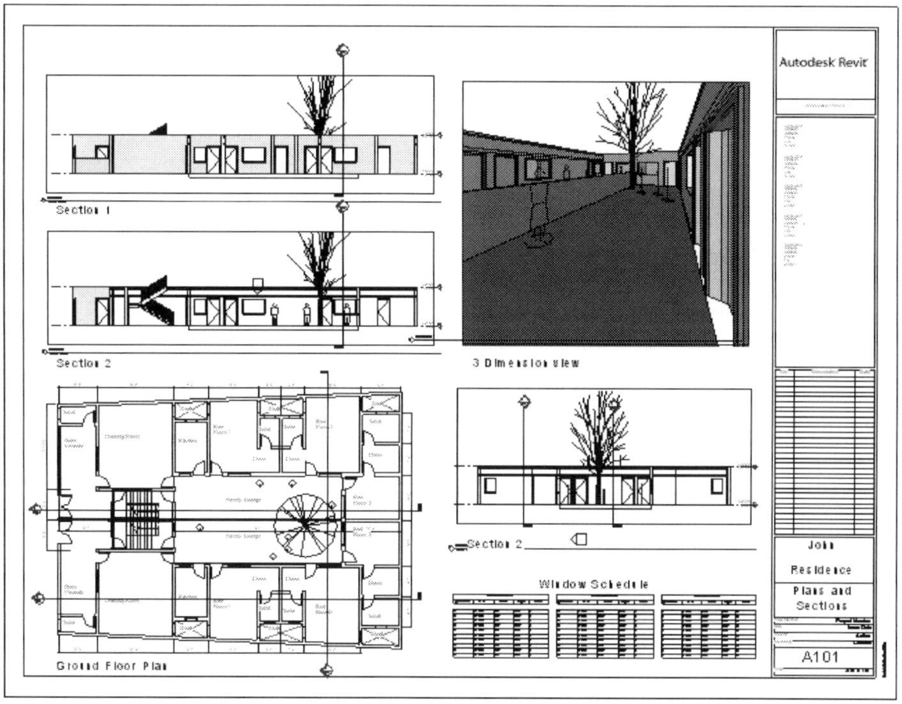

Figure 13-15 *Previewing the print settings displayed using the* ***Print Preview*** *tool*

TUTORIAL

Tutorial 1 Apartment 1

In this tutorial, you will create a drawing sheet for the *Apartment 1* project created in Tutorial 2 of Chapter 12, and then add the first floor plan, north elevation, longitudinal section, room schedule, and door schedule to the sheet, as shown in Figure 13-16. You will also add the following project and sheet information in the sheet title block:

(Expected time: 30 min)

1. Drawing sheet parameters:
   Titleblock to be used-
   For Imperial **C17 x 22 Horizontal** from the **English-Imperial > Titleblocks** folder.
   For Metric **A2 metric** from the **English > Titleblocks** folder.

2. Project information:
   Owner name: **Apartment Society**
   Project Name: **Apartment 1**
   Sheet name: **Plan, Elevation and Schedules**
   Sheet number: **Apt-1-01**

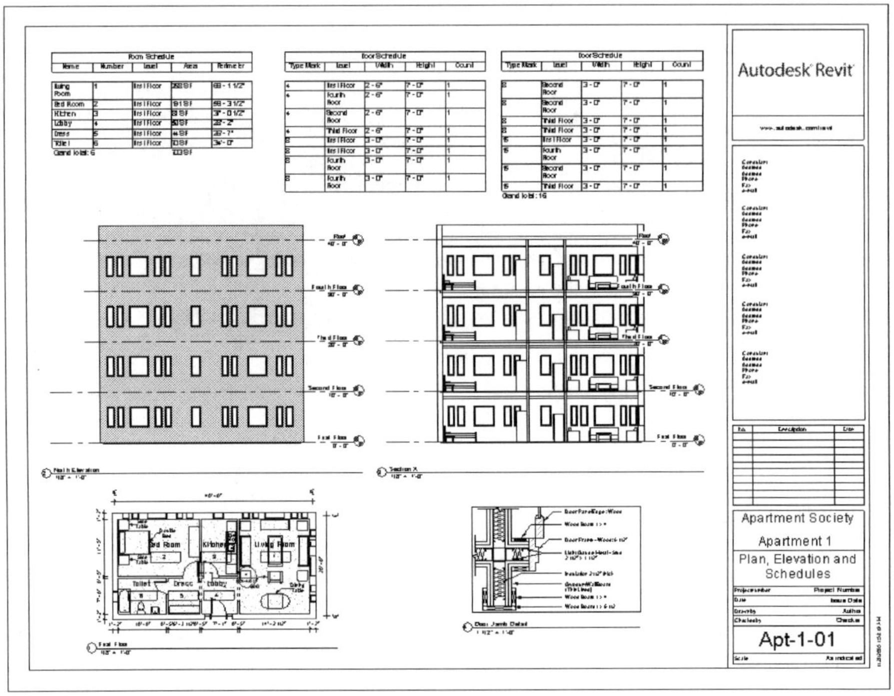

Figure 13-16 Sheet layout for creating a drawing sheet for the Apartment 1 project

The following steps are required to complete this tutorial:

a. Add a sheet by using the **New Sheet** tool and by loading the specified titleblock.
b. Add the project information to the sheet titleblock.
c. Add views to the sheet view and compose the drawing sheet.
d. Modify the visibility settings of the added views using the **Activate View** tool.
e. Add schedules to the sheet.
f. Edit schedules to compose them on the sheet.
g. Save the project using the **Save As** tool.
   For Imperial *c13_Apartment1_tut1.rvt*
   For Metric *M_c13_Apartment1_tut1.rvt*
h. Close the project.

Adding a Sheet to the Project

First, you need to open the project file and create a new sheet view. You can load the titleblock from the specified folder.

1. Choose **Open > Project** from the **File** menu and open the *c12_Apartment1_tut2.rvt* (for Metric *M_c12_Apartment1_tut2.rvt*) file created in Tutorial 2 of Chapter 12. You can also download this file from *https://www.Cadsoft.com*. The path of the file is as follows:*Textbooks > Civil/GIS > Revit Architecture > Exploring Autodesk Revit 2025 for Architecture*

2. Choose the **Sheet** tool from the **Sheet Composition** panel in the **View** tab; the **New Sheet** dialog box is displayed.

3. Choose the **Load** button; the **Load Family** dialog box is displayed. In this dialog box, choose the titleblock **C 17 x 22 Horizontal** from **English-Imperial > Titleblocks** folder (for Metric **A2 metric** from the **Metric > Titleblocks** folder). On doing so, the specified titleblock is added to the list in the **Select titleblocks** area in the **New Sheet** dialog box.

4. Choose the **OK** button to close the **New Sheet** dialog box and create the sheet view by using the loaded titleblock.

Adding the Project Information to the Sheet Titleblock

In this section, you need to add the project information to the titleblock using the **Properties** palette. Further, you need to directly enter other project parameters in the title block.

1. Select the titleblock from the drawing window. Then, in the **Properties** palette, click in the value fields for the **Sheet Name** and **Sheet Number** parameters and enter **Plan, Elevation and Schedules** and **Apt-1-01**, respectively.

2. Choose **Apply**. You will notice that the values entered in the **Properties** palette are displayed in the sheet.

3. Next in the titleblock, click in the **Owner** field and enter **Apartment Society** in the edit box. Now, press ENTER to view the new value in the sheet.

4. Similarly, you need to click on the **Project Name** field in the titleblock in the drawing area and enter **Apartment 1** in the edit box and then press ENTER. Now, the project details appear in the titleblock, as shown in Figure 13-17.

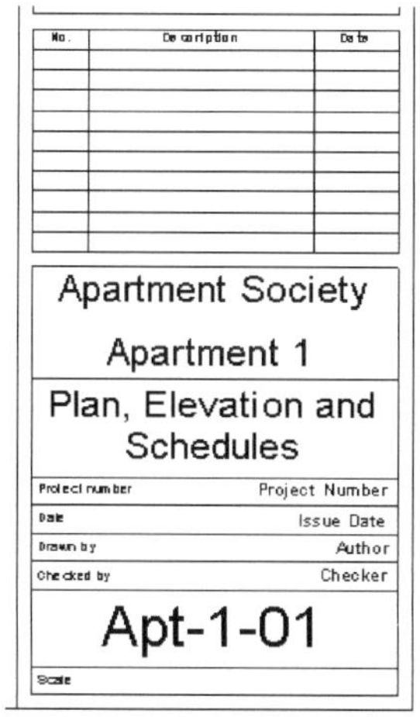

Figure 13-17 *Project information in the titleblock of the sheet view*

Adding Project Views to the Drawing Sheet

In this section, you need to add the specified project views by dragging their name from the **Project Browser** in the sheet. Further, you need to place the project views at their designated place in the sheet based on the sheet layout.

1. Drag **First Floor** from the **Floor Plans** head in the **Project Browser** into the drawing sheet. The view appears in the form of a rectangle in the sheet view. You can also use the **Place View** tool to add the view to the sheet.

2. Next, move the cursor to the lower left area of the titleblock such that the corner of the rectangle is close to the lower left corner of the drawing sheet. Next, click to place the view. The first floor plan view is added to the sheet and appears enclosed in a rectangle.

3. In the **Project Browser**, click on the **North** elevation node under the **Elevations (Building Elevation)** head and then right-click to display the shortcut menu. Choose the **Rename** option from the shortcut menu. In the **Name** edit box of this dialog box, enter **North Elevation** and then press ENTER to rename it.

4. Drag **North Elevation** and **Section X** from the **Project Browser** and add them to the specified locations in the sheet, refer to Figure 13-18. The three views are added to the sheet, as shown in Figure 13-18.

5. Next, drag the **Door Jamb Detail** view from the **Project Browser** and add it to the desired location in the sheet, refer to Figure 13-16.

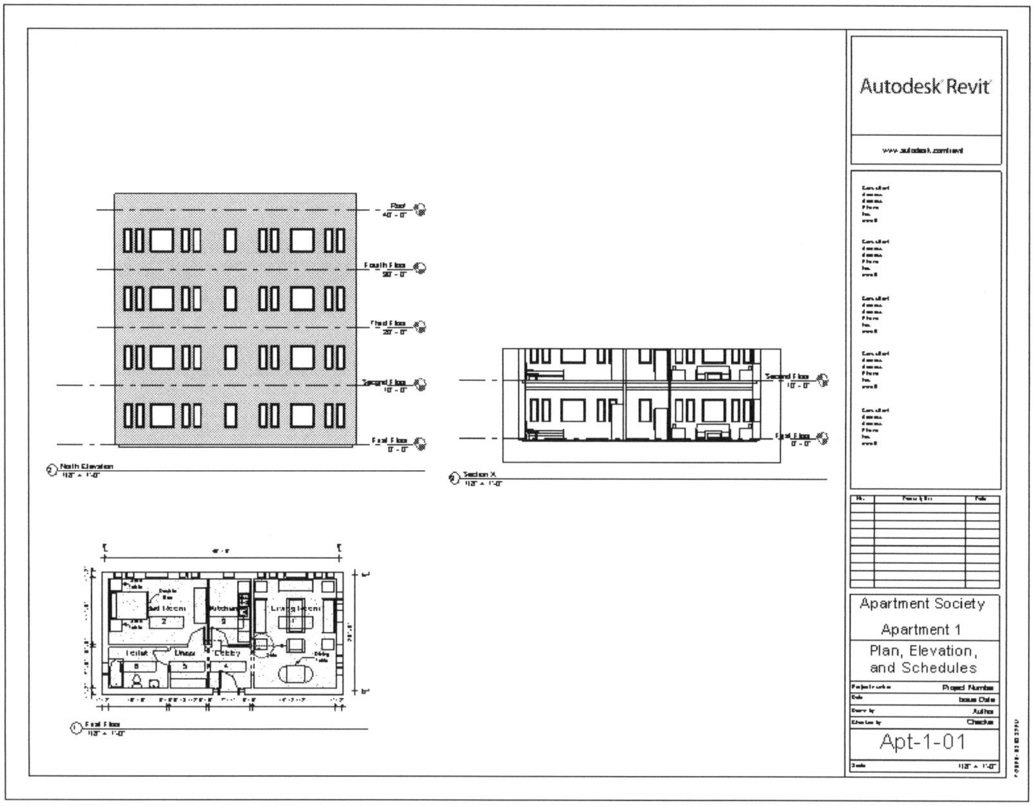

Figure 13-18 Project views added to the drawing sheet view

Modifying the Visibility of the Added Project Views

Notice that the project views in the drawing sheet given for this tutorial do not show the grids and section lines but show the complete section view. You can access each project view directly from the drawing sheet using the **Activate View** tool. Then, you can modify their visibility settings.

1. Select the **North Elevation** view from the drawing or highlight the view and then right-click to display the shortcut menu. Choose the **Activate View** option from the shortcut menu. On doing so, all the views except the selected one become grey. You can now use the editing tools to modify the building model and the visibility settings of the activated view.

2. Choose the **Visibility/ Graphics** tool from the **Graphics** panel in the **View** tab; the **Visibility/Graphic Overrides for Elevation: North Elevation** dialog box is displayed.

3. In the **Annotation Categories** tab, clear the check boxes for **Grids** and **Sections** in the **Visibility** column. Choose **OK** to close the dialog box and return to the drawing window.

4. Right-click in the drawing window, and then choose the **Deactivate View** option from the shortcut menu.

5. Repeat step 1 for activating the **Section X** view from the sheet. Next, repeat steps 2 and 3 to turn off the display of grids and sections in the activated view.

6. In the **Properties** palette for the **Section X** view, ensure that the check box corresponding to the **Crop Region Visible** parameter is selected. Next, in the sheet, select the boundary of the section view to display the view range controls.

7. Increase the size of the view range and display the entire section view by using the drag control located on the top as shown in the drawing sheet for this tutorial. You may use the **Zoom** tool to zoom in the section view. Next, right-click and choose **Deactivate View** from the shortcut menu.

Adding Schedules to the Drawing Sheet

Next, you need to add schedules to the drawing sheet in the same manner as you did in the case of the project view. Autodesk Revit, however, gives you the flexibility to divide larger schedules into smaller panels to accommodate them in the drawing sheet.

1. Drag **Room Schedule** from the **Schedules/Quantities** head in the **Project Browser**.

2. Move the cursor to the top left corner of the titleblock. The schedule is displayed and moves along with the cursor. Click to drop and add the room schedule at the location shown in the drawing sheet.

3. Similarly, drag and drop **Door Schedule** from the **Project Browser** to the drawing sheet. The door schedule is long and therefore you can use the **Split Schedule Table** tool to split it into two parts.

4. Click on the **Split Schedule Table** control available in the middle of the right edge of the door schedule, as shown in Figure 13-20

. The schedule is divided into two parts and the second part is placed on the right of the first part, refer to Figure 13-18.

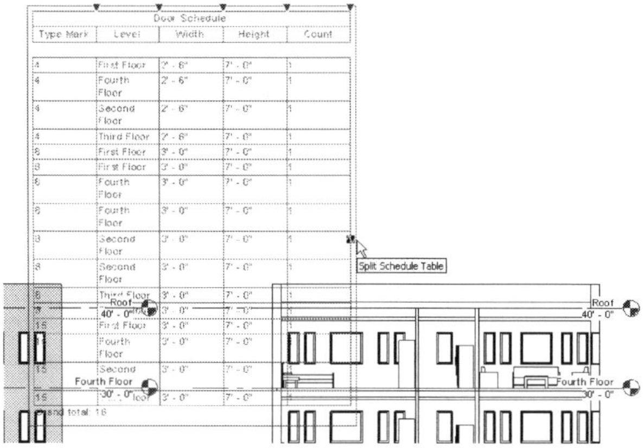

*Figure 13-20 Using the **Split Schedule Table** tool to split the table into two parts*

Saving the Project

In this section, you will save the project file using the **Save As** tool.

1. Choose **Save As > Project** from the **File** menu; the **Save As** dialog box is displayed. Enter **c13_Apartment1_tut1 (M_c13_Apartment1_tut1** for Metric) in the **File name** edit box and then choose **Save**.
2. Next, choose **Close** from the **File** menu to close the project file.

EXERCISES

Exercise 1 Club

Create a drawing sheet for the *Club* project created in Exercise 3 of Chapter 12. Add the first floor and roof plan, north elevation, sections, and room schedule to the sheet, as shown in Figure 13-21. Use the following project and sheet information parameters:

(Expected time: 30 min)

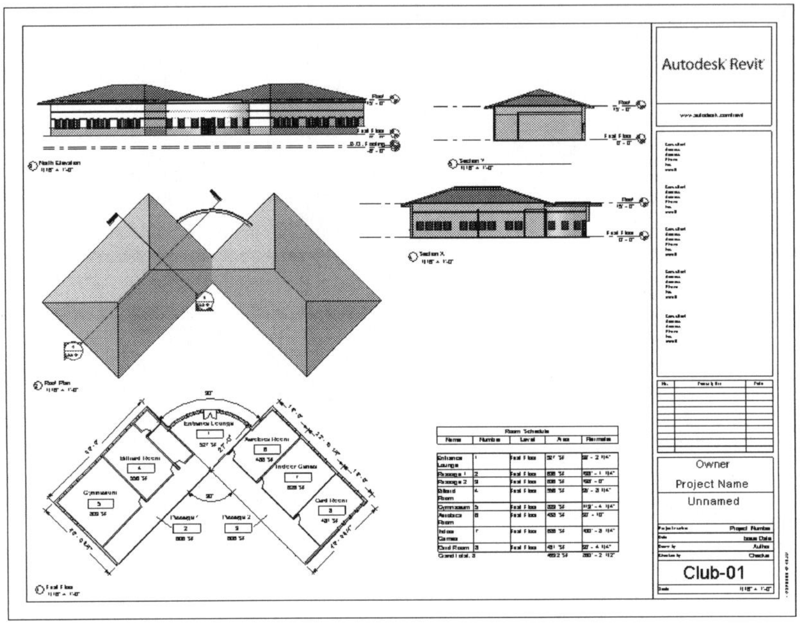

Figure 13-21 *Sheet view for creating a drawing sheet for the Club project*

1. Drawing sheet parameters:
   Title block to be used:
   For Imperial **C 17 x 22 Horizontal** from **English-Imperial > Titleblocks** folder.
   For Metric **A2 metric** from **English > Titleblocks** folder.

2. Project information:
   Owner name: **Club House**
   Sheet title: **Club Building**
   Sheet name: **Project Drawings**
   Sheet number: **Club-01**

3. Name of the file to be saved-
   For Imperial **c13_Club_ex1**
   For Metric **M_c13_Club_ex1**

Note

The Club project has been created using the Commercial-Default.rte template and therefore it already contains various sheet views. Some of the views have already been placed in them. You need to create a copy of the views and then add them to the new sheet that you create. You may also need to modify the view scale.

Chapter 14

Creating 3D Views

Learning Objectives

After completing this chapter, you will be able to:

- *Generate orthographic view of a building model*
- *Use the Navigation tools to view a building model*
- *Generate perspective view using the Camera tool*
- *Modify the properties of perspective view*
- *Lock and unlock 3D view*
- *Create 3d Section view*

INTRODUCTION

A three-dimensional view is a powerful medium of developing and presenting architectural and interior design concepts. Viewing a building model in all the three dimensions helps the architects and interior designers to visualize various building forms and shapes. It also conveys their vision of project to the clients, who, in most cases, are unable to understand the architectural drawings. The exterior 3D views of a building not only depict its overall scale, shape, and volume but can also represent the proposed exterior finishes as envisaged by the architects. The interior 3D views can be used to represent various interior elements such as interior partitions, furniture, plants, and interior finishes.

The use of computer graphics in the preparation of three-dimensional exterior and interior views gives numerous advantages to designers. The inevitable alterations in a building model during the design development can easily be studied with respect to their implications on the three-dimensional aspect of design. Software also enables the designers to develop design alternatives with different exterior and interior materials.

As described in the previous chapters, a building model of a project created using Autodesk Revit is an amalgamation of various three-dimensional elements such as walls, doors, windows, floor, roof, stairs, furniture, and so on. Each of these elements is a three-dimensional parametric entity. Autodesk Revit's powerful parametric technology enables the designers to create and edit these elements, and study them in all the three dimensions simultaneously. Autodesk Revit provides various tools to view these elements in 3D.

This chapter explains the tools used for generating 3D views of a structure. It describes the procedure for generating new orthographic and perspective views for a building model. The ease of generating 3D views in Autodesk Revit gives a major advantage to all the users. They can also easily modify the view properties to achieve the desired effect in the exterior or interior 3D views.

THREE-DIMENSIONAL (3D) VIEWS

In Autodesk Revit, you can display a variety of three-dimensional (3D) views. There are two basic types of 3D views that can be created and viewed.

Orthographic View: An orthographic view is a 3D view of a building model in which all the elements are displayed in their actual sizes, irrespective of their distance from the source. This view is displayed automatically as a default 3D view. Once this view has been displayed, you can then use various tools to modify its properties.

Perspective View: A perspective view is displayed by placing a camera and specifying its eye elevation and target position.

BIM
CONCEPTS

The three dimensional views created for the BIM project is based on various disciplines of the stakeholders involved during the worksharing process. For a project, you can create multiple duplicate views and change their disciplines and view templates based on the requirement of the project. Also, it is recommended to use specified view templates and object styles based on the EIR and BIM Execution Plan of the project.

Generating Orthographic View

Ribbon: View > Create > 3D View drop-down > Default 3D View

In Autodesk Revit, the default 3D view is an orthographic view. To view a building model in orthographic projection, invoke the **Default 3D View** tool from the **Create** panel; the current view will change into the default 3D view.

Note

*In Revit, there is no default {3D} view in the **Project Browser**. However, when you invoke the **Default 3D View** tool from the **Create** panel, the default {3D} view is added to the **Project Browser**.*

When you create an orthographic 3D view for the first time in a project, a camera is automatically placed at the default position (southeast corner) and the corresponding view is displayed in the drawing window. The default view is named as **{3D}** and is added under the **3D Views** head of the **Project Browser**. Figure 14-1 shows an example of the 3D view of a building model displayed using the **Default 3D View** tool.

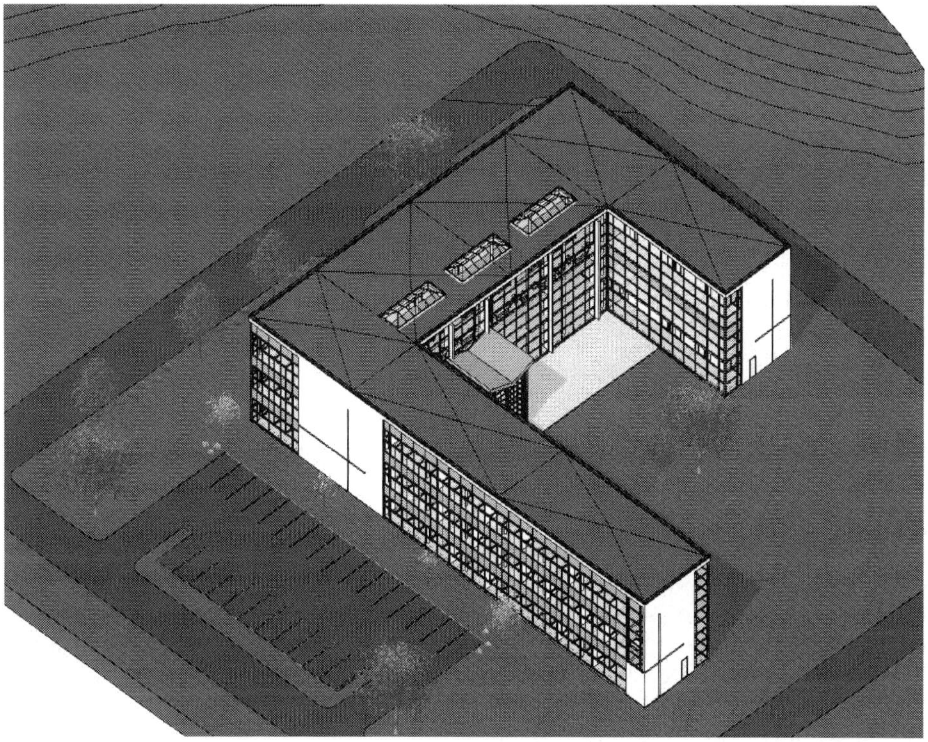

Figure 14-1 *3D view of a building model displayed using the **Default 3D View** tool*

Tip

You can rename the default 3D view in the Project Browser. After renaming it, when the ***Default 3D View*** *tool is invoked from the* **View** *tab, Autodesk Revit will create another default view under the* ***{3D}*** *head in the Project Browser.*

Dynamically Viewing Models Using Navigation Tools

The 3D environment of Revit has two important tools for navigation: SteeringWheels and ViewCube. You can use these tools to navigate through 3D views easily. These tools have already been discussed in brief in Chapter 2.

The SteeringWheels, as shown in Figure 14-2, is a tracking menu that enables you to navigate easily within the exterior and interior faces of a model. The ViewCube helps you simplify the switching of views from one point to another, as shown in Figure 14-3.

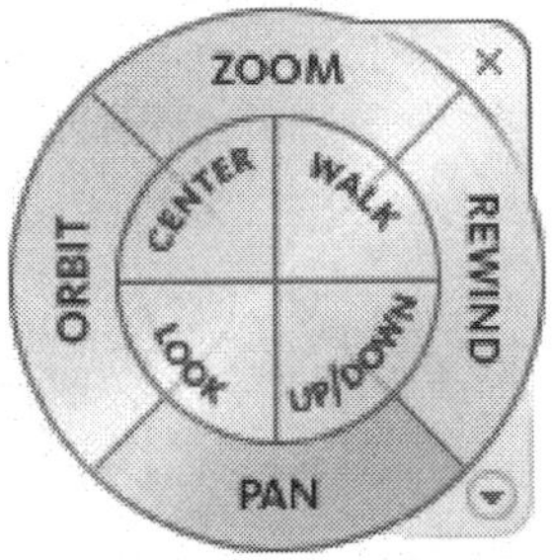

Figure 14-2 *The tools in the SteeringWheels*

Figure 14-3 *The tools in the ViewCube*

Note

If the SteeringWheels is not displayed by default, press F8 or invoke it from the ***Navigation Bar****. You can also display or hide the SteeringWheels navigation by pressing Shift+W keys.*

Using the Orient Tool

Autodesk Revit enables you to view the building model from certain predefined orientations such as north, south, east, west, northwest, southeast, and so on. These orientations are created by a camera placed in the respective directions and at a certain elevation. You can easily view the building model from these orientations using the **Orient** tool. To invoke this tool, right-click on the ViewCube; a shortcut menu will be displayed. Next, choose the **Orient to View, Orient to a Direction**, or **Orient to a Plane** option from the shortcut menu. You can choose the **Orient to View** option from the shortcut menu to transform the current view of the project to any of the views of floor plans or elevations. You can use the **Orient to a Direction** option from the shortcut menu to change the direction of the current project view to any of the orthographic or isometric view directions such as **North**, **South**, and **Northeast Isometric**. Similarly, you can use the **Orient to a Plane** option from the shortcut menu to change the orientation of the current project view to any of the specified workplanes in the project.

Note

By default, a building plan is oriented in the north direction toward the top of the screen. You can, however, choose the ***Location*** *tool from the* ***Project Location*** *panel in the* ***Manage*** *tab to modify the orientation.*

Generating Perspective Views

Perspective views provide a realistic view of the exterior or interior space of a building model. Whenever you see an exterior or an interior view of a building with naked eye, it appears as the perspective view with the wall edges of the building converging at imaginary points. Using Autodesk Revit, you can easily create perspective views of a building model by defining the camera location and the target point. In most building projects, you will find it easier to use the floor plan view to define these points (camera location and camera target). You can then modify the view parameters to achieve the desired result.

Fly Mode for View

In Autodesk Revit 2025, you can now navigate into your project using the **Fly** tool. The **Fly** tool enables you to walk through the perspective view of the model and look around. The **Fly** tool can be used in both interior or exterior perspective view of the model. The perspective view can be created by using the **Camera** tool or by switching the **Projection Mode** to **Perspective** in the **Properties** palette of the **Default 3D View.** After creating the perspective view, click on the **Fly** tool icon in the **Navigation** bar; the cursor will change into the look around icon. To navigate the direction, click to hold and drag the right mouse button to look around. The 4 arrow keys are used to move in forward, backward, left, and right directions. Alternatively, use W, S, A, and D keys to move in the forward, backward, left, and right directions, respectively. Use E and Q keys to move up and move down or else use the Page Up and Page Down buttons. The Shift and the mouse wheel is used to set the speed factor to fly through the model. The maximum speed factor to navigate through the model is 50. The current speed factor is shown in the Status Bar.

Locking and Unlocking 3D Views

In Revit, you can lock, unlock, and save the orientation of a 3D view. By locking a 3D view, you can add annotations and element tags to it. To view the options for locking, unlocking, and saving the 3D views, choose the **Unlocked 3D View** option from the **View Control Bar**. Note that the **Unlocked 3D View** option will be displayed in the 3D and camera views only. On choosing the **Unlocked 3D View** option, a flyout will be displayed, as shown in Figure 14-4. In the flyout, the following options are displayed: **Save Orientation and Lock View**, **Restore Orientation and Lock View**, and **Unlock View**. Note that the **Restore Orientation and Lock View** and **Unlock View** options are inactive by default.

***Figure 14-4** Flyout displayed on choosing the **Unlocked 3D View** option*

On choosing the **Save Orientation and Lock View** option, the **Rename Default 3D View To Lock** dialog box will be displayed. In this dialog box, enter a name in the **Name** edit box and choose the **OK** button; the current view will be renamed, saved, and locked. After you save and lock the orientation of the 3D view, you can also tag the elements. After tagging the elements you can unlock the view. To do so, choose the **Unlock View** option from the flyout; the tags added will disappear and you can re-orient the 3D view. After re-orienting the view, you can restore the previous locked view that was saved. To do so, choose the **Restore Orientation and Lock View** option from the flyout.

Note

*In the Camera view, on choosing the **Save Orientation and Lock View** option from the **View Control Bar**, the **Rename Default 3D View To Lock** dialog box will not be displayed.*

Using the Section Box

A section box is used to limit the model geometry in a file that you want to export to other program such as Autodesk 3ds Max or Autodesk 3ds Max Design. It helps you clip unwanted content in a file. By adjusting the extent of a section box using drag controls, you can clip the unwanted portion of a model in the 3D view, thereby reducing the file size. This helps in easy and fast export of heavy files. Note that a section box can be enabled in a 3D view only.

To enable a section box, open the required 3D view and in the **Properties** palette, select the check box in the **Value** column corresponding to the **Section Box** parameter. Now, choose the **Apply** button in the **Properties** palette; a section box will be displayed in the 3D view, as shown in Figure 14-5. Select the section box to display the blue color arrow controls. Drag the blue arrow controls and modify the extent of the section box to clip the unwanted portion of the model. Note that the section box option is available in the **Properties** palette for camera view also.

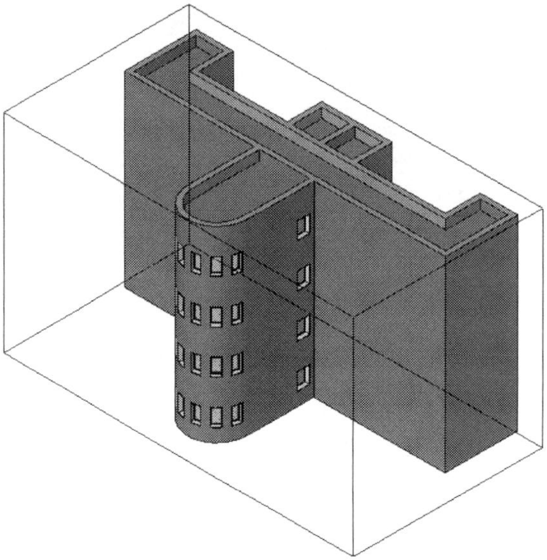

Figure 14-5 *Section box generated for the 3D view*

Note

*To hide the section box, clear the check box for the **Section Box** parameter in the **Properties** palette.*

Autodesk Revit allows you to modify the extents of a section box from the views other than the 3D view such as a plan view or elevation view. To do so, select the section box in a 3D view and then open the required plan or the elevation view. The blue arrow controls will be displayed in the respective view as well. Modify the extent of the section box and then open the 3D view again to see the change. If required, you can modify the extent of the scope box again in the 3D view. To hide the scope box, select the scope box and right-click; a shortcut menu will be displayed. Choose **Hide in view > Elements** from the shortcut menu; the scope box will become invisible in the view. To save the view, duplicate the view and rename it.

In Revit, you can also use the **Selection Box** tool to create a 3d section for the selected elements in the view. To create a 3d section, select the elements that you want to keep it in the section box and then choose the **Selection Box** tool from the **View** panel of the displayed contextual tab. On doing so, the elements selected will only be displayed inside the section box.

Tip

*To modify the camera and target locations, open the view in which the camera was placed. Right-click on the **3D view** name in the **Project Browser** that was created by using the camera. Choose the **Show Camera** option from the shortcut menu to display the location of the camera and the target points in the drawing window. The camera and target points can now be dragged to a new location. Autodesk Revit immediately regenerates the 3D view for the new location.*

Note

The extents of a section box are not affected by the crop regions.

TUTORIALS

Tutorial 1 Apartment 1

In this tutorial, you will create two interior perspective views for the first floor of the *Apartment 1* project created in Chapter 13 of this textbook, refer to Figures 14-6 and 14-7. The first figure shows the living room and the second figure shows the view of the lobby area from the living room. The exact view angle and height are not important for this tutorial.

(Expected time: 20 min)

The following steps are required to complete this tutorial:

a. Open the project file and the ground floor plan view.
b. Generate the interior perspective view of the living room using the camera tool, refer to Figure 14-8 and Figure 14-9.
c. Generate the interior perspective view of the lobby room, refer to Figures 14-10 and 14-11.
d. Save the project using the **Save As** tool.
   For Imperial *c14_Apartment1_tut1.rvt*
   For Metric *M_c14_Apartment1_tut1.rvt*
e. Close the project.

Figure 14-6 *3D view of the living room for the Apartment 1 project*

Figure 14-7 *3D view of the lobby area for the Apartment 1 project*

Opening the Project File and Invoking the Camera Tool

1. Choose **Open > Project** from the **File** menu and open the *c13_Apartment 1_tut1.rvt* file created in Tutorial 1 of Chapter 13. You can also download this file from *https://www.Cadsoft.com*. The path of the file is as follows: *Textbooks > Civil/GIS > Revit Architecture > Exploring Autodesk Revit 2025 for Architecture*.

2. Double-click on **First Floor** in the **Floor Plans** head of the **Project Browser** to open the corresponding floor plan in the drawing window.

3. Invoke the **Camera** tool from **View > Create > 3D View** drop-down; the **Options Bar** is displayed.

Generating the Interior Perspective View of the Living Room

In this section of the tutorial, you will specify the camera and the target points for the camera. You will use the default value **5'6"** (eye level) for both these points.

1. Select the **Perspective** check box in the **Options Bar**, if it is not selected by default.

2. Move the cursor to the lower right corner of the living room and click to specify the camera point.

3. Now, move the cursor toward the diagonally opposite ends of the living room and click when the field of view resembles the graphic shown in Figure 14-8; the 3D interior view is displayed in the drawing window, shown in the Figure 14-9.

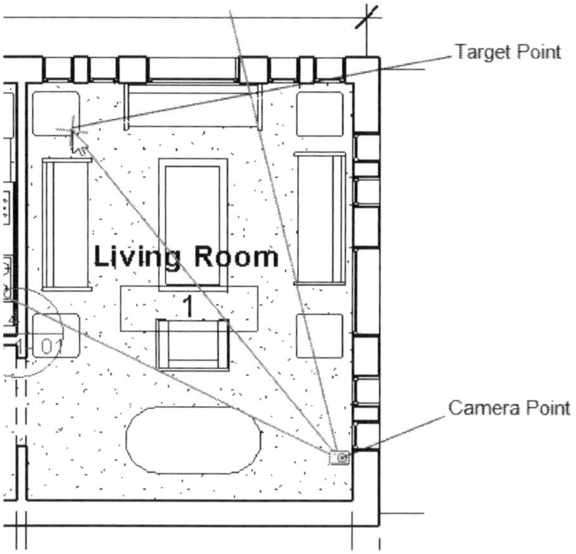

Figure 14-8 *Specifying the camera point and the target point of the interior view*

Figure 14-9 The 3D interior view of the living room

4. The selected 3D view shows the drag controls (blue dots at the midpoint of the four sides of the rectangle). Right-click in the drawing window and then choose **Zoom Out (2x)** from the shortcut menu.

5. Use the drag controls to increase the view extents based on the 3D view given for the living room. Next, press ESC to exit the **Modify | Cameras** tab.

6. In the **Project Browser**, right-click on **3D View 1** under **3D Views**; a shortcut menu is displayed. From the shortcut menu, choose the **Rename** option; an edit box is displayed. In the edit box, enter **Living Area**; the **3D View 1** view is renamed as **Living Area**.

Generating the Interior Perspective View of the Lobby

You can use similar steps to generate the 3D interior view of the lobby area.

1. Right-click on the **First Floor** view in the **Floor Plans** head in the **Project Browser** and choose **Open** from the shortcut menu to open the view.

2. Invoke the **Camera** tool from **View > Create > 3D View** drop-down. Ensure that the **Perspective** check box is selected in the **Options Bar** (default setting).

3. Move the cursor near the center of the opening between the lobby area and the living room, refer to Figure 14-10, and click to specify the camera point.

4. Move the cursor horizontally toward the left and click when the field of view resembles the graphic shown in Figure 14-10; the 3D interior view of the lobby area is displayed in the drawing window.

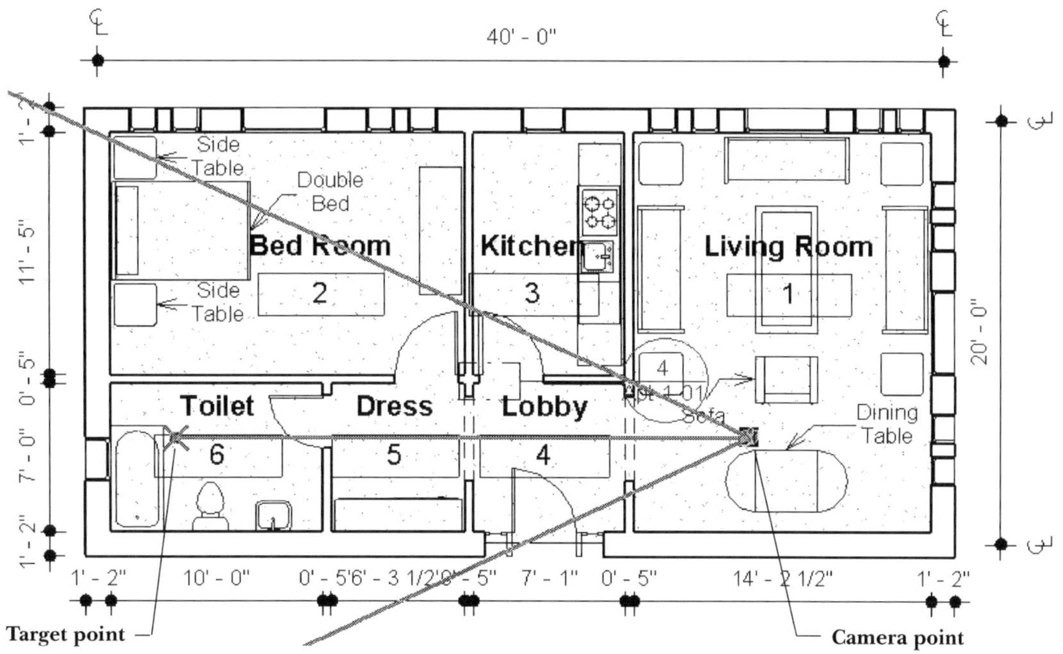

Figure 14-10 *Specifying the camera point and the target point for generating the lobby interior view*

5. Right-click in the drawing window and then choose **Zoom Out (2x)** from the shortcut menu displayed. Use the drag controls to increase the view extents.

6. Now, choose **Visual Style > Shaded** from the **View Control Bar**; the current view is shaded.

7. Use the drag controls to increase the view extents based on the 3D view given for the lobby area, refer to Figure 14-11.

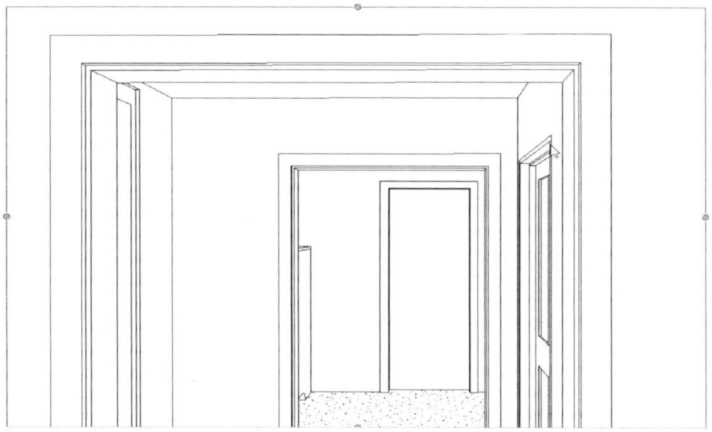

Figure 14-11 *The 3D interior view of the lobby area*

8. Right-click on **3D View 1** in the **Project Browser** under **3D Views** and rename it as **Lobby Area** by using the **Rename** option from the shortcut menu.

Saving the Project

In this section, you will save the project file using the **Save As** tool.

1. Now, choose **Save As > Project** from the **File** menu; the **Save As** dialog box is displayed. Enter **c14_Apartment1_tut1** in the **File name** edit box and choose **Save**; the project file is saved.

2. Choose **Close** from the **File** menu to close the project file.

Tutorial 2 Club

Create an exterior 3D perspective view and an orthographic view for the *Club* project created in Exercise 1 of Chapter 13, refer to Figures 14-12 and 14-13. Note that the exterior 3D perspective view is from the main entrance. The exact view and height are not important for this tutorial.

(Expected time: 20 min)

The following steps are required to complete this tutorial:

a. Open the project file and the First floor plan view.
b. Invoke the **Camera** tool to generate the exterior 3D view.
c. Generate the exterior perspective view of the Main Entrance, refer to Figure 14-14.
d. Generate the orthographic view of the building, refer to Figure 14-17.
e. Use the SteeringWheels and the **ORBIT** tool to display the orthographic view.

f. Save the project using the **Save As** tool.
   For Imperial *c14_Club_tut2.rvt*
   For Metric *M_c14_Club_tut2.rvt*
g. Close the project.

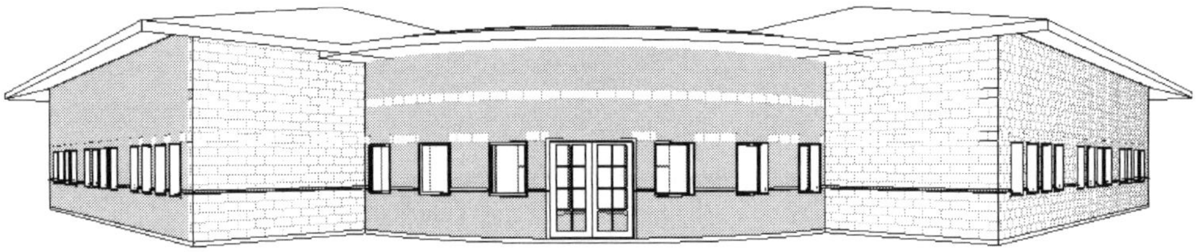

Figure 14-12 *3D exterior view of the main entrance of the club building*

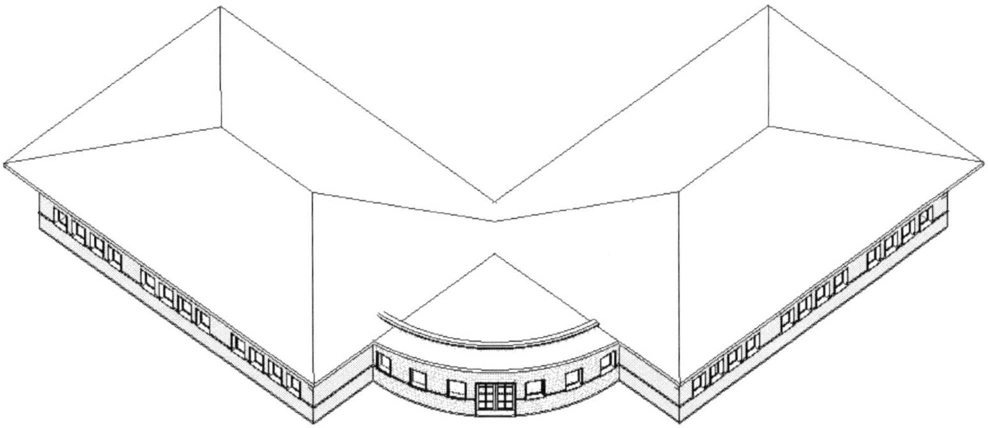

Figure 14-13 Orthographic view of the Club building

Opening the Project File and Invoking the Camera Tool

1. Choose **Open > Project** from the **File** menu and open the *c13_Club_ex1.rvt* file created in Exercise 1 of Chapter 13. You can also download this file from *https://www.Cadsoft.com*. The path of the file is as follows: *Textbooks > Civil/GIS > Revit Architecture > Exploring Autodesk Revit 2025 for Architecture*.

2. Double-click on **First Floor** in the **Floor Plans** head of the **Project Browser** to open the corresponding floor plan in the drawing window.

3. Invoke the **Camera** tool from **View > Create > 3D View** drop-down; the options for inserting the camera are displayed in the **Options Bar**.

Generating the Exterior Perspective View of the Main Entrance

You need to specify the camera and target points. Also, you will use the default value **5'6"** (**1650** mm) (eye level) for both these points.

1. Select the **Perspective** check box in the **Options Bar** if not selected by default.

2. Move the cursor vertically above the main entrance door and click to specify the camera point, as shown in Figure 14-14.

3. Move the cursor vertically downward and click when the field of view resembles the graphic shown in Figure 14-14; the drawing window shows the 3D exterior view of the main entrance, as shown in Figure 14-15.

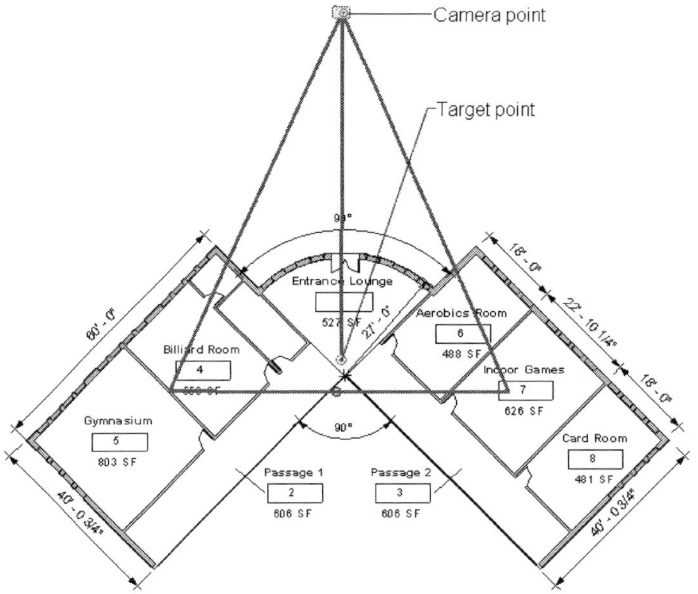

Figure 14-14 *Specifying the camera and the target points for the exterior view*

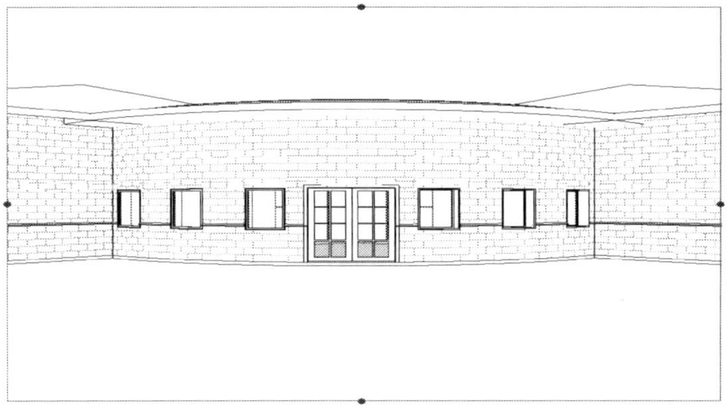

Figure 14-15 *3D exterior view of the main entrance*

4. Right-click in the drawing window and choose **Zoom Out (2x)** from the shortcut menu.

5. Ensure that the crop region of the 3D view is selected in the drawing and it displays the drag controls (blue dots at the midpoint of the four sides of the rectangle).

6. Use the drag controls to increase the view extents based on the 3D view given for the entrance view. Press ESC to exit the **Camera** tool.

7. Rename **3D View 1** to **Entrance View** by using the **Rename** option from the shortcut menu displayed on right-clicking in the view name in the **Project Browser**.

Generating the Orthographic View of the Club Building

You need to display the orthographic view of the club building using the SteeringWheels.

1. Invoke the **Default 3D View** tool from **View > Create > 3D View** drop-down.

2. Move the cursor toward the ViewCube and right-click on it; a shortcut menu is displayed.

3. Choose **Orient to a Direction > Northeast Isometric** from the shortcut menu; the northeast isometric view is displayed, as shown in Figure 14-16.

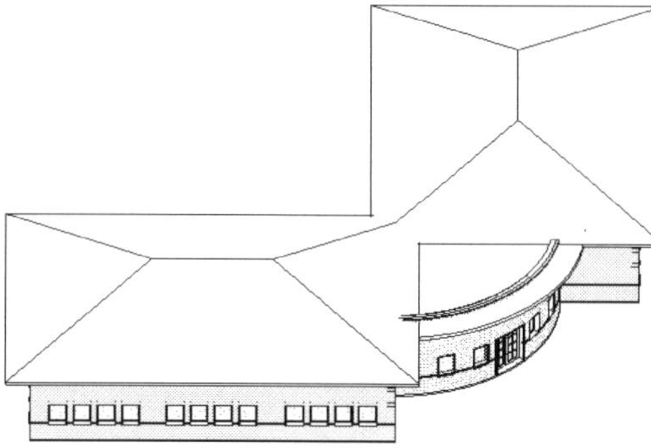

Figure 14-16 *Northeast isometric view of the Club building*

4. Press the F8 key to display SteeringWheels.

5. Next, press and hold the **ORBIT** tool in SteeringWheels and drag the cursor to display a view similar to the one specified for the project. Release the mouse button when the appropriate view is displayed, refer to Figure 14-17. Press ESC to exit.

6. In the **Project Browser**, right-click on **{3D}** under **3D Views** and rename **{3D}** to **Orthographic View** by using the **Rename** option from the shortcut menu displayed.

7. Double-click on **First Floor** in the **Floor Plans** head of the **Project Browser** to open corresponding floor plan in the drawing window.

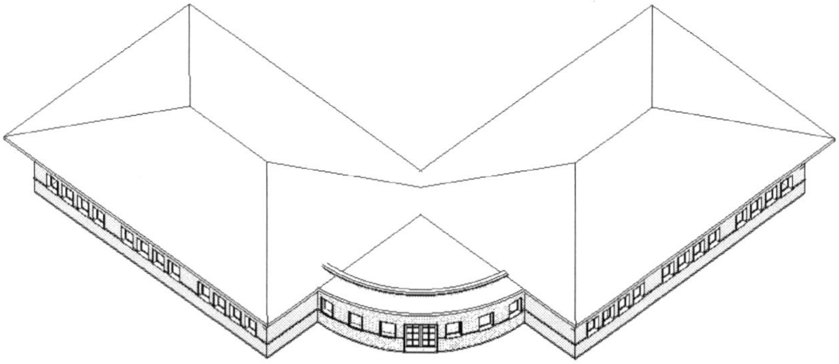

Figure 14-17 *Orthographic view of the Club project*

Saving the Project

In this section, you will save the project file using the **Save As** tool.

1. Choose **Save As > Project** from the **File** menu; the **Save As** dialog box is displayed.
2. Enter **c14_Club_tut2** in the **File name** edit box and choose **Save** to save the project file.
3. Choose **Close** from the **File** menu to close the project file.

EXERCISE

Exercise 1 — Apartment 1

Download *c14-revit-2025-exr01.rvt* (for Metric *M_ c14-revit-2025-exr01.rvt*) from the Cadsoft website and display the 3D exterior views. Also, create two exterior perspective views, one from the front and the other from the side of the building model, as shown in Figures 14-18 and 14-19, respectively. Use an appropriate camera location, target point, eye elevation, and target elevation to achieve the given perspective view. **(Expected time: 30 min)**

Save the file as *c14_Apartment-1_ex1.rvt* (for Metric *M_c14_Apartment-1_ex1.rvt*)

Figure 14-18 *3D exterior front view for Exercise 1*

Figure 14-19 *3D exterior side view for Exercise 1*

Exercise 2 Office Building 1

Download the file *c14-revit-2025-exr02.rvt* (for Metric *M_ c14-revit-2025-exr02.rvt*) from the Cadsoft website and display a 3D interior view as shown in Figure 14-20. Use an appropriate camera location, target point, eye elevation, and target elevation to achieve the given perspective view. **(Expected time: 15 min)**

Save the file as *c14_Office-1_ex2.rvt* (for Metric *M_c14_Office-1_ex2.rvt*)

Figure 14-20 *3D exterior front view for Exercise 2*

Exercise 3 Office Building 2

Open the file *c10_Office-2_tut1.rvt* (for Metric *M_ c10_Office-2_tut1.rvt*) created in Tutorial 1 of Chapter 10 and display an exterior 3D view, as shown in Figure 14-21. Use an appropriate camera location and target point. The eye elevation and target elevation for the view are 15'0" and 35'0", respectively. **(Expected time: 15 min)**

Save the file as *c14_Office-2_ex3.rvt* (for Metric *M_c14_Office-2_ex3.rvt*)

Figure 14-21 *3D exterior view for Exercise 3*

Chapter 15

Rendering Views and Creating Walkthroughs

Learning Objectives

After completing this chapter, you will be able to:

- *Create a rendered scene*
- *Set natural and artificial lighting*
- *Use different materials and textures*
- *Create and render a walkthrough*
- *Add people and vehicles to a building model*
- *Use the Decal tool*
- *Know about Autodesk Rendering*
- *Understanding rendering in Enscape*

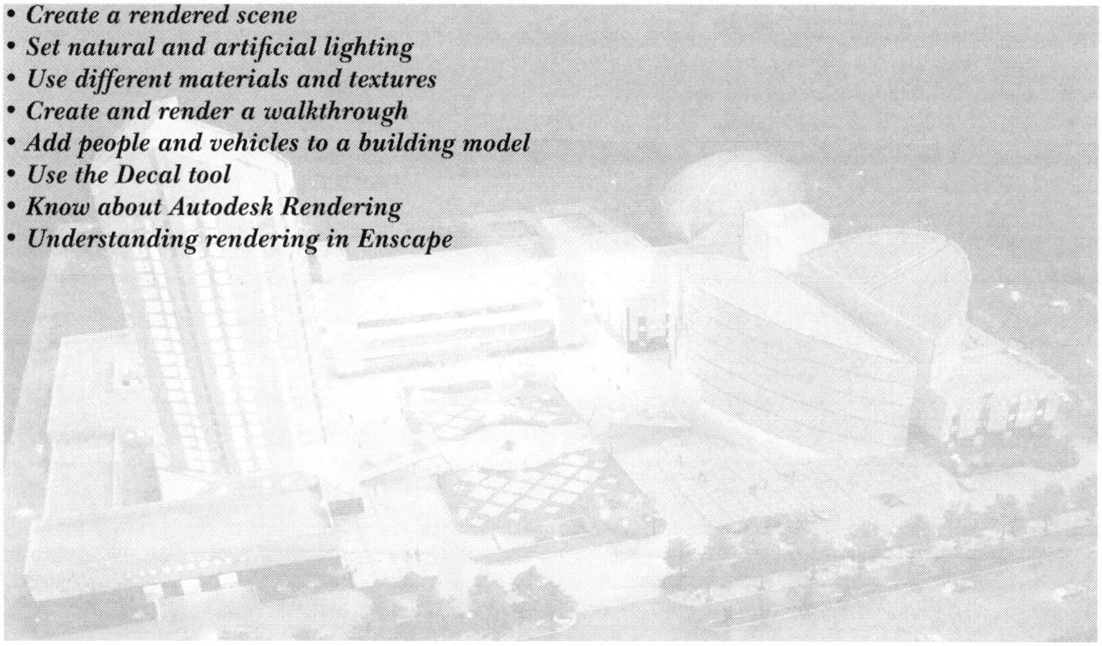

RENDERING IN REVIT

Rendering is a process of generating digital image from a three-dimensional model containing geometry, texture, light, and shading information through a computer program.

The rendering process helps you give a clear picture of the final outcome of an architectural project that you are working on. It bridges the gap between the final outcome and the two-dimensional working drawing, thereby giving you and your client the freedom to select the final finish with reduced cycle time for the approval and finalization of the design. Also, rendering with its rich photorealistic images helps you generate the actual bill of material for a project.

As described in the previous chapter, creating three-dimensional views of a building model enables you to visualize the interior and exterior of a building. However, to give a realistic effect to the view, it is necessary to apply materials to the building elements and render the view. The rendered exterior views of a building not only depict its overall scale, shape, and volume, but also represent the proposed exterior finish as envisaged by the architect. Similarly, a rendered interior three-dimensional view can be used to represent various interior elements such as interior partitions, furniture, plants, and interior finishes.

Rendering Workflow

In this chapter, you will learn about various methods and techniques used in a rendering workflow. A rendering workflow is a process chart that shows the flow of the rendering process to achieve the end result in the form of high quality rendered image, refer to Figure 15-1.

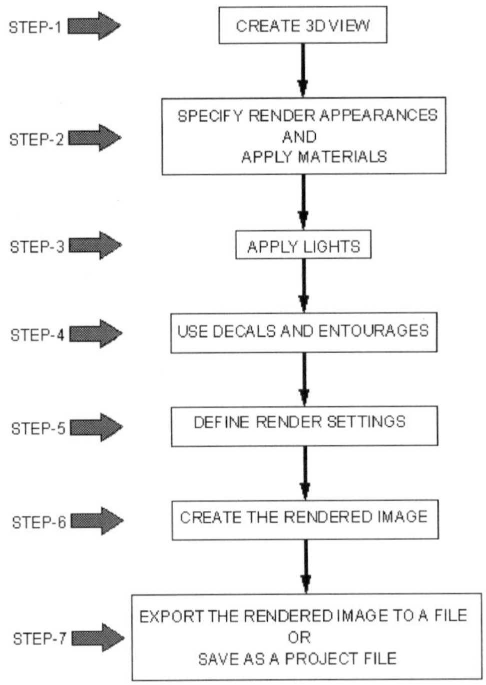

Figure 15-1 *The rendering workflow chart*

Note
In the Rendering workflow, you can alter the order of steps from step1 to step 4.

Introduction to Materials

In Revit, managing material, which is an important part of rendering process, includes creating and adding of various materials. Material specifies the actual look of the model elements as they appear in the rendered images. Materials convey the structural and descriptive information of model elements. In Revit, you can apply materials to individual elements in a building model or apply materials to elements at the time of defining their families.

Before you learn how to assign materials to the elements, you should be familiar with some of the concepts that are discussed next.

Applying Lights

Light is an important factor of the rendering process. It affects the overall appearance of elements and materials in the rendered view. It not only enables you to view the building model, but also helps you express its mood along with its ambience. For example, for the exterior of a shopping mall, a vibrant mood can be created by using bright afternoon lights with sharp shadows. On the contrary, for the exterior of a residential building, soft evening lights can be used to give a relaxing ambience. Similarly, an interior view of a showroom can be brightly lit and a bar interior view can be given an accent lighting. Autodesk Revit enables you to add and control the properties of a variety of artificial lighting fixtures based on the project requirement.

Using Decals and Entourages

The **Decal** tool is used to place an image on the face of a flat or a cylindrical element in the building model. This tool can also be used to represent signals, posters, billboards, and paintings in a rendered view. A decal is an external graphic file that is linked to the project file. The file formats that can be used as decal are *jpeg*, *targa*, *tiff*, and *bmp*. Since the decal files are linked to the project file and are not saved with it, the size of the project file does not increase. It is, however, recommended to save decals along with a project file so that the link is maintained. Entourages are the collection of people, trees, cars, parking elements, and other site components that can be inserted in the model to give a realistic look to the rendered image.

Rendering Settings

After creating three-dimensional view for rendering,
applying lights and materials to elements, and placing entourages and decals, you need to do render settings before the final rendering. To render and create the presets for the rendering process, choose the **Show Rendering Dialog** button from **View Control Bar** or type RR; the **Rendering** dialog box will be displayed.

In the **Rendering** dialog box, you can define the view area for rendering, specify the quality of output, specify the light settings, set the background of the rendered image, adjust the rendered image exposure quality, and save the rendered image to a project view or export it as an external image file. Various settings are discussed next.

CREATING A WALKTHROUGH

A well rendered walkthrough can make a project presentation impressive. Using the **Walkthrough** tool, you can easily generate a walkthrough for a building model by rendering a series of 3D views defined by camera frames. The process of generating a walkthrough involves specifying a walkthrough path, editing it, and finally recording the walkthrough.

Creating the Walkthrough Path

Ribbon:	View > Create > 3D View drop-down > Walkthrough

The first step for creating a walkthrough is to define the walkthrough path that a camera follows through the building model. In most cases, this path is defined in the plan view. But, you can also use the section, elevation, and 3D views to define the path.

The walkthrough path is defined by specifying the points that form a spline in the project view. The specified points become key frames in the walkthrough. Autodesk Revit automatically generates additional intermediate frames between these key frames.

To create a walkthrough path, first open the view in which it is to be defined. Next, invoke the **Walkthrough** tool from **View > Create> 3D View** drop-down; the **Modify | Walkthrough** tab will be displayed and the options related to the path creation will be displayed in the **Options Bar**, as shown in Figure 15-2.

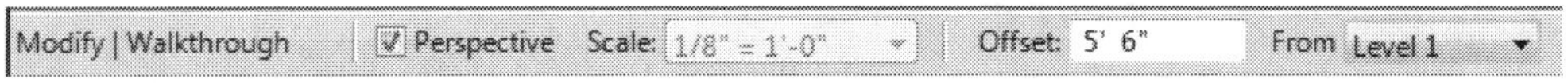

***Figure 15-2** The **Options Bar** displaying the options for creating a walkthrough*

Editing and Playing the Walkthrough

When you create a walkthrough path, Autodesk Revit assigns **Walkthrough 1** as its name by default and then it is added to the **Project Browser** under the **Walkthroughs** heading.

To view the created walkthrough, double-click on its name in the **Project Browser** or right-click over the name and then choose **Open** from the shortcut menu; the drawing window will display the 3D view of the last frame. Select the frame from the drawing window; the **Modify | Cameras** tab will be displayed and the crop boundary with the grips will also be displayed. You can use the grips to resize the crop region. Alternatively, if you choose the **Size Crop** button in the **Crop** panel of the **Modify |Cameras** tab, the **Crop Region Size** dialog box will be displayed. In this dialog box, enter the new values for the **Width** and **Height** parameters of the crop region. The **Field of view** radio button in this dialog box is selected to modify the aspect ratio of the crop region. The **Scale** radio button is selected to scale the crop region by locking its aspect ratio. Note that while scaling the crop region, you can modify its width and height without affecting the field of view of the scene. After specifying the desired values, choose **OK**; the **Crop Region Size** dialog box will be closed and the crop region will be resized according to the parameters specified.

Now, to display the walkthrough path, double-click on the level name in which the path was created. While the walkthrough path is displayed in the drawing area, choose the **Edit Walkthrough** button in the **Modify | Cameras** tab; the **Edit Walkthrough** tab will be displayed, as shown in Figure 15-3. The options in the **Edit Walkthrough** tab are used to play or edit a walkthrough. The key frames specified to create the path are highlighted with dots in the drawing.

Figure 15-3 *The options for editing and playing a walkthrough in the* ***Edit Walkthrough*** *tab*

Recording a Walkthrough

Autodesk Revit enables you to record the walkthrough as **AVI** files and to export these files to a disk. The **AVI** files can be used to export the walkthrough. To export these files, open the walkthrough and choose **Export > Images and Animations > Walkthrough** from the **File** menu; the **Length/Format** dialog box will be displayed providing you with various options for exporting the walkthrough. The **Output Length** area of this dialog box enables you to select the frames to be exported. The **All frames** radio button in this area can be selected to export all the frames in the walkthrough. Alternatively, you can select the **Frame range** radio button and specify the number of frames to be exported by entering suitable values in the **Start** and **End** edit boxes. The **Frames/second** edit box is used to specify the speed at which the frames will be recorded.

The **Format** area of the **Length/Format** dialog box provides you with the options to specify the type and quality for the recording of the walkthrough. You can select the view type to be used for generating and recording the walkthrough from the **Visual Style** drop-down list. To record the rendered view of frames, select the **Rendering** option from this drop-down list. You can also specify the quality for the recording by setting the **Dimensions** and **Zoom to** spinners. Next, choose the **OK** button to invoke the **Export Walkthrough** dialog box, as shown in Figure 15-4. This dialog box is used to save a walkthrough. Enter a suitable name in the **File name** edit box and select the **AVI Files** option from the **Files of type** drop-down list. Next, choose the **Save** button from the **Export Walkthrough** dialog box to start recording the walkthrough. Now, Autodesk Revit will display the **Video Compression** dialog box that enables you to select a video compressor from a list of compressors installed on your computer. When you choose the **OK** button, the dialog box will be closed and Autodesk Revit will start recording the walkthrough. The progress indicator and **Status Bar** at the bottom of the screen will display the progress of the processes being executed.

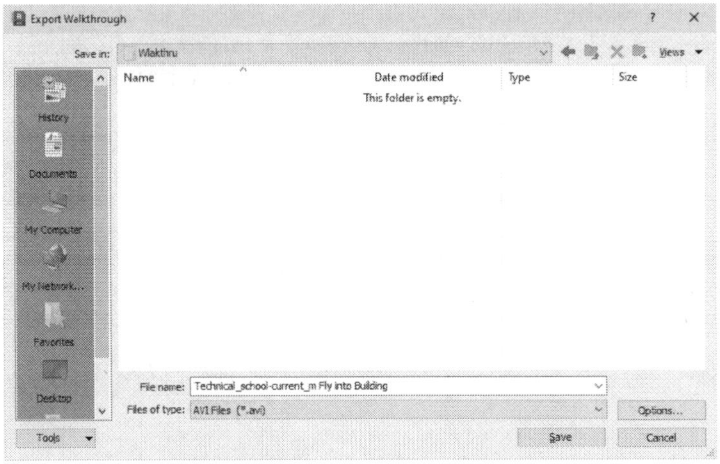

Figure 15-4 *The* ***Export Walkthrough*** *dialog box*

Note

The recorded file can be viewed using any of the media players. Note that this animation can now be viewed even without opening Autodesk Revit.

AUTODESK RENDERING

Autodesk Rendering is a cloud based rendering service that helps you to create photorealistic images and panoramic views for your project. It is a user interactive platform that provides a broad set of features including online cloud services and related products.

The main features of the **Autodesk Rendering** service are:

1. Using the **Autodesk Rendering** service, you can render your Revit projects from any computer.

2. Using the **Autodesk Rendering** service, you can access multiple versions of your renderings, render images as panoramas, change rendering quality, and apply background environments to rendered scenes.

3. Using this service, the design files can be reviewed even without the Revit software by accessing the **Autodesk** account online or even on mobile applications. The design files can be shared among users and the file updates and e-mail notifications reach the users whenever the file is edited.

4. The rendered views generated by the **Autodesk Rendering** application are grouped as collection in the render gallery. You can also create and view panoramic views. Panoramic views can be accessed from the render gallery.

5. Using the **Autodesk Rendering** service, you can adjust the exposure of the rendered image directly without having to re-render from the original desktop application.

6. The **Autodesk Rendering** service helps you to preview the rendered image of a single day or multi-day solar studies.

Rendering in Cloud

To render an image in the cloud, open a view in Revit interface. Next, choose the **Render in Cloud** tool from the **Presentation** panel of the **View** tab; the **Render in Cloud** dialog box will be displayed, as shown in Figure 15-5. Choose the **Continue** button; the **Render in Cloud** dialog box will be displayed again with the rendering options, as shown in Figure 15-6. In this dialog box, the 3D view to be rendered is to be chosen from the **3D View** swatch. You can choose the image type to be displayed, still or panorama, from the **Output Type** drop-down list. You can define the quality of rendered image as standard or final by choosing options from the **Render Quality** drop-down list. You can specify image size for rendering by selecting an option from the **Image Size** drop-down list. From the **Exposure** drop-down list, you can select an option to specify whether the exposure being adjusted for the rendered image will be in advanced or native mode. After setting the required parameters, you can choose the **Render** button to render the image in the cloud. The estimated time to render a view is displayed on the right corner of this dialog box.

Note

*If a project is not saved in Revit, and the **Render in Cloud** tool is used to render the image, then the **Render in Cloud** message box will be displayed, prompting you to save the project and proceed.*

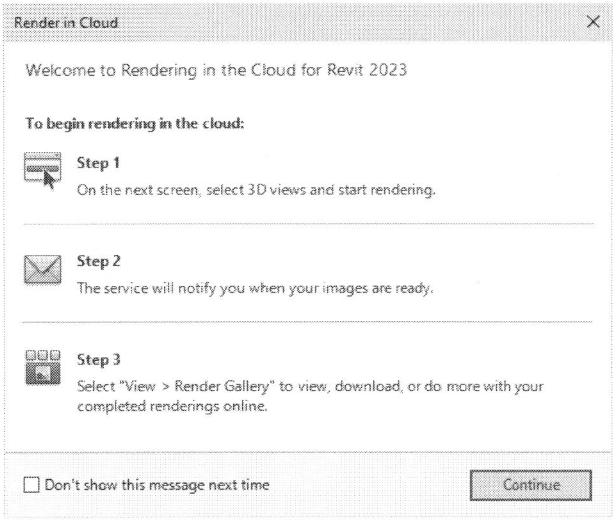

***Figure 15-5** The **Render in Cloud** dialog box*

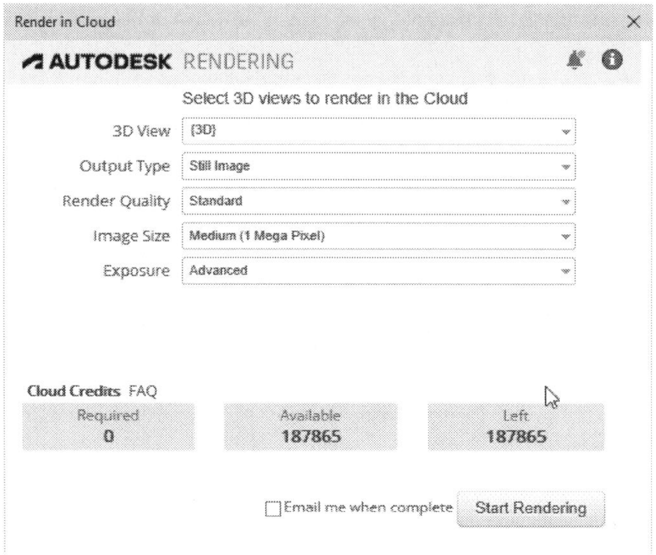

*Figure 15-6 The **Render in Cloud** dialog box with rendering options*

Tip
*The **Autodesk 360 / Energy Analysis** and **Autodesk 360 / Structural Analysis** cloud based services are available on subscription. The **Autodesk 360 / Energy Analysis** service helps in checking the energy efficiency of a building project whereas the **Autodesk 360 / Structural Analysis** service helps in performing structural analysis of the building elements.*

Accessing Render Gallery

The rendered images and the images being rendered are available online on **Render Gallery** page. The images are arranged in such a way that the newest image is displayed first. The grouping of images is done on the basis of file extension such as *.rvt* or *.dwg*. To access the Render gallery from Revit, choose the **Render Gallery** tool from the **Presentation** panel of the **View** tab; the **Autodesk 360 | Rendering** page will be displayed. You need to sign in to your account to access the Render gallery from the **Autodesk Rendering** page.

RENDERING IN ENSCAPE

Enscape is a virtual reality and real-time rendering plugin for Autodesk Revit. It has a user-friendly interface and can be downloaded from the Autodesk App Store. It is used to create walkthrough of a project. Enscape does not require uploading the Revit project to the cloud or exporting to other programs. It is a real-time rendering plugin and all changes made in a Revit project are immediately updated in it. Enscape immediately shows the changes you make to the project even in VR. With the option to create a standalone file of your architectural project, you can send an Enscape file to your clients or colleagues for quick demonstrations. Note that the Enscape version 3.02 +45914 is used for illustration and examples in this section.

Advantages of Enscape

Enscape provides a live link between Revit and real-time rendering which enables architects to incorporate rendering into the design process at all project phases. Enscape greatly shortens the time taken between designing and visualization as anyone involved in the project can quickly create a real-time walkthrough or renderings with a single click. It is simple to use and can be used with minimal training.

It can help in minimizing the project costs by using **VR Mode**. Using this mode of Enscape, a user can navigate and see the clashes in the architectural project even before it is constructed.

Enscape includes several new features such as the **BIM Mode**. With this tool, users can simply click on an element in Enscape to show the BIM data available on it such as dimensions, material types, quantities, and even prices. There is no need to go back and forth between Enscape and Revit to find the information you need. Similarly, you can search for a particular element to highlight one or all instances in Enscape. In addition, you can now create floor plans or cross-sections directly within Enscape.

Encape has its own ever-expanding **Asset Library** which will help you to fill your Revit project with various assets such as vegetation, people, furniture, accessories, vehicles, and lightings to make it even more realistic.

Encape has various options to customise renderings and visualization by providing various types of backgrounds such as forest, construction sites, mountains, and town. Also, you can maximize your visualization quality by adding details such as clouds or fog and you can also create the 360° panorama views of the project for rendering.

Starting Enscape Rendering

Once you have downloaded the Enscape (version 3.02 +45914) software from the enscape official website, it can be directly accessed inside the Revit 2025. To start rendering using the Enscape, open the desired 3D View and choose the **Encape** tab in the Revit interface; the Enscape's interface and all of its tools will be displayed, as shown in Figure 15-7. Now, click on the **Start** button from the **Control** panel; Enscape will start in a new tab of your computer and the Flymode view of the Encape will be displayed, as shown in Figure 15-8. You can switch the Flymode view to the Walkmode view by pressing the SPACEBAR from your keyboard. In both of the modes, various tools to navigate inside the view are provided at the bottom of the screen.

Figure 15-7 *The* ***Enscape*** *tab*

Figure 15-8** The **Flymode** view of the **Enscape

Navigating in Enscape

To navigate in the Enscape view, you can use the tools given at bottom of the view. To navigate both in Flymode and Walkmode, press and drag the left mouse button to look around in desired direction. The arrow keys of the keyboard are used to move in forward, backward, left, and right directions. Alternatively, use the W, S, A, and D keys to move in forward, backward, left, and right directions, respectively. To move fast, you can use a combination of the W, S, A, or D key and the corresponding arrow key. You can move even faster by pressing and holding the CTRL key along with required combination of keys as stated above. Use the E and Q key to move up and down respectively. To drag a view, press and hold the mouse wheel a drag the cursor in the desired direction. You can also look in or look out in a view by scrolling the mouse wheel forward and backward. You can reach a location or an element in Flymode or in Walkmode by double-clicking on the desired location or the element.

Creating Views Using Enscape

You can create camera view in your Revit interface using Enscape. To do so, navigate to the desired view in Enscape and then choose the **View Management** tool from the menubar, as shown in Figure 15-9. On doing so, the **View Management** window will be displayed in the right hand side, as shown in Figure 15-10. Choose the **Create View** button from the displayed window and then enter a view name in the **CAD View** name edit box. Now choose the **Create View** button; the view will be created and updated in the Revit file. In Revit the view will be added with the specified name under the **3D Views** heading of the **Project Browser**. Double-click on the view name created under the **3D Views** to open and see the created view, as shown in Figure 15-11.

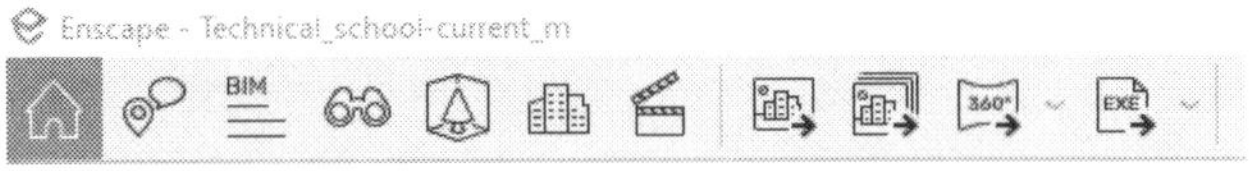

***Figure 15-9** Choosing the **View Management** tool*

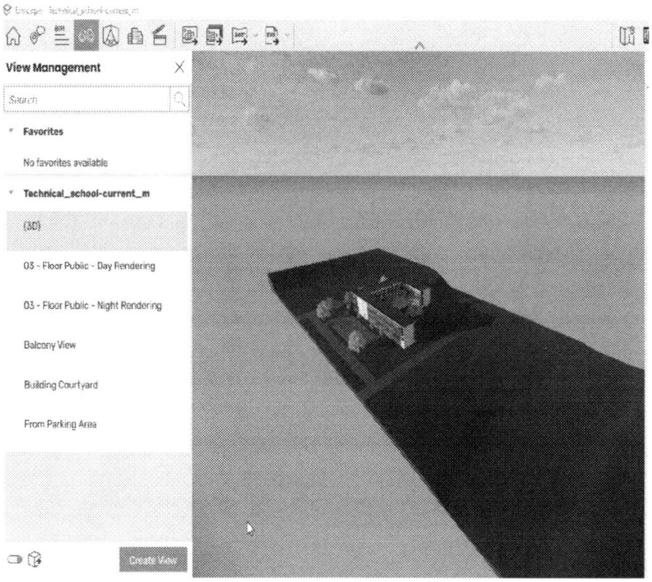

Figure 15-10 *The* ***View Management*** *window*

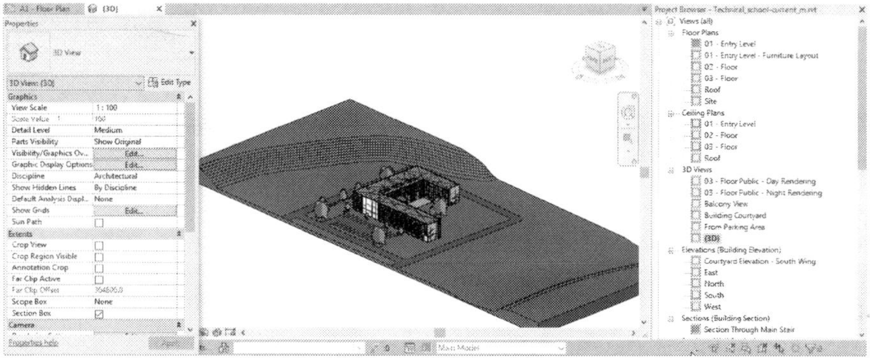

Figure 15-11 *View created in Revit using Encape*

Rendering Images Using Enscape

You can render image of your desired view using Enscape. To do so, navigate to the desired view in the Enscape and minimise the Enscape. Now from Revit interface, you can choose **Render Image** tool from the **Tools** panel, as shown in Figure 15-12; the rendered image will be exported and saved in Revit.

*Figure 15-12 The **Render Image** tool*

Placing Components Using Enscape Asset Library

Enscape comes with an in-built **Asset Library** which can be used to place new components in a Revit project. These components have **Instance Properties** and **Edit Type** properties like Revit components. Therefore, the properties such as **Level**, **Offset from Level**, **Height** and **Width** of these components can also be modified. To place an asset from Enscape in the Revit project by project, choose the **Asset Library** option from the **Tools** panel of the **Enscape** tab; the **Enscape Asset Library** dialog box will be displayed with all the available assets, which are grouped by category, as shown in Figure 15-13. On the top left in the **Enscape Asset Library** dialog box, there is a search bar to search an asset by name. Below this search bar, the asset categories are shown and include **All** category. Below the categories are the options to filter results by **Tag**. Hovering the mouse on any of the assets will reveal its name and you will also see a gray star appearing next to the asset. Clicking on the gray star next to an asset will change the color of the star from gray to gold, indicating the asset is now marked as Favorite. This asset is then listed in the Favorites category, until you click the star again which will remove it from the Favorites category.

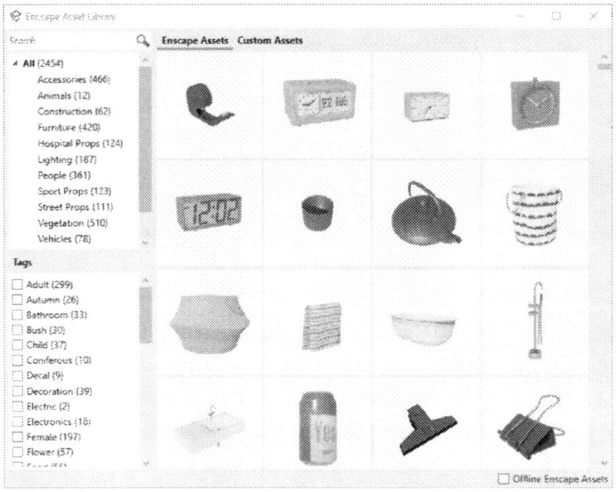

***Figure 15-13 Enscape Asset Library** dialog box*

To place an asset from Enscape, open a view in the Revit interface and select the desired asset from the **Enscape Asset Library** dialog box. Now to place the asset on a surface, hover the mouse in the Revit viewport; you will see a small cross-hair pointer. Each surface that you hit with the pointer will be highlighted in blue. Select the target surface you want to place the asset on. Once the asset is placed, press the ESC key until the **Enscape Asset Library** dialog box comes back in focus. The asset will now appear in Enscape rendered views as well as in all the Revit views.

Changing the Day time in Enscape

You can adjust the time of the day in Enscape easily. To do so, go to the Enscape screen, press and hold the SHIFT key along with the right mouse button and drag it toward right to increase the time or drag the mouse towards left to reduce the time. Alternately, you can use U and I key, to decrease or increase the time respectively. You can see the corresponding changes in the in the artificial and natural light depending on the adjusted time.

Exporting Enscape Project

You can save your Enscape project as a standalone file with ".exe" file extension and then share it with others. To do so, in the Enscape interface, click on the **Exe Standalone** from the menubar as shown in Figure 15-14; the **Create Enscape Standalone** dialog box will be displayed. Now browse to the location where you want to save the file. In the **File Name** area, assign a suitable name to the file and choose the **Save** button; the file will be saved as a standalone file at the desired location. Now this file can be shared with other users. This file opens in a default Enscape viewer in the computer. The user will have the options same as that of Enscape in this viewer such as Flymode, Walkmode and settings even if the user does not have Enscape installed in the computer. The user can change the settings of the Enscape rendering.

Figure 15-14 *Selecting the* ***Exe Standalone*** *option*

Making a Video of the Project

You can make a video in Enscape of a walkthrough of your Revit project. To do so, minimize your Revit interface and switch to the Enscape screen and enable a perspective view. Now press the V key; the **Video Editor w**indow box will be displayed, as shown in Figure 15-15. Here you have to tell the Enscape about the path that the video needs to follow. A video path is made up of multiple numbers of keyframes and these keyframes contain the information of camera position and camera rotation. To add a keyframe, first make sure that you are at the desired starting point of the video. Now click on the + button displayed under the **Keyframe** area in the **Video Editor** window; a triangular keyframe will be added in the timeline under the **Timeline** area. Similarly, navigate to other desired locations that you want to be in your video in the Enscape interface and add multiple keyframes. Before adding a keyframe for the video path, make sure you had adjusted the camera angle and the camera elevation by adjusting the view with your mouse. You can preview your video by clicking on the **Preview** button in the **Video Editor** window. After you have a finalized video, you can save this video after rendering it. To do so, click on the **Export** option in the **Timeline** area; the **Video Export** dialog box will be displayed. Specify the settings of the video in this dialog box and then choose the **Export** button; the **Save Video** dialog box will be displayed as shown in Figure 15-16. Now browse and choose the desired folder in which you want to save the rendered video and click on the **Save** button to save the rendered video.

Note

You can adjust the time of the day for all the keyframes by using U and I keys.

Figure 15-15 *The **Video Editor** window in Enscape*

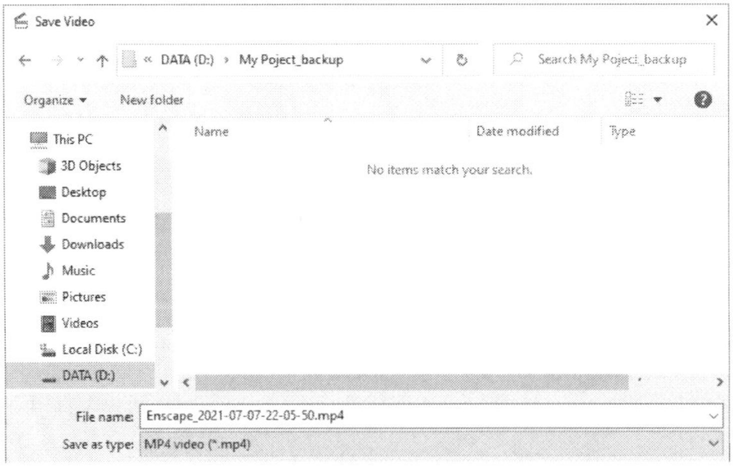

Figure 15-16 *The **Save Video** dialog box*

TUTORIALS

Tutorial 1 Apartment 1

Using the project file *c14_Apartment1_tut1.rvt,* created in Tutorial 1 of Chapter 14, render the interior view of the living room at first floor, as shown in Figure 15-17. Replace the existing finish of materials assigned to the floor, wall, ceiling, and chair with new materials/textures (given with their path). Also, create a 200-frame interior walkthrough at eye level using the path shown in Figure 15-18. **(Expected time: 1 hr)**

The rendered image can be downloaded from *https://www.cadsoft.com*. The path of the file is as follows: *Textbooks > Civil/GIS > Revit Architecture> Exploring Autodesk Revit 2025 for Architecture.*

Note
The expected time of completing the tutorial does not include the rendering time. The rendering time will depend upon the system configuration.

Figure 15-17 *The interior view of the living room after rendering*

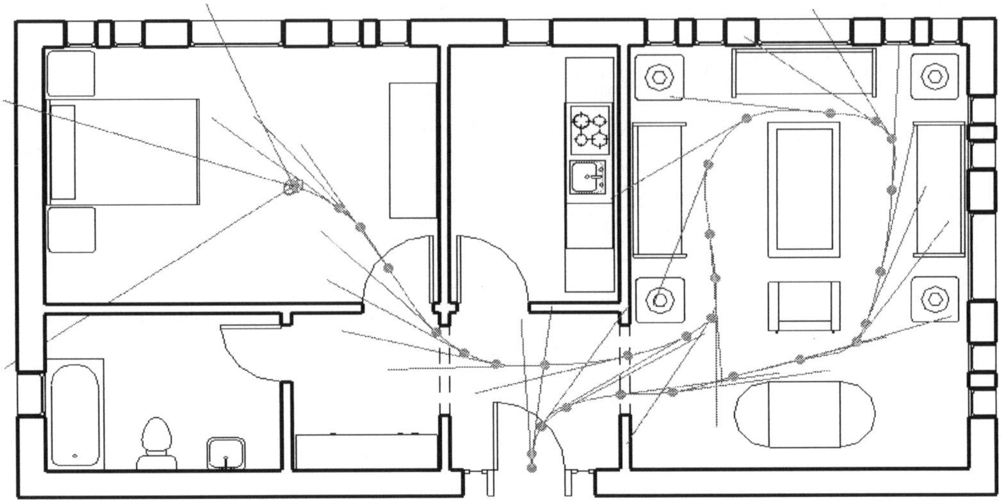

Figure 15-18 *Sketch plan showing the walkthrough path for the Apartment 1 project*

The materials used for rendering the scene are as follows:

1. New material for walls: **Finishes - Interior - Gypsum Wall Board** with **Wall-Paint Matte** render appearance.
2. New material for ceiling: **Finishes - Interior - Gypsum Wall Board - Ceiling**
3. New material for floor: **Vinyl Composition Tile**
4. New cushion material for **Chair Corbu: Textile - Brown Leather**

Add lights based on the parameters given next, refer to Figures 15-19 and 15-20. The exact location of lights is not important in this tutorial.

1. Light Fixtures to be loaded and used from the **English-Imperial > Lighting Fixture** folder:
   Table Lamp - Standard
   Pendant Light - Disk

The following steps are required to complete this tutorial:

a. Open the specified project file and the first floor plan view.
b. Apply materials to the floor using the **Paint** tool.
c. Change the finish of the ceiling, walls, and textile of the chair using the **Material Browser** dialog box.
d. Load lighting fixtures and add them to the floor and ceiling plan.
e. Add RPC to the view.
f. Set the rendering scene and generate the rendered image.
g. Specify the walkthrough path and create a walkthrough.
h. Save the project using the **Save As** tool.
   For Imperial *c15_Apartment1_tut1.rvt*
   For Metric *M_c15_Apartment1_tut1.rvt*
i. Close the project.

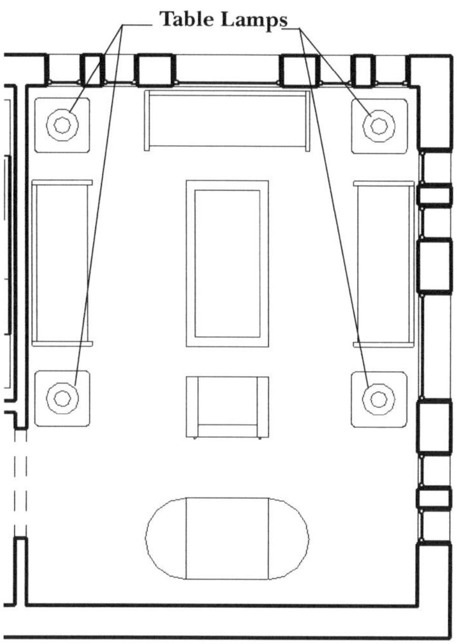

Figure 15-19 Sketch floor plan for adding table lamps to the living room

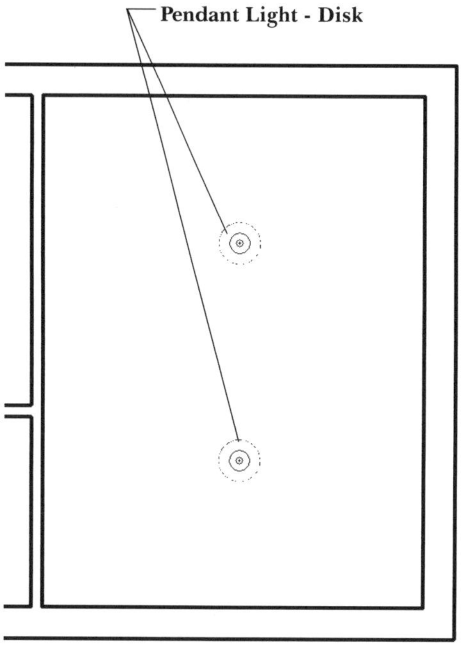

Figure 15-20 Sketch ceiling plan for adding ceiling lights to the living room

Opening the Project File and the First Floor Plan View

1. Choose **Open > Project** from the **File** menu and then open the *c14_Apartment1_tut1.rvt* for imperial and *M_c14_Apartment1_tut1.rvt* for metric file created in Tutorial 1 of Chapter 14. You can also download this file from *https://www.cadsoft.com*. The path of the file is as follows: *Textbooks > Civil/GIS > Revit Architecture > Exploring Autodesk Revit 2025 for Architecture.*

2. Now, double-click on **First Floor** in the **Floor Plans** head of the **Project Browser** to open the corresponding floor plan in the drawing window.

   As dimensions and tags are not required for this tutorial, you can turn off their visibility by using the **Visibility/ Graphics** tool.

3. Choose the **Visibility/ Graphics** tool from the **Graphics** panel of the **View** tab; the **Visibility/Graphic Overrides for Floor Plan: First Floor** dialog box is displayed. Choose the **Annotation Categories** tab and clear the check boxes for **Callouts**, **Dimensions**, **Furniture Tags**, and **Room Tags**.

4. Next, choose the **Apply** button and then the **OK** button to view the changes in the drawing area and to close the dialog box, respectively.

Modifying the Floor Material Using the Paint Tool

In this section of the tutorial, you will modify the finishes of the materials used in the building model. Also, you will use the **Paint** tool to apply the material to the floor in the section view.

1. Double-click on **Section X** under the **Sections (Building Section)** head in the **Project Browser** to display the section in the drawing window.

2. Choose the **Paint** tool from **Modify** > **Geometry > Paint** drop-down; the **Material Browser <default material name>** dialog box is displayed.

3. In the **Search** edit box of the **Material Browser** dialog box, type the **Vinyl Composition Tile** for Imperial and **Birch** for Metric; the corresponding material is displayed in the **Project Materials: All** area.

4. Select the **Vinyl Composition Tile** material.

5. Now, in the drawing area, move the cursor near the top of the floor at the **First Floor** level in the section view and click when it is highlighted; the selected material is applied to the floor top. Choose the **Done** button in the **Material Browser** dialog box to close the dialog box.

Modifying the Finish of the Materials Using the Material Editor

In this section, you will modify the finish of other elements by using the **Material Browser** dialog box.

1. Choose the **Modify** button from the **Select** panel and then move the cursor over the left exterior wall. Next, click to select the wall when it is highlighted.
2. Now, choose the **Type Properties** button in the **Properties** panel of the **Modify | Walls** tab; the **Type Properties** dialog box is displayed.

3. Next, choose the **Edit** button in the **Value** column for the **Structure** parameter to display the **Edit Assembly** dialog box, refer to Figure 15-25.

   The finishes and structure of the selected wall type are displayed in the **Layers** area and are arranged with the exterior finish as the first and the interior finish as the last layer, refer to Figure 15-21. As the material properties of the interior finish need to be modified, you will modify the last layer (**Finish 2 [5]**).

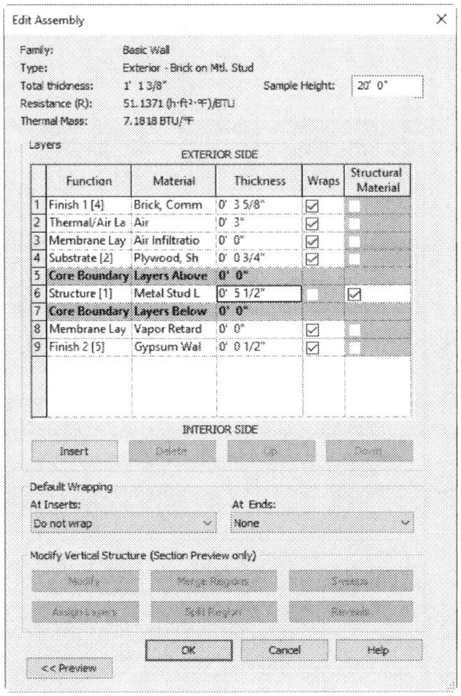

Figure 15-21 *The* ***Edit Assembly*** *dialog box for the exterior wall type*

4. Click in the **Material** column for the **Finish 2 [5]** layer and choose the browse button; the **Material Browser** dialog box along with the **Material Editor** pane is displayed.

5. In the **Material Browser** dialog box, ensure that the **Gypsum Wall Board** option is selected in the **Project Materials: All** area and its parameters are displayed in the **Material Editor** pane.

6. Choose the **Appearance** tab; the preview of the selected material and its properties are displayed. Now, choose the **Replaces this asset** button from this tab; the **Asset Browser** dialog box is displayed, as shown in Figure 15-22.

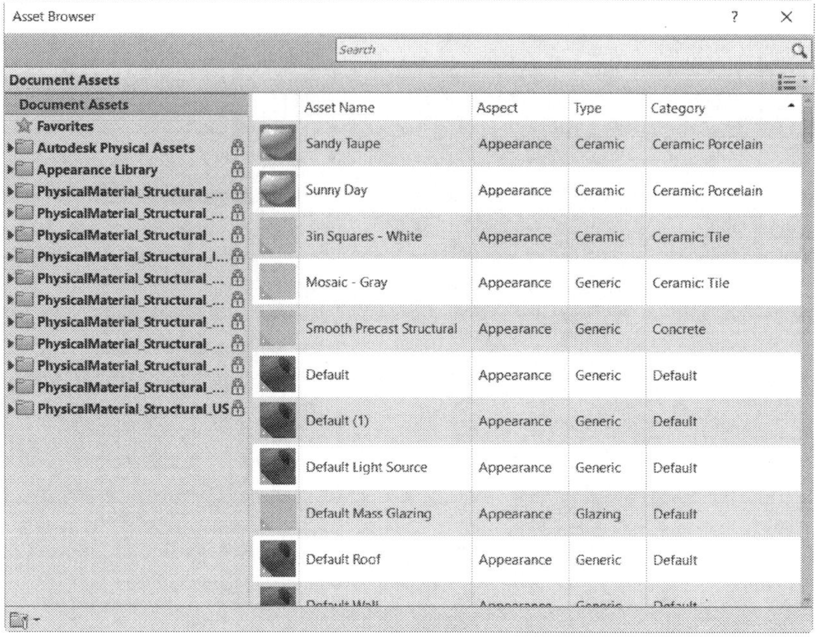

*Figure 15-22 The **Asset Browser** dialog box*

7. In the **Asset Browser** dialog box, choose **Appearance Library > Wall Paint > Matte > Biege** from **Appearance Library**; a double arrow button is displayed.

8. Choose the **Replaces the current asset in the editor with this asset** button to replace the current asset from the selected option. Close the **Asset Browser** dialog box and then choose the **OK** button in the **Material Browser** dialog box; the **Material Browser** dialog box along with the **Material Editor** pane is closed.

9. Close all the dialog boxes and return to the drawing window. Now, press ESC to exit the selection.

   Next, modify the appearance of the ceiling material.

10. Open the first floor ceiling plan and select it; the **Modify | Multi-Select** tab is displayed. Choose the **Filter** tool from the **Selection** panel of this contextual tab; the **Filter** dialog box is displayed. By default, all the check boxes are selected. Clear all the check boxes except the check box corresponding to the **Ceilings** parameter.

11. Choose the **OK** button; the **Filter** dialog box is closed; the **Modify | Ceilings** contextual tab is displayed.

12. Now, choose the **Type Properties** tool from the **Properties** panel of the **Modify | Ceilings** tab; the **Type Properties** dialog box is displayed.

13. Choose the **Edit** button in the **Value** field of the **Structure** parameter; the **Edit Assembly** dialog box is displayed with different layers of the selected ceiling.

14. Click in the **Material** column, the **Finish 2 [5]** layer for Imperial and **Structure[1]** for Metric, and then choose the browse button to display the **Material Browser** dialog box.

15. In the **Material Browser** dialog box, the **Gypsum Wall Board** option is selected in the **Project Materials: All** area. The **Material Editor** pane displays the properties of material. Choose the **Duplicate Material and Assets** option from the **Creates and duplicates materials** drop-down list located at the lower left corner of the **Material Browser** dialog box; the name **Gypsum Wall Board(1)** is displayed in the **Name** edit box of the **Identity** tab in the **Material Browser** dialog box. Enter the name **Finish - Interior - Gypsum Wall Board - Ceiling** in this edit box.

16. Now, choose the **Replaces this asset** button from the **Appearance** tab; the **Asset Browser** dialog box is displayed. In this dialog box, choose **Paint > Glossy > Paint - Enamel Glossy (White)** from **Appearance Library**. Click on double-arrow button to replace the current asset.

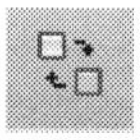

17. Close the **Asset Browser** dialog box. Choose the **Apply** button and then the **OK** button to close the **Material Browser** dialog box. Choose the **OK** button in the **Edit Assembly** and the **Type Properties** dialog boxes to close them.

    Next, you will modify the texture of the chair textile.

18. Open the first floor plan view in the drawing window using the **Project Browser**.

19. Select the **Chair-Corbu** for Imperial (**M_Chair-Corbo** for Metric) in the living room and then choose the **Edit Type** button from the **Properties** palette; the **Type Properties** dialog box is displayed.

20. In this dialog box, click in the **Value** column for the **Cushion Material** type parameter and choose the browse button; the **Material Browser** dialog box is displayed.

21. In the **Material Browser** dialog box, choose the **Duplicate Material and Assets** option from the **Creates and Duplicates Materials** drop-down list at the lower left corner of the **Material Browser** dialog box. Enter the name **Textile - Brown Leather** in the **Name** edit box and press ENTER.

22. Choose the **Replaces this asset** button from the **Appearance** tab in the **Material Editor** pane; the **Asset Browser** dialog box is displayed. In the **Asset Browser** dialog box, choose **Fabric > Leather** option from the **Appearance Library**.

23. Next, select the **Brown** material in the right pane of the **Appearance Library** area and click on the double arrow button to replace the current asset.

24. Now, close the **Asset Browser** dialog box and choose the **Graphics** tab from the **Materials Browser** dialog box; the **Graphic Properties** will be displayed in the **Material Editor** pane. Select the **Use Render Appearance** check box in the **Shading** area and then choose the **OK** button in the **Material Browser** and the **Type Properties** dialog boxes to close them. Now, press ESC to exit the selection of the chair. The **Leather** material are assigned to the chair.

Loading and Adding Light Fixtures to the Building Model

In this section of the tutorial, you will load the light fixtures and then add them to the floor and ceiling plan.

1. Choose the **Place a Component** tool from **Architecture > Build > Component** drop-down; the **Modify | Place Component** tab is displayed.

2. Now, choose the **Load Family** tool from the **Mode** panel; the **Load Family** dialog box is displayed.

3. Navigate through the **English-Imperial > Lighting > Architectural > Internal** folder and load the **Table Lamp - Standard** and **Pendant Light - Disk** for Imperial **(English > Lighting > Architectural > Internal** folder and load the M_**Table Lamp - Standard** and **M_Pendant Light - Disk** for Metric) fixtures using the CTRL key. Choose the **Open** button to return to the drawing area.

4. From the **Type Selector** drop-down list, select the **Table Lamp-Standard: 60W - 120V** fixture.

5. Move the cursor near the center of the side table placed at the lower right of the living room and click when the **Midpoint and Midpoint** object snap is displayed, as shown in Figure 15-23. Now, choose the **Modify** button to exit the selection mode.

6. Next, select the added table lamp; instance parameters for the component are displayed in the **Properties** palette.

7. Enter the value **1'3"** for Imperial (**381mm** for Metric) in the value column for the **Offset** instance parameter and then choose the **Apply** button.

8. Use the **Copy** tool in the **Modify** panel to copy the added lamp on the other three side tables, as shown in Figure 15-24 (Select the **Multiple** check box in the **Options Bar**). Now, press ESC twice to exit the current selection.

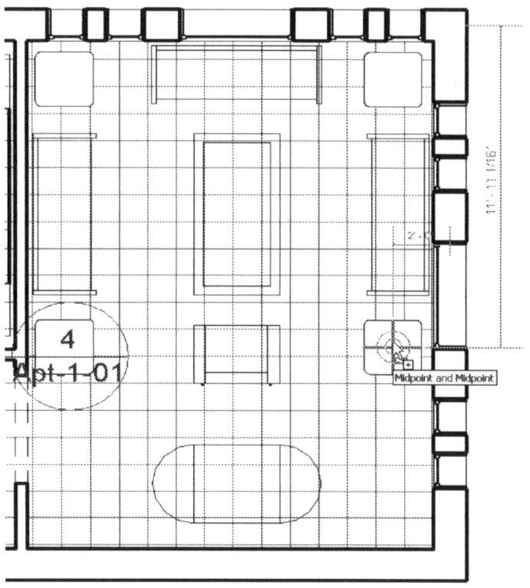

Figure 15-23 Adding lamp to the side table

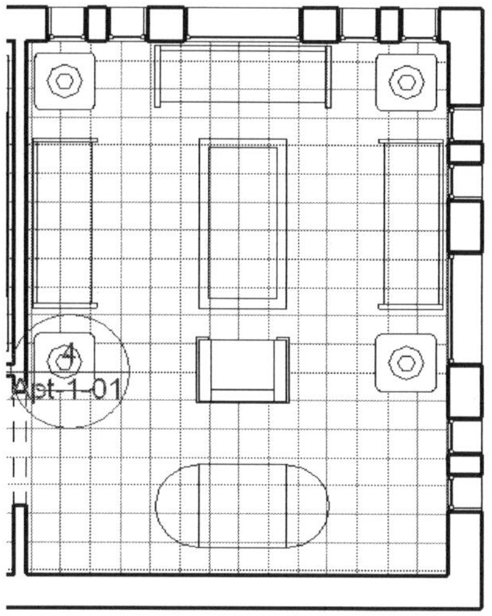

Figure 15-24 Creating multiple copies of the table lamp component

9. Double-click on **First Floor** under the **Ceiling Plans** head in the **Project Browser** to display the ceiling plan view.

10. Next, choose the **Place a Component** tool from the **Architecture > Build > Component** drop-down; the **Modify | Place Component** tab is displayed. Select the **Pendant Light - Disk: 100W - 120V** fixture from the **Type Selector** drop-down list.

11. Move the cursor over the sitting area in the living room and click to add the light fixture at the approximate location, as shown in Figure 15-25.

12. Similarly, add the same fixture to the other side of the living room, as shown in Figure 15-26 and then press ESC twice to exit the **Modify | Place Component** tab.

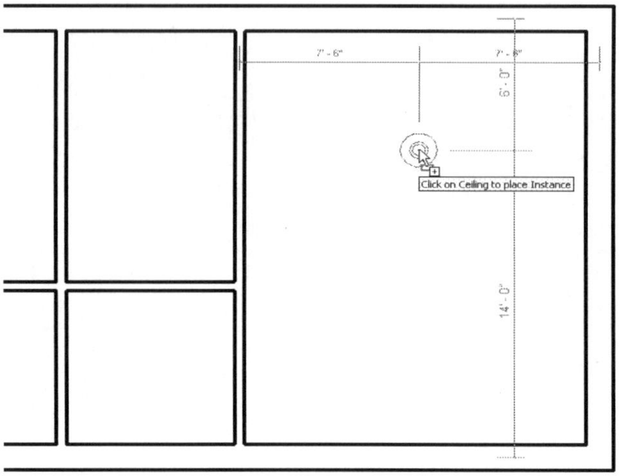

Figure 15-25 Adding the first light fixture in the ceiling plan

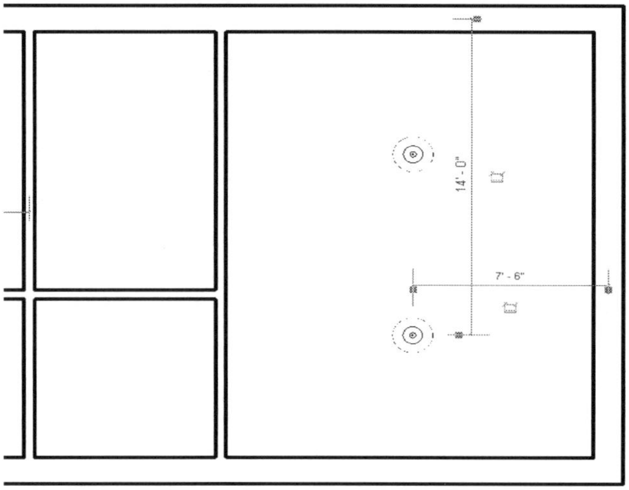

Figure 15-26 Adding the second light fixture in the ceiling plan

Adding RPC to the Interior Scene

In this section of the tutorial, you will use the **Place a Component** tool to load and add the specified RPC to the interior scene. The specified RPC (Rich Photorealistic Content) can be added at the location shown in Figure 15-13.

1. Open the first floor plan view from the **Floor Plans** heading in the drawing window using the **Project Browser**.

2. Now, choose the **Place a Component** tool from the **Architecture > Build > Component** drop-down; the **Modify | Place Component** tab is displayed.

3. Choose the **Load Family** tool from the **Mode** panel; the **Load Family** dialog box is displayed.

4. Navigate through the **English-Imperial > Entourage** folder, select the **RPC Female** for Imperial (**English > Entourage** folder, select the **M_RPC Female** for Metric) family file and then choose **Open**; the selected family file is now loaded.

5. Next, select the **RPC Female: Cynthia** option from the **Type Selector** drop-down list.

6. Now, select the **Rotate after placement** check box from the **Options Bar** and then click on the sofa and move the cursor counterclockwise. Next, click to place and align the RPC in the view, as shown in Figure 15-27.

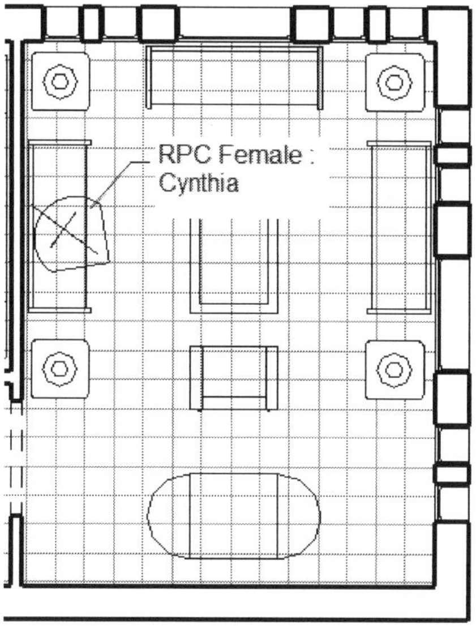

Figure 15-27 *RPC added to the view*

7. Press the ESC key twice to exit the insertion process. Double-click on **Living Area** under the **3D Views** head in the **Project Browser** to view the 3D view after adding the lighting fixtures and RPC, refer to Figure 15-28.

8. Choose **Visual Style > Shaded** from **View Control Bar**; the current view will be shaded.

Note

You can set the position of the RPCs by selecting the required elevation or section views from the ***Project Browser****.*

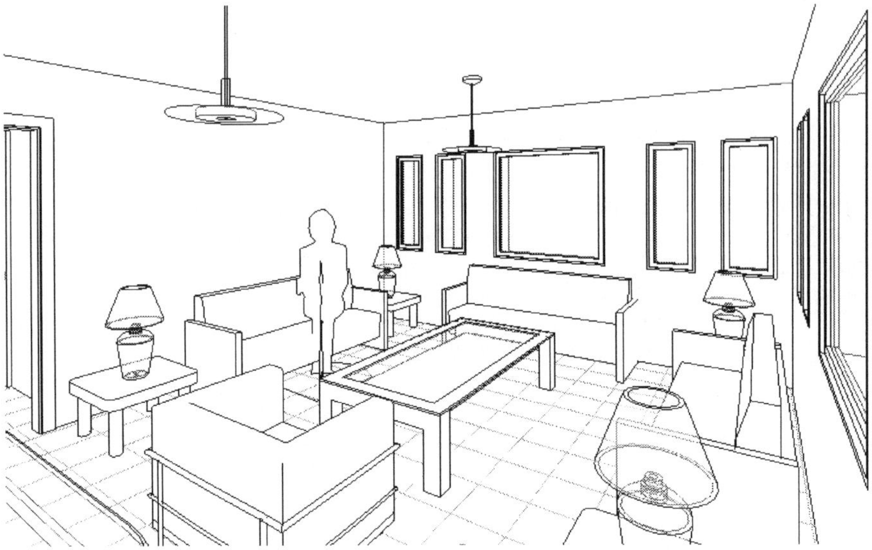

Figure 15-28 *The interior 3D view of the living room with the added lighting fixtures and RPC*

Setting Up the Interior Rendering Scene

In this section of the tutorial, you will invoke the **Rendering** dialog box and specify the settings to render the interior scene.

1. Invoke the **Rendering** dialog box by choosing the **Show Rendering Dialog** button from **View Control Bar**. Alternatively, you can press RR to display the **Rendering** dialog box.
2. In the **Quality** area of the **Rendering** dialog box, select the **Best** option from the **Setting** drop-down list.
3. In the **Lighting** area, select the **Interior: Sun and Artificial** option from the **Scheme** drop-down list.
4. Choose the Browse button displayed next to the **Sun Setting** text box; the **Sun Settings** dialog box is displayed.
5. Select the **Still** radio button in the **Solar Study** area.
6. Choose the button next to the **Location** parameter in the **Settings** area to invoke the **Location Weather and Site** dialog box.
7. In the **Location** tab, select the **Default City List** option from the **Define Location by** drop-down list. Select **Chicago, IL** from the **City** drop-down list and choose the **OK** button.
8. Set the date **14/06/2025** in the **Date** drop-down list and then set the time to **10:00 PM** in the **Time** spinner.

9. Clear the **Ground Plane at Level** check box, if selected.

10. Choose **Apply** and **OK**; the specified settings are applied to the scene and the **Sun Settings** dialog box is closed.

11. Next, in the **Image** area of the **Rendering** dialog box, choose the **Adjust Exposure** button to invoke the **Exposure Control** dialog box.

12. In the **Exposure Control** dialog box, set the values of the parameters as given below:

    Exposure Value: **6.5** Highlights: **0.45**
    Saturation: **1** Shadows: **0.2**
    White Point: **4000**

13. Choose **Apply** and **OK**; the specified settings are applied to the scene and the **Exposure Control** dialog box is closed.

14. In the **Rendering** dialog box, choose the **Render** button; Autodesk Revit starts the rendering process using the mental ray rendering engine for the interior view. The **Rendering Progress** dialog box is displayed on your screen. On completion of the process, the rendered image is displayed in the drawing area, as shown in Figure 15-29.

Figure 15-29 *The interior 3D view of the living room rendered using mental ray*

15. After the rendering process is complete, choose the **Save to Project** button in the **Image** area; the **Save to Project** dialog box is displayed. Enter **Living Area** in the **Name** edit box and choose the **OK** button. This will add the rendered image under the **Renderings** head in the **Project Browser**. Now, close the **Rendering** dialog box.

The rendered image can be downloaded from *https://www.Cadsoft.com*. The path of the file is as follows: *Textbooks > Civil/GIS > Revit Architecture > Exploring Autodesk Revit 2025 for Architecture*

Note
*Depending on the hardware configuration of the system, it might take considerable time to render. To do a quick render, select **Draft** or **Low** from the **Setting** drop-down list in the **Rendering** dialog box.*

Generating the Walkthrough

In this section, you will specify the walkthrough path in the floor plan view.

1. Double-click on **First Floor** under the **Floor Plans** head in the **Project Browser**.

2. Choose the **Walkthrough** tool from the **View > Create > 3D View** drop-down; the **Modify | Walkthrough** tab is displayed. Note that the cursor changes into a **+** mark and different options related to creating a walkthrough path are displayed in the **Options Bar**.

3. Now, start sketching the walkthrough path by clicking at the first point near the center of the main entrance door of the apartment. Ensure that the first point is specified inside the lobby area and not outside the entrance door. Move the cursor toward the living room and click to specify multiple points in such a manner that they form a counterclockwise loop around the center table in the living room, as shown in Figure 15-30.

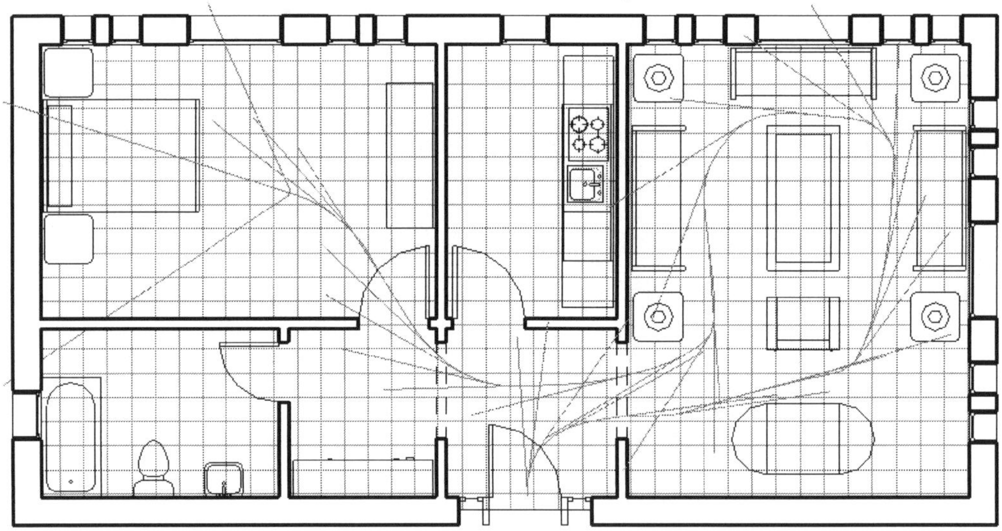

Figure 15-30 *Sketch plan showing the walkthrough path for the Apartment 1 project*

4. Similarly, move the cursor across the lobby and inside the bed room. Next, click to specify multiple points of the walkthrough path, refer to Figure 15-30. The exact location of these points is not critical.

5. After specifying the last point, choose the **Finish Walkthrough** button from the **Walkthrough** panel of the **Modify | Walkthrough** tab; walkthrough path is now completed and the **Modify | Cameras** tab is displayed.

Note

The dots indicate the specified points and are shown only for illustration. They are not displayed when the walkthrough path is being specified.

Note that the generated walkthrough is now added to the **Project Browser** as **Walkthrough 1** under the new **Walkthroughs** head. You can now modify the total frames for the walkthrough using the **Edit Walkthrough** button.

6. Double-click on **Walkthrough 1** under the **Walkthroughs** head in the **Project Browser**. Choose the **Edit Walkthrough** button from the **Walkthrough** panel of the **Modify | Cameras** tab; the **Edit Walkthrough** contextual tab is displayed.

7. In the **Options Bar**, ensure that the **Active Camera** option is selected from the **Controls** drop-down list and **200** is displayed in the **Frame** edit box. Now, choose the **Frame Settings** button next to the **Frame** edit box; the **Walkthrough Frames** dialog box is displayed.

8. In the **Walkthrough Frames** dialog box, enter **200** as the value in the **Total Frames** edit box and choose the **OK** button; the specified settings are now assigned to the walkthrough and then **Walkthrough Frames** dialog box closes. Also the total number of frames in the walkthrough gets modified from **300** to **200** as specified for this tutorial.

Recording the Walkthrough

In this section of the tutorial, you will open the walkthrough, adjust the field of view of the camera and then record it by exporting it using the **AVI** file format.

1. In the **Options Bar**, enter **1** in the **Frame** edit box. Now, choose the **Edit Walkthrough** tab and then choose the **Open Walkthrough** button from the **Walkthrough** panel. The first frame is opened in the drawing window.

2. Right-click inside the drawing window and choose **Zoom Out (2x)** from the shortcut menu displayed.

3. Drag the extents of the rectangle to modify the field of view for the camera such that the view resembles the illustration given in Figure 15-31.

4. Right-click again and choose **Zoom To Fit** to enlarge the view.

5. Next, you will export the walkthrough as an AVI file.

From the **File** menu, choose **Export > Images and Animations > Walkthrough**; the **Length Format** dialog box is displayed, as shown in Figure 15-32.

Figure 15-31 *First frame of the walkthrough with the modified field of view*

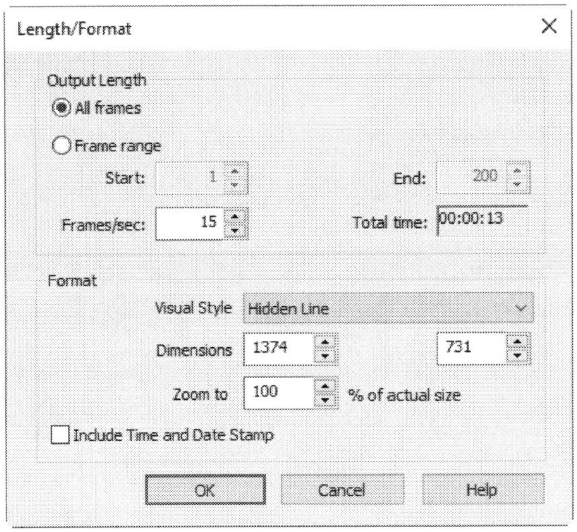

Figure 15-32 *The* ***Length/Format*** *dialog box*

6. In the **Length/Format** dialog box, ensure that the **All frames** radio button is selected and then set **10** in the **Frames/sec** spinner.

7. In the **Format** area, select **Rendering** from the **Visual Style** drop-down list.

8. Next, set the value in the **Zoom to** spinner to **43** and choose the **OK** button; the **Length/Format** dialog box closes and the **Export Walkthrough** dialog box is displayed.

9. In the **Export Walkthrough** dialog box, specify an appropriate path, enter **c15_Apartment1_tut1_Walkthrough 1** for Imperial and **M_c15_Apartment_tut1_ Walkthrough 1** for Metric in the **File name** edit box, and then make sure that AVI Files is the default file format selected in the **Files of type** drop-down list. Now, choose the **Save** button; the rendering and recording process starts. After rendering of the first frame, the **Video Compression** dialog box gets displayed. Keep the **Full Frames (Uncompressed)** option selected in the **Compressor** drop-down list and choose the **OK** button to continue.

   Now, the rendering and recording process starts for each frame of the walkthrough and the **Rendering Progress** dialog box is displayed. When the procedure is completed, the file is recorded at the specified location. You can then use any media player to view the walkthrough file.

Saving the Project

In this section, you will save the project file using the **Save As** tool.

1. Choose **Save As > Project** from the **File** menu; the **Save As** dialog box is displayed.
2. Enter **c15_Apartment1_tut1** for Imperial and **M_ c15_Apartment1_tut1** for Metric in the **File name** edit box and then choose **Save**.
3. Choose **Close** from the **File** menu to close the project file.

EXERCISES

Exercise 1 Club

Using the *c14_Club_tut2.rvt* project file created in Tutorial 2 of Chapter 14, render the entrance view, as shown in Figure 15-33. Assume all the information required for this exercise.
Save the file as *c15_Club_ex1.rvt* **(Expected time: 30 min)**

Figure 15-33 *Rendered exterior view of the Club project*

Exercise 2 — Apartment 1 - Night View

Using the project file created in Tutorial 1 of this chapter, render the night scene of the living room, as shown in Figure 15-34. You need to modify the time settings to 6:00 A.M. Save the file as *c15_Apartment-1_ex2.rvt.* **(Expected time: 30 min)**

The rendered image can be downloaded from *https://www.cadsoft.com*. The path of the file is as follows: *Textbooks > Civil/GIS > Revit Architecture > Exploring Autodesk Revit 2025 for Architecture.*

Figure 15-34 *Rendered interior view of the living room*

Chapter 16

Using Advanced Features

Learning Objectives

After completing this chapter, you will be able to:

- *Add structural components to a project*
- *Create multiple design options*
- *Create area schemes and area plans*
- *Create color schemes*
- *Use project phasing tools*
- *Link building models and share coordinates*
- *Share projects using worksets*
- *Purge unused elements from the project file*
- *Organize the Project Browser*
- *Generate solar studies*
- *Understand Point Cloud*
- *Publish views to DWF*
- *Understanding Route Analysis*

INTRODUCTION

This chapter describes the advanced features of Revit such as adding structural components, creating design options using the phasing tools, and worksharing methodology. With the help of these advanced features, you can easily add structural components to a building model using various tools and in-built components. You can also create various design options based on the project requirement, and also use the area analysis tools to analyze and generate area plans for the building model.

In this chapter, you will learn to represent the phase-wise development of a project, link independently developed projects to each other, and share their coordinates using the project sharing and linking tools. Also, the concept of worksets, which enables multiple users to work on a single project and interoperability of Revit Architecture with other program such as Autodesk 3ds Max has been discussed in the chapter.

CREATING STRUCTURAL COMPONENTS

Structural components describe the structural system of a building project. The use of some of these elements, such as walls, floors, roof, stairs, and so on has been discussed in the earlier chapters. The structural components may or may not comprise the basic structural support system of a building project. For example, in a multistory building project, steel framing components can form the basic support system and the walls can be non-load bearing. Therefore, it may be necessary to specify and differentiate between the structural and architectural components for some projects.

Creating Structural Walls

There are four types of functions that can be assigned to a parametric wall. The functions that can be assigned to the wall are interior, exterior, retaining, and foundation. The interior or exterior wall functions have been described in the earlier chapters. A structural wall differs from a non-structural wall only in its structural usage. The four structural usage options are:

Non-bearing	Walls supporting only their own weight or are used to divide spaces
Bearing	Walls supporting a vertical load in addition to their own weight
Shear	Walls supporting a lateral thrust of shear
Structural combined	Walls serving more than one purpose

Tip

*While using structural elements, you can choose the **Structural Settings** button (represented by an inclined arrow) from the **Structure** panel of the **Structure** tab; the **Structural Settings** dialog box will be displayed. In this dialog box, you can set the parameters to control the representation of structural elements in a project file. You can also use the Structural-Default.rte template file that contains the view properties and range settings specifically for working with the structural elements.*

Creating Structural Columns

Structural columns can be placed in the plan or 3D view. You can place a single structural column or multiple structural columns by selecting the grid intersection. To add a structural column, invoke the **Column** tool from the **Structure** panel of the **Structure** tab; the **Modify | Place Structural Column** tab will be displayed, as shown in Figure 16-1. In this tab, you can choose the **Load Family** button from the **Mode** panel and then load the additional structural columns from the **Structural Columns** folder in the **English-Imperial** folder. Based on the material specification, four subfolders **Concrete**, **Precast Concrete**, **Steel**, and **Wood** have been provided to load the structural columns from the column families.

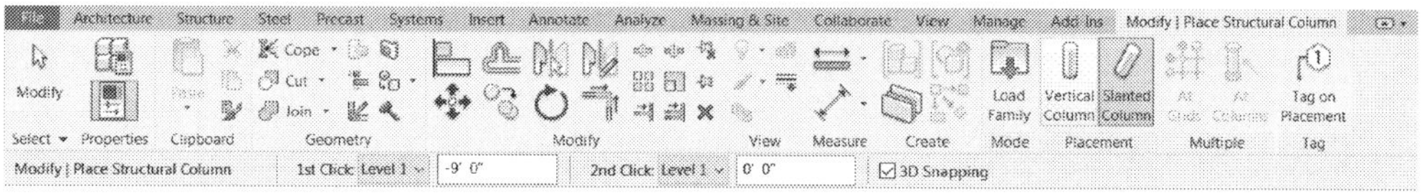

Figure 16-1 *The* ***Modify / Place Structural Column*** *tab*

Adding Structural Beams and Braces

Beams are structural components that connect the columns or the structural walls. Beams can be joined to the structural bearing walls if their structural usage property is set to bearing or structural combined. You can add beams individually or by selecting the grids along which they are required.

Cutting Openings in Beams, Braces, and Columns

Revit allows you to cut openings in structural members such as beams, braces, and columns. You can cut an opening in a structural member after selecting planes. Every structural member (Beam, Brace, and Column) shows two planes: horizontal and vertical. These planes are perpendicular to each other, as shown in Figures 16-2 and 16-3. You can also lay ducts and pipes passing through the beams and columns by creating the openings of the corresponding shape. You can create different shapes of openings using different sketching tools.

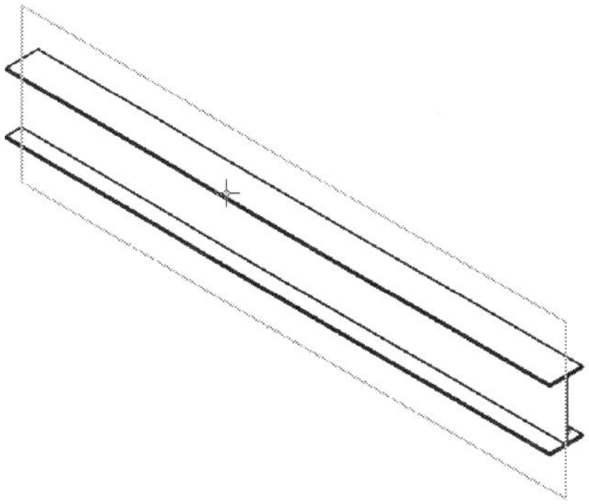

Figure 16-2 *The vertical plane of the beam*

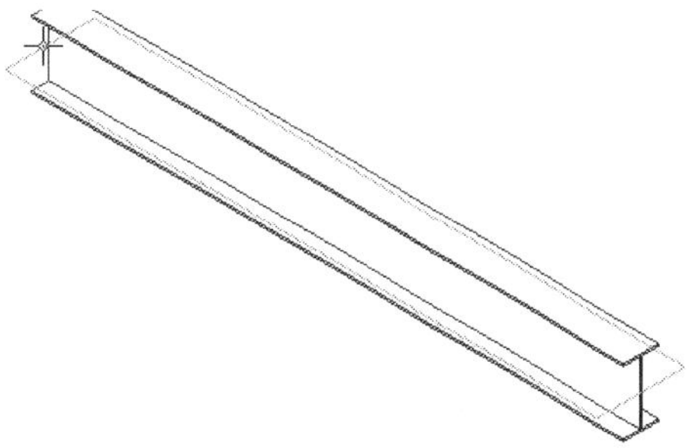

Figure 16-3 *The horizontal plane of the beam*

GENERATING MULTIPLE DESIGN OPTIONS

In Revit, the **Design Options** tool provides you with the flexibility of choosing several alternatives for your design. As a designer, you need to come up with multiple design ideas, so that you can select the best option that suits the functional, aesthetical, and economical requirement of your project. Therefore, you need to develop and evaluate several design options and then execute the best possible design option for your project.

Generating Design Options for a Project

The **Design Options** tool is used to generate design options for a project. Invoke this tool from the **Design Options** panel in the **Manage** tab; the **Design Options** dialog box will be displayed. To generate a new design option set, choose the **New** button in the **Option Set** section of the **Design Options** dialog box; a new option set will be added to the left pane. Next, to add a new option in this option set, choose the **New** button from the **Option** section. You can also rename an option set by using the **Rename** button in the **Option Set** section. Figure 16-4 shows various design options in the **Design Options** dialog box.

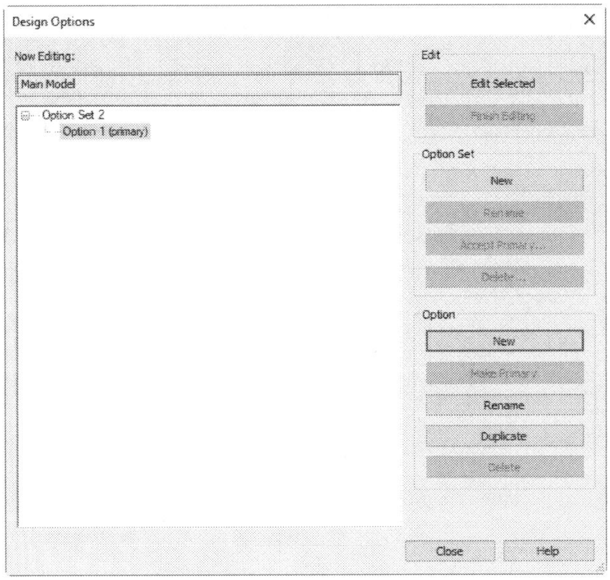

***Figure 16-4** The **Design Options** dialog box with the design option sets and design options created*

Presenting Design Options

After creating the design options, you can compare and present them to clients. To do so, create multiple copies of the view of a design option, and set the visibility setting for each view. You can use the **Visibility/Graphics** tool to control the visibility of a design option in each view. Select the view from the **Project Browser** and right-click. Next, choose **Duplicate View > Duplicate** from the shortcut menu; the copy of the view will be created. You may create a number of copies of the view based on the number of design options generated. Use the shortcut menu to rename each view based on the design option it represents.

Tip

*While using the **Design Options** tool, you need to keep in mind that the view-specific elements such as dimensions and text cannot be added to the design options. Also, you need to be careful in keeping all dependent elements in the same design option.*

USING AREA ANALYSIS TOOLS

The area analysis tools are used to analyze and represent different area types. These tools can also be used to create area schemes such as gross building area, rentable area, and area plans. You can access these tools from the **Area** drop-down in the **Room & Area** panel of the **Architecture** tab, as shown in Figure 16-5.

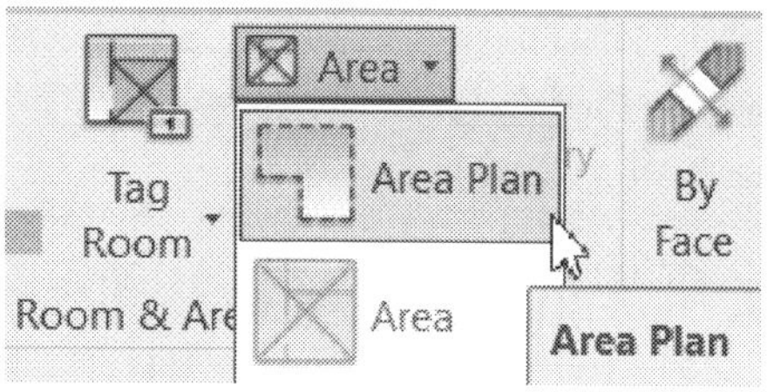

Figure 16-5 *Various tools in the **Area** drop-down*

Note

You can also generate area schemes for the design options developed for a project.

Area Schemes

Area Schemes are the standards that you set to calculate the area of an enclosed boundary with a specific space. You can use the **Area** tool from the **Area** drop-down to differentiate the spaces based on their usage. You can create and implement various types of area schemes in your current project based on the local standards that are followed for the area calculation.

You can create multiple area measurement schemes. To do so, invoke the **Area and Volume Computations** tool from the **Room & Area** panel of the **Architecture** tab; the **Area and Volume Computations** dialog box will be displayed. In this dialog box, choose the **Area Schemes** tab; the options in this tab will be displayed, as shown in Figure 16-6.

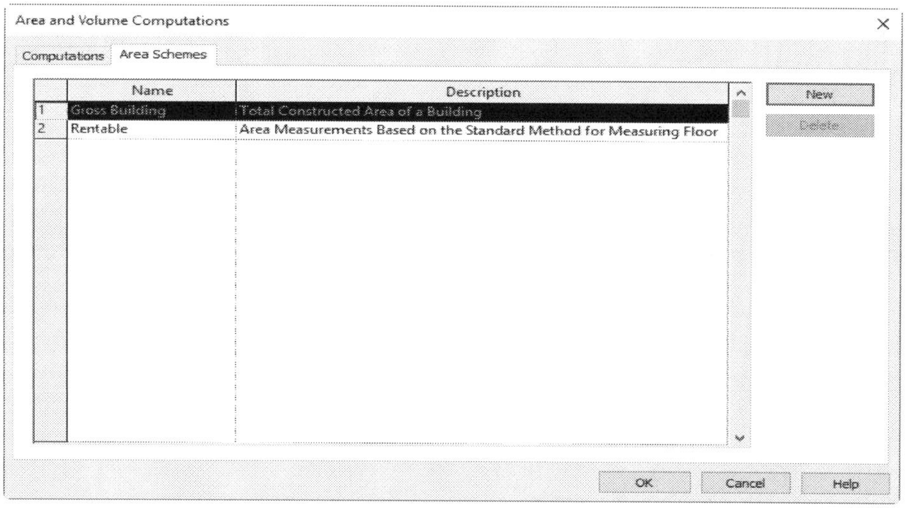

Figure 16-6 *The **Area and Volume Computations** dialog box*

Area Plans

The **Area Plan** tool is used to create area plans. Area plans are views that describe the subject of the area scheme. You can create a number of area plans for a single area scheme. For example, a rentable area scheme can be created for a commercial project and then you can create the area plans showing the rentable area for each level. Each area plan has distinct area boundaries, tags, and color fills.

Area Schedules

Once you have added area tags to the spaces and renamed them appropriately, you can proceed to create area schedules. To do so, invoke the **Schedule/Quantities** tool from **View > Create > Schedules** drop-down; the **New Schedule** dialog box will be displayed. In this dialog box, the **Category** list box displays the area schemes created in the project. Select the desired area scheme from this list and ensure that the **Schedule building components** radio button is selected. Choose the **OK** button in the **New Schedule** dialog box; the **Schedule Properties** dialog box will be displayed. Select and add the parameters to be included in the scheduled fields from the **Available fields** area, as shown in Figure 16-7.

You can also use the **Fields**, **Filter**, **Sorting/Grouping**, **Formatting**, and **Appearance** tabs available in the **Schedule Properties** dialog box to set the properties of the area schedule. Next, choose the **OK** button; the **Modify Schedule/Quantities** tab will be displayed and the created schedule will be displayed in the drawing area. Add it to the drawing sheet for the presentation of the area schemes. Figure 16-8 shows an example of an area schedule.

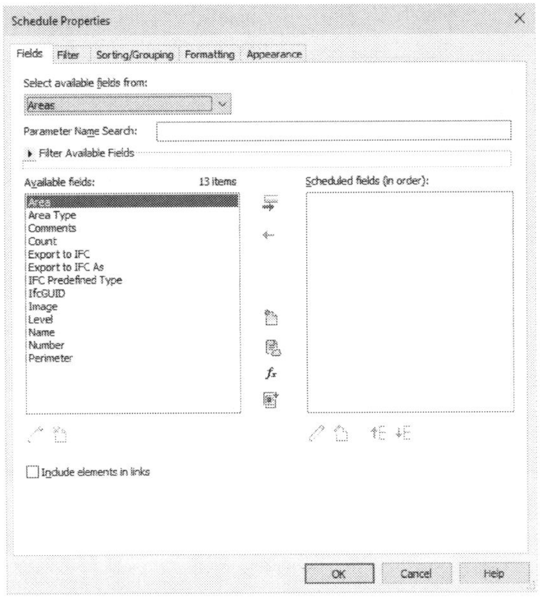

*Figure 16-7 The **Schedule Properties** dialog box*

Area Schedule (Area Scheme 1)		
Level	Name	Area
Level 1	Living Room	445 SF
Level 1	Kitchen	126 SF
Level 1	Bed Room	734 SF
Level 1	Patio	532 SF
Level 1	Front Open	273 SF
Level 1	Washroom	80 SF
Level 1	Closet	80 SF
Level 1	Bed Room 2	176 SF
Level 1	Shaft 3	47 SF
Level 1	Shaft 2	57 SF
Level 1	Shaft 1	43 SF
Level 1	Utility	45 SF
Level 1	Stairs	128 SF

Figure 16-8 *An example of a created area schedule*

COLOR SCHEMES

Color schemes are used to represent and categorize different areas in the project graphically by using color codes. In other words, color schemes are used to color the rooms and areas in a view. You can create color schemes and then apply them to different plan views. You can create different color schemes for the first and second floors of a building. A color scheme can be created based on a specific category such as Gross area, Rentable area, or an instance property of the room or area such as room perimeter, number, or name. You can also create a color scheme depending on the utilization of different rooms in a building. For example, you can assign different colors to represent areas such as office, storage, or accounts. Then, you can add a color scheme legend to help you identify different areas by the colors assigned to them.

Creating Color Schemes

Before creating a color scheme, the rooms and areas should be defined in the plan view. You can create a color scheme based on any parameter of a room or an area. For example, if you want to create a color scheme for rooms on the basis of room number, room area, or names, make sure that the rooms are numbered and named in the plan view.

Modifying Color Scheme

To modify a color scheme, select the existing color scheme legend and choose **Scheme > Edit Scheme** from the **Modify Color Fill Legends** tab; the **Edit Color Scheme** dialog box will be displayed. Modify the colors, fill patterns, title, parameter, and other values as required and choose the **Apply** button to view the changes. Choose the **OK** button to retain the changes.

MASKING REGIONS

While working in an architectural project, sometimes it becomes difficult to demarcate between different visible elements due to overlapping of lineworks. Therefore, to reduce the congestion of lineworks and interference of unwanted elements in your drawing, it is necessary to hide or obscure them in the view.

Adding Masking Regions to a Project

To mask or obscure any element in a project, open the floor plan view and choose the **Masking Region** tool from **Annotate > Detail > Region** drop-down; the screen will enter the sketch mode. Choose the required sketching tool from the **Draw** panel, select the line type from the **Line Style** drop-down list, and then sketch the masking regions in the closed loops. After sketching the masking regions, choose the **Finish Edit Mode** button from the **Mode** panel of the **Modify |Create Masking Region Boundary** tab. Figure 16-9 shows the plumbing fixtures such as tub and sink masked in the drawing.

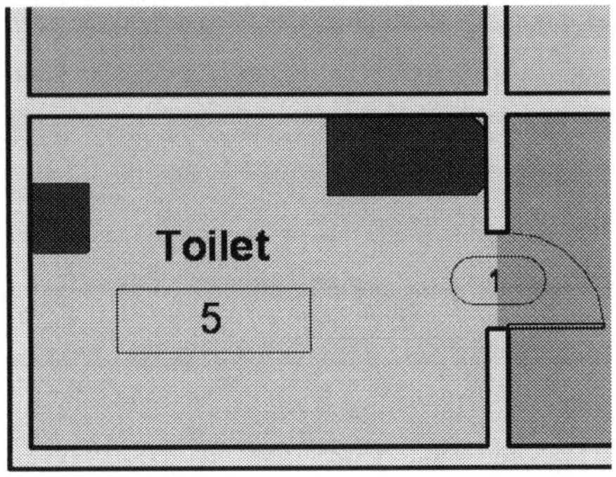

Figure 16-9 *Masking regions added to tub and sink*

Adding Masking Regions to a Detail Family

While creating and defining a detail family, you can add a masking region to it so that when you load it into a project, it can obscure or hide the elements which are interfering with it. To add a masking region to a detail family, open the floor plan view and choose the **Masking Region** tool from the **Detail** panel of the **Annotate** tab, while you are working in the **Revit Family Editor**. On doing so, the **Modify | Create Masking Region Boundary** tab will be displayed for sketching and editing the masking region. Next, choose the required sketching tool from the **Draw** panel. Then, sketch the masking region over the detail family in a closed loop. Next, in the **Properties** palette, you can clear the check box in the **Value** column for the **Visible** instance parameter to disable the masking region boundaries. As a result, when you load the family into the project, the masking region will not be visible.

Adding Masking Regions to a Model Family

You can create and define a masking region for a model family in the same way as you do for the detail family. While creating a masking region for a model family, you can control the position of the plane where it is drawn by using the **Draw in Foreground** instance parameter in the **Properties** palette of the masking region. If you clear the check box in the value column of the **Draw in Foreground** instance parameter, then you can draw the masking region at the work plane where the model family was created. If you select this check box, then you can draw the masking region at a plane of the model family that is nearest to you in the view.

CREATING DISPLACED VIEW

In Autodesk Revit 2025, the **Displace Elements** tool is used to create a displaced view of an element or group of elements. The element or group of elements of which a displaced view is created is called the displacement set. The displaced view of element is used to see the internal parts of the building which are obstructed due to roofs, and walls of the building. This helps the designer to view the 3D model of the building more closely. This tool is used to move the displacement set along the X, Y, and Z axes upto a specified distance away from a model.

USING PROJECT PHASING TOOLS

In some architectural projects, building works need to be carried out in phases. Phases represent the time periods over which a project is to be executed. One of the examples of such a project is building a large institutional complex that requires the construction work of different buildings within the stipulated time. Another example can be an interior renovation project, in which certain building components may need to be constructed and demolished over a number of stages or phases. Revit enables the architects and the interior designers to represent phases in the views of a building model. Using the phasing tools, you can generate 3D views that show the phase-wise addition or demolition of building components and the resulting building model as well.

Understanding Phasing Concepts

To use the phasing tools in Revit, you first need to understand the basic concepts of the working of these tools and their usage. You can create any number of phases for a project and specify them for the building elements. Revit allows you to specify phasing for views and also for the building components. Each new building component added to a project has the **Phase Created** and **Phase Demolished** instance parameters associated with it. These instance properties can be modified. The **Phase Created** instance parameter indicates the phase during which the element will be added and the **Phase Demolished** instance parameter shows the phase in which it will be demolished. By default, you can assign one of the following values for these instance parameters in a building element:

New	Element created in the current phase
Existing	Element created in an earlier phase and continues to exist in the current phase
Demolished	Element created in an earlier phase and demolished in the current phase
Temporary	Element created and demolished in the current phase

LINKING BUILDING MODELS AND SHARING COORDINATES

Some large projects may require you to develop independent building models and later combine them into a comprehensive project. For example, in a large educational campus project, you may need to independently develop the building models of the administrative building, academic complex, faculty residences, student accommodation, and so on, and later add them into a single host project file. You can do so by using Revit's linking and sharing coordinates tools that enable you to link models and share their coordinates within a single host project file. These tools also enable you to work on building models in other project files. When you link projects, you need to share the coordinates of the host model with the linked model so that the linked files retain their positions. To do so, invoke the **Acquire Coordinates** tool from **Manage > Project Locations > Coordinates** drop-down and then select the linked project in your drawing. On selecting the linked project, the origin of the linked project's shared coordinate becomes the origin of the host project's shared coordinates. Revit Architecture provides the flexibility and easy management of the linked models by enhancing the linking of the models and organizing the linked files in the **Project Browser**. You can easily access the linked files, the nested link files, and the Link Manager from the **Project Browser**. The nested Revit links are also listed under the **Revit Links** head with the host link in the **Project Browser**.

Linking or Importing Models

To link or import one project to the other, open the host project file and choose the **Insert** tab, options in this tab will be displayed, as shown in Figure 16-10. You can choose the options from the **Import** or **Link** panel of the **Insert** tab depending upon the file format to be imported or linked. Before you import or link any file, make sure that the file contains the data and geometry that are compatible with Revit Architecture. To link *.dwg, .dwf, .skp, .dgn,* and other files, choose the **Link CAD** tool from the **Link** panel of the **Insert** tab; the **Link CAD Formats** dialog box will be displayed, as shown in Figure 16-11. In this dialog box, select the file type from the **Files of type** drop-down list. Next, browse to the required location and select the file. The options in this dialog box are discussed next.

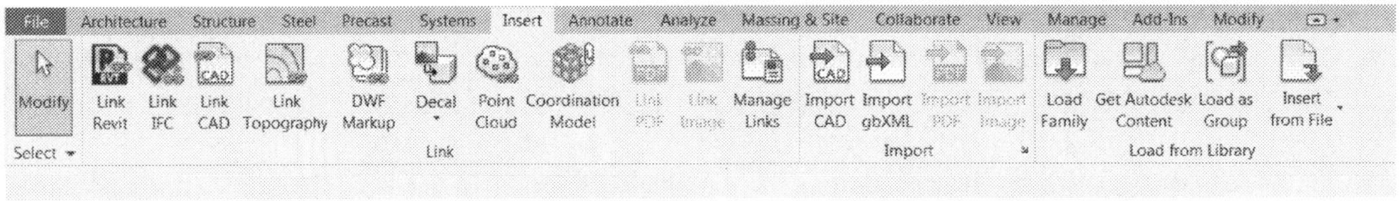

Figure 16-10 The linking and the importing options in the ***Insert*** *tab*

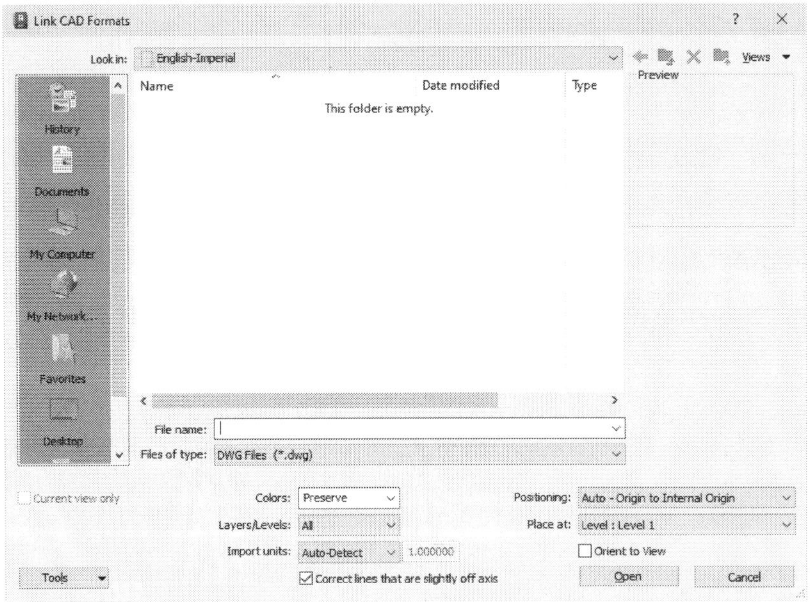

Figure 16-11 The ***Link CAD Formats*** *dialog box*

Linking Revit Models

To link or import the Revit files, choose the **Link Revit** tool from the **Link** panel of the **Insert** tab; the **Import/Link RVT** dialog box will be displayed, as shown in Figure 16-12. In this dialog box, navigate to the desired folder and select the file to be linked to the host project; the preview image of the selected project file will be displayed in the **Preview** pane. Choose the **Open** button to open the file. Select the appropriate positioning option from the **Positioning** area to position the model automatically or manually in the host project.

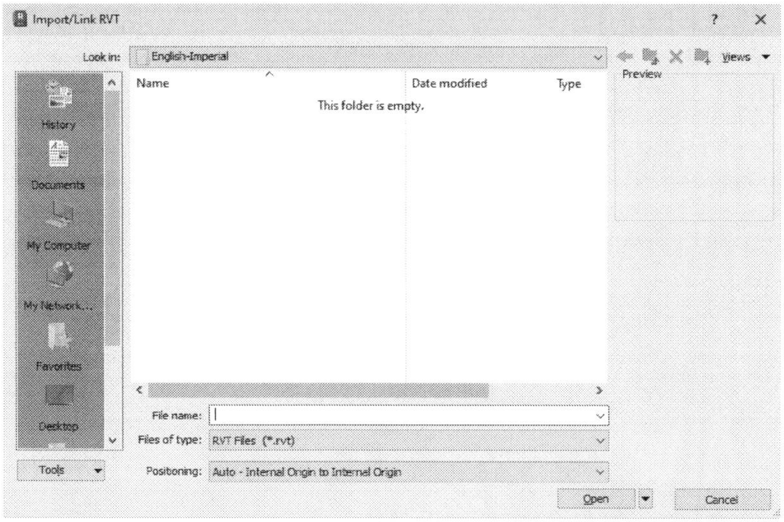

***Figure 16-12** The **Import/Link RVT** dialog box*

Nested Linked Models

If you import a Revit model that already contains a linked file, the linked file becomes a nested link file. You can control the visibility of the nested linked models in the host project file. When you link a file, by default, the nested link model will not be displayed in the host project file and the **Nested Links Invisible** message box will be displayed, as shown in Figure 16-13, indicating that the elements in the linked model will not be visible in the project view. To display the nested link model in the host model, open the parent link view in which the file is linked and choose the **Manage Links** tool in the **Import** panel of **Insert** tab; the **Manage Links** dialog box will be displayed. Choose the **Revit** tab from the dialog box; the name of the linked file will be displayed in the **Linked Name** column. Click in the **Reference Type** column and change the value in this column to **Attachment**. Now, link the parent model file to the host project; the nested linked model will be displayed in the project. If you do not want to display the nested model elements in the host project, accept the default **Overlay** option. When you select the **Overlay** option, Revit does not load the linked model when its host is linked to another model. The following example explains the use of the **Overlay** and **Attachment** options in linking models. Example: Project A consists of a single horizontal wall, as shown in Figure 16-14. The project A is imported and linked to the project B that consists of a single vertical wall. Figure 16-15 shows the Project B and Figure 16-16 shows the project A linked to the project B. The project B becomes a host for project A and project A becomes a link for project B. Now, the project B along with the linked project A (consisting of both horizontal and vertical walls) is imported to another project, Project C, which consists of a single inclined wall. In this case, the project A (consisting of a horizontal wall) becomes a nested link. Figure 16-17 shows the project C after linking with the project B. You will notice that the horizontal wall of project A will not be displayed in the project C. This is because the **Overlay** option is selected in the **Reference Type** column of the **Manage Links** dialog box. To display the horizontal wall of the project A (nested model) in the project C, open the project B parent view, consisting of vertical and horizontal walls. Next, choose the **Manage Links** tool from the **Link** panel in the **Insert** tab; the **Manage Links** dialog box will be displayed. Choose the **Revit** tab in the dialog box; the project A will be displayed in the **Linked File** column. Select the **Attachment** option for the linked project A in the **Reference Type** column and choose the **OK** button. Close the parent view and import the project B into the project C. Now, the horizontal wall of the project A (nested link) will be displayed in the project C, as shown in Figure 16-18. The project A will also be added in the **Project Browser** as the nested link under the parent project B, as shown in Figure 16-19.

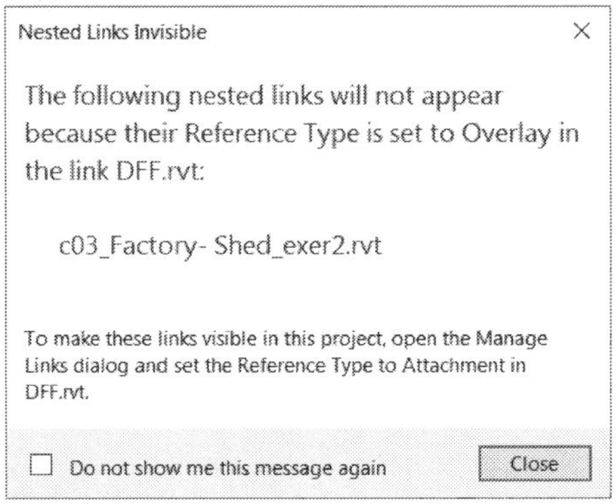

Figure 16-13 *The **Nested Links Invisible** message box*

Figure 16-14 *Project A consisting of a horizontal wall*

Figure 16-15 *Project B consisting of a vertical wall*

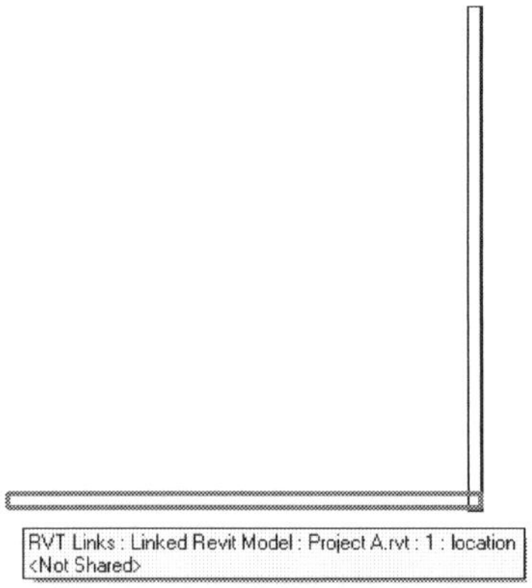

Figure 16-16 *Project A linked to project B*

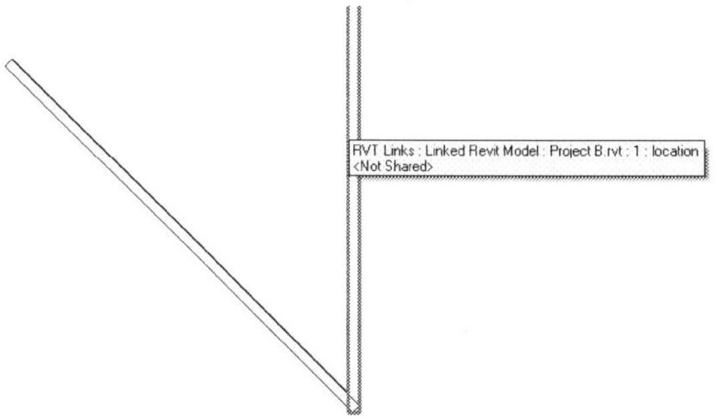

Figure 16-17 *The host project B linked to project C with the* ***Overlay*** *option selected*

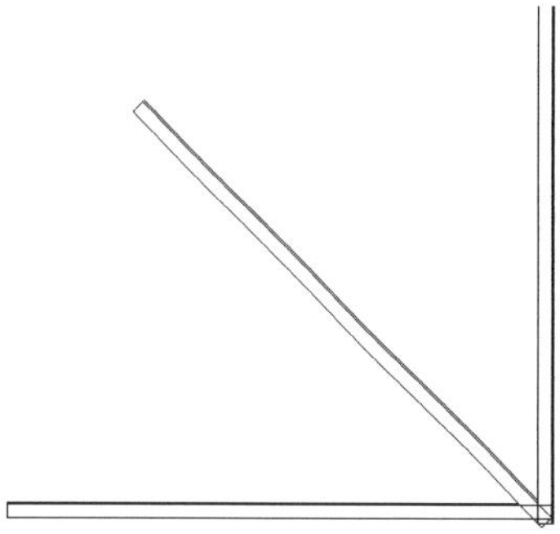

Figure 16-18 *Project B linked to Project C after selecting the* ***Attachment*** *option for project A*

Figure 16-19 *Project A added as a nested link with parent project B in the* ***Project Browser***

Converting Linked Models to Groups - Binding Links

You can convert the linked Revit models into groups in the host project. You can do so by binding the linked model with the host project, thereby making it a part of the host project. After binding the linked model, the model geometry will be transformed into a group, and therefore making changes in the host project will be easier for you. To bind and group a linked model, select the linked model in the drawing; the **Modify | RVT Links** tab will be displayed. Choose the **Bind Link** tool from the **Link** panel; the **Bind Link Options** dialog box will be displayed, as shown in Figure 16-20. In the dialog box, select the **Attached Details**, **Levels**, and **Grids** check boxes to include them in the group and then choose the **OK** button; the linked model will be converted into a group. If there is any group in the project with the same name as that of the linked Revit model, a dialog box with a message will be displayed prompting you to replace the group. Choose **Yes** to replace the existing group or choose **No** to rename the group. On choosing **Yes**, another dialog box with a message to remove the link will be displayed, as shown in Figure 16-21. Choose the **Remove Link** button in the dialog box to remove the link or choose the **OK** button to remove it later; the linked model will be transformed into a group and listed in the **Project Browser** under the **Groups** head.

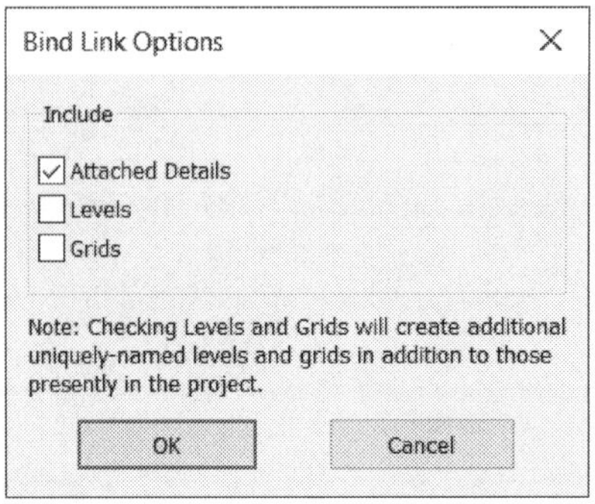

***Figure 16-20** The **Bind Link Options** dialog box*

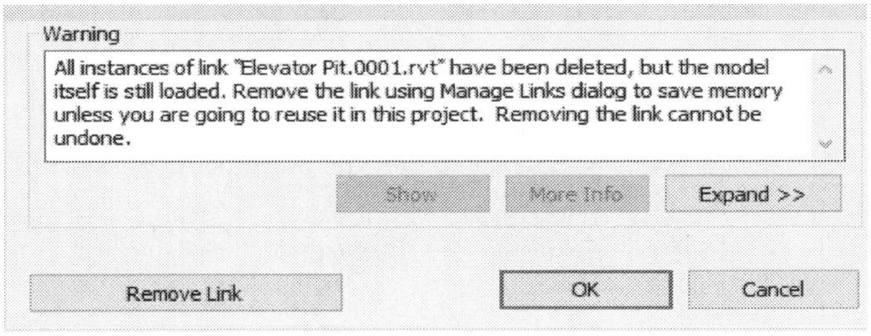

***Figure 16-21** The message box displayed on binding the link*

Managing Links

Revit enables you to manage links between the host and the linked projects. To do so, choose the **Manage Links** tool from the **Link** panel of the **Insert** tab; the **Manage Links** dialog box will be displayed with a list of linked projects, as shown in Figure 16-22.

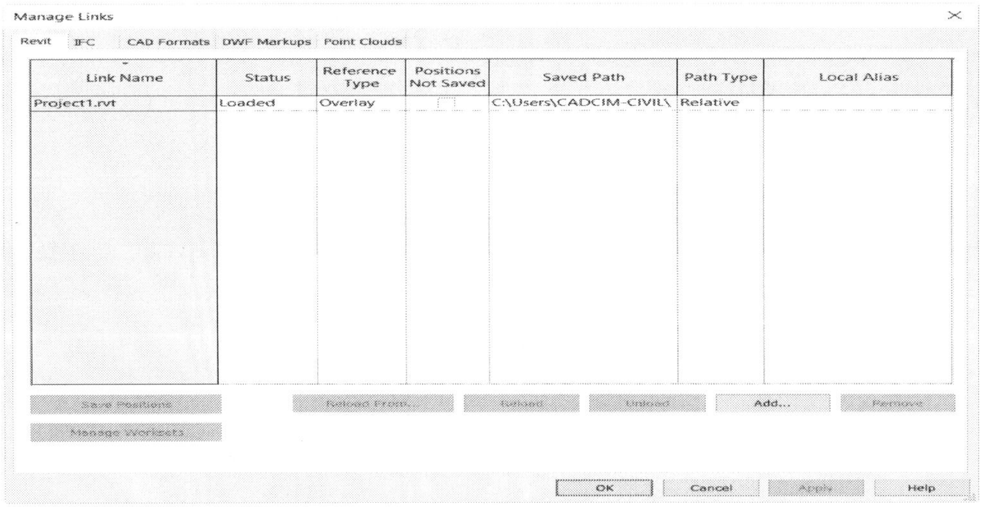

***Figure 16-22** The **Manage Links** dialog box*

Including Elements of Linked Models in Schedules

In Autodesk Revit, you can include different model elements from a linked file such as walls, doors, windows, and so on into a schedule. To do so, select the existing schedule of the current project from the **Schedules/Quantities(all)** head in the **Project Browser**; the instance properties of the selected schedule will be displayed in the **Properties** palette. In the value column of the palette, choose **Edit** in the **Fields** parameter; the **Schedule Properties** dialog box will be displayed. Select the **Include elements in Linked** files check box to include the elements from a linked file, if required. To include project information from a linked file, select **Project Information** from the **Select available fields from** drop-down list. Select the example explained earlier, project A will be the top-level nested model. Choose the **Apply** button to view the changes in the project and then choose the **OK** button.

Applying the Color Schemes of the Host Model to Rooms and Areas of the Linked Models

In Autodesk Revit, you can apply the color scheme assigned to the rooms and areas in the host file to the rooms and areas in the linked file. To do so, choose the **Color Schemes** tool from the **Room & Area** panel of the **Architecture** tab; the **Edit Color Scheme** dialog box will be displayed. In this dialog box, make sure that the **Include Elements from linked** files check box is selected in the **Options** area. Choose the **OK** button to close the dialog box. Next, choose the **Visibility/ Graphics** tool from the **Graphics** panel of the **View** tab; the **Visibility/Graphic Override** dialog box will be displayed for the current view. Next, choose the **Revit Links** tab from the dialog box. Select the linked model to which you want to apply the color scheme of the host file. Choose the default **By Host View** button in the **Display Settings** column; the **RVT Links Display Settings** dialog box will be displayed. You can select the **By host view** or **Custom** radio button in this dialog box if the linked model consists of rooms. However, if the linked model consists of only areas, select the **Custom** radio button. On selecting the **Custom** radio button, all options in the **Basics** tab will be enabled. From the **Linked view** drop-down list, select the view of the linked model to which you want to apply the color scheme. Next, from the **Color Fill** drop-down list, select the **By host view** option. Choose the **Apply** button to apply the changes. Next, choose the **OK** button to close the dialog box.

Copying Linked Model Elements

You can copy the elements of the linked model to the host model. To do so, move the cursor over the linked model and press TAB to highlight the required elements, in case the file has a nested link. Click on the element when it is highlighted and choose the **Copy to Clipboard** tool from the **Clipboard** panel of the **Modify | RVT Links** tab. Next, open the host file in which you want to paste the element and choose the **Paste from Clipboard** tool from the **Clipboard** panel; the linked model file from which you have selected the element will be displayed in the **Project Browser** under the **Revit Links** head. You can place the element either by clicking at the required location in the drawing area or by dragging the linked model from the **Project Browser** and dropping it in the drawing area.

WORKSHARING CONCEPTS

Worksharing is a method of distributing work among people involved in a project, and accomplishing it within the stipulated period of time. In Worksharing, each person involved in the project is assigned a task that has to be accomplished through proper planning and by coordinating with the other members of the team.

Worksharing Using Workset Tools

In Autodesk Revit, you work in an integrated single file model. This implies that all the construction related information and documents are available in a single model file and this forms the basis of the worksharing process.

Process of Worksharing

With the start of the new project, you start creating the basic building geometry like the shell of the model with basic detailing. But as the project progresses, you underscore the need of sharing the project with multiple users for which you need to speed up the process of work toward completion. To invite multiple users to work in your project, you save the file that you have worked as a central file in a network location and make copies of it in the local hard drives of individual team members. By doing so, you divide the work into different parts. Everybody will start working with the copied central file from his or her individual workstation. Before starting with the file, each member working in the project is instructed to rename the copied central file in their individual hard drive location and then open it in Revit Architecture to proceed further. The flow diagram in Figure 16-23 explains in brief the process of worksharing in brief.

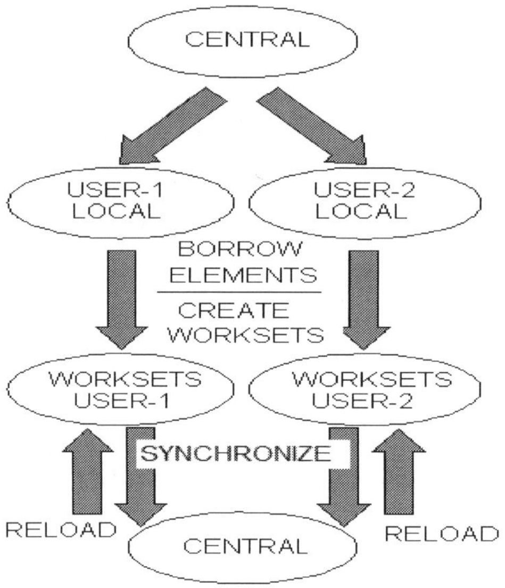

Figure 16-23 *Flow diagram explaining the Worksharing process*

Saving Methodology in Worksharing

While working in a worksharing environment, it is important to learn the methods of saving your files. There are two methods of saving a file and sharing your files with others by saving a local file and synchronizing with the central file. You can save your file as a local file by choosing **Save** from the **File** menu. The **Save** option is generally used to save the local file. Using this method, you can just secure the work that you have done without affecting the worksharing process. The other method is to synchronize the file with central. The **Synchronize with Central** option is used to publish the work in the central file. The moment you synchronize your file to central, you acquire the changes made by other team members to their local models, provided they had also saved their files to the central file at that particular interval.

Element Ownership Concepts

In a worksharing environment, you need to know about the status of ownership of a particular element or workset in the project. To check the status of ownership of an element, select it and check whether you can modify or reposition it. If you own the element, it will be highlighted in blue and can be modified and repositioned. In case, you do not own the element that you have selected, it will be highlighted in red, but with a blue icon attached to it indicating it cannot be modified.

PURGING UNUSED ELEMENT FAMILIES

The **Purge Unused** tool enables you to unload families and family types from the project file. This reduces the file size to a considerable extent. However, in case the worksets have been enabled, they must be opened to use this tool.

Invoke the **Purge Unused** tool from the **Settings** panel of the **Manage** tab; the **Purged unused** dialog box will be displayed, as shown in Figure 16-24. It displays the list of element categories that can be unloaded. You can select the families, groups, or annotation types and choose the **OK** button to purge them. You can also use the **Check All** and **Check None** buttons for the selection. In Autodesk Revit, you can also purge unused materials.

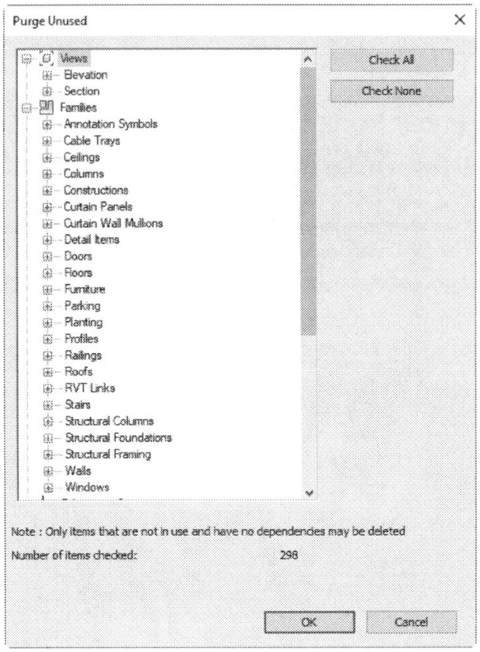

Figure 16-24 *The **Purge unused** dialog box*

TRANSFERRING PROJECT STANDARDS

The **Transfer Project Standards** tool allows you to copy and paste project standards from one project to another. The project standards that can be transferred include family types, materials, view templates, and object types.

ORGANIZING THE PROJECT BROWSER

Autodesk Revit also enables you to organize the **Project Browser** based on the documents to be prepared for the project. Invoke the **Browser Organization** tool from **View > Windows > User Interface** drop-down; the **Browser Organization** dialog box is displayed. The **Views** tab of this dialog box displays the parameters that can be selected for organizing the browser.

The **Sheets** tab can be used to view the sheets created for a project in the **Project Browser**. You can select an option to display the sheets. For example, you can rename the sheets created for a construction document package based on their sheet numbers. You can also select the **Sheet Prefix** check box in the **Sheets** tab to arrange the sheet display in the **Project Browser**. Figure 16-25 shows the **Project Browser** with such an arrangement.

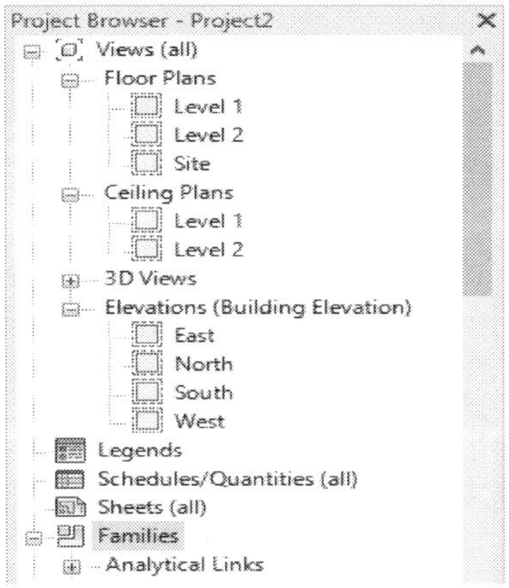

Figure 16-25 *An example of organised **Project Browser***

GENERATING SHADOWS

Another powerful feature provided in Autodesk Revit is the automatic generation of shadows based on the location of the sun with respect to the building. Shadows can be generated in the plan, elevation, section, or 3D view. To use this feature, open the view and click on the inclined arrow symbol (Graphic Display Options) in the **Graphics** panel of the **View** tab. On doing so, the **Graphic Display Options** dialog box will be displayed. In this dialog box, select the **Cast Shadows** check box. The intensity and direction of the shadow can also be controlled by using various options from **Graphic Display Options** dialog box. You can also choose the **Shadow On** button from the **View Control Bar** to generate shadows in the view.

AUTODESK REVIT SOLAR STUDIES

Revit Architecture's solar study helps you visualize and analyze the position of Sun and solar effects on buildings and sites in the real world at a specified time, date, duration, and location. You can analyze the effect of the Sun on the same building at different locations, such as Boston or Melbourne, by changing the location of the building from Boston to Melbourne. The **Solar Study** tools in Revit Architecture provide you with a wide range of locations to view the Sun and shadow effects on the same building or site but at different locations by generating the solar studies. Revit Architecture allows you to generate a still solar study in a single frame depicting the shadows at a specified date and time. It can also generate an animated three-dimensional solar study consisting of multiple frames depicting the shadow movements for a specified duration of time.

Generating Still Solar Study

You can generate a still solar study to see the shadow patterns based on the location of the building, or the azimuth and altitude of the Sun from the horizontal surface. To generate a still solar study, open the floor plan view or the site plan view. In the **Properties** palette of the opened view, set the value of the **Orientation** parameter to **True North** and choose the **Apply** button. Next, choose the **Rotate True North** tool from **Manage > Project Location > Position** drop-down; a rotation symbol will be displayed in the drawing area. In the **Options Bar**, specify the rotation angle in the **Angle from Project to True North** edit box. This angle specified in the edit box will be the angle difference of the building from the Project North direction to the **True North** direction. You can rotate the building to the **East** or **West** of the **True North** direction. Open the 3D view of the building and adjust the view of the building as required. Choose the **View** tab and then choose the inclined arrow symbol in the **Graphics** panel; the **Graphic Display Options** dialog box will be displayed, as shown in Figure 16-26.

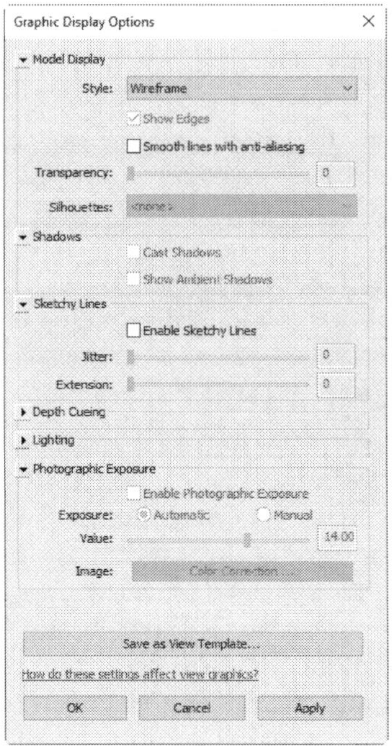

*Figure 16-26 The **Graphic Display Options** dialog box*

Creating an Animated Solar Study

In Revit Architecture, you can also create an animated single-day or multi-day solar study. The single-day solar study generates the animation of shadow movements for a specified period of time in a single day and at a particular location. For example, you can create a solar study from 6 AM to 7 PM on August 15 in Boston. The multi-day solar study helps you create an animation of shadow movements at a fixed time of a day, but for specified number of days at a particular location. For example, you can create a solar study at 2 PM from August 15 to August 20, 2012 in Boston.

POINT CLOUD

Point Cloud is a collection of point data in X, Y, and Z ordinates. This data is collected by 3D scanners. These 3D scanned point clouds are then used to create 3D revit models. These point cloud files are inserted in the Revit project to modify the existing condition of a building.

Inserting a Point Cloud File

Point cloud files are .rcp and .rcs type format files. If the file is in raw format such as .pcg or .las format, then you will be prompted to index the data. The .rcp and .rcs format files are indexed point cloud files which are inserted in a Revit project.

Point Cloud Visibility

To view a point cloud in a project, invoke the **Visibility/ Graphic** tool from the **Graphics** panel of the **View** tab; the **Visibility/Graphics Overrides for <current view>** dialog box will be displayed. In this dialog box, choose the **Point Cloud** tab. Next, in the **Visibility** column of this tab, you can select or clear the checkbox corresponding to the selected point cloud file to show or hide the point cloud in the view, respectively. You can also assign the color for each point cloud in the model. To do so, choose the **RGB** button; the **Point Cloud Color Mode** dialog box will be displayed. In the **Point display** area of this dialog box, by default, the **RGB** option is selected in the **Color Mode** drop-down list. On selecting any other color option in the drop-down list, the **Settings** area, and the **Clear Overrides** button at the bottom of the dialog box will be activated. Specify the required settings in the **Settings** area. You can also choose the **Clear Overrides** button to restore the previous settings. Next, choose the **OK** button; the dialog box will be closed.

Using Point Cloud File in a Project

In Autodesk Revit, after inserting the point cloud file, you will use that file in the project. This file helps the user to provide a higher level of visual appearance of existing condition of a building. To capture the existing building, laser scanners are used to sample point from the surface of the existing building and saves that data as point cloud. The data collected by the laser scanner is huge. Thus, Revit links a point cloud as a reference rather than embedding the complete file. Now, you can modify the existing building condition easily and the work performance will be also be improved due to better visibility.

REVIT INTEROPERABILITY

Autodesk Revit is quite compatible with CAD-based programs. You can create your own designs in Revit Architecture and then export them into other programs for visualization and creating photorealistic renderings of your designs. Revit is user-friendly, parametric, flexible, and compatible features help you create and implement your designs better and faster. Revit Architecture supports a wide range of file formats, thereby enabling the files to be imported and exported to other programs such as Autodesk 3ds Max, Autodesk Inventor, Revit Structures, Revit MEP, SketchUp, Microstation, and so on.

Interoperability with Autodesk 3ds Max

You can export a 3D view from Revit to Autodesk 3ds Max through DWG format to create photorealistic renderings and visualizations. The interoperability of Revit with 3ds Max has improved in the following ways:

1. Before exporting the views, you can use the section box to restrict the unwanted content in the file to be exported to Autodesk 3ds Max. The section box has been enhanced to enable you to switch to 2D view while modifying the section box size for the 3D view. Reducing the amount of data from the file reduces the size, thus making the importing and exporting of data faster and easier.

2. The objects that have different materials assigned to different surfaces in Revit Architecture and the objects that have a complex geometry are exported in 3D view as ACIS solids. When such objects are imported in 3ds Max, they display different materials assigned to different surfaces unlike the previous releases that showed only one material for the entire ACIS solid.

3. The files exported from Revit and imported into 3ds Max show the Revit Material Names, unlike the Material names that used to be displayed for the materials in the previous releases.

4. To export drawings from Revit to 3ds Max, you do not need to use the plug-in anymore, as this functionality has been incorporated in 3ds Max.

Interoperability with Trimble SketchUp

Revit Architecture supports SKP files. You can create your design in Revit and then export it to Trimble SketchUp. Similarly, you can import the design from Trimble SketchUp and link it to Revit. Trimble SketchUp is best suited for modeling and visualization. You can work on your designs using both Revit and SketchUp. If you want to quickly create a model of your design, assign materials and visualize it, then you can use SketchUp and then Revit Architecture to modify the designs. If you want to create a design, and at the same time, want to analyze and document the results, you should start your design in Revit and then use SketchUp at a later stage when you need to add materials and visualization. Trimble SketchUp along with Google Earth helps you place your models on the web using the real world coordinates and share it with world using the **3D Ware house** feature of SketchUp that helps you to search, share and store 3D models.

Publishing Tips in Revit 2025

Revit Architecture allows you to export the 2D and 3D project views to Design Web Format (DWF). DWF provides you with an alternative of Portable Drawing Format (PDF) to publish your project views. The advantages of using the DWF are given below:

1. You can print the same DWF file using any printer since the DWF file is not device specific.

2. You can share the DWF files electronically through emails because of their smaller size.

3. The DWF files are generally one-twentieth the size of the original DWG files.

4. You can publish all your data in one DWF file instead of multiple DWG files.

5. You can place your designs on web using the Google search engine.

6. You can share your files irrespective of the versions of AutoCAD.

7. You can decide what to publish in DWF file and avoid the sharing of the original files.

8. You can see models clearly in a DWF file by zooming in the DWF files.

9. You can use a 3D viewer to view designs in DWF files.

10. Using the DWF files, you can share your designs in a much secure and easy way. You can share your designs even with people who do not use Revit Architecture.

ROUTE ANALYSIS

The route analysis allows you to find out the shortest path and the quickest time of traversing between two defined points in a model. Using this analysis, you can design your space to prevent any obstacle in the path of traverse between spaces in the model.

Creating a Travel Path

Open a plan view and then choose the **Path of Travel** tool in the **Route Analysis** panel of the **Analyze** tab; the **Modify | Place Path of Travel** contextual tab will be displayed. In this tab, you can select an option from the **Line Style** drop down in the **Line Style** panel to specify a line type to the path of travel. Also, you can choose the **Tag on Placement** button from the **Tag** panel to enable the display of tag along with the path of travel. After choosing an appropriate option click to specify the start point of the path and then click to specify the end of the path. On doing so, a path of travel will be generated as the shortest travel distance between the two specified points avoiding model element obstacles in the path of travel, as shown in Figure 16-27. After the path is generated, you can make changes to the route by modifying the model categories that are considered obstacles to the path of travel. The settings for the route analysis can be modified by using the **Route Analysis Settings** dialog box. After the setting is modified, the travel path can be updated by using the **Update** tool that can be applied by selecting the travel path.

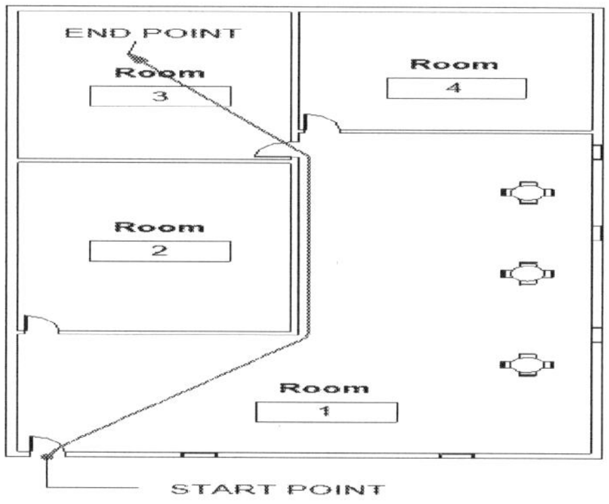

Figure 16-27 *Path of travel displayed in a model*

Modifying Properties of the Path

After creating the path you can view and modify the instance proprieties of the path. To do so, select the path created; the instance properties of the path will be displayed in the **Properties** palette, as shown in Figure 16-28. In this palette, you can select an option from the drop-down list corresponding to the **Line Style** parameter to change the line pattern of the selected path. The **Speed**, **View Name**, **Level, Time**, and **Length** are read-only parameters and cannot be modified. You can assign a value to the **Mark** parameter that can be used for reference in the schedules and tags created for the path of travel.

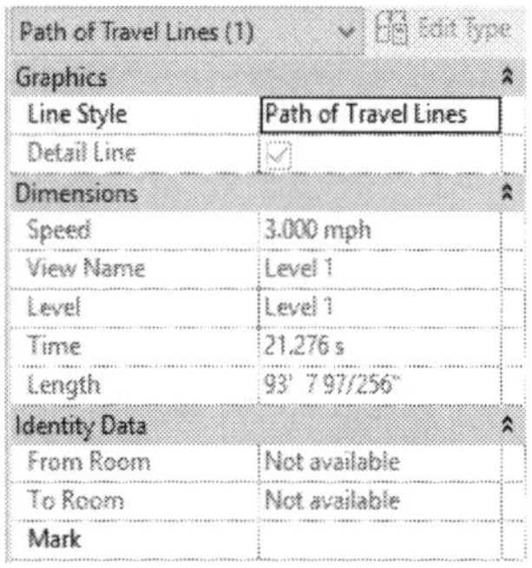

***Figure 16-28** Partial view of the **Properties** pallete displaying the instance properties of the travel path*

Setting the Travel Path

After creating the path you can modify the route of the path based on the obstacles that you choose to come in its traverse. To modify the settings of the obstacles in the path, choose the **Route Analysis Settings** button in the **Route Analysis** panel of the **Analyze** tab. On doing so, the **Route Analysis Settings** dialog box is displayed as shown in Figure 16-29.

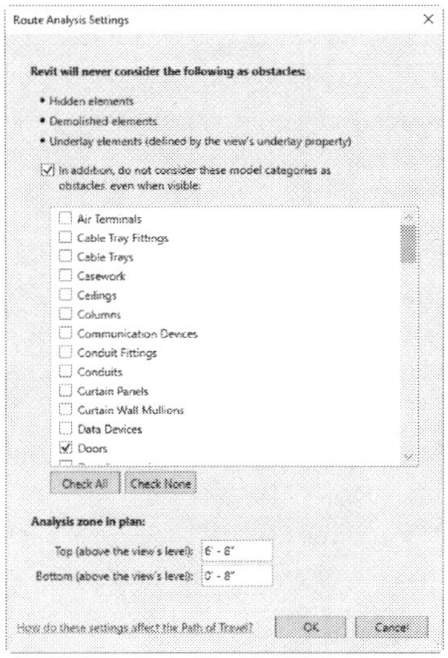

***Figure 16-29** The **Route Analysis Settings** dialog box*

TUTORIALS

In the following two tutorials, you will combine the projects created in this book. In the first tutorial, you will create two facade options for the *Apartment 1* project. In the second tutorial, you will combine the *Elevator and Stair Lobby* and the *Apartment 2* project to create a cluster of apartments. You will then link the cluster to the *Site Layout* project created earlier in Chapter 9.

Tutorial 1 Apartment 1

In this tutorial, you will create two design options for the exterior facade. The first option uses 18" x 18" (for Metric 450 x 450 mm) square structural columns and beams, as shown in Figure 16-30. The distance between these members is shown in Figure 16-31 and Figure 16-32. The second design option is created using 12" X 12" (for Metric 300 x 300mm) square concrete columns and 12" X 24" rectangular concrete beams, as shown in Figure 16-63. The distances between the structural members and some other dimensions are shown in Figures 16-64. You can assume the missing dimensions for this tutorial. **(Expected time: 45 min)**

Note

In this tutorial, the usage of structural elements has been given only as an example for creating the design options for a facade. Their general usage in a project should, however, depend on their placement.

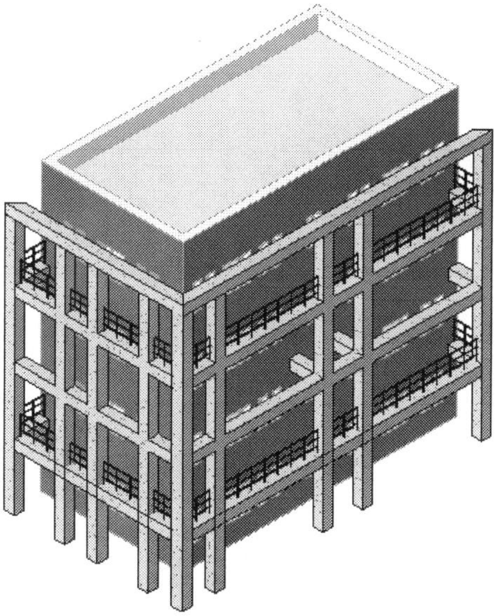

Figure 16-30 *The 3D view for creating structural elements for Design Option 1*

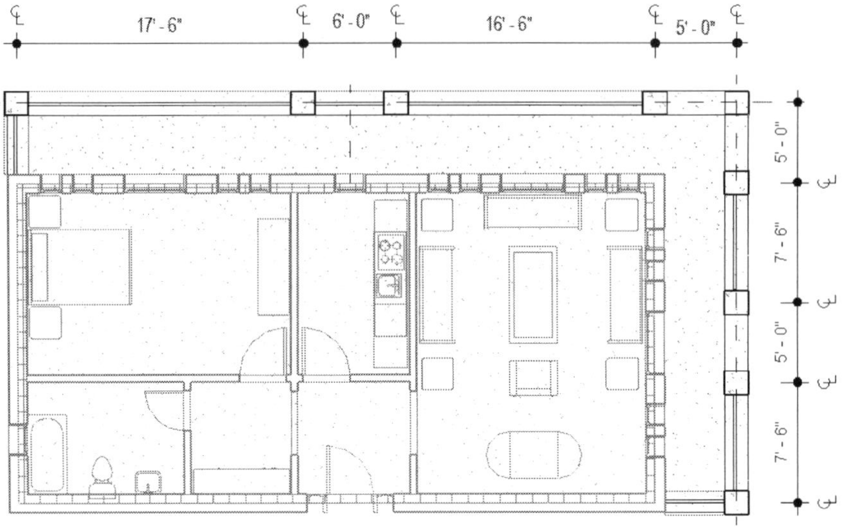

Figure 16-31 *The sketch plan for creating structural elements for Design Option 1*

Figure 16-32 *The north elevation view for creating structural elements for Design Option 1*

The following steps are required to complete this tutorial:

a. Create the design options in the *Apartment 1* project.
b. Add structural elements to the Design Option 1, refer to Figures 16-30 through 16-32.
c. Add structural elements to the Design Option 2.
d. Save the project using the **Save As** tool.
   For Imperial *c16_Apartment1_tut1.rvt*
   For Metric **M**_*c16_Apartment1_tut1.rvt*
e. Close the project using the **Close** tool.

Creating Design Options

In this section of the tutorial, you will first create the two design options and then add elements to each of them.

1. Choose **Open > Project** from the **File** menu and open the *c15_Apartment1_tut1.rvt* file (for Metric *M_c15_Apartment1_tut1.rvt* file) created in Tutorial 1 of Chapter 15. You can also download this file from *https://www.cadsoft.com*. The path of the file is as follows: *Textbooks > Civil/GIS > Revit Architecture > Exploring Autodesk Revit 2025 for Architecture.*

2. Double-click on **First Floor** in the **Floor Plans** head of the **Project Browser** to open the corresponding floor plan in the drawing window.

3. Choose the **Manage** tab and then choose the **Design Options** tool from the **Design Options** panel; the **Design Options** dialog box is displayed.

4. In the **Design Options** dialog box, choose the **New** button from the **Option Set** area; the **Option 1** node with **Option 1(primary)** subhead is displayed in the left side pane of **Design Option** dialog box.

Note

*The headings for the **Option Set**, **Edit**, and **Option** areas in the **Design Options** dialog box will disappear on creating a new option set. To locate these areas in the dialog box, refer to Figure 16-4 in this chapter.*

5. In the **Option** area, choose the **New** button to create **Option 2** under the **Option Set 1**.

6. Select **Option Set 1** and choose the **Rename** button in the **Option Set** area; the **Rename** dialog box is displayed. Enter **Facade Options** in the **New** edit box and choose **OK**; the **Option Set 1** node is renamed to **Facade Options**.

7. Similarly, rename **Option 1(primary)** as **18"X18" columns** (**450 x 450 mm** for metric) and **Option 2** as **12"X12"** (**300 x 300 mm** for metric) **columns**.

8. Select the option **18"X18" columns (primary)** and choose the **Edit Selected** button from the **Edit** area; the drawing area turns into edit mode for the selected design option.

9. Now, choose the **Close** button to close the **Design Options** dialog box.

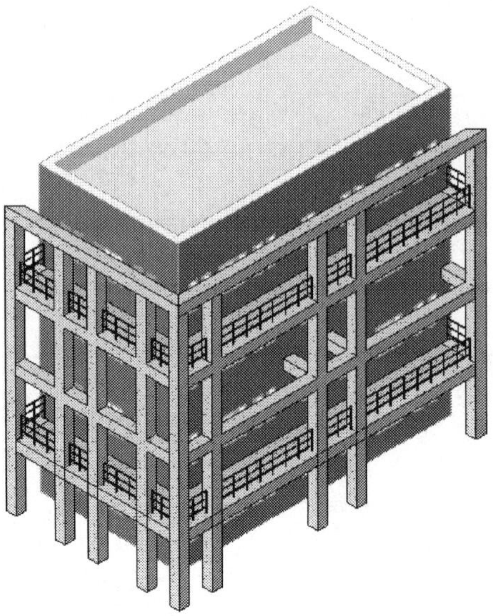

Figure 16-33 The 3D view for creating structural elements for Design Option 2

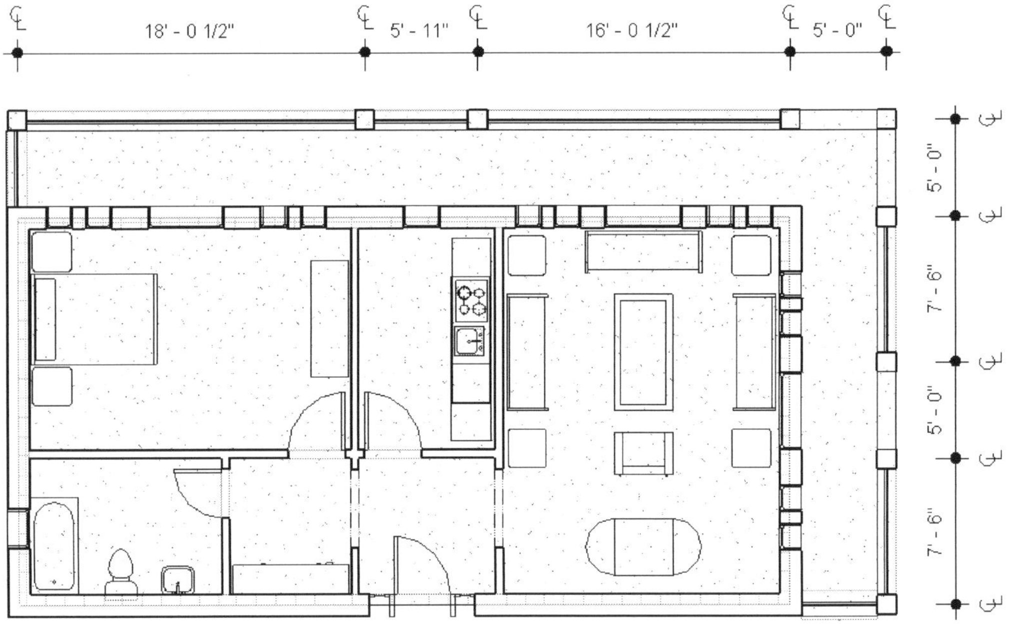

Figure 16-34 The sketch plan for creating structural elements for Design Option 2

Adding Structural Components to Design Option 1

In this section of the tutorial, you will load and add structural columns in Design Option 1.

1. You need to add the structural columns to the drawing. To do so, choose the **Structural Column** tool from **Architecture > Build > Column** drop-down; the **Modify | Place Structural Column** tab is displayed.

2. From the **Mode** panel, choose the **Load Family** tool; the **Load Family** dialog box is displayed.

3. In the **Load Family** dialog box, choose **English-Imperial > Structural Columns > Concrete** folder (for Metric **English > Structural Columns > Concrete** folder) and select the *Concrete-Square-Column.rfa* file (for Metric *M_Concrete-Square-Column.rfa* file). Next, choose **Open**; the **Concrete-Square-Column** family (for Metric **M_Concrete-Square-Column** family) is loaded into the project file.

4. From the **Type Selector** drop-down list in the **Properties** palette, select **Concrete-Square-Column: 18 x 18** (for Metric **M_Concrete-Square-Column: 450 x 450mm**).

5. Move the cursor near the top right corner of the floor plan and click when the vertical dimension in the upward direction displays **5'-0"** (1500 mm) and the wall centerline is displayed, as shown in Figure 16-35. Now, exit from the **Modify | Place Structural Column** tab by choosing the **Modify** button from the **Select** panel.

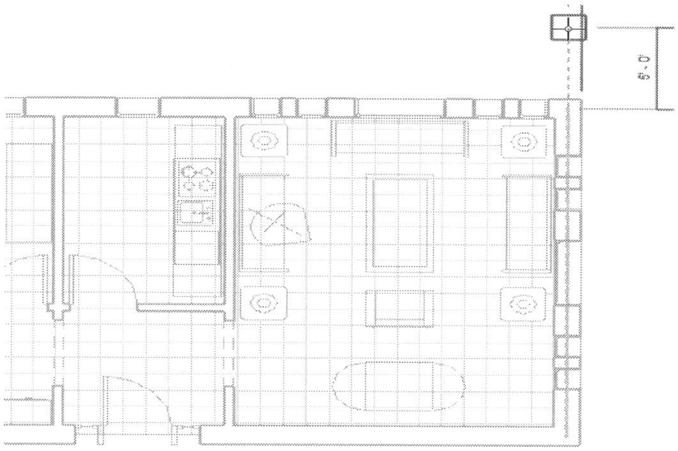

Figure 16-35 *Adding structural elements to the Design Option 1*

6. In the **3D Views** head in the **Project Browser**, double-click on **{3D}**.

7. Select the added structural column to display its instance properties in the **Properties** palette. In the **Properties** palette, click in the value field of the **Base Level** instance parameter and select the **First Floor** option from the drop-down list displayed, if it is not selected by default.

8. Click on the value field of the **Top Level** instance parameter and select **Roof** from the drop-down list displayed. In the value fields of the **Base Offset** and **Top Offset** instance parameters, enter **0'0"** (**0 mm**) and **-6"** (**-150 mm**), respectively.

9. Choose the **Apply** button from the **Properties** palette; the specified instance parameters are assigned to the selected column.

10. Open the **First Floor** plan view. Select the structural column and make its copies, refer to Figure 16-28. To make copies, choose the **Copy** tool from the **Modify** panel of the **Modify | Structural Columns** tab and select the **Multiple** checkbox from the **Options Bar** and create copies in the horizontal direction at the specified dimension, refer to Figure 16-28.

11. Now, choose the **Modify** button from the **Select** panel to clear the selection.

12. Repeat the procedure followed in step 10 to draw the column in vertical direction, refer to Figure 16-28. After adding all columns as per the sketch plan, choose the **Modify** button from the **Select** panel.

13. To add beams to the columns, choose the **Beam** tool from the **Structure** panel of the **Structure** tab.

14. On invoking the **Beam** tool, the **Modify | Place Beam** tab is displayed. In this tab, choose the **Load Family** tool from the **Mode** panel; the **Load Family** dialog box is displayed.

15. In this dialog box, choose **English-Imperial > Structural Framing > Concrete** folder (for Metric **English > Structural Framing > Concrete** folder) and then select the *Concrete-Rectangular Beam.rfa* file (for Metric *M_Concrete-Rectangular Beam.rfa* file), and then choose **Open**; the **Concrete - Rectangular Beam** family (for Metric **M_Concrete - Rectangular Beam**) is loaded in the project.

16. In the **Properties** palette, notice that the **Concrete-Rectangular Beam 16 x 32** type (for Metric **M_Concrete-Rectangular Beam 400 x 800mm**) is selected by default in the **Type Selector** drop-down list. Next, choose the **Edit Type** button; the **Type Properties** dialog box for the selected beam type is displayed.

17. Choose the **Duplicate** button; the **Name** dialog box is displayed. Enter **18 x 18** (for Metric **450 x 450mm**) in the **Name** edit box and choose **OK**; the **Name** dialog box closes.

18. In the **Type Properties** dialog box, modify the values for **b** and **h** type parameters to **1' 6"** (**450 mm**) respectively and choose the **OK** button to assign the specified value to the new type and close the **Type Properties** box.

19. In the **Options Bar**, select the **Level: Second Floor** option from the **Placement Plane** drop-down list.

20. Sketch the horizontal beam between the two columns, as shown in Figure 16-36. Ignore the warning displayed.

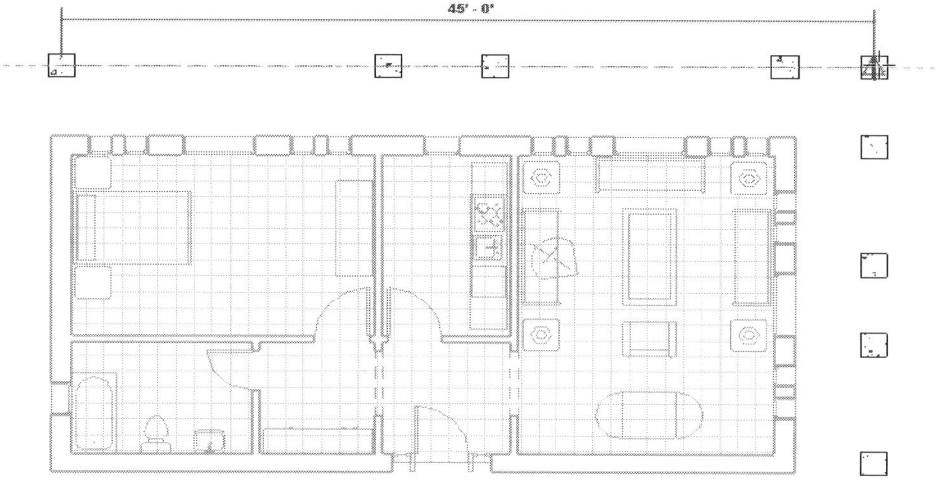

Figure 16-36 *Adding the horizontal structural beam to the Design Option 1*

21. Similarly, sketch the vertical beam between the two columns. Again, ignore the warning. After sketching the vertical beam, choose the **Modify** button; the current selection is cleared.

22. In the **Project Browser**, double-click on **{3D}** under the **3D Views** head; the default 3D view of the project is displayed in the drawing area.

23. Right-click on the **ViewCube** tool; a shortcut menu is displayed. Choose **Orient to a Direction > Northeast Isometric** from the shortcut menu; the current view reorients toward the northeast isometric view, as shown in Figure 16-37.

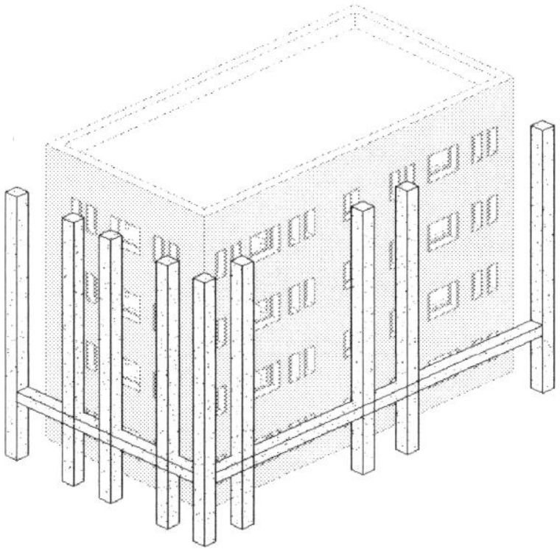

Figure 16-37 *The 3D view of the added structural elements*

24. Select the two beams using the CTRL key; the **Modify | Structural Framing** tab is displayed. Now, choose the **Copy to Clipboard** tool from the **Clipboard** panel; the selected beams are copied to the clipboard.

25. To paste the copied beams, choose the **Aligned to Selected Levels** tool from the **Paste** drop-down in the **Clipboard** panel; the **Select Levels** dialog box is displayed.

26. In this dialog box, select the **Third Floor**, **Fourth Floor**, and **Roof** options using the CTRL key and then choose the **OK** button to paste the beams at the selected levels. Now, choose the **Modify** button to exit the current selection.

27. Select the beams at the roof level and use the drag controls to complete the beam. You will use the **East** and **North Elevation** views to drag and extend the beams. After this step, the 3D view should appear similar to the illustration shown in Figure 16-38.

Similarly, you can add other smaller beams between the levels and the beams perpendicular to the walls, based on the given sketch plan and 3D View for Option 1. You may need to convert the exterior wall to bearing.

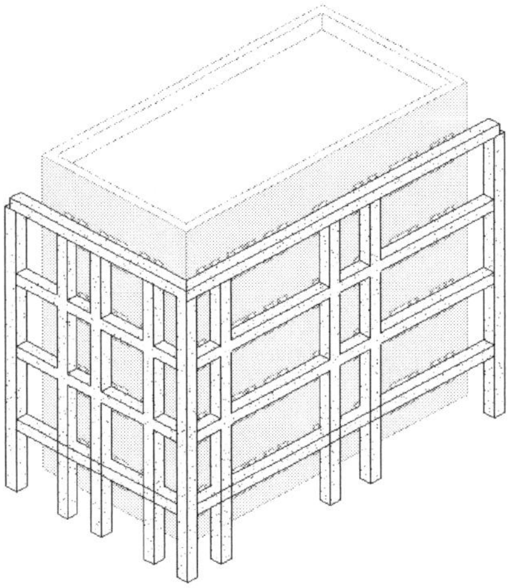

Figure 16-38 *The 3D view of the copied structural beams*

28. In the **Project Browser**, double-click on the **Second Floor**; to display the corresponding view.

29. Invoke the **Floor: Architectural** tool from **Architecture >Build > Floor** drop-down and add the floor slab between the exterior wall and the columns on the two sides at the second floor level. After you create the floor, Autodesk Revit displays a prompt box. Choose the **No** button to keep the walls detached from the slabs created in this design option.

30. Choose the **Sketch Path** tool from the **Architecture** > **Circulation** > **Railing** drop-down and then create the default railing on the second floor between the structural columns, as shown in the 3D view for Design Option 1, refer to Figure 16-30.

31. After creating the railing and the floor, select them; the **Modify | Multi-Select** tab is displayed. Choose the **Copy to Clipboard** tool from the **Clipboard** panel of the **Modify | Multi-Select** tab; the selected elements are copied to the clipboard.

32. Choose the **Aligned to Selected Levels** tool from the **Modify | Multi-Select > Clipboard > Paste** drop-down; the **Select Levels** dialog box is displayed. From this dialog box, select the **Fourth Floor** option and then choose **OK**; the **Select Levels** dialog box closes and the selected element gets pasted. Choose the **Modify** button to exit the current selection.

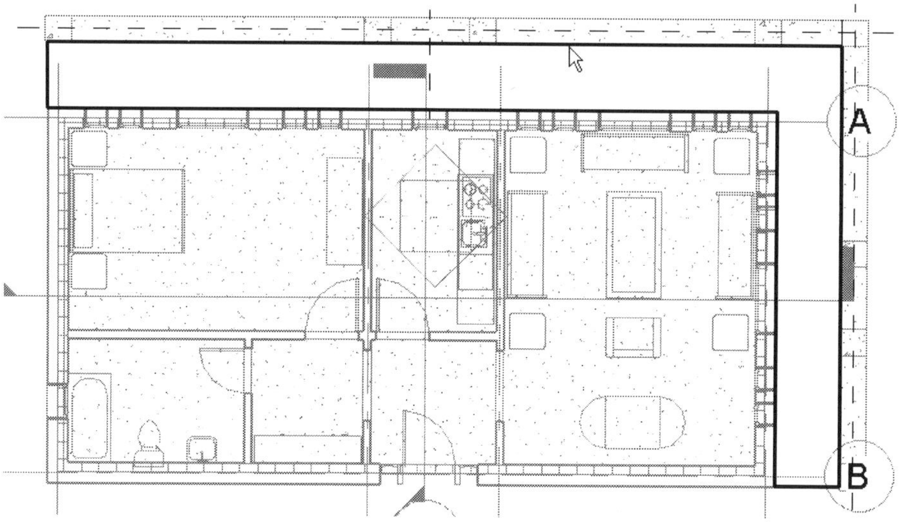

***Figure 16-39** A foor slab added using the **Floor** tool*

33. Choose the **Manage** tab and then choose the **Design Options** tool from the **Design Options** panel; the **Design Options** dialog box is displayed.

34. In this dialog box, choose the **Finish Editing** button from the **Edit** area and then choose the **Close** button; the **Design Options** dialog box closes and the 3D view appears similar to the view given for Option 1, refer to Figure 16-38. Notice that the **Main Model** option is selected in the **Active Design Option** drop-down list that is displayed below the **Pick to Edit** button in the **Design Options** panel of the **Manage** tab.

Adding Structural Components to Design Option 2

In this section of the tutorial, you will create elements to Design Option 2. In general, you can use two methods to create elements to a design option. The first method is to create the elements as they were created in the previous option. The second method is to copy elements from Option 1 to Option 2 and then modify them based on the given project parameters. You will use the second method in this tutorial.

1. To start adding element to the design option, choose the **Design Options** tool from the **Design Options** panel of the **Manage** tab; the **Design Options** dialog box is displayed; as shown in Figure 16-40. In this dialog box, select the **18" X 18" columns (primary)** (for

Metric **450 x 450 mm columns(primary)**) choose the **Edit Selected** button. Next, choose the **Close** button; the **Design Option** dialog box is closed.

2. Select all elements created in Design Option 1, including columns, beams, floors, and railings and choose the **Copy to Clipboard** tool from the **Clipboard** panel of the **Modify | Multi-Select** tab and then choose the **Manage** tab.

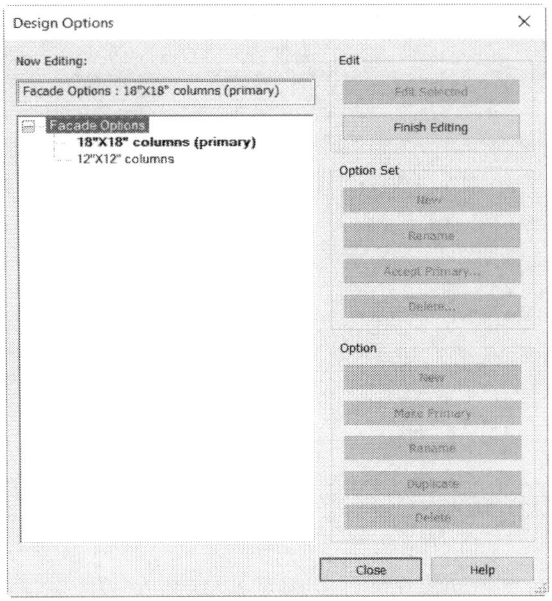

Figure 16-40 The ***Design Option*** *dialog box*

3. In this tab, select the **12"X 12" columns** (for Metric **300 x 300mm columns**) option from the **Active Design Option** drop-down list in the **Design Options** panel. Notice that from the current view, the elements of the previous design option disappear.

4. Choose the **Modify** tab and then choose the **Aligned to Same Place** tool from the **Paste** drop-down in the **Clipboard** panel. Notice that the copied elements of Design Option 1 are pasted in Design Option 2.

5. Select all the column and by using the **Type Selector** drop-down list in the **Properties** palette, replace all **Concrete-Square-Column 18" X 18"** (for Metric **M_Concrete-Square-Column 450 x 450mm**) types with **Concrete-Square-Column 12" X 12"**types (for Metric **Concrete-Square-Column 300 x 300mm** type). Similarly, select all the beams and replace all **Concrete-Rectangular Beam : 18" X 18"** beams with **Concrete-Rectangular Beam: 12" X 24"** beams.

6. Use the drag control to extend the beams, wherever necessary, such that they completely lie over the column. You may need to work in various views to select and extend the beams based on the given 3D view for Design Option 2.

7. Similarly, select the railings and edit them such that they span between the columns. Also, modify their properties to match the given railing profile. Assume the dimensions for the railing and baluster placement.

8. Select and extend the floor slabs at the second and fourth floors up to the beams. Now, the 3D view of the project should resemble the views given for creating Design Option 2, refer to Figures 16-33 through 16-35.

9. Choose the **Design Options** tool from the **Design Options** panel in the **Manage** tab to invoke the **Design Options** dialog box. Choose the **Finish Editing** button from this dialog box. Choose the **Close** button to close the dialog box.

Saving the Project

In this section, you will save the project file using the **Save As** tool.

1. Now, choose **Save As > Project** from the **File** menu; the **Save As** dialog box is displayed. Enter **c16_Apartment1_tut1.rvt** (for Metric **M_c16_Apartment1_tut1.rvt**) in the **File name** edit box of the **Save As** dialog box.

2. Choose **Save**; the **Save As** dialog closes and the file is saved.

Note

You will not close the project file for Apartment 1 project at this stage as this file will be required for the next tutorial for building the Apartment Complex.

EXERCISE

Exercise 1 Apartment 2

Modify the *Apartment Cluster* project (for Metric *M_Apartment_Cluster* project) created in the Tutorial 1 of this chapter. Add structural elements for the *Apartment 2* exterior wall, based on the sketch plan shown in Figure 16-55 and the sketch elevation, as shown in Figure 16-56. Use 18"x18" columns to generate the facade. The 3D view should appear similar to the illustration given in Figure 16-57. Modify the *Apartment 2* project file and reinsert the elements into the *Apartment Cluster* file. **(Expected time: 45 min)**

Save the file as-

For Imperial *c16_apartment_cluster_ex1.rvt*

For Metric *M_c16_apartment_cluster_ex1.rvt)*

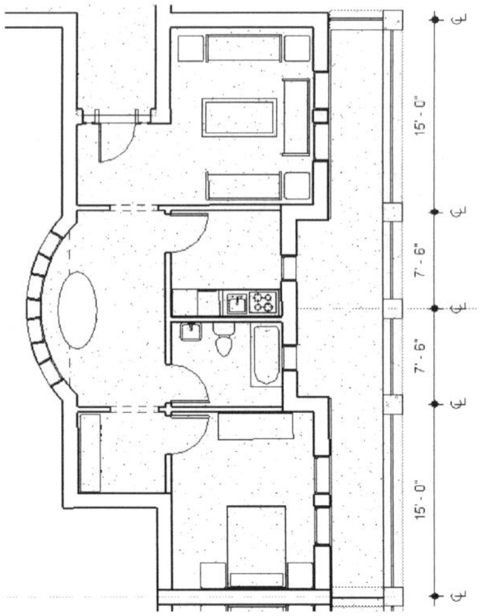

Figure 16-55 *The sketch plan view for adding structural elements to the Apartment 2 project*

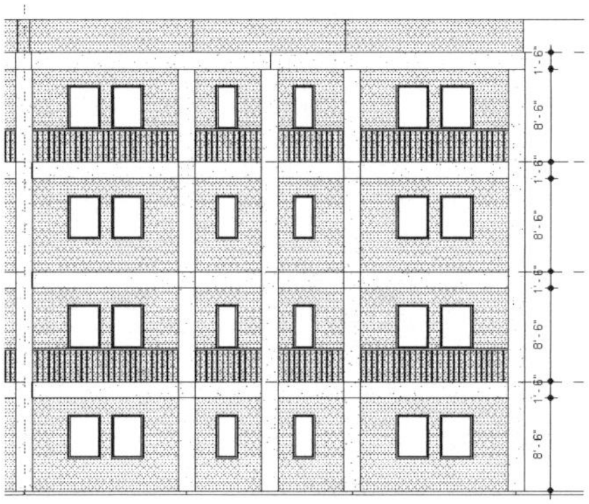

Figure 16-56 *Sketch elevation for adding structural elements to the Apartment 2 project*

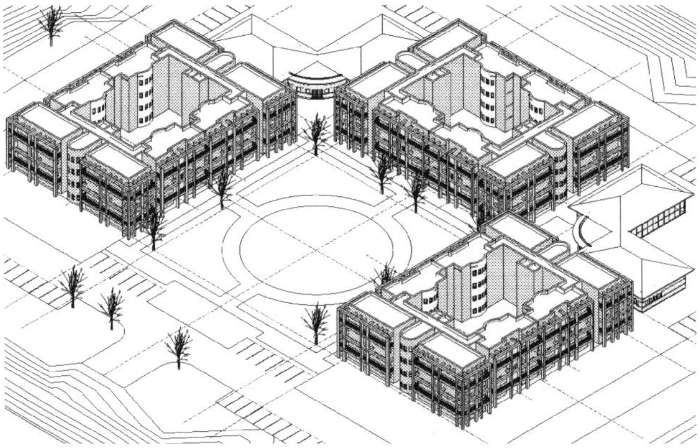

Figure 16-57 *The 3D view of the completed Apartment Complex project*

Save the file as-

For Imperial *c16_apartment_cluster_ex1.rvt*
For Metric *M_c16_apartment_cluster_ex1.rvt*)

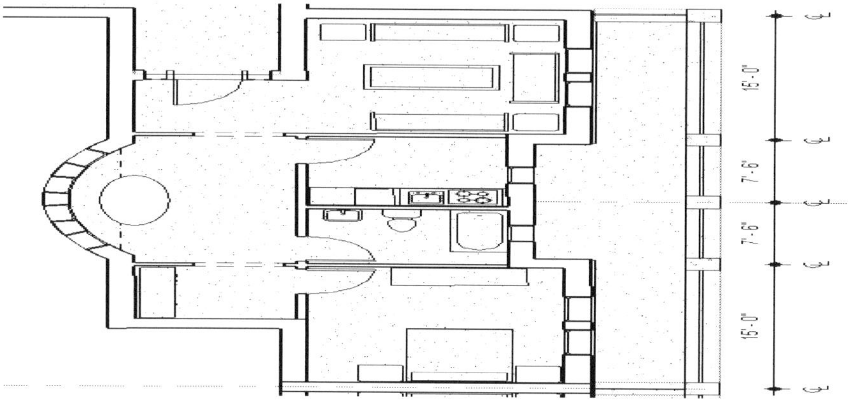

Figure 16-55 *The sketch plan view for adding structural elements to the Apartment 2 project*

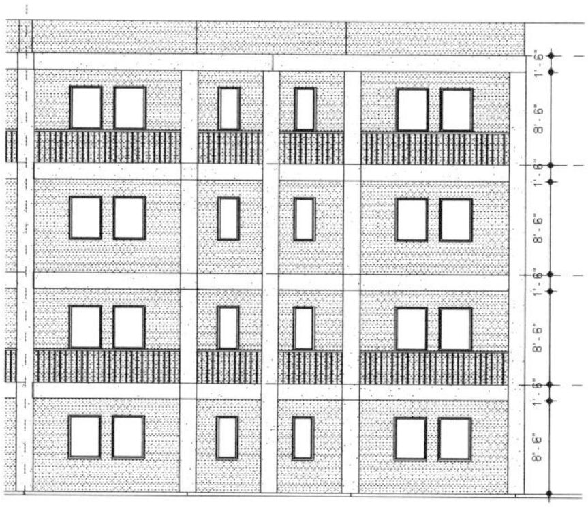

Figure 16-56 *Sketch elevation for adding structural elements to the Apartment 2 project*

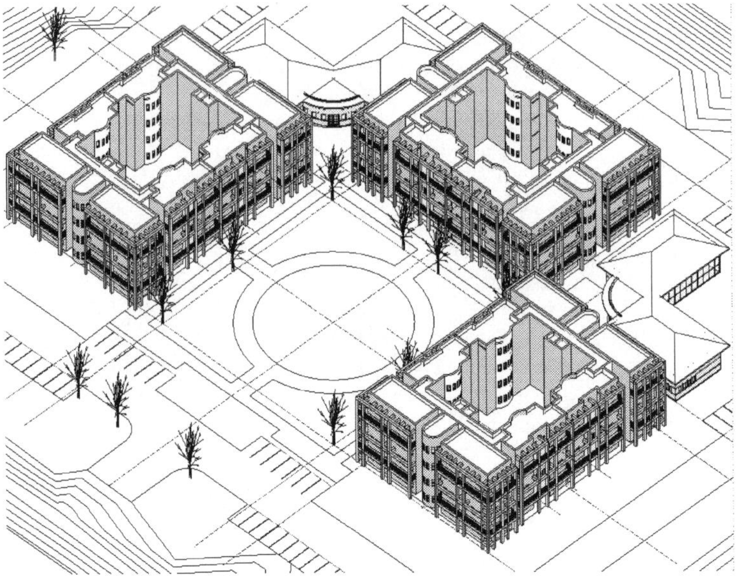

Figure 16-57 *The 3D view of the completed Apartment Complex project*

Index

D

E

F

G

H

I

K

L

M

N

O

P

R

S

T

U

V

W

Z

Other Publications by CADSoft Technologies

The following is the list of some of the publications by CADSOFT Technologies. Please visit *www.cadsofttech.com* for the complete listing.

Revit Architecture for Novices Textbooks

- Revit Architecture 2020 for Novices: Learn By Doing
- Revit Architecture 2019 for Novices: Learn By Doing
- Revit Architecture 2018 for Novices: Learn By Doing

Revit Structure for Novices Textbooks

- Revit Structure 2023 for Novices: Learn By Doing
- Revit Structure 2020 for Novices: Learn By Doing
- Revit Structure 2019 for Novices: Learn By Doing

Revit MEP for Novices Textbooks

- Revit MEP 2019 for Novices: Learn By Doing

AutoCAD Civil 3D for Novices Textbooks

- AutoCAD Civil 3D 2023 for Novices: Learn By Doing

Autodesk 3dx Max for Novices Textbooks

- Autodesk 3ds Max 2020 for Novices: Learn By Doing
- Autodesk 3ds Max 2019 for Novices: Learn By Doing

Autodesk Maya for Novices Textbook

- Autodesk Maya 2019 for Novices: Learn By Doing

Pixologic ZBrush for Novices Textbook

- Pixologic ZBrush 2018 for Novices: Learn By Doing

MAXON CINEMA 4D for Novices Textbook

- MAXON CINEMA 4D R20 Studio for Novices: Learn By Doing

AutoCAD for Novices Textbooks

- AutoCAD 2020 for Novices: Learn By Doing
- AutoCAD 2019 for Novices: Learn By Doing
- AutoCAD 2018 for Novices: Learn By Doing

Inventor Professional for Novices Textbook

- Inventor 2020 for Novices: Learn By Doing

Siemens NX for Novices Textbook

- Siemens NX 2019 for Novices: (Continous Release) Learn By Doing

Creo Parametric for Novices Textbooks

- Creo Parametric 6.0 for Novices: Learn By Doing
- Creo Parametric 5.0 for Novices: Learn By Doing

SOLIDWORKS for Novices Textbook

- SOLIDWORKS 2018 for Novices: Learn By Doing

Made in United States
Orlando, FL
04 November 2026

90203564R00258